Mathematics for Machine Learning

The fundamental mathematical tools needed to understand machine learning include linear algebra, analytic geometry, matrix decompositions, vector calculus, optimization, probability, and statistics. These topics are traditionally taught in disparate courses, making it hard for data science or computer science students, or professionals, to efficiently learn the mathematics.

This self-contained textbook bridges the gap between mathematical and machine learning texts, introducing the mathematical concepts with a minimum of prerequisites. It uses these concepts to derive four central machine learning methods: linear regression, principal component analysis, Gaussian mixture models, and support vector machines. For students and others with a mathematical background, these derivations provide a starting point to machine learning texts. For those learning the mathematics for the first time, the methods help build intuition and practical experience with applying mathematical concepts.

Every chapter includes worked examples and exercises to test understanding. Programming tutorials are offered on the book's web site.

MARC PETER DEISENROTH is the DeepMind Chair in Artificial Intelligence at University College London. Prior to this, Marc was a faculty member at Imperial College London. His research areas include data-efficient learning, probabilistic modeling, and autonomous decision making. His research received Best Paper Awards at the ICRA 2014 and the ICCAS 2016. Marc has been awarded the President's Award for Outstanding Early Career Researcher at Imperial College London, a Google Faculty Research Award, and a Microsoft PhD grant.

A. ALDO FAISAL leads the Brain & Behaviour Lab at Imperial College London, where he is faculty at the Departments of Bioengineering and Computing and a Fellow of the Data Science Institute. He is the director of the 20Mio£ United Kingdom Research and Innovation (UKRI) Center for Doctoral Training in AI for Healthcare. He obtained a PhD in computational neuroscience at Cambridge University and became Junior Research Fellow in the Computational and Biological Learning Lab. His research is at the interface of neuroscience and machine learning to understand and reverse engineer brains and behavior.

CHENG SOON ONG is Principal Research Scientist at the Machine Learning Research Group, Data61, CSIRO and Adjunct Associate Professor at the Australian National University. His research focuses on enabling scientific discovery by extending statistical machine learning methods. He received his PhD in computer science at Australian National University in 2005. He has been a lecturer in the Department of Computer Science at ETH Zürich, and has worked in the Diagnostic Genomics Team at NICTA in Melbourne.

Mathematics for Machine Learning

Marc Peter Deisenroth
University College London

A. Aldo Faisal
Imperial College London

Cheng Soon Ong
Data61, CSIRO

CAMBRIDGE
UNIVERSITY PRESS

University Printing House, Cambridge CB2 8BS, United Kingdom

One Liberty Plaza, 20th Floor, New York, NY 10006, USA

477 Williamstown Road, Port Melbourne, VIC 3207, Australia

314–321, 3rd Floor, Plot 3, Splendor Forum, Jasola District Centre, New Delhi – 110025, India

79 Anson Road, #06–04/06, Singapore 079906

Cambridge University Press is part of the University of Cambridge.

It furthers the University's mission by disseminating knowledge in the pursuit of education, learning, and research at the highest international levels of excellence.

www.cambridge.org
Information on this title: www.cambridge.org/9781108470049
DOI: 10.1017/9781108679930

First published 2020

A catalogue record for this publication is available from the British Library.

Library of Congress Cataloging-in-Publication Data
Names: Deisenroth, Marc Peter, author. | Faisal, A. Aldo, author. | Ong, Cheng Soon, author.
Title: Mathematics for machine learning / Marc Peter Deisenroth, A. Aldo Faisal, Cheng Soon Ong.
Description: Cambridge ; New York, NY : Cambridge University Press, 2020. |
Includes bibliographical references and index.
Identifiers: LCCN 2019040762 (print) | LCCN 2019040763 (ebook) |
ISBN 9781108470049 (hardback) | ISBN 9781108455145 (paperback) | ISBN 9781108679930 (epub)
Subjects: LCSH: Machine learning–Mathematics.
Classification: LCC Q325.5 .D45 2020 (print) | LCC Q325.5 (ebook) | DDC 006.3/1–dc23
LC record available at https://lccn.loc.gov/2019040762
LC ebook record available at https://lccn.loc.gov/2019040763

ISBN 978-1-108-47004-9 Hardback
ISBN 978-1-108-45514-5 Paperback

Additional resources for this publication at https://mml-book.com.

Contents

List of Symbols

Symbol	Typical meaning
$a, b, c, \alpha, \beta, \gamma$	Scalars are lowercase
$\boldsymbol{x}, \boldsymbol{y}, \boldsymbol{z}$	Vectors are bold lowercase
$\boldsymbol{A}, \boldsymbol{B}, \boldsymbol{C}$	Matrices are bold uppercase
$\boldsymbol{x}^\top, \boldsymbol{A}^\top$	Transpose of a vector or matrix
$\boldsymbol{A}^{-1}$	Inverse of a matrix
$\langle \boldsymbol{x}, \boldsymbol{y} \rangle$	Inner product of $\boldsymbol{x}$ and $\boldsymbol{y}$
$\boldsymbol{x}^\top \boldsymbol{y}$	Dot product of $\boldsymbol{x}$ and $\boldsymbol{y}$
$B = (\boldsymbol{b}_1, \boldsymbol{b}_2, \boldsymbol{b}_3)$	(Ordered) tuple
$\boldsymbol{B} = [\boldsymbol{b}_1, \boldsymbol{b}_2, \boldsymbol{b}_3]$	Matrix of column vectors stacked horizontally
$\mathcal{B} = \{\boldsymbol{b}_1, \boldsymbol{b}_2, \boldsymbol{b}_3\}$	Set of vectors (unordered)
$\mathbb{Z}, \mathbb{N}$	Integers and natural numbers, respectively
$\mathbb{R}, \mathbb{C}$	Real and complex numbers, respectively
$\mathbb{R}^n$	n-dimensional vector space of real numbers
$\forall x$	Universal quantifier: for all x
$\exists x$	Existential quantifier: there exists x
$a := b$	a is defined as b
$a =: b$	b is defined as a
$a \propto b$	a is proportional to b, i.e., $a = \text{constant} \cdot b$
$g \circ f$	Function composition: "g after f"
$\iff$	If and only if
$\implies$	Implies
$\mathcal{A}, \mathcal{C}$	Sets
$a \in \mathcal{A}$	a is an element of the set $\mathcal{A}$
$\emptyset$	Empty set
D	Number of dimensions; indexed by $d = 1, \dots, D$
N	Number of data points; indexed by $n = 1, \dots, N$
$\boldsymbol{I}_m$	Identity matrix of size $m \times m$
$\boldsymbol{0}_{m,n}$	Matrix of zeros of size $m \times n$
$\boldsymbol{1}_{m,n}$	Matrix of ones of size $m \times n$
$\boldsymbol{e}_i$	Standard/canonical vector (where i is the component that is 1)
dim(V)	Dimensionality of vector space V

Symbol	Typical meaning
$\text{rk}(\boldsymbol{A})$	Rank of matrix $\boldsymbol{A}$
$\text{Im}(\Phi)$	Image of linear mapping Φ
$\ker(\Phi)$	Kernel (null space) of a linear mapping Φ
$\text{span}[\boldsymbol{b}_1]$	Span (generating set) of $\boldsymbol{b}_1$
$\text{tr}(\boldsymbol{A})$	Trace of $\boldsymbol{A}$
$\det(\boldsymbol{A})$	Determinant of $\boldsymbol{A}$
$\lvert\cdot\rvert$	Absolute value or determinant (depending on context)
$\lVert\cdot\rVert$	Norm; Euclidean unless specified
λ	Eigenvalue or Lagrange multiplier
E_λ	Eigenspace corresponding to eigenvalue λ
$\boldsymbol{\theta}$	Parameter vector
$\frac{\partial f}{\partial x}$	Partial derivative of f with respect to x
$\frac{\mathrm{d}f}{\mathrm{d}x}$	Total derivative of f with respect to x
∇	Gradient
$\mathfrak{L}$	Lagrangian
$\mathcal{L}$	Negative log-likelihood
$\binom{n}{k}$	Binomial coefficient, n choose k
$\mathbb{V}_X[\boldsymbol{x}]$	Variance of $\boldsymbol{x}$ with respect to the random variable X
$\mathbb{E}_X[\boldsymbol{x}]$	Expectation of $\boldsymbol{x}$ with respect to the random variable X
$\text{Cov}_{X,Y}[\boldsymbol{x}, \boldsymbol{y}]$	Covariance between $\boldsymbol{x}$ and $\boldsymbol{y}$.
$X \perp\!\!\!\perp Y \mid Z$	X is conditionally independent of Y given Z
$X \sim p$	Random variable X is distributed according to p
$\mathcal{N}(\boldsymbol{\mu}, \boldsymbol{\Sigma})$	Gaussian distribution with mean $\boldsymbol{\mu}$ and covariance $\boldsymbol{\Sigma}$
$\text{Ber}(\mu)$	Bernoulli distribution with parameter μ
$\text{Bin}(N, \mu)$	Binomial distribution with parameters N, μ
$\text{Beta}(\alpha, \beta)$	Beta distribution with parameters α, β

List of Abbreviations and Acronyms

Acronym	Meaning
e.g.	Exempli gratia (Latin: for example)
GMM	Gaussian mixture model
i.e.	Id est (Latin: this means)
i.i.d.	Independent, identically distributed
MAP	Maximum a posteriori
MLE	Maximum likelihood estimation/estimator
ONB	Orthonormal basis
PCA	Principal component analysis
PPCA	Probabilistic principal component analysis
REF	Row-echelon form
SPD	Symmetric, positive definite
SVM	Support vector machine

Preface

Machine learning is the latest in a long line of attempts to distill human knowledge and reasoning into a form that is suitable for constructing machines and engineering automated systems. As machine learning becomes more ubiquitous and its software packages become easier to use, it is natural and desirable that the low-level technical details are abstracted away and hidden from the practitioner. However, this brings with it the danger that a practitioner becomes unaware of the design decisions and, hence, the limits of machine learning algorithms.

The enthusiastic practitioner who is interested to learn more about the magic behind successful machine learning algorithms currently faces a daunting set of prerequisite knowledge:

- Programming languages and data analysis tools
- Large-scale computation and the associated frameworks
- Mathematics and statistics and how machine learning builds on it

At universities, introductory courses on machine learning tend to spend early parts of the course covering some of these prerequisites. For historical reasons, courses in machine learning tend to be taught in the computer science department, where students are often trained in the first two areas of knowledge, but not so much in mathematics and statistics.

Current machine learning textbooks primarily focus on machine learning algorithms and methodologies and assume that the reader is competent in mathematics and statistics. Therefore, these books only spend one or two chapters of background mathematics, either at the beginning of the book or as appendices. We have found many people who want to delve into the foundations of basic machine learning methods who struggle with the mathematical knowledge required to read a machine learning textbook. Having taught undergraduate and graduate courses at universities, we find that the gap between high school mathematics and the mathematics level required to read a standard machine learning textbook is too big for many people.

This book brings the mathematical foundations of basic machine learning concepts to the fore and collects the information in a single place so that this skills gap is narrowed or even closed.

Why Another Book on Machine Learning?

Machine learning builds upon the language of mathematics to express concepts that seem intuitively obvious but that are surprisingly difficult to formalize. Once formalized properly, we can gain insights into the task we want to solve. One common complaint of students of mathematics around the globe is that the topics covered seem to have little relevance to practical problems. We believe that machine learning is an obvious and direct motivation for people to learn mathematics.

This book is intended to be a guidebook to the vast mathematical literature that forms the foundations of modern machine learning. We motivate the need for mathematical concepts by directly pointing out their usefulness in the context of fundamental machine learning problems. In the interest of keeping the book short, many details and more advanced concepts have been left out. Equipped with the basic concepts presented here, and how they fit into the larger context of machine learning, the reader can find numerous resources for further study, which we provide at the end of the respective chapters. For readers with a mathematical background, this book provides a brief but precisely stated glimpse of machine learning. In contrast to other books that focus on methods and models of machine learning (MacKay, 2003; Bishop, 2006; Alpaydin, 2010; Murphy, 2012; Barber, 2012; Shalev-Shwartz and Ben-David, 2014; Rogers and Girolami, 2016) or programmatic aspects of machine learning (Müller and Guido, 2016; Raschka and Mirjalili, 2017; Chollet and Allaire, 2018), we provide only four representative examples of machine learning algorithms. Instead, we focus on the mathematical concepts behind the models themselves. We hope that readers will be able to gain a deeper understanding of the basic questions in machine learning and connect practical questions arising from the use of machine learning with fundamental choices in the mathematical model.

"Math is linked in the popular mind with phobia and anxiety. You'd think we're discussing spiders." (Strogatz, 2014, 281)

We do not aim to write a classical machine learning book. Instead, our intention is to provide the mathematical background, applied to four central machine learning problems, to make it easier to read other machine learning textbooks.

Who Is the Target Audience?

As applications of machine learning become widespread in society, we believe that everybody should have some understanding of its underlying principles. This book is written in an academic mathematical style, which enables us to be precise about the concepts behind machine learning. We encourage readers unfamiliar with this seemingly terse style to persevere and to keep the goals of each topic in mind. We sprinkle comments and remarks throughout the text, in the hope that it provides useful guidance with respect to the big picture.

The book assumes the reader to have mathematical knowledge commonly covered in high school mathematics and physics. For example, the reader should have seen derivatives and integrals before, and geometric vectors in two or three dimensions. Starting from there, we generalize these concepts. Therefore, the target audience of the book includes undergraduate university students, evening learners and learners participating in online machine learning courses.

In analogy to music, there are three types of interaction that people have with machine learning:

Astute Listener The democratization of machine learning by the provision of open-source software, online tutorials and cloud-based tools allows users to not worry about the specifics of pipelines. Users can focus on extracting insights from data using off-the-shelf tools. This enables non-tech-savvy domain experts to benefit from machine learning. This is similar to listening to music; the user is able to choose and discern between different types of machine learning, and benefits from it. More experienced users are like music critics, asking important questions about the application of machine learning in society such as ethics, fairness and privacy of the individual. We hope that this book provides a foundation for thinking about the certification and risk management of machine learning systems and allows them to use their domain expertise to build better machine learning systems.

Experienced Artist Skilled practitioners of machine learning can plug and play different tools and libraries into an analysis pipeline. The stereotypical practitioner would be a data scientist or engineer who understands machine learning interfaces and their use cases and is able to perform wonderful feats of prediction from data. This is similar to a virtuoso playing music, where highly skilled practitioners can bring existing instruments to life and bring enjoyment to their audience. Using the mathematics presented here as a primer, practitioners would be able to understand the benefits and limits of their favourite method, and to extend and generalize existing machine learning algorithms. We hope that this book provides the impetus for more rigorous and principled development of machine learning methods.

Fledgling Composer As machine learning is applied to new domains, developers of machine learning need to develop new methods and extend existing algorithms. They are often researchers who need to understand the mathematical basis of machine learning and uncover relationships between different tasks. This is similar to composers of music who, within the rules and structure of musical theory, create new and amazing pieces. We hope this book provides a high-level overview of other technical books for people who want to become composers of machine learning. There is a great need in society for new researchers who are able to propose and explore novel approaches for attacking the many challenges of learning from data.

Acknowledgments

We are grateful to many people who looked at early drafts of the book and suffered through painful expositions of concepts. We tried to implement their ideas that we did not vehemently disagree with. We would like to especially acknowledge Christfried Webers for his careful reading of many parts of the book, and his detailed suggestions on structure and presentation. Many friends and colleagues have also been kind enough to provide their time and energy on different versions of each chapter. We have been lucky to benefit from the generosity of the online community, who have suggested improvements via github.com, which greatly improved the book.

The following people have found bugs, proposed clarifications and suggested relevant literature, either via github.com or personal communication. Their names are sorted alphabetically.

Abdul-Ganiy Usman
Adam Gaier
Adele Jackson
Aditya Menon
Alasdair Tran
Aleksandar Krnjaic
Alexander Makrigiorgos
Alfredo Canziani
Ali Shafti
Amr Khalifa
Andrew Tanggara
Angus Gruen
Antal A. Buss
Antoine Toisoul Le Cann
Areg Sarvazyan
Artem Artemev
Artyom Stepanov
Bill Kromydas
Bob Williamson
Boon Ping Lim
Chao Qu
Cheng Li
Chris Sherlock
Christopher Gray
Daniel McNamara
Daniel Wood
Darren Siegel
David Johnston
Dawei Chen
Ellen Broad
Fengkuangtian Zhu
Fiona Condon
Georgios Theodorou
He Xin
Irene Raissa Kameni
Jakub Nabaglo
James Hensman
Jamie Liu
Jean Kaddour
Jean-Paul Ebejer
Jerry Qiang
Jitesh Sindhare
John Lloyd
Jonas Ngnawe
Jon Martin
Justin Hsi

Kai Arulkumaran
Kamil Dreczkowski
Lily Wang
Lionel Tondji Ngoupeyou
Lydia Knüfing
Mahmoud Aslan
Mark Hartenstein
Mark van der Wilk
Markus Hegland
Martin Hewing
Matthew Alger
Matthew Lee
Maximus McCann
Mengyan Zhang
Michael Bennett
Michael Pedersen
Minjeong Shin
Mohammad Malekzadeh
Naveen Kumar
Nico Montali
Oscar Armas
Patrick Henriksen
Patrick Wieschollek
Pattarawat Chormai
Paul Kelly
Petros Christodoulou
Piotr Januszewski
Pranav Subramani
Quyu Kong
Ragib Zaman
Rui Zhang
Ryan-Rhys Griffiths
Salomon Kabongo
Samuel Ogunmola
Sandeep Mavadia
Sarvesh Nikumbh
Sebastian Raschka
Senanayak Sesh Kumar Karri
Seung-Heon Baek
Shahbaz Chaudhary
Shakir Mohamed
Shawn Berry
Sheikh Abdul Raheem Ali
Sheng Xue
Sridhar Thiagarajan
Syed Nouman Hasany
Szymon Brych
Thomas Bühler
Timur Sharapov
Tom Melamed
Vincent Adam
Vincent Dutordoir
Vu Minh
Wasim Aftab
Wen Zhi
Wojciech Stokowiec
Xiaonan Chong
Xiaowei Zhang
Yazhou Hao
Yicheng Luo
Young Lee
Yu Lu
Yun Cheng
Yuxiao Huang
Zac Cranko
Zijian Cao
Zoe Nolan

Contributors through github, whose real names were not listed on their github profile, are the following:

SamDataMad
bumptiousmonkey
idoamihai
deepakiim
insad
HorizonP
cs-maillist
kudo23
empet
victorBigand
17SKYE
jessjing1995

We are also very grateful to Parameswaran Raman and the many anonymous reviewers, organized by Cambridge University Press, who read one or more chapters of earlier versions of the manuscript, and provided constructive criticism that led to considerable improvements. A special mention goes to Dinesh Singh Negi, our LaTeX support for detailed and prompt advice about LaTeX-related issues. Last but not least, we are very grateful to our editor Lauren Cowles, who has been patiently guiding us through the gestation process of this book.

Part I

Mathematical Foundations

1

Introduction and Motivation

Machine learning is about designing algorithms that automatically extract valuable information from data. The emphasis here is on "automatic," i.e., machine learning is concerned about general-purpose methodologies that can be applied to many datasets, while producing something that is meaningful. There are three concepts that are at the core of machine learning: data, a model, and learning.

Since machine learning is inherently data driven, *data* is at the core of machine learning. The goal of machine learning is to design general-purpose methodologies to extract valuable patterns from data, ideally without much domain-specific expertise. For example, given a large corpus of documents (e.g., books in many libraries), machine learning methods can be used to automatically find relevant topics that are shared across documents (Hoffman et al., 2010). To achieve this goal, we design *models* that are typically related to the process that generates data, similar to the dataset we are given. For example, in a regression setting, the model would describe a function that maps inputs to real-valued outputs. To paraphrase Mitchell (1997): A model is said to learn from data if its performance on a given task improves after the data is taken into account. The goal is to find good models that generalize well to yet unseen data, which we may care about in the future. *Learning* can be understood as a way to automatically find patterns and structure in data by optimizing the parameters of the model.

data

model

learning

While machine learning has seen many success stories, and software is readily available to design and train rich and flexible machine learning systems, we believe that the mathematical foundations of machine learning are important in order to understand fundamental principles upon which more complicated machine learning systems are built. Understanding these principles can facilitate creating new machine learning solutions, understanding and debugging existing approaches, and learning about the inherent assumptions and limitations of the methodologies we are working with.

1.1 Finding Words for Intuitions

A challenge we face regularly in machine learning is that concepts and words are slippery, and a particular component of the machine learning system can be abstracted to different mathematical concepts. For example, the word "algorithm" is used in at least two different senses in the context of machine learning. In the first sense, we use the phrase "machine learning algorithm" to mean a system that makes predictions based on input data. We refer to these algorithms as *predictor*.

predictor

In the second sense, we use the exact same phrase "machine learning algorithm" to mean a system that adapts some internal parameters of the predictor so that it performs well on future unseen input data. Here we refer to this adaptation as *training* a system.

training

This book will not resolve the issue of ambiguity, but we want to highlight upfront that, depending on the context, the same expressions can mean different things. However, we attempt to make the context sufficiently clear to reduce the level of ambiguity.

The first part of this book introduces the mathematical concepts and foundations needed to talk about the three main components of a machine learning system: data, models, and learning. We will briefly outline these components here, and we will revisit them again in Chapter 8 once we have discussed the necessary mathematical concepts.

While not all data is numerical, it is often useful to consider data in a number format. In this book, we assume that *data* has already been appropriately converted into a numerical representation suitable for reading into a computer program. Therefore, we think of data as vectors. As another illustration of how subtle words are, there are (at least) three different ways to think about vectors: a vector as an array of numbers (a computer science view), a vector as an arrow with a direction and magnitude (a physics view), and a vector as an object that obeys addition and scaling (a mathematical view).

data as vectors

A *model* is typically used to describe a process for generating data, similar to the dataset at hand. Therefore, good models can also be thought of as simplified versions of the real (unknown) data-generating process, capturing aspects that are relevant for modeling the data and extracting hidden patterns from them. A good model can then be used to predict what would happen in the real world without performing real-world experiments.

model

We now come to the crux of the matter, the *learning* component of machine learning. Assume we are given a dataset and a suitable model. *Training* the model means to use the data available to optimize some parameters of the model with respect to a utility function that evaluates how well the model predicts the training data. Most training methods can be thought of as an approach analogous to climbing a hill to reach its peak. In this analogy, the peak of the hill corresponds to a maximum of some desired performance measure. However, in practice, we are interested in the model to perform well on unseen data. Performing well on data that we have already seen (training data) may only mean that we found a good way to memorize the data. However, this may not generalize well to unseen data, and, in practical applications, we often need to expose our machine learning system to situations that it has not encountered before.

learning

Let us summarize the main concepts of machine learning that we cover in this book:

- We represent data as vectors.
- We choose an appropriate model, either using the probabilistic or optimization view.
- We learn from available data by using numerical optimization methods with the aim that the model performs well on data not used for training.

1.2 Two Ways to Read This Book

We can consider two strategies for understanding the mathematics for machine learning:

- **Bottom-up:** Building up the concepts from foundational to more advanced. This is often the preferred approach in more technical fields, such as mathematics. This strategy has the advantage that the reader at all times is able to rely on their previously learned concepts. Unfortunately, for a practitioner many of the foundational concepts are not particularly interesting by themselves, and the lack of motivation means that most foundational definitions are quickly forgotten.
- **Top-down:** Drilling down from practical needs to more basic requirements. This goal-driven approach has the advantage that the readers know at all times why they need to work on a particular concept, and there is a clear path of required knowledge. The downside of this strategy is that the knowledge is built on potentially shaky foundations, and the readers have to remember a set of words that they do not have any way of understanding.

We decided to write this book in a modular way to separate foundational (mathematical) concepts from applications so that this book can be read in both ways. The book is split into two parts, where Part I lays the mathematical foundations and Part II applies the concepts from Part I to a set of fundamental machine learning problems, which form four pillars of machine learning as illustrated in Figure 1.1: regression, dimensionality reduction, density estimation, and classification. Chapters in Part I mostly build upon the previous ones, but it is possible to skip a chapter and work backward if necessary. Chapters in Part II are only loosely coupled and can be read in any order. There are many pointers forward and backward between the two parts of the book to link mathematical concepts with machine learning algorithms.

Of course there are more than two ways to read this book. Most readers learn using a combination of top-down and bottom-up approaches, sometimes building up basic mathematical skills before attempting more complex concepts, but also choosing topics based on applications of machine learning.

Part I Is about Mathematics

The four pillars of machine learning we cover in this book (see Figure 1.1) require a solid mathematical foundation, which is laid out in Part I.

We represent numerical data as vectors and represent a table of such data as a matrix. The study of vectors and matrices is called *linear algebra*, which we introduce in Chapter 2. The collection of vectors as a matrix is also described there.

linear algebra

Given two vectors representing two objects in the real world, we want to make statements about their similarity. The idea is that vectors that are similar should be predicted to have similar outputs by our machine learning algorithm (our predictor). To formalize the idea of similarity between vectors, we need to introduce operations that take two vectors as input and return a numerical

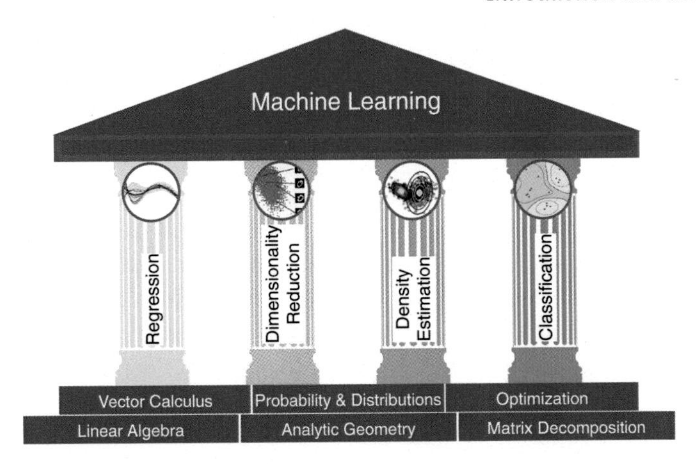

Figure 1.1 The foundations and four pillars of machine learning.

value representing their similarity. The construction of similarity and distances is central to *analytic geometry* and is discussed in Chapter 3.

analytic geometry

In Chapter 4, we introduce some fundamental concepts about matrices and *matrix decomposition*. Some operations on matrices are extremely useful in machine learning, and they allow for an intuitive interpretation of the data and more efficient learning.

matrix decomposition

We often consider data to be noisy observations of some true underlying signal. We hope that by applying machine learning we can identify the signal from the noise. This requires us to have a language for quantifying what "noise" means. We often would also like to have predictors that allow us to express some sort of uncertainty, e.g., to quantify the confidence we have about the value of the prediction at a particular test data point. Quantification of uncertainty is the realm of *probability theory* and is covered in Chapter 6.

probability theory

To train machine learning models, we typically find parameters that maximize some performance measure. Many optimization techniques require the concept of a gradient, which tells us the direction in which to search for a solution. Chapter 5 is about *vector calculus* and details the concept of gradients, which we subsequently use in Chapter 7, where we talk about *optimization* to find maxima/minima of functions.

vector calculus

optimization

Part II Is about Machine Learning

The second part of the book introduces *four pillars of machine learning* as shown in Figure 1.1. We illustrate how the mathematical concepts introduced in the first part of the book are the foundation for each pillar. Broadly speaking, chapters are ordered by difficulty (in ascending order).

In Chapter 8, we restate the three components of machine learning (data, models, and parameter estimation) in a mathematical fashion. In addition, we provide some guidelines for building experimental setups that guard against overly optimistic evaluations of machine learning systems. Recall that the goal is to build a predictor that performs well on unseen data.

In Chapter 9, we will have a close look at *linear regression*, where our objective is to find functions that map inputs $\boldsymbol{x} \in \mathbb{R}^D$ to corresponding observed

linear regression

function values $y \in \mathbb{R}$, which we can interpret as the labels of their respective inputs. We will discuss classical model fitting (parameter estimation) via maximum likelihood and maximum a posteriori estimation, as well as Bayesian linear regression, where we integrate the parameters out instead of optimizing them.

Chapter 10 focuses on *dimensionality reduction*, the second pillar in Figure 1.1, using principal component analysis. The key objective of dimensionality reduction is to find a compact, lower-dimensional representation of high-dimensional data $\boldsymbol{x} \in \mathbb{R}^D$, which is often easier to analyze than the original data. Unlike regression, dimensionality reduction is only concerned about modeling the data – there are no labels associated with a data point $\boldsymbol{x}$.

dimensionality reduction

In Chapter 11, we will move to our third pillar: *density estimation*. The objective of density estimation is to find a probability distribution that describes a given dataset. We will focus on Gaussian mixture models for this purpose, and we will discuss an iterative scheme to find the parameters of this model. As in dimensionality reduction, there are no labels associated with the data points $\boldsymbol{x} \in \mathbb{R}^D$. However, we do not seek a low-dimensional representation of the data. Instead, we are interested in a density model that describes the data.

density estimation

Chapter 12 concludes the book with an in-depth discussion of the fourth pillar: *classification*. We will discuss classification in the context of support vector machines. Similar to regression (Chapter 9), we have inputs $\boldsymbol{x}$ and corresponding labels y. However, unlike regression, where the labels were real-valued, the labels in classification are integers, which requires special care.

classification

1.3 Exercises and Feedback

We provide some exercises in Part I, which can be done mostly by pen and paper. For Part II, we provide programming tutorials (jupyter notebooks) to explore some properties of the machine learning algorithms we discuss in this book.

We appreciate that Cambridge University Press strongly supports our aim to democratize education and learning by making this book freely available for download at

`https://mml-book.com`

where tutorials, errata, and additional materials can be found. Mistakes can be reported and feedback provided using the preceding URL.

2

Linear Algebra

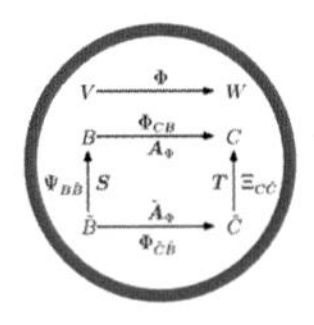

algebra

When formalizing intuitive concepts, a common approach is to construct a set of objects (symbols) and a set of rules to manipulate these objects. This is known as an *algebra*. Linear algebra is the study of vectors and certain rules to manipulate vectors. The vectors many of us know from school are called "geometric vectors," which are usually denoted by a small arrow above the letter, e.g., $\overrightarrow{x}$ and $\overrightarrow{y}$. In this book, we discuss more general concepts of vectors and use a bold letter to represent them, e.g., $\boldsymbol{x}$ and $\boldsymbol{y}$.

In general, vectors are special objects that can be added together and multiplied by scalars to produce another object of the same kind. From an abstract mathematical viewpoint, any object that satisfies these two properties can be considered a vector. Here are some examples of such vector objects:

1. Geometric vectors. This example of a vector may be familiar from high school mathematics and physics. Geometric vectors – see Figure 2.1(a) – are directed segments, which can be drawn (at least in two dimensions). Two geometric vectors $\overrightarrow{\boldsymbol{x}}, \overrightarrow{\boldsymbol{y}}$ can be added, such that $\overrightarrow{\boldsymbol{x}} + \overrightarrow{\boldsymbol{y}} = \overrightarrow{\boldsymbol{z}}$ is another geometric vector. Furthermore, multiplication by a scalar $\lambda \overrightarrow{\boldsymbol{x}}$, $\lambda \in \mathbb{R}$, is also a geometric vector. In fact, it is the original vector scaled by λ. Therefore, geometric vectors are instances of the vector concepts introduced previously. Interpreting vectors as geometric vectors enables us to use our intuitions about direction and magnitude to reason about mathematical operations.
2. Polynomials are also vectors; see Figure 2.1(b): Two polynomials can be added together, which results in another polynomial; and they can be multiplied by a scalar $\lambda \in \mathbb{R}$, and the result is a polynomial as well. Therefore, polynomials are (rather unusual) instances of vectors. Note that polynomials

Figure 2.1 Different types of vectors. Vectors can be surprising objects, including (a) geometric vectors and (b) polynomials.

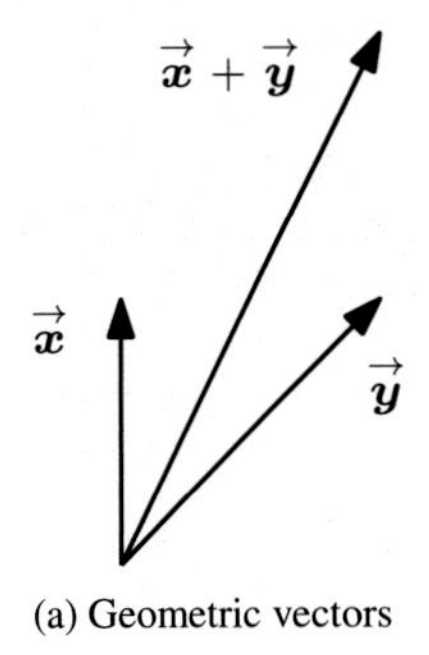

(a) Geometric vectors

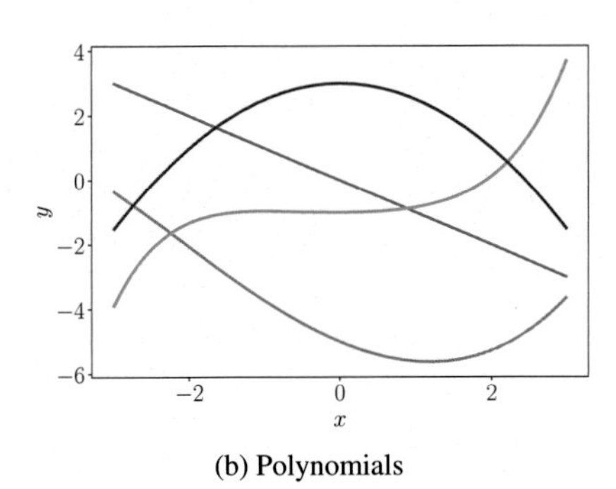

(b) Polynomials

are very different from geometric vectors. While geometric vectors are concrete "drawings," polynomials are abstract concepts. However, they are both vectors in the sense previously described.

3. Audio signals are vectors. Audio signals are represented as a series of numbers. We can add audio signals together, and their sum is a new audio signal. If we scale an audio signal, we also obtain an audio signal. Therefore, audio signals are a type of vector, too.
4. Elements of $\mathbb{R}^n$ (tuples of n real numbers) are vectors. $\mathbb{R}^n$ is more abstract than polynomials, and it is the concept we focus on in this book. For instance,

$$\boldsymbol{a} = \begin{bmatrix} 1 \\ 2 \\ 3 \end{bmatrix} \in \mathbb{R}^3 \tag{2.1}$$

is an example of a triplet of numbers. Adding two vectors $\boldsymbol{a}, \boldsymbol{b} \in \mathbb{R}^n$ componentwise results in another vector: $\boldsymbol{a} + \boldsymbol{b} = \boldsymbol{c} \in \mathbb{R}^n$. Moreover, multiplying $\boldsymbol{a} \in \mathbb{R}^n$ by $\lambda \in \mathbb{R}$ results in a scaled vector $\lambda\boldsymbol{a} \in \mathbb{R}^n$. Considering vectors as elements of $\mathbb{R}^n$ has an additional benefit that it loosely corresponds to arrays of real numbers on a computer. Many programming languages support array operations, which allow for convenient implementation of algorithms that involve vector operations.

Be careful to check whether array operations actually perform vector operations when implementing on a computer.

Linear algebra focuses on the similarities between these vector concepts. We can add them together and multiply them by scalars. We will largely focus on vectors in $\mathbb{R}^n$ since most algorithms in linear algebra are formulated in $\mathbb{R}^n$. We will see in Chapter 8 that we often consider data to be represented as vectors in $\mathbb{R}^n$. In this book, we will focus on finite-dimensional vector spaces, in which case there is a 1:1 correspondence between any kind of vector and $\mathbb{R}^n$. When it is convenient, we will use intuitions about geometric vectors and consider array-based algorithms.

One major idea in mathematics is the idea of "closure." This is the question: What is the set of all things that can result from my proposed operations? In the case of vectors: What is the set of vectors that can result by starting with a small set of vectors, and adding them to each other and scaling them? This results in a vector space (Section 2.4). The concept of a vector space and its properties underlie much of machine learning. The concepts introduced in this chapter are summarized in Figure 2.2.

Pavel Grinfeld's series on linear algebra: `http://tinyurl.com/nahclwm`
Gilbert Strang's course on linear algebra: `http://tinyurl.com/29p5q8j`
3Blue1Brown series on linear algebra: `https://tinyurl.com/h5g4kps`

This chapter is mostly based on the lecture notes and books by Drumm and Weil (2001), Strang (2003), Hogben (2013), Liesen and Mehrmann (2015), as well as Pavel Grinfeld's Linear Algebra series. Other excellent resources are Gilbert Strang's Linear Algebra course at MIT and the Linear Algebra Series by 3Blue1Brown.

Linear algebra plays an important role in machine learning and general mathematics. The concepts introduced in this chapter are further expanded to include the idea of geometry in Chapter 3. In Chapter 5, we will discuss vector calculus, where a principled knowledge of matrix operations is essential. In Chapter 10,

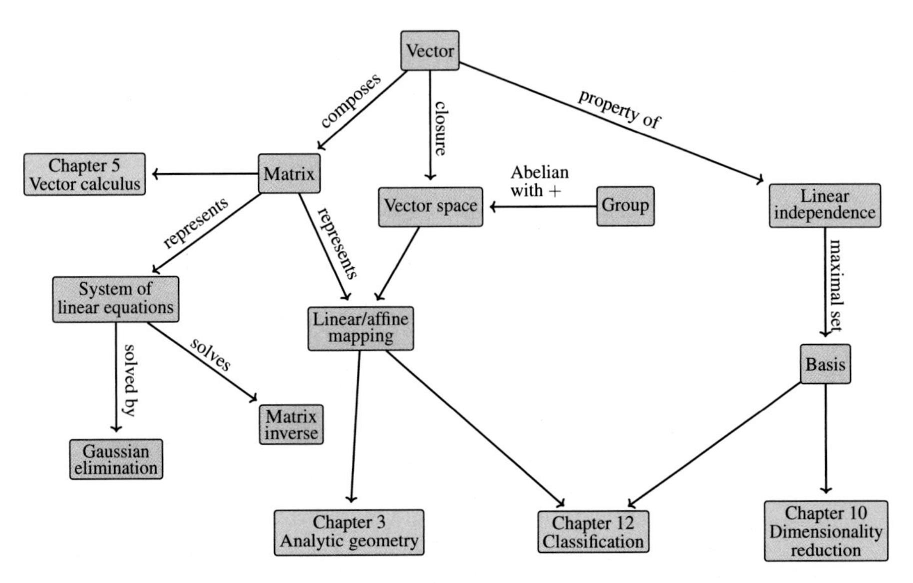

Figure 2.2 A mind map of the concepts introduced in this chapter, along with where they are used in other parts of the book.

we will use projections (to be introduced in Section 3.8) for dimensionality reduction with principal component analysis (PCA). In Chapter 9, we will discuss linear regression, where linear algebra plays a central role for solving least-squares problems.

2.1 Systems of Linear Equations

Systems of linear equations play a central part of linear algebra. Many problems can be formulated as systems of linear equations, and linear algebra gives us the tools for solving them.

Example 2.1

A company produces products $N_1, \dots, N_n$ for which resources $R_1, \dots, R_m$ are required. To produce a unit of product N_j, a_{ij} units of resource R_i are needed, where $i = 1, \dots, m$ and $j = 1, \dots, n$.

The objective is to find an optimal production plan, i.e., a plan of how many units x_j of product N_j should be produced if a total of b_i units of resource R_i are available and (ideally) no resources are left over.

If we produce $x_1, \dots, x_n$ units of the corresponding products, we need a total of

$$a_{i1}x_1 + \cdots + a_{in}x_n \tag{2.2}$$

many units of resource R_i. An optimal production plan $(x_1, \dots, x_n) \in \mathbb{R}^n$, therefore, has to satisfy the following system of equations:

$$\begin{aligned} a_{11}x_1 + \cdots + a_{1n}x_n &= b_1 \\ &\vdots \\ a_{m1}x_1 + \cdots + a_{mn}x_n &= b_m \end{aligned} \, , \tag{2.3}$$

where $a_{ij} \in \mathbb{R}$ and $b_i \in \mathbb{R}$.

Equation (2.3) is the general form of a *system of linear equations*, and $x_1, \dots, x_n$ are the *unknowns* of this system. Every n-tuple $(x_1, \dots, x_n) \in \mathbb{R}^n$ that satisfies (2.3) is a *solution* of the linear equation system.

system of linear equations

solution

Example 2.2
The system of linear equations

$$\begin{array}{ccccccc} x_1 & + & x_2 & + & x_3 & = & 3 \quad (1) \\ x_1 & - & x_2 & + & 2x_3 & = & 2 \quad (2) \\ 2x_1 & & & + & 3x_3 & = & 1 \quad (3) \end{array} \tag{2.4}$$

has *no solution:* Adding the first two equations yields $2x_1 + 3x_3 = 5$, which contradicts the third equation (3).

Let us have a look at the system of linear equations

$$\begin{array}{ccccccc} x_1 & + & x_2 & + & x_3 & = & 3 \quad (1) \\ x_1 & - & x_2 & + & 2x_3 & = & 2 \quad (2) \\ & & x_2 & + & x_3 & = & 2 \quad (3) \end{array}. \tag{2.5}$$

From the first and third equation, it follows that $x_1 = 1$. From (1) + (2), we get $2x_1 + 3x_3 = 5$, i.e., $x_3 = 1$. From (3), we then get that $x_2 = 1$. Therefore, $(1, 1, 1)$ is the only possible and *unique solution* (verify that $(1, 1, 1)$ is a solution by plugging in).

As a third example, we consider

$$\begin{array}{ccccccc} x_1 & + & x_2 & + & x_3 & = & 3 \quad (1) \\ x_1 & - & x_2 & + & 2x_3 & = & 2 \quad (2) \\ 2x_1 & & & + & 3x_3 & = & 5 \quad (3) \end{array}. \tag{2.6}$$

Since $(1) + (2) = (3)$, we can omit the third equation (redundancy). From (1) and (2), we get $2x_1 = 5 - 3x_3$ and $2x_2 = 1 + x_3$. We define $x_3 = a \in \mathbb{R}$ as a free variable, such that any triplet

$$\left(\frac{5}{2} - \frac{3}{2}a, \frac{1}{2} + \frac{1}{2}a, a\right), \quad a \in \mathbb{R} \tag{2.7}$$

is a solution of the system of linear equations, i.e., we obtain a solution set that contains *infinitely many* solutions.

In general, for a real-valued system of linear equations we obtain either no, exactly one, or infinitely many solutions. Linear regression (Chapter 9) solves a version of Example 2.1 when we cannot solve the system of linear equations.

Remark (Geometric Interpretation of Systems of Linear Equations). In a system of linear equations with two variables x_1, x_2, each linear equation defines a line on the x_1x_2-plane. Since a solution to a system of linear equations must satisfy all equations simultaneously, the solution set is the intersection of these lines. This intersection set can be a line (if the linear equations describe the same line), a point, or empty (when the lines are parallel). An illustration is given in Figure 2.3 for the system

$$\begin{aligned} 4x_1 + 4x_2 &= 5 \\ 2x_1 - 4x_2 &= 1 \end{aligned} \tag{2.8}$$

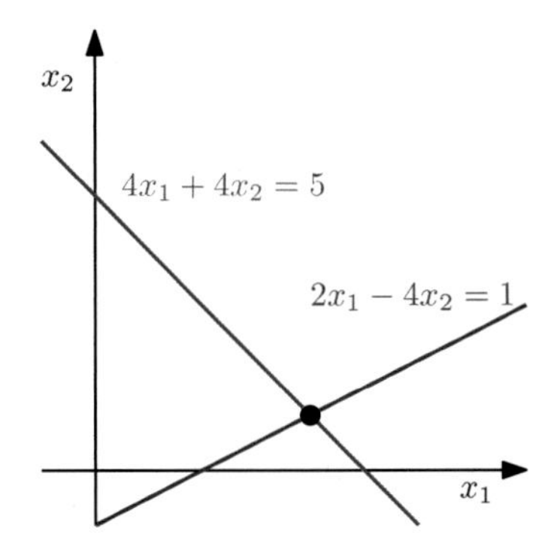

Figure 2.3 The solution space of a system of two linear equations with two variables can be geometrically interpreted as the intersection of two lines. Every linear equation represents a line.

where the solution space is the point $(x_1, x_2) = (1, \frac{1}{4})$. Similarly, for three variables, each linear equation determines a plane in three-dimensional space. When we intersect these planes, i.e., satisfy all linear equations at the same time, we can obtain a solution set that is a plane, a line, a point, or empty (when the planes have no common intersection). ♢

For a systematic approach to solving systems of linear equations, we will introduce a useful compact notation. We collect the coefficients a_{ij} into vectors and collect the vectors into matrices. In other words, we write the system from (2.3) in the following form:

$$x_1 \begin{bmatrix} a_{11} \\ \vdots \\ a_{m1} \end{bmatrix} + x_2 \begin{bmatrix} a_{12} \\ \vdots \\ a_{m2} \end{bmatrix} + \cdots + x_n \begin{bmatrix} a_{1n} \\ \vdots \\ a_{mn} \end{bmatrix} = \begin{bmatrix} b_1 \\ \vdots \\ b_m \end{bmatrix} \tag{2.9}$$

$$\iff \begin{bmatrix} a_{11} & \cdots & a_{1n} \\ \vdots & & \vdots \\ a_{m1} & \cdots & a_{mn} \end{bmatrix} \begin{bmatrix} x_1 \\ \vdots \\ x_n \end{bmatrix} = \begin{bmatrix} b_1 \\ \vdots \\ b_m \end{bmatrix}. \tag{2.10}$$

In the following, we will have a close look at these *matrices* and define computation rules. We will return to solving linear equations in Section 2.3.

2.2 Matrices

Matrices play a central role in linear algebra. They can be used to compactly represent systems of linear equations, but they also represent linear functions (linear mappings), as we will see later in Section 2.7. Before we discuss some of these interesting topics, let us first define what a matrix is and what kind of operations we can do with matrices. We will see more properties of matrices in Chapter 4.

matrix

Definition 2.1 (Matrix). With $m, n \in \mathbb{N}$ a real-valued (m, n) *matrix* $\boldsymbol{A}$ is an $m \cdot n$-tuple of elements a_{ij}, $i = 1, \ldots, m$, $j = 1, \ldots, n$, which is ordered according to a rectangular scheme consisting of m rows and n columns:

$$\boldsymbol{A} = \begin{bmatrix} a_{11} & a_{12} & \cdots & a_{1n} \\ a_{21} & a_{22} & \cdots & a_{2n} \\ \vdots & \vdots & & \vdots \\ a_{m1} & a_{m2} & \cdots & a_{mn} \end{bmatrix}, \quad a_{ij} \in \mathbb{R}. \tag{2.11}$$

By convention $(1, n)$-matrices are called *rows*, and $(m, 1)$-matrices are called *columns*. These special matrices are also called *row/column vectors*.

row

column

row vector

column vector

$\mathbb{R}^{m\times n}$ is the set of all real-valued (m, n)-matrices. $\boldsymbol{A} \in \mathbb{R}^{m\times n}$ can be equivalently represented as $\boldsymbol{a} \in \mathbb{R}^{mn}$ by stacking all n columns of the matrix into a long vector; see Figure 2.4.

Figure 2.4 By stacking its columns, a matrix $\boldsymbol{A}$ can be represented as a long vector $\boldsymbol{a}$.

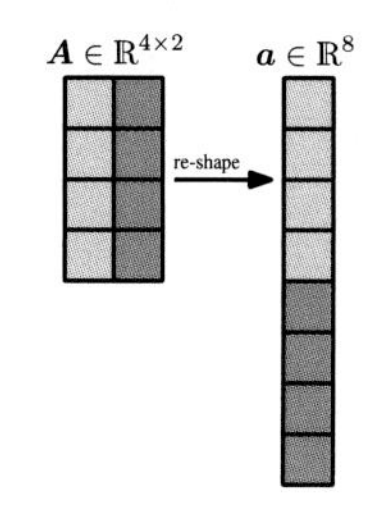

2.2.1 Matrix Addition and Multiplication

The sum of two matrices $\boldsymbol{A} \in \mathbb{R}^{m\times n}, \boldsymbol{B} \in \mathbb{R}^{m\times n}$ is defined as the element wise sum, i.e.,

$$\boldsymbol{A} + \boldsymbol{B} := \begin{bmatrix} a_{11} + b_{11} & \cdots & a_{1n} + b_{1n} \\ \vdots & & \vdots \\ a_{m1} + b_{m1} & \cdots & a_{mn} + b_{mn} \end{bmatrix} \in \mathbb{R}^{m\times n}. \tag{2.12}$$

For matrices $\boldsymbol{A} \in \mathbb{R}^{m\times n}, \boldsymbol{B} \in \mathbb{R}^{n\times k}$ the elements c_{ij} of the product $\boldsymbol{C} = \boldsymbol{AB} \in \mathbb{R}^{m\times k}$ are computed as

Note the size of the matrices.

```
C =
np.einsum('il,
lj', A, B)
```

$$c_{ij} = \sum_{l=1}^{n} a_{il} b_{lj}, \qquad i = 1, \dots, m, \quad j = 1, \dots, k. \tag{2.13}$$

This means, to compute element c_{ij} we multiply the elements of the ith row of $\boldsymbol{A}$ with the jth column of $\boldsymbol{B}$ and sum them up. Later in Section 3.2, we will call this the *dot product* of the corresponding row and column. In cases where we need to be explicit that we are performing multiplication, we use the notation $\boldsymbol{A} \cdot \boldsymbol{B}$ to denote multiplication (explicitly showing "$\cdot$").

There are n columns in $\boldsymbol{A}$ and n rows in $\boldsymbol{B}$ so that we can compute $a_{il}b_{lj}$ for $l = 1, \dots, n$.

Commonly, the dot product between two vectors $\boldsymbol{a}, \boldsymbol{b}$ is denoted by $\boldsymbol{a}^\top \boldsymbol{b}$ or $\langle \boldsymbol{a}, \boldsymbol{b} \rangle$.

Remark. Matrices can only be multiplied if their "neighboring" dimensions match. For instance, an $n \times k$-matrix $\boldsymbol{A}$ can be multiplied with a $k \times m$-matrix $\boldsymbol{B}$, but only from the left side:

$$\underbrace{\boldsymbol{A}}_{n\times k} \underbrace{\boldsymbol{B}}_{k\times m} = \underbrace{\boldsymbol{C}}_{n\times m} \tag{2.14}$$

The product $\boldsymbol{BA}$ is not defined if $m \neq n$ since the neighboring dimensions do not match. ◇

Remark. Matrix multiplication is *not* defined as an elementwise operation on matrix elements, i.e., $c_{ij} \neq a_{ij}b_{ij}$ (even if the size of $\boldsymbol{A}, \boldsymbol{B}$ was chosen appropriately). This kind of elementwise multiplication often appears in programming languages when we multiply (multidimensional) arrays with each other, and is called a *Hadamard product*. ◇

Hadamard product

Example 2.3

For $\boldsymbol{A} = \begin{bmatrix} 1 & 2 & 3 \\ 3 & 2 & 1 \end{bmatrix} \in \mathbb{R}^{2\times 3}$, $\boldsymbol{B} = \begin{bmatrix} 0 & 2 \\ 1 & -1 \\ 0 & 1 \end{bmatrix} \in \mathbb{R}^{3\times 2}$, we obtain

$$\boldsymbol{AB} = \begin{bmatrix} 1 & 2 & 3 \\ 3 & 2 & 1 \end{bmatrix} \begin{bmatrix} 0 & 2 \\ 1 & -1 \\ 0 & 1 \end{bmatrix} = \begin{bmatrix} 2 & 3 \\ 2 & 5 \end{bmatrix} \in \mathbb{R}^{2\times 2}, \tag{2.15}$$

$$\boldsymbol{BA} = \begin{bmatrix} 0 & 2 \\ 1 & -1 \\ 0 & 1 \end{bmatrix} \begin{bmatrix} 1 & 2 & 3 \\ 3 & 2 & 1 \end{bmatrix} = \begin{bmatrix} 6 & 4 & 2 \\ -2 & 0 & 2 \\ 3 & 2 & 1 \end{bmatrix} \in \mathbb{R}^{3\times 3}. \tag{2.16}$$

Figure 2.5 Even if both matrix multiplications $\boldsymbol{AB}$ and $\boldsymbol{BA}$ are defined, the dimensions of the results can be different.

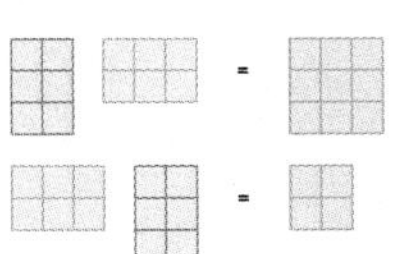

From this example, we can already see that matrix multiplication is not commutative, i.e., $\boldsymbol{AB} \neq \boldsymbol{BA}$; see also Figure 2.5 for an illustration.

identity matrix

Definition 2.2 (Identity Matrix)**.** In $\mathbb{R}^{n\times n}$, we define the *identity matrix*

$$\boldsymbol{I}_n := \begin{bmatrix} 1 & 0 & \cdots & 0 & \cdots & 0 \\ 0 & 1 & \cdots & 0 & \cdots & 0 \\ \vdots & \vdots & \ddots & \vdots & \ddots & \vdots \\ 0 & 0 & \cdots & 1 & \cdots & 0 \\ \vdots & \vdots & \ddots & \vdots & \ddots & \vdots \\ 0 & 0 & \cdots & 0 & \cdots & 1 \end{bmatrix} \in \mathbb{R}^{n\times n} \tag{2.17}$$

as the $n \times n$-matrix containing 1 on the diagonal and 0 everywhere else.

Now that we defined matrix multiplication, matrix addition, and the identity matrix, let us have a look at some properties of matrices:

associativity

- *Associativity:*

$$\forall \boldsymbol{A} \in \mathbb{R}^{m\times n}, \boldsymbol{B} \in \mathbb{R}^{n\times p}, \boldsymbol{C} \in \mathbb{R}^{p\times q} : (\boldsymbol{AB})\boldsymbol{C} = \boldsymbol{A}(\boldsymbol{BC}) \tag{2.18}$$

distributivity

- *Distributivity:*

$$\forall \boldsymbol{A}, \boldsymbol{B} \in \mathbb{R}^{m\times n}, \boldsymbol{C}, \boldsymbol{D} \in \mathbb{R}^{n\times p} : (\boldsymbol{A} + \boldsymbol{B})\boldsymbol{C} = \boldsymbol{AC} + \boldsymbol{BC} \tag{2.19a}$$

$$\boldsymbol{A}(\boldsymbol{C} + \boldsymbol{D}) = \boldsymbol{AC} + \boldsymbol{AD} \tag{2.19b}$$

- Multiplication with the identity matrix:

$$\forall \boldsymbol{A} \in \mathbb{R}^{m\times n} : \boldsymbol{I}_m \boldsymbol{A} = \boldsymbol{A}\boldsymbol{I}_n = \boldsymbol{A} \tag{2.20}$$

Note that $\boldsymbol{I}_m \neq \boldsymbol{I}_n$ for $m \neq n$.

2.2.2 Inverse and Transpose

A square matrix possesses the same number of columns and rows.

inverse

Definition 2.3 (Inverse)**.** Consider a square matrix $\boldsymbol{A} \in \mathbb{R}^{n\times n}$. Let matrix $\boldsymbol{B} \in \mathbb{R}^{n\times n}$ have the property that $\boldsymbol{AB} = \boldsymbol{I}_n = \boldsymbol{BA}$. $\boldsymbol{B}$ is called the *inverse* of $\boldsymbol{A}$ and denoted by $\boldsymbol{A}^{-1}$.

regular
invertible
nonsingular
singular
noninvertible

Unfortunately, not every matrix $\boldsymbol{A}$ possesses an inverse $\boldsymbol{A}^{-1}$. If this inverse does exist, $\boldsymbol{A}$ is called *regular/invertible/nonsingular*, otherwise *singular/noninvertible*. When the matrix inverse exists, it is unique. In Section 2.3, we will discuss a general way to compute the inverse of a matrix by solving a system of linear equations.

Remark (Existence of the Inverse of a 2×2-matrix). Consider a matrix

$$\boldsymbol{A} := \begin{bmatrix} a_{11} & a_{12} \\ a_{21} & a_{22} \end{bmatrix} \in \mathbb{R}^{2\times 2}. \tag{2.21}$$

If we multiply $\boldsymbol{A}$ with

$$\boldsymbol{B} := \begin{bmatrix} a_{22} & -a_{12} \\ -a_{21} & a_{11} \end{bmatrix} \tag{2.22}$$

we obtain

$$\boldsymbol{AB} = \begin{bmatrix} a_{11}a_{22} - a_{12}a_{21} & 0 \\ 0 & a_{11}a_{22} - a_{12}a_{21} \end{bmatrix} = (a_{11}a_{22} - a_{12}a_{21})\boldsymbol{I}. \tag{2.23}$$

Therefore,

$$\boldsymbol{A}^{-1} = \frac{1}{a_{11}a_{22} - a_{12}a_{21}} \begin{bmatrix} a_{22} & -a_{12} \\ -a_{21} & a_{11} \end{bmatrix} \tag{2.24}$$

if and only if $a_{11}a_{22} - a_{12}a_{21} \neq 0$. In Section 4.1, we will see that $a_{11}a_{22} - a_{12}a_{21}$ is the determinant of a 2×2-matrix. Furthermore, we can generally use the determinant to check whether a matrix is invertible. ◇

Example 2.4 (Inverse Matrix)
The matrices

$$\boldsymbol{A} = \begin{bmatrix} 1 & 2 & 1 \\ 4 & 4 & 5 \\ 6 & 7 & 7 \end{bmatrix}, \quad \boldsymbol{B} = \begin{bmatrix} -7 & -7 & 6 \\ 2 & 1 & -1 \\ 4 & 5 & -4 \end{bmatrix} \tag{2.25}$$

are inverse to each other since $\boldsymbol{AB} = \boldsymbol{I} = \boldsymbol{BA}$.

Definition 2.4 (Transpose). For $\boldsymbol{A} \in \mathbb{R}^{m \times n}$ the matrix $\boldsymbol{B} \in \mathbb{R}^{n \times m}$ with $b_{ij} = a_{ji}$ is called the *transpose* of $\boldsymbol{A}$. We write $\boldsymbol{B} = \boldsymbol{A}^\top$.

transpose

In general, $\boldsymbol{A}^\top$ can be obtained by writing the columns of $\boldsymbol{A}$ as the rows of $\boldsymbol{A}^\top$. The following are some important properties of inverses and transposes:

The main diagonal (sometimes called "principal diagonal," "primary diagonal," "leading diagonal," or "major diagonal") of a matrix $\boldsymbol{A}$ is the collection of entries A_{ij} where $i = j$.

The scalar case of (2.28) is $\frac{1}{2+4} = \frac{1}{6} \neq \frac{1}{2} + \frac{1}{4}$.

$$\boldsymbol{AA}^{-1} = \boldsymbol{I} = \boldsymbol{A}^{-1}\boldsymbol{A} \tag{2.26}$$

$$(\boldsymbol{AB})^{-1} = \boldsymbol{B}^{-1}\boldsymbol{A}^{-1} \tag{2.27}$$

$$(\boldsymbol{A} + \boldsymbol{B})^{-1} \neq \boldsymbol{A}^{-1} + \boldsymbol{B}^{-1} \tag{2.28}$$

$$(\boldsymbol{A}^\top)^\top = \boldsymbol{A} \tag{2.29}$$

$$(\boldsymbol{A} + \boldsymbol{B})^\top = \boldsymbol{A}^\top + \boldsymbol{B}^\top \tag{2.30}$$

$$(\boldsymbol{AB})^\top = \boldsymbol{B}^\top\boldsymbol{A}^\top \tag{2.31}$$

Definition 2.5 (Symmetric Matrix). A matrix $\boldsymbol{A} \in \mathbb{R}^{n \times n}$ is *symmetric* if $\boldsymbol{A} = \boldsymbol{A}^\top$.

symmetric matrix

Note that only (n, n)-matrices can be symmetric. Generally, we call (n, n)-matrices also *square matrices* because they possess the same number of rows and columns. Moreover, if $\boldsymbol{A}$ is invertible, then so is $\boldsymbol{A}^\top$, and $(\boldsymbol{A}^{-1})^\top = (\boldsymbol{A}^\top)^{-1} =: \boldsymbol{A}^{-\top}$.

square matrix

Remark (Sum and Product of Symmetric Matrices). The sum of symmetric matrices $\boldsymbol{A}, \boldsymbol{B} \in \mathbb{R}^{n \times n}$ is always symmetric. However, although their product is always defined, it is generally not symmetric:

$$\begin{bmatrix} 1 & 0 \\ 0 & 0 \end{bmatrix} \begin{bmatrix} 1 & 1 \\ 1 & 1 \end{bmatrix} = \begin{bmatrix} 1 & 1 \\ 0 & 0 \end{bmatrix}. \tag{2.32}$$

2.2.3 Multiplication by a Scalar

Let us look at what happens to matrices when they are multiplied by a scalar $\lambda \in \mathbb{R}$. Let $\boldsymbol{A} \in \mathbb{R}^{m \times n}$ and $\lambda \in \mathbb{R}$. Then $\lambda \boldsymbol{A} = \boldsymbol{K}$, $K_{ij} = \lambda\, a_{ij}$. Practically, λ scales each element of $\boldsymbol{A}$. For $\lambda, \psi \in \mathbb{R}$, the following holds:

associativity

- *Associativity:*
  $(\lambda\psi)\boldsymbol{C} = \lambda(\psi\boldsymbol{C}), \quad \boldsymbol{C} \in \mathbb{R}^{m \times n}$
- $\lambda(\boldsymbol{BC}) = (\lambda\boldsymbol{B})\boldsymbol{C} = \boldsymbol{B}(\lambda\boldsymbol{C}) = (\boldsymbol{BC})\lambda, \quad \boldsymbol{B} \in \mathbb{R}^{m \times n}, \boldsymbol{C} \in \mathbb{R}^{n \times k}$.
  Note that this allows us to move scalar values around.

distributivity

- $(\lambda\boldsymbol{C})^\top = \boldsymbol{C}^\top\lambda^\top = \boldsymbol{C}^\top\lambda = \lambda\boldsymbol{C}^\top$ since $\lambda = \lambda^\top$ for all $\lambda \in \mathbb{R}$.
- *Distributivity:*
  $(\lambda + \psi)\boldsymbol{C} = \lambda\boldsymbol{C} + \psi\boldsymbol{C}, \quad \boldsymbol{C} \in \mathbb{R}^{m \times n}$
  $\lambda(\boldsymbol{B} + \boldsymbol{C}) = \lambda\boldsymbol{B} + \lambda\boldsymbol{C}, \quad \boldsymbol{B}, \boldsymbol{C} \in \mathbb{R}^{m \times n}$

Example 2.5 (Distributivity)
If we define

$$\boldsymbol{C} := \begin{bmatrix} 1 & 2 \\ 3 & 4 \end{bmatrix}, \tag{2.33}$$

then for any $\lambda, \psi \in \mathbb{R}$ we obtain

$$(\lambda + \psi)\boldsymbol{C} = \begin{bmatrix} (\lambda+\psi)1 & (\lambda+\psi)2 \\ (\lambda+\psi)3 & (\lambda+\psi)4 \end{bmatrix} = \begin{bmatrix} \lambda+\psi & 2\lambda+2\psi \\ 3\lambda+3\psi & 4\lambda+4\psi \end{bmatrix} \tag{2.34a}$$

$$= \begin{bmatrix} \lambda & 2\lambda \\ 3\lambda & 4\lambda \end{bmatrix} + \begin{bmatrix} \psi & 2\psi \\ 3\psi & 4\psi \end{bmatrix} = \lambda\boldsymbol{C} + \psi\boldsymbol{C}\,. \tag{2.34b}$$

2.2.4 Compact Representations of Systems of Linear Equations

If we consider the system of linear equations

$$\begin{aligned} 2x_1 + 3x_2 + 5x_3 &= 1 \\ 4x_1 - 2x_2 - 7x_3 &= 8 \\ 9x_1 + 5x_2 - 3x_3 &= 2 \end{aligned} \tag{2.35}$$

and use the rules for matrix multiplication, we can write this equation system in a more compact form as

$$\begin{bmatrix} 2 & 3 & 5 \\ 4 & -2 & -7 \\ 9 & 5 & -3 \end{bmatrix} \begin{bmatrix} x_1 \\ x_2 \\ x_3 \end{bmatrix} = \begin{bmatrix} 1 \\ 8 \\ 2 \end{bmatrix}. \tag{2.36}$$

Note that x_1 scales the first column, x_2 the second one, and x_3 the third one.

Generally, a system of linear equations can be compactly represented in their matrix form as $\boldsymbol{Ax} = \boldsymbol{b}$; see (2.3), and the product $\boldsymbol{Ax}$ is a (linear) combination of the columns of $\boldsymbol{A}$. We will discuss linear combinations in more detail in Section 2.5.

2.3 Solving Systems of Linear Equations

In (2.3), we introduced the general form of an equation system, i.e.,

$$\begin{aligned} a_{11}x_1 + \cdots + a_{1n}x_n &= b_1 \\ &\vdots \\ a_{m1}x_1 + \cdots + a_{mn}x_n &= b_m \,, \end{aligned} \tag{2.37}$$

where $a_{ij} \in \mathbb{R}$ and $b_i \in \mathbb{R}$ are known constants and x_j are unknowns, $i = 1, \ldots, m$, $j = 1, \ldots, n$. Thus far, we saw that matrices can be used as a compact way of formulating systems of linear equations so that we can write $\boldsymbol{A}\boldsymbol{x} = \boldsymbol{b}$; see (2.10). Moreover, we defined basic matrix operations, such as addition and multiplication of matrices. In the following, we will focus on solving systems of linear equations and provide an algorithm for finding the inverse of a matrix.

2.3.1 Particular and General Solution

Before discussing how to generally solve systems of linear equations, let us have a look at an example. Consider the system of equations

$$\begin{bmatrix} 1 & 0 & 8 & -4 \\ 0 & 1 & 2 & 12 \end{bmatrix} \begin{bmatrix} x_1 \\ x_2 \\ x_3 \\ x_4 \end{bmatrix} = \begin{bmatrix} 42 \\ 8 \end{bmatrix} . \tag{2.38}$$

The system has two equations and four unknowns. Therefore, in general we would expect infinitely many solutions. This system of equations is in a particularly easy form, where the first two columns consist of a 1 and a 0. Remember that we want to find scalars $x_1, \ldots, x_4$, such that $\sum_{i=1}^{4} x_i \boldsymbol{c}_i = \boldsymbol{b}$, where we define $\boldsymbol{c}_i$ to be the ith column of the matrix and $\boldsymbol{b}$ the right-hand side of (2.38). A solution to the problem in (2.38) can be found immediately by taking 42 times the first column and 8 times the second column so that

$$\boldsymbol{b} = \begin{bmatrix} 42 \\ 8 \end{bmatrix} = 42 \begin{bmatrix} 1 \\ 0 \end{bmatrix} + 8 \begin{bmatrix} 0 \\ 1 \end{bmatrix} . \tag{2.39}$$

Therefore, a solution is $[42, 8, 0, 0]^\top$. This solution is called a *particular solution* or *special solution*. However, this is not the only solution of this system of linear equations. To capture all the other solutions, we need to be creative in generating $\boldsymbol{0}$ in a nontrivial way using the columns of the matrix: Adding $\boldsymbol{0}$ to our special solution does not change the special solution. To do so, we express the third column using the first two columns (which are of this very simple form)

particular solution
special solution

$$\begin{bmatrix} 8 \\ 2 \end{bmatrix} = 8 \begin{bmatrix} 1 \\ 0 \end{bmatrix} + 2 \begin{bmatrix} 0 \\ 1 \end{bmatrix} \tag{2.40}$$

so that $\boldsymbol{0} = 8\boldsymbol{c}_1 + 2\boldsymbol{c}_2 - 1\boldsymbol{c}_3 + 0\boldsymbol{c}_4$ and $(x_1, x_2, x_3, x_4) = (8, 2, -1, 0)$. In fact, any scaling of this solution by $\lambda_1 \in \mathbb{R}$ produces the $\boldsymbol{0}$ vector, i.e.,

$$\begin{bmatrix} 1 & 0 & 8 & -4 \\ 0 & 1 & 2 & 12 \end{bmatrix} \left(\lambda_1 \begin{bmatrix} 8 \\ 2 \\ -1 \\ 0 \end{bmatrix} \right) = \lambda_1 (8\boldsymbol{c}_1 + 2\boldsymbol{c}_2 - \boldsymbol{c}_3) = \boldsymbol{0} . \tag{2.41}$$

Following the same line of reasoning, we express the fourth column of the matrix in (2.38) using the first two columns and generate another set of nontrivial versions of $\mathbf{0}$ as

$$\begin{bmatrix} 1 & 0 & 8 & -4 \\ 0 & 1 & 2 & 12 \end{bmatrix} \left(\lambda_2 \begin{bmatrix} -4 \\ 12 \\ 0 \\ -1 \end{bmatrix} \right) = \lambda_2(-4\boldsymbol{c}_1 + 12\boldsymbol{c}_2 - \boldsymbol{c}_4) = \mathbf{0} \tag{2.42}$$

for any $\lambda_2 \in \mathbb{R}$. Putting everything together, we obtain all solutions of the equation system in (2.38), which is called the *general solution*, as the set

general solution

$$\left\{ \boldsymbol{x} \in \mathbb{R}^4 : \boldsymbol{x} = \begin{bmatrix} 42 \\ 8 \\ 0 \\ 0 \end{bmatrix} + \lambda_1 \begin{bmatrix} 8 \\ 2 \\ -1 \\ 0 \end{bmatrix} + \lambda_2 \begin{bmatrix} -4 \\ 12 \\ 0 \\ -1 \end{bmatrix}, \lambda_1, \lambda_2 \in \mathbb{R} \right\}. \tag{2.43}$$

Remark. The general approach we followed consisted of the following three steps:

1. Find a particular solution to $\boldsymbol{Ax} = \boldsymbol{b}$.
2. Find all solutions to $\boldsymbol{Ax} = \mathbf{0}$.
3. Combine the solutions from steps 1 and 2 to the general solution.

Neither the general nor the particular solution is unique. ◇

The system of linear equations in the preceding example was easy to solve because the matrix in (2.38) has this particularly convenient form, which allowed us to find the particular and the general solution by inspection. However, general equation systems are not of this simple form. Fortunately, there exists a constructive algorithmic way of transforming any system of linear equations into this particularly simple form: Gaussian elimination. Key to Gaussian elimination are elementary transformations of systems of linear equations, which transform the equation system into a simple form. Then we can apply the three steps to the simple form that we just discussed in the context of the example in (2.38).

2.3.2 Elementary Transformations

elementary transformations

Key to solving a system of linear equations are *elementary transformations* that keep the solution set the same, but that transform the equation system into a simpler form:

- Exchange of two equations (rows in the matrix representing the system of equations)
- Multiplication of an equation (row) with a constant $\lambda \in \mathbb{R}\backslash\{0\}$
- Addition of two equations (rows)

Example 2.6

For $a \in \mathbb{R}$, we seek all solutions of the following system of equations:

$$\begin{array}{rcrcrcrcrcr} -2x_1 & + & 4x_2 & - & 2x_3 & - & x_4 & + & 4x_5 & = & -3 \\ 4x_1 & - & 8x_2 & + & 3x_3 & - & 3x_4 & + & x_5 & = & 2 \\ x_1 & - & 2x_2 & + & x_3 & - & x_4 & + & x_5 & = & 0 \\ x_1 & - & 2x_2 & & & - & 3x_4 & + & 4x_5 & = & a \end{array}. \tag{2.44}$$

We start by converting this system of equations into the compact matrix notation $\boldsymbol{Ax} = \boldsymbol{b}$. We no longer mention the variables $\boldsymbol{x}$ explicitly and build the *augmented matrix* (in the form $[\boldsymbol{A} \mid \boldsymbol{b}]$)

augmented matrix

$$\left[\begin{array}{ccccc|c} -2 & 4 & -2 & -1 & 4 & -3 \\ 4 & -8 & 3 & -3 & 1 & 2 \\ 1 & -2 & 1 & -1 & 1 & 0 \\ 1 & -2 & 0 & -3 & 4 & a \end{array}\right] \begin{array}{l} \text{Swap with } R_3 \\ \\ \text{Swap with } R_1 \\ \\ \end{array}$$

where we used the vertical line to separate the left-hand side from the right-hand side in (2.44). We use $\rightsquigarrow$ to indicate a transformation of the augmented matrix using elementary transformations.

The augmented matrix $[\boldsymbol{A} \mid \boldsymbol{b}]$ compactly represents the system of linear equations $\boldsymbol{Ax} = \boldsymbol{b}$.

Swapping Rows 1 and 3 leads to

$$\left[\begin{array}{ccccc|c} 1 & -2 & 1 & -1 & 1 & 0 \\ 4 & -8 & 3 & -3 & 1 & 2 \\ -2 & 4 & -2 & -1 & 4 & -3 \\ 1 & -2 & 0 & -3 & 4 & a \end{array}\right] \begin{array}{l} \\ -4R_1 \\ +2R_1 \\ -R_1 \end{array}$$

When we now apply the indicated transformations (e.g., subtract Row 1 four times from Row 2), we obtain

$$\left[\begin{array}{ccccc|c} 1 & -2 & 1 & -1 & 1 & 0 \\ 0 & 0 & -1 & 1 & -3 & 2 \\ 0 & 0 & 0 & -3 & 6 & -3 \\ 0 & 0 & -1 & -2 & 3 & a \end{array}\right] \begin{array}{l} \\ \\ \\ -R_2 - R_3 \end{array}$$

$$\rightsquigarrow \left[\begin{array}{ccccc|c} 1 & -2 & 1 & -1 & 1 & 0 \\ 0 & 0 & -1 & 1 & -3 & 2 \\ 0 & 0 & 0 & -3 & 6 & -3 \\ 0 & 0 & 0 & 0 & 0 & a+1 \end{array}\right] \begin{array}{l} \\ \cdot(-1) \\ \cdot(-\frac{1}{3}) \\ \\ \end{array}$$

$$\rightsquigarrow \left[\begin{array}{ccccc|c} 1 & -2 & 1 & -1 & 1 & 0 \\ 0 & 0 & 1 & -1 & 3 & -2 \\ 0 & 0 & 0 & 1 & -2 & 1 \\ 0 & 0 & 0 & 0 & 0 & a+1 \end{array}\right]$$

This (augmented) matrix is in a convenient form, the *row-echelon form* (REF). Reverting this compact notation back into the explicit notation with the variables we seek, we obtain

row-echelon form

$$\begin{array}{rcrcrcrcrcr} x_1 & - & 2x_2 & + & x_3 & - & x_4 & + & x_5 & = & 0 \\ & & & & x_3 & - & x_4 & + & 3x_5 & = & -2 \\ & & & & & & x_4 & - & 2x_5 & = & 1 \\ & & & & & & & & 0 & = & a+1 \end{array} . \tag{2.45}$$

Only for $a = -1$ this system can be solved. A *particular solution* is

particular solution

$$\begin{bmatrix} x_1 \\ x_2 \\ x_3 \\ x_4 \\ x_5 \end{bmatrix} = \begin{bmatrix} 2 \\ 0 \\ -1 \\ 1 \\ 0 \end{bmatrix} . \tag{2.46}$$

general solution

The *general solution*, which captures the set of all possible solutions, is

$$\left\{ \boldsymbol{x} \in \mathbb{R}^5 : \boldsymbol{x} = \begin{bmatrix} 2 \\ 0 \\ -1 \\ 1 \\ 0 \end{bmatrix} + \lambda_1 \begin{bmatrix} 2 \\ 1 \\ 0 \\ 0 \\ 0 \end{bmatrix} + \lambda_2 \begin{bmatrix} 2 \\ 0 \\ -1 \\ 2 \\ 1 \end{bmatrix}, \quad \lambda_1, \lambda_2 \in \mathbb{R} \right\}. \tag{2.47}$$

In the following, we will detail a constructive way to obtain a particular and general solution of a system of linear equations.

pivot

Remark (Pivots and Staircase Structure). The leading coefficient of a row (first nonzero number from the left) is called the *pivot* and is always strictly to the right of the pivot of the row above it. Therefore, any equation system in row-echelon form always has a "staircase" structure. ♢

row-echelon form

Definition 2.6 (Row-Echelon Form). A matrix is in *row-echelon form* if

- All rows that contain only zeros are at the bottom of the matrix; correspondingly, all rows that contain at least one nonzero element are on top of rows that contain only zeros.
- Looking at nonzero rows only, the first nonzero number from the left (also called the *pivot* or the *leading coefficient*) is always strictly to the right of the pivot of the row above it.

pivot
leading coefficient
In other texts, it is sometimes required that the pivot is 1.
basic variable
free variable

Remark (Basic and Free Variables). The variables corresponding to the pivots in the row-echelon form are called *basic variable*, and the other variables are *free variable*. For example, in (2.45), x_1, x_3, x_4 are basic variables, whereas x_2, x_5 are free variables. ♢

Remark (Obtaining a Particular Solution). The row-echelon form makes our lives easier when we need to determine a particular solution. To do this, we express the right-hand side of the equation system using the pivot columns, such that $\boldsymbol{b} = \sum_{i=1}^{P} \lambda_i \boldsymbol{p}_i$, where $\boldsymbol{p}_i$, $i = 1, \ldots, P$, are the pivot columns. The λ_i are determined easiest if we start with the rightmost pivot column and work our way to the left.

In the previous example, we would try to find $\lambda_1, \lambda_2, \lambda_3$ so that

$$\lambda_1 \begin{bmatrix} 1 \\ 0 \\ 0 \\ 0 \end{bmatrix} + \lambda_2 \begin{bmatrix} 1 \\ 1 \\ 0 \\ 0 \end{bmatrix} + \lambda_3 \begin{bmatrix} -1 \\ -1 \\ 1 \\ 0 \end{bmatrix} = \begin{bmatrix} 0 \\ -2 \\ 1 \\ 0 \end{bmatrix}. \tag{2.48}$$

From here, we find relatively directly that $\lambda_3 = 1, \lambda_2 = -1, \lambda_1 = 2$. When we put everything together, we must not forget the nonpivot columns for which we set the coefficients implicitly to 0. Therefore, we get the particular solution $\boldsymbol{x} = [2, 0, -1, 1, 0]^\top$. ♢

reduced row-echelon form

Remark (Reduced Row-Echelon Form). An equation system is in *reduced row-echelon form* (also: *row-reduced-echelon form* or *row canonical form*) if

- It is in row-echelon form.
- Every pivot is 1.
- The pivot is the only nonzero entry in its column.

The reduced row-echelon form will play an important role later in Section 2.3.3 because it allows us to determine the general solution of a system of linear equations in a straightforward way.

Gaussian elimination

Remark (Gaussian Elimination). *Gaussian elimination* is an algorithm that performs elementary transformations to bring a system of linear equations into reduced row-echelon form. ◇

Example 2.7 (Reduced Row-Echelon Form)
Verify that the following matrix is in reduced row-echelon form (the pivots are in **bold**):

$$\boldsymbol{A} = \begin{bmatrix} \mathbf{1} & 3 & 0 & 0 & 3 \\ 0 & 0 & \mathbf{1} & 0 & 9 \\ 0 & 0 & 0 & \mathbf{1} & -4 \end{bmatrix}. \tag{2.49}$$

The key idea for finding the solutions of $\boldsymbol{Ax} = \boldsymbol{0}$ is to look at the *nonpivot columns*, which we will need to express as a (linear) combination of the pivot columns. The reduced row-echelon form makes this relatively straightforward, and we express the nonpivot columns in terms of sums and multiples of the pivot columns that are on their left: The second column is three times the first column (we can ignore the pivot columns on the right of the second column). Therefore, to obtain $\boldsymbol{0}$, we need to subtract the second column from three times the first column. Now we look at the fifth column, which is our second nonpivot column. The fifth column can be expressed as 3 times the first pivot column, 9 times the second pivot column, and -4 times the third pivot column. We need to keep track of the indices of the pivot columns and translate this into 3 times the first column, 0 times the second column (which is a nonpivot column), 9 times the third column (which is our second pivot column), and -4 times the fourth column (which is the third pivot column). Then we need to subtract the fifth column to obtain $\boldsymbol{0}$. In the end, we are still solving a homogeneous equation system.

To summarize, all solutions of $\boldsymbol{Ax} = \boldsymbol{0}, \boldsymbol{x} \in \mathbb{R}^5$ are given by

$$\left\{ \boldsymbol{x} \in \mathbb{R}^5 : \boldsymbol{x} = \lambda_1 \begin{bmatrix} 3 \\ -1 \\ 0 \\ 0 \\ 0 \end{bmatrix} + \lambda_2 \begin{bmatrix} 3 \\ 0 \\ 9 \\ -4 \\ -1 \end{bmatrix}, \quad \lambda_1, \lambda_2 \in \mathbb{R} \right\}. \tag{2.50}$$

2.3.3 The Minus-1 Trick

In the following, we introduce a practical trick for reading out the solutions $\boldsymbol{x}$ of a homogeneous system of linear equations $\boldsymbol{A}\boldsymbol{x} = \boldsymbol{0}$, where $\boldsymbol{A} \in \mathbb{R}^{k \times n}, \boldsymbol{x} \in \mathbb{R}^n$.

To start, we assume that $\boldsymbol{A}$ is in reduced row-echelon form without any rows that just contain zeros, i.e.,

$$\boldsymbol{A} = \begin{bmatrix} 0 & \cdots & 0 & \mathbf{1} & * & \cdots & * & 0 & * & \cdots & * & 0 & * & \cdots & * \\ \vdots & & \vdots & 0 & 0 & \cdots & 0 & \mathbf{1} & * & \cdots & * & \vdots & \vdots & & \vdots \\ \vdots & & \vdots & \vdots & \vdots & & \vdots & 0 & \vdots & & \vdots & \vdots & \vdots & & \vdots \\ \vdots & & \vdots & \vdots & \vdots & & \vdots & \vdots & \vdots & & \vdots & 0 & \vdots & & \vdots \\ 0 & \cdots & 0 & 0 & 0 & \cdots & 0 & 0 & 0 & \cdots & 0 & \mathbf{1} & * & \cdots & * \end{bmatrix}, \tag{2.51}$$

where $*$ can be an arbitrary real number, with the constraints that the first nonzero entry per row must be 1 and all other entries in the corresponding column must be 0. The columns $j_1, \ldots, j_k$ with the pivots (marked in **bold**) are the standard unit vectors $\boldsymbol{e}_1, \ldots, \boldsymbol{e}_k \in \mathbb{R}^k$. We extend this matrix to an $n \times n$-matrix $\tilde{\boldsymbol{A}}$ by adding $n - k$ rows of the form

$$\begin{bmatrix} 0 & \cdots & 0 & -1 & 0 & \cdots & 0 \end{bmatrix} \tag{2.52}$$

so that the diagonal of the augmented matrix $\tilde{\boldsymbol{A}}$ contains either 1 or -1. Then the columns of $\tilde{\boldsymbol{A}}$ that contain the -1 as pivots are solutions of the homogeneous equation system $\boldsymbol{A}\boldsymbol{x} = \boldsymbol{0}$. To be more precise, these columns form a basis (Section 2.6.1) of the solution space of $\boldsymbol{A}\boldsymbol{x} = \boldsymbol{0}$, which we will later call the *kernel* or *null space* (see Section 2.7.3).

kernel
null space

Example 2.8 (Minus-1 Trick)

Let us revisit the matrix in (2.49), which is already in REF:

$$\boldsymbol{A} = \begin{bmatrix} 1 & 3 & 0 & 0 & 3 \\ 0 & 0 & 1 & 0 & 9 \\ 0 & 0 & 0 & 1 & -4 \end{bmatrix}. \tag{2.53}$$

We now augment this matrix to a 5×5 matrix by adding rows of the form (2.52) at the places where the pivots on the diagonal are missing and obtain

$$\tilde{\boldsymbol{A}} = \begin{bmatrix} 1 & 3 & 0 & 0 & 3 \\ 0 & -1 & 0 & 0 & 0 \\ 0 & 0 & 1 & 0 & 9 \\ 0 & 0 & 0 & 1 & -4 \\ 0 & 0 & 0 & 0 & -1 \end{bmatrix}. \tag{2.54}$$

From this form, we can immediately read out the solutions of $\boldsymbol{A}\boldsymbol{x} = \boldsymbol{0}$ by taking the columns of $\tilde{\boldsymbol{A}}$, which contain -1 on the diagonal:

$$\left\{ \boldsymbol{x} \in \mathbb{R}^5 : \boldsymbol{x} = \lambda_1 \begin{bmatrix} 3 \\ -1 \\ 0 \\ 0 \\ 0 \end{bmatrix} + \lambda_2 \begin{bmatrix} 3 \\ 0 \\ 9 \\ -4 \\ -1 \end{bmatrix}, \quad \lambda_1, \lambda_2 \in \mathbb{R} \right\}, \tag{2.55}$$

which is identical to the solution in (2.50) that we obtained by "insight."

Calculating the Inverse

To compute the inverse A^{-1} of $A \in \mathbb{R}^{n\times n}$, we need to find a matrix X that satisfies $AX = I_n$. Then $X = A^{-1}$. We can write this down as a set of simultaneous linear equations $AX = I_n$, where we solve for $X = [x_1|\cdots|x_n]$. We use the augmented matrix notation for a compact representation of this set of systems of linear equations and obtain

$$\left[A|I_n\right] \quad \rightsquigarrow \cdots \rightsquigarrow \quad \left[I_n|A^{-1}\right]. \tag{2.56}$$

This means that if we bring the augmented equation system into reduced row-echelon form, we can read out the inverse on the right-hand side of the equation system. Hence, determining the inverse of a matrix is equivalent to solving systems of linear equations.

Example 2.9 (Calculating an Inverse Matrix by Gaussian Elimination)
To determine the inverse of

$$A = \begin{bmatrix} 1 & 0 & 2 & 0 \\ 1 & 1 & 0 & 0 \\ 1 & 2 & 0 & 1 \\ 1 & 1 & 1 & 1 \end{bmatrix} \tag{2.57}$$

we write down the augmented matrix

$$\left[\begin{array}{cccc|cccc} 1 & 0 & 2 & 0 & 1 & 0 & 0 & 0 \\ 1 & 1 & 0 & 0 & 0 & 1 & 0 & 0 \\ 1 & 2 & 0 & 1 & 0 & 0 & 1 & 0 \\ 1 & 1 & 1 & 1 & 0 & 0 & 0 & 1 \end{array}\right]$$

and use Gaussian elimination to bring it into reduced row-echelon form

$$\left[\begin{array}{cccc|cccc} 1 & 0 & 0 & 0 & -1 & 2 & -2 & 2 \\ 0 & 1 & 0 & 0 & 1 & -1 & 2 & -2 \\ 0 & 0 & 1 & 0 & 1 & -1 & 1 & -1 \\ 0 & 0 & 0 & 1 & -1 & 0 & -1 & 2 \end{array}\right],$$

such that the desired inverse is given as its right-hand side:

$$A^{-1} = \begin{bmatrix} -1 & 2 & -2 & 2 \\ 1 & -1 & 2 & -2 \\ 1 & -1 & 1 & -1 \\ -1 & 0 & -1 & 2 \end{bmatrix}. \tag{2.58}$$

We can verify that (2.58) is indeed the inverse by performing the multiplication AA^{-1} and observing that we recover I_4.

2.3.4 Algorithms for Solving a System of Linear Equations

In the following, we briefly discuss approaches to solving a system of linear equations of the form $Ax = b$. We make the assumption that a solution exists. Should there be no solution, we need to resort to approximate solutions, which we do not cover in this chapter. One way to solve the approximate problem is using the approach of linear regression, which we discuss in detail in Chapter 9.

In special cases, we may be able to determine the inverse $\boldsymbol{A}^{-1}$, such that the solution of $\boldsymbol{A}\boldsymbol{x} = \boldsymbol{b}$ is given as $\boldsymbol{x} = \boldsymbol{A}^{-1}\boldsymbol{b}$. However, this is only possible if $\boldsymbol{A}$ is a square matrix and invertible, which is often not the case. Otherwise, under mild assumptions (i.e., $\boldsymbol{A}$ needs to have linearly independent columns) we can use the transformation

$$\boldsymbol{A}\boldsymbol{x} = \boldsymbol{b} \iff \boldsymbol{A}^\top \boldsymbol{A}\boldsymbol{x} = \boldsymbol{A}^\top \boldsymbol{b} \iff \boldsymbol{x} = (\boldsymbol{A}^\top \boldsymbol{A})^{-1}\boldsymbol{A}^\top \boldsymbol{b} \tag{2.59}$$

Moore–Penrose pseudo-inverse

and use the *Moore–Penrose pseudo-inverse* $(\boldsymbol{A}^\top \boldsymbol{A})^{-1}\boldsymbol{A}^\top$ to determine the solution (2.59) that solves $\boldsymbol{A}\boldsymbol{x} = \boldsymbol{b}$, which also corresponds to the minimum norm least-squares solution. A disadvantage of this approach is that it requires many computations for the matrix-matrix product and computing the inverse of $\boldsymbol{A}^\top \boldsymbol{A}$. Moreover, for reasons of numerical precision it is generally not recommended to compute the inverse or pseudo-inverse. In the following, we therefore briefly discuss alternative approaches to solving systems of linear equations.

Gaussian elimination plays an important role when computing determinants (Section 4.1), checking whether a set of vectors is linearly independent (Section 2.5), computing the inverse of a matrix (Section 2.2.2), computing the rank of a matrix (Section 2.6.2), and determining a basis of a vector space (Section 2.6.1). Gaussian elimination is an intuitive and constructive way to solve a system of linear equations with thousands of variables. However, for systems with millions of variables, it is impractical as the required number of arithmetic operations scales cubically in the number of simultaneous equations.

In practice, systems of many linear equations are solved indirectly, by either stationary iterative methods, such as the Richardson method, the Jacobi method, the Gauß–Seidel method, and the successive overrelaxation method, or Krylov subspace methods, such as conjugate gradients, generalized minimal residual, or biconjugate gradients. We refer to the books by Stoer and Burlirsch (2002), Strang (2003), and Liesen and Mehrmann (2015) for further details.

Let $\boldsymbol{x}_*$ be a solution of $\boldsymbol{A}\boldsymbol{x} = \boldsymbol{b}$. The key idea of these iterative methods is to set up an iteration of the form

$$\boldsymbol{x}^{(k+1)} = \boldsymbol{C}\boldsymbol{x}^{(k)} + \boldsymbol{d} \tag{2.60}$$

for suitable $\boldsymbol{C}$ and $\boldsymbol{d}$ that reduces the residual error $\|\boldsymbol{x}^{(k+1)} - \boldsymbol{x}_*\|$ in every iteration and converges to $\boldsymbol{x}_*$. We will introduce norms $\|\cdot\|$, which allow us to compute similarities between vectors, in Section 3.1.

2.4 Vector Spaces

Thus far, we have looked at systems of linear equations and how to solve them (Section 2.3). We saw that systems of linear equations can be compactly represented using matrix-vector notation (2.10). In the following, we will have a closer look at vector spaces, i.e., a structured space in which vectors live.

In the beginning of this chapter, we informally characterized vectors as objects that can be added together and multiplied by a scalar, and they remain objects of the same type. Now we are ready to formalize this, and we will start by introducing the concept of a group, which is a set of elements and an operation defined on these elements that keeps some structure of the set intact.

2.4.1 Groups

Groups play an important role in computer science. Besides providing a fundamental framework for operations on sets, they are heavily used in cryptography, coding theory, and graphics.

Definition 2.7 (Group). Consider a set $\mathcal{G}$ and an operation $\otimes : \mathcal{G} \times \mathcal{G} \to \mathcal{G}$ defined on $\mathcal{G}$. Then $G := (\mathcal{G}, \otimes)$ is called a *group* if the following hold:

group

closure

associativity

neutral element

inverse element

1. *Closure* of $\mathcal{G}$ under $\otimes$: $\forall x, y \in \mathcal{G} : x \otimes y \in \mathcal{G}$
2. *Associativity:* $\forall x, y, z \in \mathcal{G} : (x \otimes y) \otimes z = x \otimes (y \otimes z)$
3. *Neutral element:* $\exists e \in \mathcal{G}\, \forall x \in \mathcal{G} : x \otimes e = x$ and $e \otimes x = x$
4. *Inverse element:* $\forall x \in \mathcal{G}\, \exists y \in \mathcal{G} : x \otimes y = e$ and $y \otimes x = e$. We often write x^{-1} to denote the inverse element of x.

Remark. The inverse element is defined with respect to the operation $\otimes$ and does not necessarily mean $\frac{1}{x}$. ♢

If additionally $\forall x, y \in \mathcal{G} : x \otimes y = y \otimes x$, then $G = (\mathcal{G}, \otimes)$ is an *Abelian group* (commutative).

Abelian group

Example 2.10 (Groups)
Let us have a look at some examples of sets with associated operations and see whether they are groups:

- $(\mathbb{Z}, +)$ is a group.
- $(\mathbb{N}_0, +)$ is not a group: Although $(\mathbb{N}_0, +)$ possesses a neutral element (0), the inverse elements are missing.

$\mathbb{N}_0 := \mathbb{N} \cup \{0\}$

- $(\mathbb{Z}, \cdot)$ is not a group: Although $(\mathbb{Z}, \cdot)$ contains a neutral element (1), the inverse elements for any $z \in \mathbb{Z}, z \neq \pm 1$, are missing.
- $(\mathbb{R}, \cdot)$ is not a group since 0 does not possess an inverse element.
- $(\mathbb{R}\backslash\{0\}, \cdot)$ is Abelian.
- $(\mathbb{R}^n, +), (\mathbb{Z}^n, +), n \in \mathbb{N}$ are Abelian if $+$ is defined componentwise, i.e.,

$$(x_1, \cdots, x_n) + (y_1, \cdots, y_n) = (x_1 + y_1, \cdots, x_n + y_n). \tag{2.61}$$

  Then, $(x_1, \cdots, x_n)^{-1} := (-x_1, \cdots, -x_n)$ is the inverse element and $e = (0, \cdots, 0)$ is the neutral element.
- $(\mathbb{R}^{m \times n}, +)$, the set of $m \times n$-matrices is Abelian (with componentwise addition as defined in (2.61)).
- Let us have a closer look at $(\mathbb{R}^{n \times n}, \cdot)$, i.e., the set of $n \times n$-matrices with matrix multiplication as defined in (2.13).

- Closure and associativity follow directly from the definition of matrix multiplication.
- Neutral element: The identity matrix $\boldsymbol{I}_n$ is the neutral element with respect to matrix multiplication "$\cdot$" in $(\mathbb{R}^{n\times n}, \cdot)$.
- Inverse element: If the inverse exists ($\boldsymbol{A}$ is regular), then $\boldsymbol{A}^{-1}$ is the inverse element of $\boldsymbol{A} \in \mathbb{R}^{n\times n}$, and in exactly this case $(\mathbb{R}^{n\times n}, \cdot)$ is a group, called the *general linear group*.

general linear group

Definition 2.8 (General Linear Group)**.** The set of regular (invertible) matrices $\boldsymbol{A} \in \mathbb{R}^{n\times n}$ is a group with respect to matrix multiplication as defined in (2.13) and is called *general linear group* $GL(n, \mathbb{R})$. However, since matrix multiplication is not commutative, the group is not Abelian.

2.4.2 Vector Spaces

When we discussed groups, we looked at sets $\mathcal{G}$ and inner operations on $\mathcal{G}$, i.e., mappings $\mathcal{G} \times \mathcal{G} \to \mathcal{G}$ that only operate on elements in $\mathcal{G}$. In the following, we will consider sets that in addition to an inner operation $+$ also contain an outer operation $\cdot$, the multiplication of a vector $\boldsymbol{x} \in \mathcal{G}$ by a scalar $\lambda \in \mathbb{R}$. We can think of the inner operation as a form of addition, and the outer operation as a form of scaling. Note that the inner/outer operations have nothing to do with inner/outer products.

vector space

Definition 2.9 (Vector Space)**.** A real-valued *vector space* $V = (\mathcal{V}, +, \cdot)$ is a set $\mathcal{V}$ with two operations

$$+ : \mathcal{V} \times \mathcal{V} \to \mathcal{V} \tag{2.62}$$

$$\cdot : \mathbb{R} \times \mathcal{V} \to \mathcal{V} \tag{2.63}$$

where

1. $(\mathcal{V}, +)$ is an Abelian group
2. Distributivity:
   a. $\forall \lambda \in \mathbb{R}, \boldsymbol{x}, \boldsymbol{y} \in \mathcal{V} : \lambda \cdot (\boldsymbol{x} + \boldsymbol{y}) = \lambda \cdot \boldsymbol{x} + \lambda \cdot \boldsymbol{y}$
   b. $\forall \lambda, \psi \in \mathbb{R}, \boldsymbol{x} \in \mathcal{V} : (\lambda + \psi) \cdot \boldsymbol{x} = \lambda \cdot \boldsymbol{x} + \psi \cdot \boldsymbol{x}$
3. Associativity (outer operation): $\forall \lambda, \psi \in \mathbb{R}, \boldsymbol{x} \in \mathcal{V} : \lambda \cdot (\psi \cdot \boldsymbol{x}) = (\lambda\psi) \cdot \boldsymbol{x}$
4. Neutral element with respect to the outer operation: $\forall \boldsymbol{x} \in \mathcal{V} : 1 \cdot \boldsymbol{x} = \boldsymbol{x}$

vector
vector addition
scalar
multiplication by scalars

The elements $\boldsymbol{x} \in V$ are called *vectors*. The neutral element of $(\mathcal{V}, +)$ is the zero vector $\boldsymbol{0} = [0, \dots, 0]^\top$, and the inner operation $+$ is called *vector addition*. The elements $\lambda \in \mathbb{R}$ are called *scalars* and the outer operation $\cdot$ is a *multiplication by scalars*. Note that a scalar product is something different, and we will get to this in Section 3.2.

Remark. A "vector multiplication" $\boldsymbol{ab}$, $\boldsymbol{a}, \boldsymbol{b} \in \mathbb{R}^n$, is not defined. Theoretically, we could define an elementwise multiplication, such that $\boldsymbol{c} = \boldsymbol{ab}$ with $c_j = a_j b_j$. This "array multiplication" is common to many programming languages but makes mathematically limited sense using the standard rules for matrix multiplication: By treating vectors as $n \times 1$ matrices (which we usually do), we can use the matrix multiplication as defined in (2.13). However, then the dimensions of the vectors do not match. Only the following multiplications for vectors are defined: $\boldsymbol{a}\boldsymbol{b}^\top \in \mathbb{R}^{n\times n}$ (*outer product*), $\boldsymbol{a}^\top\boldsymbol{b} \in \mathbb{R}$ (inner/scalar/dot product). ♢

outer product

Example 2.11 (Vector Spaces)

Let us have a look at some important examples:

- $\mathcal{V} = \mathbb{R}^n, n \in \mathbb{N}$ is a vector space with operations defined as follows:
  - Addition: $\boldsymbol{x}+\boldsymbol{y} = (x_1,\dots,x_n)+(y_1,\dots,y_n) = (x_1+y_1,\dots,x_n+y_n)$ for all $\boldsymbol{x},\boldsymbol{y} \in \mathbb{R}^n$
  - Multiplication by scalars: $\lambda\boldsymbol{x} = \lambda(x_1,\dots,x_n) = (\lambda x_1,\dots,\lambda x_n)$ for all $\lambda \in \mathbb{R}, \boldsymbol{x} \in \mathbb{R}^n$
- $\mathcal{V} = \mathbb{R}^{m\times n}, m,n \in \mathbb{N}$ is a vector space with
  - Addition: $\boldsymbol{A}+\boldsymbol{B} = \begin{bmatrix} a_{11}+b_{11} & \cdots & a_{1n}+b_{1n} \\ \vdots & & \vdots \\ a_{m1}+b_{m1} & \cdots & a_{mn}+b_{mn} \end{bmatrix}$ is defined elementwise for all $\boldsymbol{A},\boldsymbol{B} \in \mathcal{V}$
  - Multiplication by scalars: $\lambda\boldsymbol{A} = \begin{bmatrix} \lambda a_{11} & \cdots & \lambda a_{1n} \\ \vdots & & \vdots \\ \lambda a_{m1} & \cdots & \lambda a_{mn} \end{bmatrix}$ as defined in Section 2.2. Remember that $\mathbb{R}^{m\times n}$ is equivalent to $\mathbb{R}^{mn}$.
- $\mathcal{V} = \mathbb{C}$, with the standard definition of addition of complex numbers.

Remark. In the following, we will denote a vector space $(\mathcal{V},+,\cdot)$ by V when $+$ and $\cdot$ are the standard vector addition and scalar multiplication. Moreover, we will use the notation $\boldsymbol{x} \in V$ for vectors in $\mathcal{V}$ to simplify notation. ♢

Remark. The vector spaces $\mathbb{R}^n, \mathbb{R}^{n\times 1}, \mathbb{R}^{1\times n}$ are only different in the way we write vectors. In the following, we will not make a distinction between $\mathbb{R}^n$ and $\mathbb{R}^{n\times 1}$, which allows us to write n-tuples as *column vectors*

column vector

$$\boldsymbol{x} = \begin{bmatrix} x_1 \\ \vdots \\ x_n \end{bmatrix}. \tag{2.64}$$

This simplifies the notation regarding vector space operations. However, we do distinguish between $\mathbb{R}^{n\times 1}$ and $\mathbb{R}^{1\times n}$ (the *row vectors*) to avoid confusion with matrix multiplication. By default, we write $\boldsymbol{x}$ to denote a column vector, and a row vector is denoted by $\boldsymbol{x}^\top$, the *transpose* of $\boldsymbol{x}$. ♢

row vector

transpose

2.4.3 Vector Subspaces

In the following, we will introduce vector subspaces. Intuitively, they are sets contained in the original vector space with the property that when we perform vector space operations on elements within this subspace, we will never leave it. In this sense, they are "closed." Vector subspaces are a key idea in machine learning. For example, Chapter 10 demonstrates how to use vector subspaces for dimensionality reduction.

vector subspace
linear subspace

Definition 2.10 (Vector Subspace)**.** Let $V = (\mathcal{V}, +, \cdot)$ be a vector space and $\mathcal{U} \subseteq \mathcal{V}, \mathcal{U} \neq \emptyset$. Then $U = (\mathcal{U}, +, \cdot)$ is called *vector subspace* of V (or *linear subspace*) if U is a vector space with the vector space operations $+$ and $\cdot$ restricted to $\mathcal{U} \times \mathcal{U}$ and $\mathbb{R} \times \mathcal{U}$. We write $U \subseteq V$ to denote a subspace U of V.

If $\mathcal{U} \subseteq \mathcal{V}$ and V is a vector space, then U naturally inherits many properties directly from V because they hold for all $\boldsymbol{x} \in \mathcal{V}$, and in particular for all $\boldsymbol{x} \in \mathcal{U} \subseteq \mathcal{V}$. This includes the Abelian group properties, the distributivity, the associativity, and the neutral element. To determine whether $(\mathcal{U}, +, \cdot)$ is a subspace of V, we still do need to show

1. $\mathcal{U} \neq \emptyset$, in particular: $\mathbf{0} \in \mathcal{U}$
2. Closure of U:
   a. With respect to the outer operation: $\forall \lambda \in \mathbb{R} \, \forall \boldsymbol{x} \in \mathcal{U} : \lambda \boldsymbol{x} \in \mathcal{U}$.
   b. With respect to the inner operation: $\forall \boldsymbol{x}, \boldsymbol{y} \in \mathcal{U} : \boldsymbol{x} + \boldsymbol{y} \in \mathcal{U}$.

Example 2.12 (Vector Subspaces)
Let us have a look at some examples:

- For every vector space V, the trivial subspaces are V itself and $\{\mathbf{0}\}$.
- Only example D in Figure 2.6 is a subspace of $\mathbb{R}^2$ (with the usual inner/outer operations). In A and C, the closure property is violated; B does not contain $\mathbf{0}$.
- The solution set of a homogeneous system of linear equations $\boldsymbol{A}\boldsymbol{x} = \mathbf{0}$ with n unknowns $\boldsymbol{x} = [x_1, \ldots, x_n]^\top$ is a subspace of $\mathbb{R}^n$.
- The solution of an inhomogeneous system of linear equations $\boldsymbol{A}\boldsymbol{x} = \boldsymbol{b}$, $\boldsymbol{b} \neq \mathbf{0}$ is not a subspace of $\mathbb{R}^n$.
- The intersection of arbitrarily many subspaces is a subspace itself.

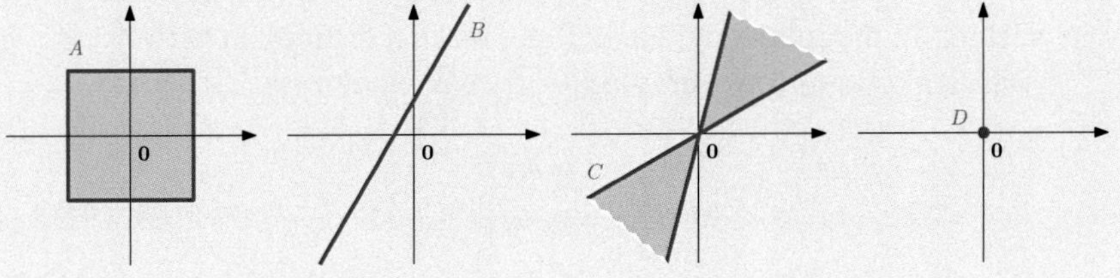

Figure 2.6 Not all subsets of $\mathbb{R}^2$ are subspaces. In A and C, the closure property is violated; B does not contain $\mathbf{0}$. Only D is a subspace.

Remark. Every subspace $U \subseteq (\mathbb{R}^n, +, \cdot)$ is the solution space of a homogeneous system of homogeneous linear equations $\boldsymbol{A}\boldsymbol{x} = \mathbf{0}$ for $\boldsymbol{x} \in \mathbb{R}^n$. ◊

2.5 Linear Independence

In the following, we will have a close look at what we can do with vectors (elements of the vector space). In particular, we can add vectors together and multiply them with scalars. The closure property guarantees that we end up with another vector in the same vector space. It is possible to find a set of vectors with which we can represent every vector in the vector space by adding them together and scaling them. This set of vectors is a *basis*, and we will discuss them in Section 2.6.1. Before we get there, we will need to introduce the concepts of linear combinations and linear independence.

Definition 2.11 (Linear Combination)**.** Consider a vector space V and a finite number of vectors $\boldsymbol{x}_1, \ldots, \boldsymbol{x}_k \in V$. Then, every $\boldsymbol{v} \in V$ of the form

$$\boldsymbol{v} = \lambda_1 \boldsymbol{x}_1 + \cdots + \lambda_k \boldsymbol{x}_k = \sum_{i=1}^{k} \lambda_i \boldsymbol{x}_i \in V \tag{2.65}$$

with $\lambda_1, \ldots, \lambda_k \in \mathbb{R}$ is a *linear combination* of the vectors $\boldsymbol{x}_1, \ldots, \boldsymbol{x}_k$.

linear combination

The $\boldsymbol{0}$-vector can always be written as the linear combination of k vectors $\boldsymbol{x}_1, \ldots, \boldsymbol{x}_k$ because $\boldsymbol{0} = \sum_{i=1}^{k} 0 \boldsymbol{x}_i$ is always true. In the following, we are interested in nontrivial linear combinations of a set of vectors to represent $\boldsymbol{0}$, i.e., linear combinations of vectors $\boldsymbol{x}_1, \ldots, \boldsymbol{x}_k$, where not all coefficients λ_i in (2.65) are 0.

Definition 2.12 (Linear (In)dependence)**.** Let us consider a vector space V with $k \in \mathbb{N}$ and $\boldsymbol{x}_1, \ldots, \boldsymbol{x}_k \in V$. If there is a non-trivial linear combination, such that $\boldsymbol{0} = \sum_{i=1}^{k} \lambda_i \boldsymbol{x}_i$ with at least one $\lambda_i \neq 0$, the vectors $\boldsymbol{x}_1, \ldots, \boldsymbol{x}_k$ are *linearly dependent*. If only the trivial solution exists, i.e., $\lambda_1 = \ldots = \lambda_k = 0$ the vectors $\boldsymbol{x}_1, \ldots, \boldsymbol{x}_k$ are *linearly independent*.

linearly dependent

linearly independent

Linear independence is one of the most important concepts in linear algebra. Intuitively, a set of linearly independent vectors consists of vectors that have no redundancy, i.e., if we remove any of those vectors from the set, we will lose something. Throughout the next sections, we will formalize this intuition more.

Example 2.13 (Linearly Dependent Vectors)

A geographic example may help to clarify the concept of linear independence. A person in Nairobi (Kenya) describing where Kigali (Rwanda) is might say, “You can get to Kigali by first going 506 km Northwest to Kampala (Uganda) and then 374 km Southwest.” This is sufficient information to describe the location of Kigali because the geographic coordinate system may be considered a two-dimensional vector space (ignoring altitude and the Earth’s curved surface). The person may add, “It is about 751 km West of here.” Although this last statement is true, it is not necessary to find Kigali given the previous information (see Figure 2.7 for an illustration). In this example, the “506 km Northwest” vector (blue) and the “374 km Southwest” vector (purple) are linearly independent. This means the Southwest vector cannot be described in terms of the Northwest vector, and vice versa. However, the third "751 km West" vector (black) is a linear combination of the other two vectors, and it makes the set of vectors linearly dependent. Equivalently, given "751 km West" and "374 km Southwest" can be linearly combined to obtain "506 km Northwest".

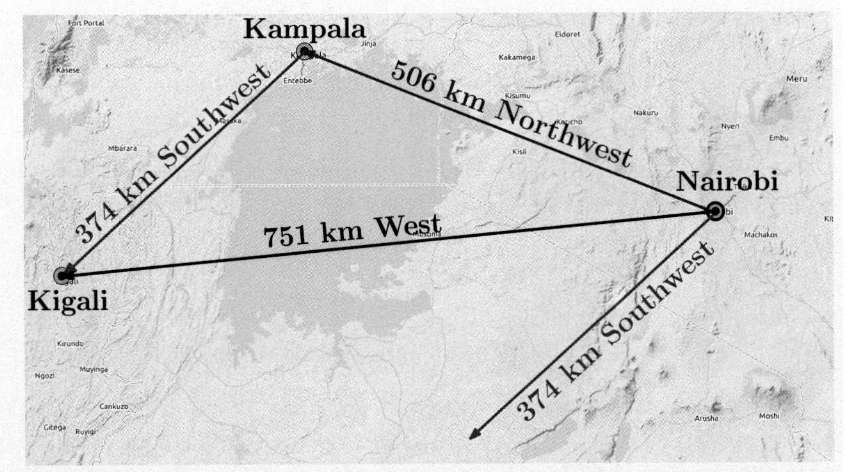

Figure 2.7 Geographic example (with crude approximations to cardinal directions) of linearly dependent vectors in a two-dimensional space (plane).

Remark. The following properties are useful to find out whether vectors are linearly independent:

- k vectors are either linearly dependent or linearly independent. There is no third option.
- If at least one of the vectors $\boldsymbol{x}_1, \ldots, \boldsymbol{x}_k$ is $\boldsymbol{0}$ then they are linearly dependent. The same holds if two vectors are identical.
- The vectors $\{\boldsymbol{x}_1, \ldots, \boldsymbol{x}_k : \boldsymbol{x}_i \neq \boldsymbol{0}, i = 1, \ldots, k\}$, $k \geqslant 2$, are linearly dependent if and only if (at least) one of them is a linear combination of the others. In particular, if one vector is a multiple of another vector, i.e., $\boldsymbol{x}_i = \lambda \boldsymbol{x}_j$, $\lambda \in \mathbb{R}$, then the set $\{\boldsymbol{x}_1, \ldots, \boldsymbol{x}_k : \boldsymbol{x}_i \neq \boldsymbol{0}, i = 1, \ldots, k\}$ is linearly dependent.
- A practical way of checking whether vectors $\boldsymbol{x}_1, \ldots, \boldsymbol{x}_k \in V$ are linearly independent is to use Gaussian elimination: Write all vectors as columns of a matrix $\boldsymbol{A}$ and perform Gaussian elimination until the matrix is in row-echelon form (the reduced row-echelon form is unnecessary here):
  - The pivot columns indicate the vectors, which are linearly independent of the vectors on the left. Note that there is an ordering of vectors when the matrix is built.
  - The nonpivot columns can be expressed as linear combinations of the pivot columns on their left. For instance, the row-echelon form

$$\begin{bmatrix} 1 & 3 & 0 \\ 0 & 0 & 2 \end{bmatrix} \tag{2.66}$$

tells us that the first and third columns are pivot columns. The second column is a nonpivot column because it is three times the first column.

All column vectors are linearly independent if and only if all columns are pivot columns. If there is at least one nonpivot column, the columns (and, therefore, the corresponding vectors) are linearly dependent.

Example 2.14

Consider $\mathbb{R}^4$ with

$$\boldsymbol{x}_1 = \begin{bmatrix} 1 \\ 2 \\ -3 \\ 4 \end{bmatrix}, \quad \boldsymbol{x}_2 = \begin{bmatrix} 1 \\ 1 \\ 0 \\ 2 \end{bmatrix}, \quad \boldsymbol{x}_3 = \begin{bmatrix} -1 \\ -2 \\ 1 \\ 1 \end{bmatrix}. \tag{2.67}$$

To check whether they are linearly dependent, we follow the general approach and solve

$$\lambda_1 \boldsymbol{x}_1 + \lambda_2 \boldsymbol{x}_2 + \lambda_3 \boldsymbol{x}_3 = \lambda_1 \begin{bmatrix} 1 \\ 2 \\ -3 \\ 4 \end{bmatrix} + \lambda_2 \begin{bmatrix} 1 \\ 1 \\ 0 \\ 2 \end{bmatrix} + \lambda_3 \begin{bmatrix} -1 \\ -2 \\ 1 \\ 1 \end{bmatrix} = \boldsymbol{0} \tag{2.68}$$

for $\lambda_1, \ldots, \lambda_3$. We write the vectors $\boldsymbol{x}_i$, $i = 1, 2, 3$, as the columns of a matrix and apply elementary row operations until we identify the pivot columns:

$$\begin{bmatrix} 1 & 1 & -1 \\ 2 & 1 & -2 \\ -3 & 0 & 1 \\ 4 & 2 & 1 \end{bmatrix} \rightsquigarrow \cdots \rightsquigarrow \begin{bmatrix} 1 & 1 & -1 \\ 0 & 1 & 0 \\ 0 & 0 & 1 \\ 0 & 0 & 0 \end{bmatrix}. \tag{2.69}$$

Here, every column of the matrix is a pivot column. Therefore, there is no nontrivial solution, and we require $\lambda_1 = 0, \lambda_2 = 0, \lambda_3 = 0$ to solve the equation system. Hence, the vectors $\boldsymbol{x}_1, \boldsymbol{x}_2, \boldsymbol{x}_3$ are linearly independent.

Remark. Consider a vector space V with k linearly independent vectors $\boldsymbol{b}_1, \ldots, \boldsymbol{b}_k$ and m linear combinations

$$\begin{aligned} \boldsymbol{x}_1 &= \sum_{i=1}^{k} \lambda_{i1} \boldsymbol{b}_i \,, \\ &\vdots \\ \boldsymbol{x}_m &= \sum_{i=1}^{k} \lambda_{im} \boldsymbol{b}_i \,. \end{aligned} \tag{2.70}$$

Defining $B = [b_1, \ldots, b_k]$ as the matrix whose columns are the linearly independent vectors $b_1, \ldots, b_k$, we can write

$$x_j = B\lambda_j, \quad \lambda_j = \begin{bmatrix} \lambda_{1j} \\ \vdots \\ \lambda_{kj} \end{bmatrix}, \quad j = 1, \ldots, m, \tag{2.71}$$

in a more compact form.

We want to test whether $x_1, \ldots, x_m$ are linearly independent. For this purpose, we follow the general approach of testing when $\sum_{j=1}^{m} \psi_j x_j = 0$. With (2.71), we obtain

$$\sum_{j=1}^{m} \psi_j x_j = \sum_{j=1}^{m} \psi_j B\lambda_j = B \sum_{j=1}^{m} \psi_j \lambda_j. \tag{2.72}$$

This means that $\{x_1, \ldots, x_m\}$ are linearly independent if and only if the column vectors $\{\lambda_1, \ldots, \lambda_m\}$ are linearly independent.

◇

Remark. In a vector space V, m linear combinations of k vectors $x_1, \ldots, x_k$ are linearly dependent if $m > k$. ◇

Example 2.15

Consider a set of linearly independent vectors $b_1, b_2, b_3, b_4 \in \mathbb{R}^n$ and

$$\begin{array}{rcrcrcrcr} x_1 & = & b_1 & - & 2b_2 & + & b_3 & - & b_4 \\ x_2 & = & -4b_1 & - & 2b_2 & & & + & 4b_4 \\ x_3 & = & 2b_1 & + & 3b_2 & - & b_3 & - & 3b_4 \\ x_4 & = & 17b_1 & - & 10b_2 & + & 11b_3 & + & b_4 \end{array}. \tag{2.73}$$

Are the vectors $x_1, \ldots, x_4 \in \mathbb{R}^n$ linearly independent? To answer this question, we investigate whether the column vectors

$$\left\{ \begin{bmatrix} 1 \\ -2 \\ 1 \\ -1 \end{bmatrix}, \begin{bmatrix} -4 \\ -2 \\ 0 \\ 4 \end{bmatrix}, \begin{bmatrix} 2 \\ 3 \\ -1 \\ -3 \end{bmatrix}, \begin{bmatrix} 17 \\ -10 \\ 11 \\ 1 \end{bmatrix} \right\} \tag{2.74}$$

are linearly independent. The reduced row-echelon form of the corresponding linear equation system with coefficient matrix

$$A = \begin{bmatrix} 1 & -4 & 2 & 17 \\ -2 & -2 & 3 & -10 \\ 1 & 0 & -1 & 11 \\ -1 & 4 & -3 & 1 \end{bmatrix} \tag{2.75}$$

is given as

$$\begin{bmatrix} 1 & 0 & 0 & -7 \\ 0 & 1 & 0 & -15 \\ 0 & 0 & 1 & -18 \\ 0 & 0 & 0 & 0 \end{bmatrix}. \tag{2.76}$$

We see that the corresponding linear equation system is nontrivially solvable: The last column is not a pivot column, and $\boldsymbol{x}_4 = -7\boldsymbol{x}_1 - 15\boldsymbol{x}_2 - 18\boldsymbol{x}_3$. Therefore, $\boldsymbol{x}_1, \ldots, \boldsymbol{x}_4$ are linearly dependent as $\boldsymbol{x}_4$ can be expressed as a linear combination of $\boldsymbol{x}_1, \ldots, \boldsymbol{x}_3$.

2.6 Basis and Rank

In a vector space V, we are particularly interested in sets of vectors $\mathcal{A}$ that possess the property that any vector $\boldsymbol{v} \in V$ can be obtained by a linear combination of vectors in $\mathcal{A}$. These vectors are special vectors, and in the following, we will characterize them.

2.6.1 Generating Set and Basis

Definition 2.13 (Generating Set and Span)**.** Consider a vector space $V = (\mathcal{V}, +, \cdot)$ and set of vectors $\mathcal{A} = \{\boldsymbol{x}_1, \ldots, \boldsymbol{x}_k\} \subseteq \mathcal{V}$. If every vector $\boldsymbol{v} \in \mathcal{V}$ can be expressed as a linear combination of $\boldsymbol{x}_1, \ldots, \boldsymbol{x}_k$, $\mathcal{A}$ is called a *generating set* of V. The set of all linear combinations of vectors in $\mathcal{A}$ is called the *span* of $\mathcal{A}$. If $\mathcal{A}$ spans the vector space V, we write $V = \text{span}[\mathcal{A}]$ or $V = \text{span}[\boldsymbol{x}_1, \ldots, \boldsymbol{x}_k]$.

generating set

span

Generating sets are sets of vectors that span vector (sub)spaces, i.e., every vector can be represented as a linear combination of the vectors in the generating set. Now we will be more specific and characterize the smallest generating set that spans a vector (sub)space.

Definition 2.14 (Basis)**.** Consider a vector space $V = (\mathcal{V}, +, \cdot)$ and $\mathcal{A} \subseteq \mathcal{V}$. A generating set $\mathcal{A}$ of V is called *minimal* if there exists no smaller set $\tilde{\mathcal{A}} \subseteq \mathcal{A} \subseteq \mathcal{V}$ that spans V. Every linearly independent generating set of V is minimal and is called a *basis* of V.

minimal

basis

Let $V = (\mathcal{V}, +, \cdot)$ be a vector space and $\mathcal{B} \subseteq \mathcal{V}, \mathcal{B} \neq \emptyset$. Then, the following statements are equivalent:

A basis is a minimal generating set and a maximal linearly independent set of vectors.

- $\mathcal{B}$ is a basis of V.
- $\mathcal{B}$ is a minimal generating set.
- $\mathcal{B}$ is a maximal linearly independent set of vectors in V, i.e., adding any other vector to this set will make it linearly dependent.
- Every vector $\boldsymbol{x} \in V$ is a linear combination of vectors from $\mathcal{B}$, and every linear combination is unique, i.e., with

$$\boldsymbol{x} = \sum_{i=1}^{k} \lambda_i \boldsymbol{b}_i = \sum_{i=1}^{k} \psi_i \boldsymbol{b}_i \tag{2.77}$$

  and $\lambda_i, \psi_i \in \mathbb{R}$, $\boldsymbol{b}_i \in \mathcal{B}$ it follows that $\lambda_i = \psi_i,\ i = 1, \ldots, k$.

Example 2.16

canonical basis

- In $\mathbb{R}^3$, the *canonical/standard basis* is

$$\mathcal{B} = \left\{ \begin{bmatrix} 1 \\ 0 \\ 0 \end{bmatrix}, \begin{bmatrix} 0 \\ 1 \\ 0 \end{bmatrix}, \begin{bmatrix} 0 \\ 0 \\ 1 \end{bmatrix} \right\}. \tag{2.78}$$

- Different bases in $\mathbb{R}^3$ are

$$\mathcal{B}_1 = \left\{ \begin{bmatrix} 1 \\ 0 \\ 0 \end{bmatrix}, \begin{bmatrix} 1 \\ 1 \\ 0 \end{bmatrix}, \begin{bmatrix} 1 \\ 1 \\ 1 \end{bmatrix} \right\}, \mathcal{B}_2 = \left\{ \begin{bmatrix} 0.5 \\ 0.8 \\ 0.4 \end{bmatrix}, \begin{bmatrix} 1.8 \\ 0.3 \\ 0.3 \end{bmatrix}, \begin{bmatrix} -2.2 \\ -1.3 \\ 3.5 \end{bmatrix} \right\}. \tag{2.79}$$

- The set

$$\mathcal{A} = \left\{ \begin{bmatrix} 1 \\ 2 \\ 3 \\ 4 \end{bmatrix}, \begin{bmatrix} 2 \\ -1 \\ 0 \\ 2 \end{bmatrix}, \begin{bmatrix} 1 \\ 1 \\ 0 \\ -4 \end{bmatrix} \right\} \tag{2.80}$$

  is linearly independent, but not a generating set (and no basis) of $\mathbb{R}^4$: For instance, the vector $[1, 0, 0, 0]^\top$ cannot be obtained by a linear combination of elements in $\mathcal{A}$.

basis vector

Remark. Every vector space V possesses a basis $\mathcal{B}$. The preceding examples show that there can be many bases of a vector space V, i.e., there is no unique basis. However, all bases possess the same number of elements, the *basis vectors*. ◇

dimension

We only consider finite-dimensional vector spaces V. In this case, the *dimension* of V is the number of basis vectors of V, and we write $\dim(V)$. If $U \subseteq V$ is a subspace of V, then $\dim(U) \leqslant \dim(V)$ and $\dim(U) = \dim(V)$ if and only if $U = V$. Intuitively, the dimension of a vector space can be thought of as the number of independent directions in this vector space.

The dimension of a vector space corresponds to the number of its basis vectors.

Remark. The dimension of a vector space is not necessarily the number of elements in a vector. For instance, the vector space $V = \text{span}[\begin{bmatrix} 0 \\ 1 \end{bmatrix}]$ is one-dimensional, although the basis vector possesses two elements. ◇

Remark. A basis of a subspace $U = \text{span}[\boldsymbol{x}_1, \ldots, \boldsymbol{x}_m] \subseteq \mathbb{R}^n$ can be found by executing the following steps:

1. Write the spanning vectors as columns of a matrix $\boldsymbol{A}$.
2. Determine the row-echelon form of $\boldsymbol{A}$.
3. The spanning vectors associated with the pivot columns are a basis of U.

Example 2.17 (Determining a Basis)
For a vector subspace $U \subseteq \mathbb{R}^5$, spanned by the vectors

$$\boldsymbol{x}_1 = \begin{bmatrix} 1 \\ 2 \\ -1 \\ -1 \\ -1 \end{bmatrix}, \quad \boldsymbol{x}_2 = \begin{bmatrix} 2 \\ -1 \\ 1 \\ 2 \\ -2 \end{bmatrix}, \quad \boldsymbol{x}_3 = \begin{bmatrix} 3 \\ -4 \\ 3 \\ 5 \\ -3 \end{bmatrix}, \quad \boldsymbol{x}_4 = \begin{bmatrix} -1 \\ 8 \\ -5 \\ -6 \\ 1 \end{bmatrix} \in \mathbb{R}^5, \tag{2.81}$$

we are interested in finding out which vectors $\boldsymbol{x}_1, \ldots, \boldsymbol{x}_4$ are a basis for U. For this, we need to check whether $\boldsymbol{x}_1, \ldots, \boldsymbol{x}_4$ are linearly independent. Therefore, we need to solve

$$\sum_{i=1}^{4} \lambda_i \boldsymbol{x}_i = \boldsymbol{0}, \tag{2.82}$$

which leads to a homogeneous system of equations with matrix

$$\begin{bmatrix} \boldsymbol{x}_1, \boldsymbol{x}_2, \boldsymbol{x}_3, \boldsymbol{x}_4 \end{bmatrix} = \begin{bmatrix} 1 & 2 & 3 & -1 \\ 2 & -1 & -4 & 8 \\ -1 & 1 & 3 & -5 \\ -1 & 2 & 5 & -6 \\ -1 & -2 & -3 & 1 \end{bmatrix}. \tag{2.83}$$

With the basic transformation rules for systems of linear equations, we obtain the row-echelon form

$$\begin{bmatrix} 1 & 2 & 3 & -1 \\ 2 & -1 & -4 & 8 \\ -1 & 1 & 3 & -5 \\ -1 & 2 & 5 & -6 \\ -1 & -2 & -3 & 1 \end{bmatrix} \rightsquigarrow \cdots \rightsquigarrow \begin{bmatrix} 1 & 2 & 3 & -1 \\ 0 & 1 & 2 & -2 \\ 0 & 0 & 0 & 1 \\ 0 & 0 & 0 & 0 \\ 0 & 0 & 0 & 0 \end{bmatrix}.$$

Since the pivot columns indicate which set of vectors is linearly independent, we see from the row-echelon form that $\boldsymbol{x}_1, \boldsymbol{x}_2, \boldsymbol{x}_4$ are linearly independent (because the system of linear equations $\lambda_1 \boldsymbol{x}_1 + \lambda_2 \boldsymbol{x}_2 + \lambda_4 \boldsymbol{x}_4 = \boldsymbol{0}$ can only be solved with $\lambda_1 = \lambda_2 = \lambda_4 = 0$). Therefore, $\{\boldsymbol{x}_1, \boldsymbol{x}_2, \boldsymbol{x}_4\}$ is a basis of U.

2.6.2 Rank

The number of linearly independent columns of a matrix $\boldsymbol{A} \in \mathbb{R}^{m \times n}$ equals the number of linearly independent rows and is called the *rank* of $\boldsymbol{A}$ and is denoted by $\text{rk}(\boldsymbol{A})$.

rank

Remark. The rank of a matrix has some important properties:

- $\text{rk}(\boldsymbol{A}) = \text{rk}(\boldsymbol{A}^\top)$, i.e., the column rank equals the row rank.
- The columns of $\boldsymbol{A} \in \mathbb{R}^{m \times n}$ span a subspace $U \subseteq \mathbb{R}^m$ with $\dim(U) = \text{rk}(\boldsymbol{A})$. Later we will call this subspace the *image* or *range*. A basis of U

can be found by applying Gaussian elimination to $\boldsymbol{A}$ to identify the pivot columns.

- The rows of $\boldsymbol{A} \in \mathbb{R}^{m\times n}$ span a subspace $W \subseteq \mathbb{R}^n$ with $\dim(W) = \text{rk}(\boldsymbol{A})$. A basis of W can be found by applying Gaussian elimination to $\boldsymbol{A}^\top$.
- For all $\boldsymbol{A} \in \mathbb{R}^{n\times n}$, it holds that $\boldsymbol{A}$ is regular (invertible) if and only if $\text{rk}(\boldsymbol{A}) = n$.
- For all $\boldsymbol{A} \in \mathbb{R}^{m\times n}$ and all $\boldsymbol{b} \in \mathbb{R}^m$, it holds that the linear equation system $\boldsymbol{Ax} = \boldsymbol{b}$ can be solved if and only if $\text{rk}(\boldsymbol{A}) = \text{rk}(\boldsymbol{A}|\boldsymbol{b})$, where $\boldsymbol{A}|\boldsymbol{b}$ denotes the augmented system.
- For $\boldsymbol{A} \in \mathbb{R}^{m\times n}$ the subspace of solutions for $\boldsymbol{Ax} = \boldsymbol{0}$ possesses dimension $n - \text{rk}(\boldsymbol{A})$. Later we will call this subspace the *kernel* or the *null space*.
- A matrix $\boldsymbol{A} \in \mathbb{R}^{m\times n}$ has *full rank* if its rank equals the largest possible rank for a matrix of the same dimensions. This means that the rank of a full-rank matrix is the lesser of the number of rows and columns, i.e., $\text{rk}(\boldsymbol{A}) = \min(m, n)$. A matrix is said to be *rank deficient* if it does not have full rank.

kernel

null space

full rank

rank deficient

◇

Example 2.18 (Rank)

- $\boldsymbol{A} = \begin{bmatrix} 1 & 0 & 1 \\ 0 & 1 & 1 \\ 0 & 0 & 0 \end{bmatrix}$.

  $\boldsymbol{A}$ has two linearly independent rows/columns so that $\text{rk}(\boldsymbol{A}) = 2$.
- $\boldsymbol{A} = \begin{bmatrix} 1 & 2 & 1 \\ -2 & -3 & 1 \\ 3 & 5 & 0 \end{bmatrix}$.

  We use Gaussian elimination to determine the rank:

$$\begin{bmatrix} 1 & 2 & 1 \\ -2 & -3 & 1 \\ 3 & 5 & 0 \end{bmatrix} \rightsquigarrow \cdots \rightsquigarrow \begin{bmatrix} 1 & 2 & 1 \\ 0 & 1 & 3 \\ 0 & 0 & 0 \end{bmatrix}. \tag{2.84}$$

  Here we see that the number of linearly independent rows and columns is 2, such that $\text{rk}(\boldsymbol{A}) = 2$.

2.7 Linear Mappings

In the following, we will study mappings on vector spaces that preserve their structure, which will allow us to define the concept of a coordinate. In the beginning of the chapter, we said that vectors are objects that can be added together and multiplied by a scalar, and the resulting object is still a vector. We wish to preserve this property when applying the mapping: Consider two real vector spaces V, W. A mapping $\Phi : V \to W$ preserves the structure of the vector space if

$$\Phi(\boldsymbol{x} + \boldsymbol{y}) = \Phi(\boldsymbol{x}) + \Phi(\boldsymbol{y}) \tag{2.85}$$

$$\Phi(\lambda \boldsymbol{x}) = \lambda\Phi(\boldsymbol{x}) \tag{2.86}$$

for all $\boldsymbol{x}, \boldsymbol{y} \in V$ and $\lambda \in \mathbb{R}$. We can summarize this in the following definition:

Definition 2.15 (Linear Mapping)**.** For vector spaces V, W, a mapping $\Phi : V \to W$ is called a *linear mapping* (or *vector space homomorphism*/*linear transformation*) if

linear mapping
vector space homomorphism
linear transformation

$$\forall \boldsymbol{x}, \boldsymbol{y} \in V \, \forall \lambda, \psi \in \mathbb{R} : \Phi(\lambda \boldsymbol{x} + \psi \boldsymbol{y}) = \lambda \Phi(\boldsymbol{x}) + \psi \Phi(\boldsymbol{y}) . \tag{2.87}$$

It turns out that we can represent linear mappings as matrices (Section 2.7.1). Recall that we can also collect a set of vectors as columns of a matrix. When working with matrices, we have to keep in mind what the matrix represents: a linear mapping or a collection of vectors. We will see more about linear mappings in Chapter 4. Before we continue, we will briefly introduce special mappings.

Definition 2.16 (Injective, Surjective, Bijective)**.** Consider a mapping $\Phi : \mathcal{V} \to \mathcal{W}$, where $\mathcal{V}, \mathcal{W}$ can be arbitrary sets. Then Φ is called

injective
surjective
bijective

- *Injective* if $\forall \boldsymbol{x}, \boldsymbol{y} \in \mathcal{V} : \Phi(\boldsymbol{x}) = \Phi(\boldsymbol{y}) \implies \boldsymbol{x} = \boldsymbol{y}$.
- *Surjective* if $\Phi(\mathcal{V}) = \mathcal{W}$.
- *Bijective* if it is injective and surjective.

If Φ is surjective, then every element in $\mathcal{W}$ can be "reached" from $\mathcal{V}$ using Φ. A bijective Φ can be "undone," i.e., there exists a mapping $\Psi : \mathcal{W} \to \mathcal{V}$ so that $\Psi \circ \Phi(\boldsymbol{x}) = \boldsymbol{x}$. This mapping Ψ is then called the inverse of Φ and normally denoted by Φ^{-1}.

With these definitions, we introduce the following special cases of linear mappings between vector spaces V and W:

isomorphism
endomorphism
automorphism

- *Isomorphism:* $\Phi : V \to W$ linear and bijective
- *Endomorphism:* $\Phi : V \to V$ linear
- *Automorphism:* $\Phi : V \to V$ linear and bijective
- We define $\mathrm{id}_V : V \to V$, $\boldsymbol{x} \mapsto \boldsymbol{x}$ as the *identity mapping* or *identity automorphism* in V.

identity mapping
identity automorphism

Example 2.19 (Homomorphism)

The mapping $\Phi : \mathbb{R}^2 \to \mathbb{C}$, $\Phi(\boldsymbol{x}) = x_1 + ix_2$, is a homomorphism:

$$\begin{aligned}
\Phi\left(\begin{bmatrix} x_1 \\ x_2 \end{bmatrix} + \begin{bmatrix} y_1 \\ y_2 \end{bmatrix}\right) &= (x_1 + y_1) + i(x_2 + y_2) = x_1 + ix_2 + y_1 + iy_2 \\
&= \Phi\left(\begin{bmatrix} x_1 \\ x_2 \end{bmatrix}\right) + \Phi\left(\begin{bmatrix} y_1 \\ y_2 \end{bmatrix}\right) \\
\Phi\left(\lambda \begin{bmatrix} x_1 \\ x_2 \end{bmatrix}\right) &= \lambda x_1 + \lambda i x_2 = \lambda(x_1 + ix_2) = \lambda \Phi\left(\begin{bmatrix} x_1 \\ x_2 \end{bmatrix}\right) .
\end{aligned} \tag{2.88}$$

This also justifies why complex numbers can be represented as tuples in $\mathbb{R}^2$: There is a bijective linear mapping that converts the elementwise addition of tuples in $\mathbb{R}^2$ into the set of complex numbers with the corresponding addition. Note that we only showed linearity, but not the bijection.

Theorem 2.17 (Theorem 3.59 in Axler (2015)). *Finite-dimensional vector spaces V and W are isomorphic if and only if* $\dim(V) = \dim(W)$.

Theorem 2.17 states that there exists a linear, bijective mapping between two vector spaces of the same dimension. Intuitively, this means that vector spaces of the same dimension are kind of the same thing, as they can be transformed into each other without incurring any loss.

Theorem 2.17 also gives us the justification to treat $\mathbb{R}^{m\times n}$ (the vector space of $m \times n$-matrices) and $\mathbb{R}^{mn}$ (the vector space of vectors of length mn) the same, as their dimensions are mn, and there exists a linear, bijective mapping that transforms one into the other.

Remark. Consider vector spaces V, W, X. Then:

- For linear mappings $\Phi : V \to W$ and $\Psi : W \to X$, the mapping $\Psi \circ \Phi : V \to X$ is also linear.
- If $\Phi : V \to W$ is an isomorphism, then $\Phi^{-1} : W \to V$ is an isomorphism, too.
- If $\Phi : V \to W$, $\Psi : V \to W$ are linear, then $\Phi + \Psi$ and $\lambda\Phi$, $\lambda \in \mathbb{R}$, are linear, too.

2.7.1 Matrix Representation of Linear Mappings

Any n-dimensional vector space is isomorphic to $\mathbb{R}^n$ (Theorem 2.17). We consider a basis $\{\boldsymbol{b}_1, \ldots, \boldsymbol{b}_n\}$ of an n-dimensional vector space V. In the following, the order of the basis vectors will be important. Therefore, we write

$$B = (\boldsymbol{b}_1, \ldots, \boldsymbol{b}_n) \tag{2.89}$$

ordered basis

and call this n-tuple an *ordered basis* of V.

Remark (Notation). We are at the point where notation gets a bit tricky. Therefore, we summarize some parts here. $B = (\boldsymbol{b}_1, \ldots, \boldsymbol{b}_n)$ is an ordered basis, $\mathcal{B} = \{\boldsymbol{b}_1, \ldots, \boldsymbol{b}_n\}$ is an (unordered) basis, and $\boldsymbol{B} = [\boldsymbol{b}_1, \ldots, \boldsymbol{b}_n]$ is a matrix whose columns are the vectors $\boldsymbol{b}_1, \ldots, \boldsymbol{b}_n$. ♢

Definition 2.18 (Coordinates). Consider a vector space V and an ordered basis $B = (\boldsymbol{b}_1, \ldots, \boldsymbol{b}_n)$ of V. For any $\boldsymbol{x} \in V$, we obtain a unique representation (linear combination)

$$\boldsymbol{x} = \alpha_1 \boldsymbol{b}_1 + \ldots + \alpha_n \boldsymbol{b}_n \tag{2.90}$$

coordinate

of $\boldsymbol{x}$ with respect to B. Then $\alpha_1, \ldots, \alpha_n$ are the *coordinates* of $\boldsymbol{x}$ with respect to B, and the vector

$$\boldsymbol{\alpha} = \begin{bmatrix} \alpha_1 \\ \vdots \\ \alpha_n \end{bmatrix} \in \mathbb{R}^n \tag{2.91}$$

coordinate vector
coordinate representation

is the *coordinate vector/coordinate representation* of $\boldsymbol{x}$ with respect to the ordered basis B.

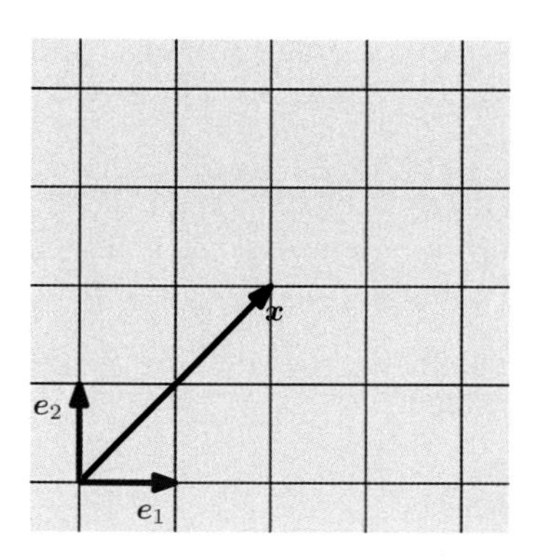

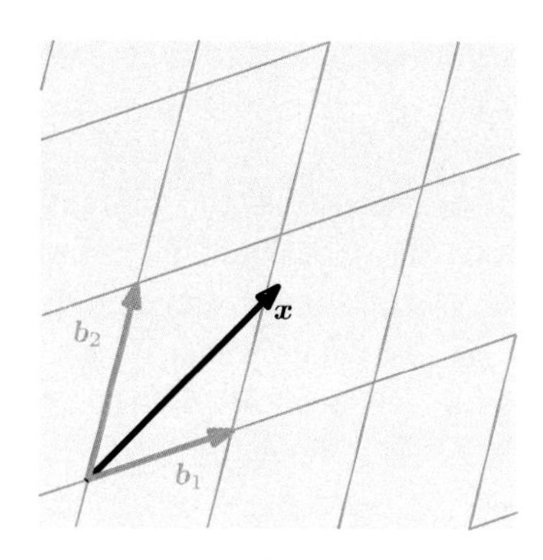

Figure 2.8 Two different coordinate systems defined by two sets of basis vectors. A vector $\boldsymbol{x}$ has different coordinate representations depending on which coordinate system is chosen.

A basis effectively defines a coordinate system. We are familiar with the Cartesian coordinate system in two dimensions, which is spanned by the canonical basis vectors $\boldsymbol{e}_1, \boldsymbol{e}_2$. In this coordinate system, a vector $\boldsymbol{x} \in \mathbb{R}^2$ has a representation that tells us how to linearly combine $\boldsymbol{e}_1$ and $\boldsymbol{e}_2$ to obtain $\boldsymbol{x}$. However, any basis of $\mathbb{R}^2$ defines a valid coordinate system, and the same vector $\boldsymbol{x}$ from before may have a different coordinate representation in the $(\boldsymbol{b}_1, \boldsymbol{b}_2)$ basis. In Figure 2.8, the coordinates of $\boldsymbol{x}$ with respect to the standard basis $(\boldsymbol{e}_1, \boldsymbol{e}_2)$ is $[2, 2]^\top$. However, with respect to the basis $(\boldsymbol{b}_1, \boldsymbol{b}_2)$ the same vector $\boldsymbol{x}$ is represented as $[1.09, 0.72]^\top$, i.e., $\boldsymbol{x} = 1.09\boldsymbol{b}_1 + 0.72\boldsymbol{b}_2$. In the following sections, we will discover how to obtain this representation.

Example 2.20

Let us have a look at a geometric vector $\boldsymbol{x} \in \mathbb{R}^2$ with coordinates $[2, 3]^\top$ with respect to the standard basis $(\boldsymbol{e}_1, \boldsymbol{e}_2)$ of $\mathbb{R}^2$. This means, we can write $\boldsymbol{x} = 2\boldsymbol{e}_1 + 3\boldsymbol{e}_2$. However, we do not have to choose the standard basis to represent this vector. If we use the basis vectors $\boldsymbol{b}_1 = [1, -1]^\top, \boldsymbol{b}_2 = [1, 1]^\top$, we will obtain the coordinates $\frac{1}{2}[-1, 5]^\top$ to represent the same vector with respect to $(\boldsymbol{b}_1, \boldsymbol{b}_2)$ (see Figure 2.9).

Figure 2.9 Different coordinate representations of a vector $\boldsymbol{x}$, depending on the choice of basis.

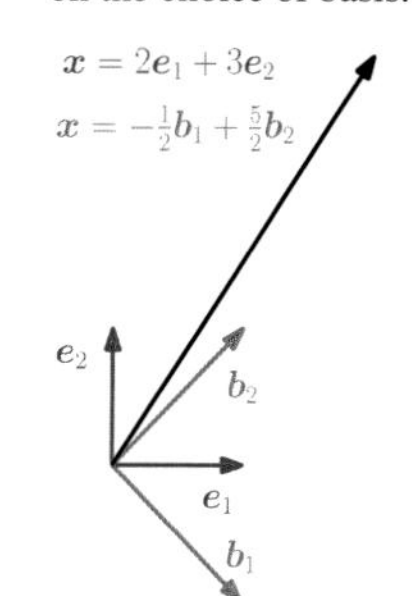

Remark. For an n-dimensional vector space V and an ordered basis B of V, the mapping $\Phi : \mathbb{R}^n \to V$, $\Phi(\boldsymbol{e}_i) = \boldsymbol{b}_i$, $i = 1, \dots, n$, is linear (and because of Theorem 2.17 an isomorphism), where $(\boldsymbol{e}_1, \dots, \boldsymbol{e}_n)$ is the standard basis of $\mathbb{R}^n$. ◇

Now we are ready to make an explicit connection between matrices and linear mappings between finite-dimensional vector spaces.

Definition 2.19 (Transformation Matrix)**.** Consider vector spaces V, W with corresponding (ordered) bases $B = (\boldsymbol{b}_1, \dots, \boldsymbol{b}_n)$ and $C = (\boldsymbol{c}_1, \dots, \boldsymbol{c}_m)$. Moreover, we consider a linear mapping $\Phi : V \to W$. For $j \in \{1, \dots, n\}$,

$$\Phi(\boldsymbol{b}_j) = \alpha_{1j}\boldsymbol{c}_1 + \cdots + \alpha_{mj}\boldsymbol{c}_m = \sum_{i=1}^{m} \alpha_{ij}\boldsymbol{c}_i \tag{2.92}$$

is the unique representation of $\Phi(\boldsymbol{b}_j)$ with respect to C. Then, we call the $m \times n$-matrix $\boldsymbol{A}_\Phi$, whose elements are given by

$$A_\Phi(i, j) = \alpha_{ij}\,, \tag{2.93}$$

transformation matrix

the *transformation matrix* of Φ (with respect to the ordered bases B of V and C of W).

The coordinates of $\Phi(\boldsymbol{b}_j)$ with respect to the ordered basis C of W are the jth column of $\boldsymbol{A}_\Phi$. Consider (finite-dimensional) vector spaces V, W with ordered bases B, C and a linear mapping $\Phi : V \to W$ with transformation matrix $\boldsymbol{A}_\Phi$. If $\hat{\boldsymbol{x}}$ is the coordinate vector of $\boldsymbol{x} \in V$ with respect to B and $\hat{\boldsymbol{y}}$ the coordinate vector of $\boldsymbol{y} = \Phi(\boldsymbol{x}) \in W$ with respect to C, then

$$\hat{\boldsymbol{y}} = \boldsymbol{A}_\Phi \hat{\boldsymbol{x}} \,. \tag{2.94}$$

This means that the transformation matrix can be used to map coordinates with respect to an ordered basis in V to coordinates with respect to an ordered basis in W.

Example 2.21 (Transformation Matrix)

Consider a homomorphism $\Phi : V \to W$ and ordered bases $B = (\boldsymbol{b}_1, \ldots, \boldsymbol{b}_3)$ of V and $C = (\boldsymbol{c}_1, \ldots, \boldsymbol{c}_4)$ of W. With

$$\begin{aligned} \Phi(\boldsymbol{b}_1) &= \boldsymbol{c}_1 - \boldsymbol{c}_2 + 3\boldsymbol{c}_3 - \boldsymbol{c}_4 \\ \Phi(\boldsymbol{b}_2) &= 2\boldsymbol{c}_1 + \boldsymbol{c}_2 + 7\boldsymbol{c}_3 + 2\boldsymbol{c}_4 \\ \Phi(\boldsymbol{b}_3) &= 3\boldsymbol{c}_2 + \boldsymbol{c}_3 + 4\boldsymbol{c}_4 \end{aligned} \tag{2.95}$$

the transformation matrix $\boldsymbol{A}_\Phi$ with respect to B and C satisfies $\Phi(\boldsymbol{b}_k) = \sum_{i=1}^{4} \alpha_{ik} \boldsymbol{c}_i$ for $k = 1, \ldots, 3$ and is given as

$$\boldsymbol{A}_\Phi = [\boldsymbol{\alpha}_1, \boldsymbol{\alpha}_2, \boldsymbol{\alpha}_3] = \begin{bmatrix} 1 & 2 & 0 \\ -1 & 1 & 3 \\ 3 & 7 & 1 \\ -1 & 2 & 4 \end{bmatrix}, \tag{2.96}$$

where the $\boldsymbol{\alpha}_j$, $j = 1, 2, 3$, are the coordinate vectors of $\Phi(\boldsymbol{b}_j)$ with respect to C.

Example 2.22 (Linear Transformations of Vectors)

We consider three linear transformations of a set of vectors in $\mathbb{R}^2$ with the transformation matrices

$$\boldsymbol{A}_1 = \begin{bmatrix} \cos(\frac{\pi}{4}) & -\sin(\frac{\pi}{4}) \\ \sin(\frac{\pi}{4}) & \cos(\frac{\pi}{4}) \end{bmatrix}, \; \boldsymbol{A}_2 = \begin{bmatrix} 2 & 0 \\ 0 & 1 \end{bmatrix}, \; \boldsymbol{A}_3 = \frac{1}{2}\begin{bmatrix} 3 & -1 \\ 1 & -1 \end{bmatrix}. \tag{2.97}$$

Figure 2.10 gives three examples of linear transformations of a set of vectors. Figure 2.10(a) shows 400 vectors in $\mathbb{R}^2$, each of which is represented by a dot at the corresponding (x_1, x_2)-coordinates. The vectors are arranged in a square. When we use matrix $\boldsymbol{A}_1$ in (2.97) to linearly transform each of these vectors, we obtain the rotated square in Figure 2.10(b). If we apply the linear mapping represented by $\boldsymbol{A}_2$, we obtain the rectangle in Figure 2.10(c) where each x_1-coordinate is stretched by 2. Figure 2.10(d) shows the original square from Figure 2.10(a) when linearly transformed using $\boldsymbol{A}_3$, which is a combination of a reflection, a rotation, and a stretch.

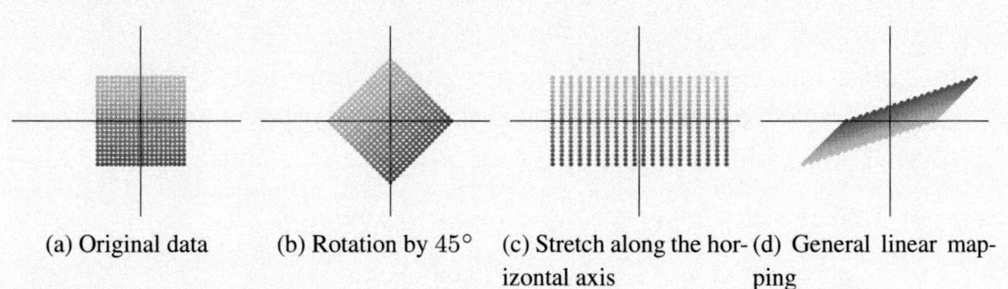

(a) Original data (b) Rotation by $45°$ (c) Stretch along the horizontal axis (d) General linear mapping

Figure 2.10 Three examples of linear transformations of the vectors shown as dots in (a); (b) rotation by $45°$; (c) stretching of the horizontal coordinates by 2; and (d) combination of reflection, rotation, and stretching.

2.7.2 Basis Change

In the following, we will have a closer look at how transformation matrices of a linear mapping $\Phi : V \to W$ change if we change the bases in V and W. Consider two ordered bases

$$B = (\boldsymbol{b}_1, \ldots, \boldsymbol{b}_n), \quad \tilde{B} = (\tilde{\boldsymbol{b}}_1, \ldots, \tilde{\boldsymbol{b}}_n) \tag{2.98}$$

of V and two ordered bases

$$C = (\boldsymbol{c}_1, \ldots, \boldsymbol{c}_m), \quad \tilde{C} = (\tilde{\boldsymbol{c}}_1, \ldots, \tilde{\boldsymbol{c}}_m) \tag{2.99}$$

of W. Moreover, $\boldsymbol{A}_\Phi \in \mathbb{R}^{m \times n}$ is the transformation matrix of the linear mapping $\Phi : V \to W$ with respect to the bases B and C, and $\tilde{\boldsymbol{A}}_\Phi \in \mathbb{R}^{m \times n}$ is the corresponding transformation mapping with respect to $\tilde{B}$ and $\tilde{C}$. In the following, we will investigate how $\boldsymbol{A}$ and $\tilde{\boldsymbol{A}}$ are related, i.e., how/whether we can transform $\boldsymbol{A}_\Phi$ into $\tilde{\boldsymbol{A}}_\Phi$ if we choose to perform a basis change from B, C to $\tilde{B}, \tilde{C}$.

Remark. We effectively get different coordinate representations of the identity mapping id_V. In the context of Figure 2.9, this would mean to map coordinates with respect to $(\boldsymbol{e}_1, \boldsymbol{e}_2)$ onto coordinates with respect to $(\boldsymbol{b}_1, \boldsymbol{b}_2)$ without changing the vector $\boldsymbol{x}$. By changing the basis and correspondingly the representation of vectors, the transformation matrix with respect to this new basis can have a particularly simple form that allows for straightforward computation. $\diamondsuit$

Example 2.23 (Basis Change)

Consider a transformation matrix

$$\boldsymbol{A} = \begin{bmatrix} 2 & 1 \\ 1 & 2 \end{bmatrix} \tag{2.100}$$

with respect to the canonical basis in $\mathbb{R}^2$. If we define a new basis

$$B = \left(\begin{bmatrix} 1 \\ 1 \end{bmatrix}, \begin{bmatrix} 1 \\ -1 \end{bmatrix} \right) \tag{2.101}$$

we obtain a diagonal transformation matrix

$$\tilde{A} = \begin{bmatrix} 3 & 0 \\ 0 & 1 \end{bmatrix} \tag{2.102}$$

with respect to B, which is easier to work with than A.

In the following, we will look at mappings that transform coordinate vectors with respect to one basis into coordinate vectors with respect to a different basis. We will state our main result first and then provide an explanation.

Theorem 2.20 (Basis Change). *For a linear mapping $\Phi : V \to W$, ordered bases*

$$B = (\boldsymbol{b}_1, \dots, \boldsymbol{b}_n), \quad \tilde{B} = (\tilde{\boldsymbol{b}}_1, \dots, \tilde{\boldsymbol{b}}_n) \tag{2.103}$$

of V and

$$C = (\boldsymbol{c}_1, \dots, \boldsymbol{c}_m), \quad \tilde{C} = (\tilde{\boldsymbol{c}}_1, \dots, \tilde{\boldsymbol{c}}_m) \tag{2.104}$$

of W, and a transformation matrix $\boldsymbol{A}_\Phi$ of Φ with respect to B and C, the corresponding transformation matrix $\tilde{\boldsymbol{A}}_\Phi$ with respect to the bases $\tilde{B}$ and $\tilde{C}$ is given as

$$\tilde{\boldsymbol{A}}_\Phi = \boldsymbol{T}^{-1} \boldsymbol{A}_\Phi \boldsymbol{S}\,. \tag{2.105}$$

Here, $\boldsymbol{S} \in \mathbb{R}^{n \times n}$ is the transformation matrix of id_V that maps coordinates with respect to $\tilde{B}$ onto coordinates with respect to B, and $\boldsymbol{T} \in \mathbb{R}^{m \times m}$ is the transformation matrix of id_W that maps coordinates with respect to $\tilde{C}$ onto coordinates with respect to C.

Proof Following Drumm and Weil (2001), we can write the vectors of the new basis $\tilde{B}$ of V as a linear combination of the basis vectors of B, such that

$$\tilde{\boldsymbol{b}}_j = s_{1j}\boldsymbol{b}_1 + \cdots + s_{nj}\boldsymbol{b}_n = \sum_{i=1}^{n} s_{ij}\boldsymbol{b}_i\,, \quad j = 1, \dots, n\,. \tag{2.106}$$

Similarly, we write the new basis vectors $\tilde{C}$ of W as a linear combination of the basis vectors of C, which yields

$$\tilde{\boldsymbol{c}}_k = t_{1k}\boldsymbol{c}_1 + \cdots + t_{mk}\boldsymbol{c}_m = \sum_{l=1}^{m} t_{lk}\boldsymbol{c}_l\,, \quad k = 1, \dots, m\,. \tag{2.107}$$

We define $\boldsymbol{S} = ((s_{ij})) \in \mathbb{R}^{n \times n}$ as the transformation matrix that maps coordinates with respect to $\tilde{B}$ onto coordinates with respect to B and $\boldsymbol{T} = ((t_{lk})) \in \mathbb{R}^{m \times m}$ as the transformation matrix that maps coordinates with respect to $\tilde{C}$ onto coordinates with respect to C. In particular, the jth column of $\boldsymbol{S}$ is the coordinate representation of $\tilde{\boldsymbol{b}}_j$ with respect to B and the kth column of $\boldsymbol{T}$ is the coordinate representation of $\tilde{\boldsymbol{c}}_k$ with respect to C. Note that both $\boldsymbol{S}$ and $\boldsymbol{T}$ are regular.

We are going to look at $\Phi(\tilde{\boldsymbol{b}}_j)$ from two perspectives. First, applying the mapping Φ, we get that for all $j = 1, \dots, n$

$$\Phi(\tilde{\boldsymbol{b}}_j) = \sum_{k=1}^{m} \underbrace{\tilde{a}_{kj}\tilde{\boldsymbol{c}}_k}_{\in W} \overset{(2.107)}{=} \sum_{k=1}^{m} \tilde{a}_{kj} \sum_{l=1}^{m} t_{lk}\boldsymbol{c}_l = \sum_{l=1}^{m} \left(\sum_{k=1}^{m} t_{lk}\tilde{a}_{kj} \right) \boldsymbol{c}_l\,, \tag{2.108}$$

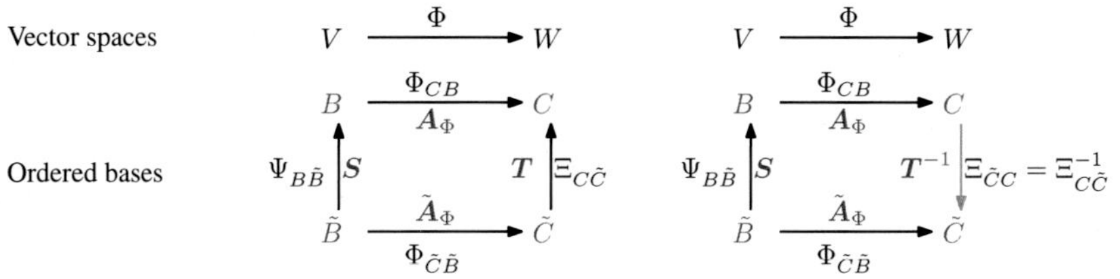

Figure 2.11 For a homomorphism $\Phi : V \to W$ and ordered bases $B, \tilde{B}$ of V and $C, \tilde{C}$ of W (marked in blue), we can express the mapping $\Phi_{\tilde{C}\tilde{B}}$ with respect to the bases $\tilde{B}, \tilde{C}$ equivalently as a composition of the homomorphisms $\Phi_{\tilde{C}\tilde{B}} = \Xi_{\tilde{C}C} \circ \Phi_{CB} \circ \Psi_{B\tilde{B}}$ with respect to the bases in the subscripts. The corresponding transformation matrices are in red.

where we first expressed the new basis vectors $\tilde{\boldsymbol{c}}_k \in W$ as linear combinations of the basis vectors $\boldsymbol{c}_l \in W$ and then swapped the order of summation.

Alternatively, when we express the $\tilde{\boldsymbol{b}}_j \in V$ as linear combinations of $\boldsymbol{b}_j \in V$, we arrive at

$$\Phi(\tilde{\boldsymbol{b}}_j) \overset{(2.106)}{=} \Phi\left(\sum_{i=1}^{n} s_{ij}\boldsymbol{b}_i\right) = \sum_{i=1}^{n} s_{ij}\Phi(\boldsymbol{b}_i) = \sum_{i=1}^{n} s_{ij}\sum_{l=1}^{m} a_{li}\boldsymbol{c}_l \tag{2.109a}$$

$$= \sum_{l=1}^{m}\left(\sum_{i=1}^{n} a_{li}s_{ij}\right)\boldsymbol{c}_l\,, \quad j = 1, \ldots, n\,, \tag{2.109b}$$

where we exploited the linearity of Φ. Comparing (2.108) and (2.109b), it follows for all $j = 1, \ldots, n$ and $l = 1, \ldots, m$ that

$$\sum_{k=1}^{m} t_{lk}\tilde{a}_{kj} = \sum_{i=1}^{n} a_{li}s_{ij} \tag{2.110}$$

and therefore

$$\boldsymbol{T}\tilde{\boldsymbol{A}}_\Phi = \boldsymbol{A}_\Phi\boldsymbol{S} \in \mathbb{R}^{m\times n}\,, \tag{2.111}$$

such that

$$\tilde{\boldsymbol{A}}_\Phi = \boldsymbol{T}^{-1}\boldsymbol{A}_\Phi\boldsymbol{S}\,, \tag{2.112}$$

which proves Theorem 2.20. □

Theorem 2.20 tells us that with a basis change in V (B is replaced with $\tilde{B}$) and W (C is replaced with $\tilde{C}$), the transformation matrix $\boldsymbol{A}_\Phi$ of a linear mapping $\Phi : V \to W$ is replaced by an equivalent matrix $\tilde{\boldsymbol{A}}_\Phi$ with

$$\tilde{\boldsymbol{A}}_\Phi = \boldsymbol{T}^{-1}\boldsymbol{A}_\Phi\boldsymbol{S}. \tag{2.113}$$

Figure 2.11 illustrates this relation: Consider a homomorphism $\Phi : V \to W$ and ordered bases $B, \tilde{B}$ of V and $C, \tilde{C}$ of W. The mapping Φ_{CB} is an instantiation of Φ and maps basis vectors of B onto linear combinations of basis vectors of C. Assume that we know the transformation matrix $\boldsymbol{A}_\Phi$ of Φ_{CB} with respect to the ordered bases B, C. When we perform a basis change from B to $\tilde{B}$ in V and from C to $\tilde{C}$ in W, we can determine the corresponding transformation matrix $\tilde{\boldsymbol{A}}_\Phi$ as follows: First, we find the matrix representation of the linear mapping $\Psi_{B\tilde{B}} : V \to V$ that maps coordinates with respect to the new basis $\tilde{B}$ onto the (unique) coordinates with respect to the "old" basis B (in V). Then we use the transformation matrix $\boldsymbol{A}_\Phi$ of $\Phi_{CB} : V \to W$ to map these coordinates onto the coordinates with respect to C in W. Finally, we use a linear mapping $\Xi_{\tilde{C}C} : W \to W$ to map the coordinates with respect to C onto coordinates with respect to $\tilde{C}$. Therefore, we can express the linear mapping $\Phi_{\tilde{C}\tilde{B}}$ as a composition of linear mappings that involve the "old" basis:

$$\Phi_{\tilde{C}\tilde{B}} = \Xi_{\tilde{C}C} \circ \Phi_{CB} \circ \Psi_{B\tilde{B}} = \Xi^{-1}_{C\tilde{C}} \circ \Phi_{CB} \circ \Psi_{B\tilde{B}} . \tag{2.114}$$

Concretely, we use $\Psi_{B\tilde{B}} = \mathrm{id}_V$ and $\Xi_{C\tilde{C}} = \mathrm{id}_W$, i.e., the identity mappings that map vectors onto themselves, but with respect to a different basis.

equivalent

Definition 2.21 (Equivalence)**.** Two matrices $\boldsymbol{A}, \tilde{\boldsymbol{A}} \in \mathbb{R}^{m\times n}$ are *equivalent* if there exist regular matrices $\boldsymbol{S} \in \mathbb{R}^{n\times n}$ and $\boldsymbol{T} \in \mathbb{R}^{m\times m}$, such that $\tilde{\boldsymbol{A}} = \boldsymbol{T}^{-1}\boldsymbol{A}\boldsymbol{S}$.

similar

Definition 2.22 (Similarity)**.** Two matrices $\boldsymbol{A}, \tilde{\boldsymbol{A}} \in \mathbb{R}^{n\times n}$ are *similar* if there exists a regular matrix $\boldsymbol{S} \in \mathbb{R}^{n\times n}$ with $\tilde{\boldsymbol{A}} = \boldsymbol{S}^{-1}\boldsymbol{A}\boldsymbol{S}$

Remark. Similar matrices are always equivalent. However, equivalent matrices are not necessarily similar. ◊

Remark. Consider vector spaces V, W, X. From the remark that follows Theorem 2.17, we already know that for linear mappings $\Phi : V \to W$ and $\Psi : W \to X$ the mapping $\Psi \circ \Phi : V \to X$ is also linear. With transformation matrices $\boldsymbol{A}_\Phi$ and $\boldsymbol{A}_\Psi$ of the corresponding mappings, the overall transformation matrix is $\boldsymbol{A}_{\Psi\circ\Phi} = \boldsymbol{A}_\Psi\boldsymbol{A}_\Phi$. ◊

In light of this remark, we can look at basis changes from the perspective of composing linear mappings:

- $\boldsymbol{A}_\Phi$ is the transformation matrix of a linear mapping $\Phi_{CB} : V \to W$ with respect to the bases B, C.
- $\tilde{\boldsymbol{A}}_\Phi$ is the transformation matrix of the linear mapping $\Phi_{\tilde{C}\tilde{B}} : V \to W$ with respect to the bases $\tilde{B}, \tilde{C}$.
- $\boldsymbol{S}$ is the transformation matrix of a linear mapping $\Psi_{B\tilde{B}} : V \to V$ (automorphism) that represents $\tilde{B}$ in terms of B. Normally, $\Psi = \mathrm{id}_V$ is the identity mapping in V.
- $\boldsymbol{T}$ is the transformation matrix of a linear mapping $\Xi_{C\tilde{C}} : W \to W$ (automorphism) that represents $\tilde{C}$ in terms of C. Normally, $\Xi = \mathrm{id}_W$ is the identity mapping in W.

If we (informally) write down the transformations just in terms of bases, then $\boldsymbol{A}_\Phi : B \to C$, $\tilde{\boldsymbol{A}}_\Phi : \tilde{B} \to \tilde{C}$, $\boldsymbol{S} : \tilde{B} \to B$, $\boldsymbol{T} : \tilde{C} \to C$ and $\boldsymbol{T}^{-1} : C \to \tilde{C}$, and

$$\tilde{B} \to \tilde{C} = \tilde{B} \to B \to C \to \tilde{C} \tag{2.115}$$

$$\tilde{\boldsymbol{A}}_\Phi = \boldsymbol{T}^{-1}\boldsymbol{A}_\Phi\boldsymbol{S} . \tag{2.116}$$

Note that the execution order in (2.116) is from right to left because vectors are multiplied at the right-hand side so that $\boldsymbol{x} \mapsto \boldsymbol{S}\boldsymbol{x} \mapsto \boldsymbol{A}_\Phi(\boldsymbol{S}\boldsymbol{x}) \mapsto \boldsymbol{T}^{-1}\big(\boldsymbol{A}_\Phi(\boldsymbol{S}\boldsymbol{x})\big) = \tilde{\boldsymbol{A}}_\Phi\boldsymbol{x}$.

Example 2.24 (Basis Change)

Consider a linear mapping $\Phi : \mathbb{R}^3 \to \mathbb{R}^4$ whose transformation matrix is

$$\boldsymbol{A}_\Phi = \begin{bmatrix} 1 & 2 & 0 \\ -1 & 1 & 3 \\ 3 & 7 & 1 \\ -1 & 2 & 4 \end{bmatrix} \tag{2.117}$$

with respect to the standard bases

$$B = \left(\begin{bmatrix} 1 \\ 0 \\ 0 \end{bmatrix}, \begin{bmatrix} 0 \\ 1 \\ 0 \end{bmatrix}, \begin{bmatrix} 0 \\ 0 \\ 1 \end{bmatrix}\right), \quad C = \left(\begin{bmatrix} 1 \\ 0 \\ 0 \\ 0 \end{bmatrix}, \begin{bmatrix} 0 \\ 1 \\ 0 \\ 0 \end{bmatrix}, \begin{bmatrix} 0 \\ 0 \\ 1 \\ 0 \end{bmatrix}, \begin{bmatrix} 0 \\ 0 \\ 0 \\ 1 \end{bmatrix}\right). \tag{2.118}$$

We seek the transformation matrix $\tilde{\boldsymbol{A}}_\Phi$ of Φ with respect to the new bases

$$\tilde{B} = \left(\begin{bmatrix} 1 \\ 1 \\ 0 \end{bmatrix}, \begin{bmatrix} 0 \\ 1 \\ 1 \end{bmatrix}, \begin{bmatrix} 1 \\ 0 \\ 1 \end{bmatrix}\right) \in \mathbb{R}^3, \quad \tilde{C} = \left(\begin{bmatrix} 1 \\ 1 \\ 0 \\ 0 \end{bmatrix}, \begin{bmatrix} 1 \\ 0 \\ 1 \\ 0 \end{bmatrix}, \begin{bmatrix} 0 \\ 1 \\ 1 \\ 0 \end{bmatrix}, \begin{bmatrix} 1 \\ 0 \\ 0 \\ 1 \end{bmatrix}\right). \tag{2.119}$$

Then,

$$\boldsymbol{S} = \begin{bmatrix} 1 & 0 & 1 \\ 1 & 1 & 0 \\ 0 & 1 & 1 \end{bmatrix}, \quad \boldsymbol{T} = \begin{bmatrix} 1 & 1 & 0 & 1 \\ 1 & 0 & 1 & 0 \\ 0 & 1 & 1 & 0 \\ 0 & 0 & 0 & 1 \end{bmatrix}, \tag{2.120}$$

where the ith column of $\boldsymbol{S}$ is the coordinate representation of $\tilde{\boldsymbol{b}}_i$ in terms of the basis vectors of B. Since B is the standard basis, the coordinate representation is straightforward to find. For a general basis B, we would need to solve a linear equation system to find the λ_i such that $\sum_{i=1}^{3} \lambda_i \boldsymbol{b}_i = \tilde{\boldsymbol{b}}_j$, $j = 1, \ldots, 3$. Similarly, the jth column of $\boldsymbol{T}$ is the coordinate representation of $\tilde{\boldsymbol{c}}_j$ in terms of the basis vectors of C.

Therefore, we obtain

$$\tilde{\boldsymbol{A}}_\Phi = \boldsymbol{T}^{-1}\boldsymbol{A}_\Phi\boldsymbol{S} = \frac{1}{2}\begin{bmatrix} 1 & 1 & -1 & -1 \\ 1 & -1 & 1 & -1 \\ -1 & 1 & 1 & 1 \\ 0 & 0 & 0 & 2 \end{bmatrix}\begin{bmatrix} 3 & 2 & 1 \\ 0 & 4 & 2 \\ 10 & 8 & 4 \\ 1 & 6 & 3 \end{bmatrix} \tag{2.121a}$$

$$= \begin{bmatrix} -4 & -4 & -2 \\ 6 & 0 & 0 \\ 4 & 8 & 4 \\ 1 & 6 & 3 \end{bmatrix}. \tag{2.121b}$$

In Chapter 4, we will be able to exploit the concept of a basis change to find a basis with respect to which the transformation matrix of an endomorphism has a particularly simple (diagonal) form. In Chapter 10, we will look at a data compression problem and find a convenient basis onto which we can project the data while minimizing the compression loss.

2.7.3 Image and Kernel

The image and kernel of a linear mapping are vector subspaces with certain important properties. In the following, we will characterize them more carefully.

Definition 2.23 (Image and Kernel).
For $\Phi : V \to W$, we define the *kernel/null space*

kernel
null space

$$\ker(\Phi) := \Phi^{-1}(\mathbf{0}_W) = \{\boldsymbol{v} \in V : \Phi(\boldsymbol{v}) = \mathbf{0}_W\} \tag{2.122}$$

and the *image/range*

image
range

$$\mathrm{Im}(\Phi) := \Phi(V) = \{\boldsymbol{w} \in W | \exists \boldsymbol{v} \in V : \Phi(\boldsymbol{v}) = \boldsymbol{w}\} . \tag{2.123}$$

We also call V and W the *domain* and *codomain* of Φ, respectively.

domain
codomain

Intuitively, the kernel is the set of vectors in $\boldsymbol{v} \in V$ that Φ maps onto the neutral element $\mathbf{0}_W \in W$. The image is the set of vectors $\boldsymbol{w} \in W$ that can be "reached" by Φ from any vector in V. An illustration is given in Figure 2.12.

Remark. Consider a linear mapping $\Phi : V \to W$, where V, W are vector spaces.

- It always holds that $\Phi(\mathbf{0}_V) = \mathbf{0}_W$ and, therefore, $\mathbf{0}_V \in \ker(\Phi)$. In particular, the null space is never empty.
- $\mathrm{Im}(\Phi) \subseteq W$ is a subspace of W, and $\ker(\Phi) \subseteq V$ is a subspace of V.
- Φ is injective (one-to-one) if and only if $\ker(\Phi) = \{\mathbf{0}\}$.

◇

Remark (Null Space and Column Space). Let us consider $\boldsymbol{A} \in \mathbb{R}^{m\times n}$ and a linear mapping $\Phi : \mathbb{R}^n \to \mathbb{R}^m$, $\boldsymbol{x} \mapsto \boldsymbol{A}\boldsymbol{x}$.

- For $\boldsymbol{A} = [\boldsymbol{a}_1, \dots, \boldsymbol{a}_n]$, where $\boldsymbol{a}_i$ are the columns of $\boldsymbol{A}$, we obtain

$$\mathrm{Im}(\Phi) = \{\boldsymbol{A}\boldsymbol{x} : \boldsymbol{x} \in \mathbb{R}^n\} = \left\{\sum_{i=1}^{n} x_i \boldsymbol{a}_i : x_1, \dots, x_n \in \mathbb{R}\right\} \tag{2.124a}$$

$$= \mathrm{span}[\boldsymbol{a}_1, \dots, \boldsymbol{a}_n] \subseteq \mathbb{R}^m , \tag{2.124b}$$

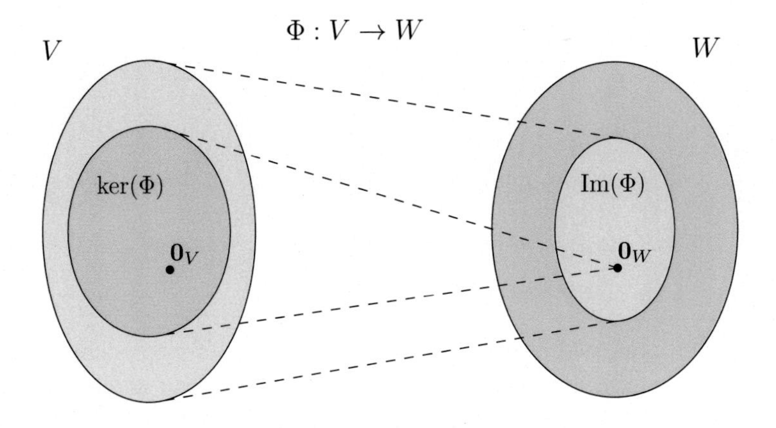

Figure 2.12 Kernel and image of a linear mapping $\Phi : V \to W$.

i.e., the image is the span of the columns of $\boldsymbol{A}$, also called the *column space*. Therefore, the column space (image) is a subspace of $\mathbb{R}^m$, where m is the "height" of the matrix.

column space

- $\text{rk}(\boldsymbol{A}) = \dim(\text{Im}(\Phi))$.
- The kernel/null space $\ker(\Phi)$ is the general solution to the homogeneous system of linear equations $\boldsymbol{Ax} = \boldsymbol{0}$ and captures all possible linear combinations of the elements in $\mathbb{R}^n$ that produce $\boldsymbol{0} \in \mathbb{R}^m$.
- The kernel is a subspace of $\mathbb{R}^n$, where n is the "width" of the matrix.
- The kernel focuses on the relationship among the columns, and we can use it to determine whether/how we can express a column as a linear combination of other columns.

Example 2.25 (Image and Kernel of a Linear Mapping)
The mapping

$$\Phi : \mathbb{R}^4 \to \mathbb{R}^2, \quad \begin{bmatrix} x_1 \\ x_2 \\ x_3 \\ x_4 \end{bmatrix} \mapsto \begin{bmatrix} 1 & 2 & -1 & 0 \\ 1 & 0 & 0 & 1 \end{bmatrix} \begin{bmatrix} x_1 \\ x_2 \\ x_3 \\ x_4 \end{bmatrix} = \begin{bmatrix} x_1 + 2x_2 - x_3 \\ x_1 + x_4 \end{bmatrix} \tag{2.125a}$$

$$= x_1 \begin{bmatrix} 1 \\ 1 \end{bmatrix} + x_2 \begin{bmatrix} 2 \\ 0 \end{bmatrix} + x_3 \begin{bmatrix} -1 \\ 0 \end{bmatrix} + x_4 \begin{bmatrix} 0 \\ 1 \end{bmatrix} \tag{2.125b}$$

is linear. To determine $\text{Im}(\Phi)$, we can take the span of the columns of the transformation matrix and obtain

$$\text{Im}(\Phi) = \text{span}[\begin{bmatrix} 1 \\ 1 \end{bmatrix}, \begin{bmatrix} 2 \\ 0 \end{bmatrix}, \begin{bmatrix} -1 \\ 0 \end{bmatrix}, \begin{bmatrix} 0 \\ 1 \end{bmatrix}]. \tag{2.126}$$

To compute the kernel (null space) of Φ, we need to solve $\boldsymbol{Ax} = \boldsymbol{0}$, i.e., we need to solve a homogeneous equation system. To do this, we use Gaussian elimination to transform $\boldsymbol{A}$ into reduced row-echelon form:

$$\begin{bmatrix} 1 & 2 & -1 & 0 \\ 1 & 0 & 0 & 1 \end{bmatrix} \rightsquigarrow \cdots \rightsquigarrow \begin{bmatrix} 1 & 0 & 0 & 1 \\ 0 & 1 & -\frac{1}{2} & -\frac{1}{2} \end{bmatrix}. \tag{2.127}$$

This matrix is in reduced row-echelon form, and we can use the Minus-1 Trick to compute a basis of the kernel (see Section 2.3.3). Alternatively, we can express the nonpivot columns (columns 3 and 4) as linear combinations of the pivot columns (columns 1 and 2). The third column $\boldsymbol{a}_3$ is equivalent to $-\frac{1}{2}$ times the second column $\boldsymbol{a}_2$. Therefore, $\boldsymbol{0} = \boldsymbol{a}_3 + \frac{1}{2}\boldsymbol{a}_2$. In the same way, we see that $\boldsymbol{a}_4 = \boldsymbol{a}_1 - \frac{1}{2}\boldsymbol{a}_2$, and therefore $\boldsymbol{0} = \boldsymbol{a}_1 - \frac{1}{2}\boldsymbol{a}_2 - \boldsymbol{a}_4$. Overall, this gives us the kernel (null space) as

$$\ker(\Phi) = \text{span}[\begin{bmatrix} 0 \\ \frac{1}{2} \\ 1 \\ 0 \end{bmatrix}, \begin{bmatrix} -1 \\ \frac{1}{2} \\ 0 \\ 1 \end{bmatrix}]. \tag{2.128}$$

rank-nullity theorem

Theorem 2.24 (Rank-Nullity Theorem). *For vector spaces V, W, and a linear mapping $\Phi : V \to W$ it holds that*

$$\dim(\ker(\Phi)) + \dim(\text{Im}(\Phi)) = \dim(V)\,. \tag{2.129}$$

fundamental theorem of linear mappings

The rank-nullity theorem is also referred to as the *fundamental theorem of linear mappings* (Axler, 2015, theorem 3.22). The following are direct consequences of Theorem 2.24:

- If $\dim(\text{Im}(\Phi)) < \dim(V)$, then $\ker(\Phi)$ is nontrivial, i.e., the kernel contains more than $\mathbf{0}_V$ and $\dim(\ker(\Phi)) \geqslant 1$.
- If $\boldsymbol{A}_\Phi$ is the transformation matrix of Φ with respect to an ordered basis and $\dim(\text{Im}(\Phi)) < \dim(V)$, then the system of linear equations $\boldsymbol{A}_\Phi \boldsymbol{x} = \mathbf{0}$ has infinitely many solutions.
- If $\dim(V) = \dim(W)$, then the following three-way equivalence holds:
  - Φ is injective
  - Φ is surjective
  - Φ is bijective

  since $\text{Im}(\Phi) \subseteq W$.

2.8 Affine Spaces

In the following, we will have a closer look at spaces that are offset from the origin, i.e., spaces that are no longer vector subspaces. Moreover, we will briefly discuss properties of mappings between these affine spaces, which resemble linear mappings.

Remark. In the machine learning literature, the distinction between linear and affine is sometimes not clear so that we can find references to affine spaces/ mappings as linear spaces/mappings. ◇

2.8.1 Affine Subspaces

Definition 2.25 (Affine Subspace). Let V be a vector space, $\boldsymbol{x}_0 \in V$ and $U \subseteq V$ a subspace. Then the subset

$$L = \boldsymbol{x}_0 + U := \{\boldsymbol{x}_0 + \boldsymbol{u} : \boldsymbol{u} \in U\} \tag{2.130a}$$

$$= \{\boldsymbol{v} \in V | \exists \boldsymbol{u} \in U : \boldsymbol{v} = \boldsymbol{x}_0 + \boldsymbol{u}\} \subseteq V \tag{2.130b}$$

affine subspace
linear manifold
direction
direction space
support point
hyperplane

is called *affine subspace* or *linear manifold* of V. U is called *direction* or *direction space*, and $\boldsymbol{x}_0$ is called *support point*. In Chapter 12, we refer to such a subspace as a *hyperplane*.

Note that the definition of an affine subspace excludes $\mathbf{0}$ if $\boldsymbol{x}_0 \notin U$. Therefore, an affine subspace is not a (linear) subspace (vector subspace) of V for $\boldsymbol{x}_0 \notin U$.

Examples of affine subspaces are points, lines, and planes in $\mathbb{R}^3$, which do not (necessarily) go through the origin.

Remark. Consider two affine subspaces $L = \boldsymbol{x}_0 + U$ and $\tilde{L} = \tilde{\boldsymbol{x}}_0 + \tilde{U}$ of a vector space V. Then $L \subseteq \tilde{L}$ if and only if $U \subseteq \tilde{U}$ and $\boldsymbol{x}_0 - \tilde{\boldsymbol{x}}_0 \in \tilde{U}$.

Affine subspaces are often described by *parameters*: Consider a k-dimensional affine space $L = \boldsymbol{x}_0 + U$ of V. If $(\boldsymbol{b}_1, \ldots, \boldsymbol{b}_k)$ is an ordered basis of U, then every element $\boldsymbol{x} \in L$ can be uniquely described as

$$\boldsymbol{x} = \boldsymbol{x}_0 + \lambda_1 \boldsymbol{b}_1 + \ldots + \lambda_k \boldsymbol{b}_k \,, \tag{2.131}$$

where $\lambda_1, \ldots, \lambda_k \in \mathbb{R}$. This representation is called *parametric equation* of L with directional vectors $\boldsymbol{b}_1, \ldots, \boldsymbol{b}_k$ and *parameters* $\lambda_1, \ldots, \lambda_k$. ♢

parametric equation

parameters

Example 2.26 (Affine Subspaces)

- One-dimensional affine subspaces are called *lines* and can be written as $\boldsymbol{y} = \boldsymbol{x}_0 + \lambda \boldsymbol{x}_1$, where $\lambda \in \mathbb{R}$, where $U = \text{span}[\boldsymbol{x}_1] \subseteq \mathbb{R}^n$ is a one-dimensional subspace of $\mathbb{R}^n$. This means that a line is defined by a support point $\boldsymbol{x}_0$ and a vector $\boldsymbol{x}_1$ that defines the direction. See Figure 2.13 for an illustration.
- Two-dimensional affine subspaces of $\mathbb{R}^n$ are called *planes*. The parametric equation for planes is $\boldsymbol{y} = \boldsymbol{x}_0 + \lambda_1 \boldsymbol{x}_1 + \lambda_2 \boldsymbol{x}_2$, where $\lambda_1, \lambda_2 \in \mathbb{R}$ and $U = [\boldsymbol{x}_1, \boldsymbol{x}_2] \subseteq \mathbb{R}^n$. This means that a plane is defined by a support point $\boldsymbol{x}_0$ and two linearly independent vectors $\boldsymbol{x}_1, \boldsymbol{x}_2$ that span the direction space.
- In $\mathbb{R}^n$, the $(n-1)$-dimensional affine subspaces are called *hyperplanes*, and the corresponding parametric equation is $\boldsymbol{y} = \boldsymbol{x}_0 + \sum_{i=1}^{n-1} \lambda_i \boldsymbol{x}_i$, where $\boldsymbol{x}_1, \ldots, \boldsymbol{x}_{n-1}$ form a basis of an $(n-1)$-dimensional subspace U of $\mathbb{R}^n$. This means that a hyperplane is defined by a support point $\boldsymbol{x}_0$ and $(n-1)$ linearly independent vectors $\boldsymbol{x}_1, \ldots, \boldsymbol{x}_{n-1}$ that span the direction space. In $\mathbb{R}^2$, a line is also a hyperplane. In $\mathbb{R}^3$, a plane is also a hyperplane.

line

plane

hyperplane

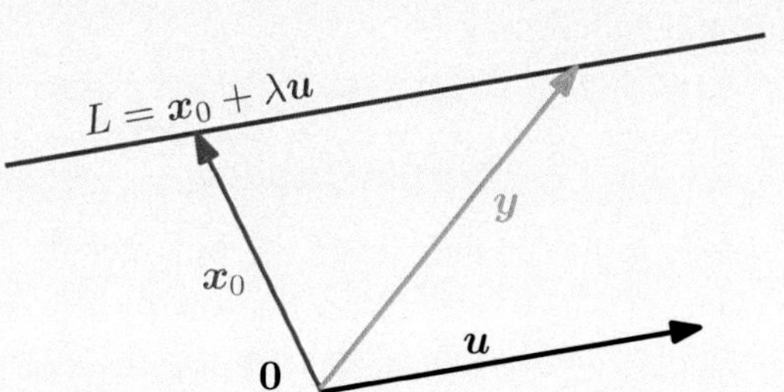

Figure 2.13 Vectors $\boldsymbol{y}$ on a line lie in an affine subspace L with support point $\boldsymbol{x}_0$ and direction $\boldsymbol{u}$.

Remark (Inhomogeneous systems of linear equations and affine subspaces). For $\boldsymbol{A} \in \mathbb{R}^{m \times n}$ and $\boldsymbol{b} \in \mathbb{R}^m$, the solution of the linear equation system $\boldsymbol{A}\boldsymbol{x} = \boldsymbol{b}$ is either the empty set or an affine subspace of $\mathbb{R}^n$ of dimension $n - \text{rk}(\boldsymbol{A})$. In particular, the solution of the linear equation $\lambda_1 \boldsymbol{x}_1 + \ldots + \lambda_n \boldsymbol{x}_n = \boldsymbol{b}$, where $(\lambda_1, \ldots, \lambda_n) \neq (0, \ldots, 0)$, is a hyperplane in $\mathbb{R}^n$.

In $\mathbb{R}^n$, every k-dimensional affine subspace is the solution of a linear inhomogeneous equation system $\boldsymbol{A}\boldsymbol{x} = \boldsymbol{b}$, where $\boldsymbol{A} \in \mathbb{R}^{m \times n}, \boldsymbol{b} \in \mathbb{R}^m$ and $\text{rk}(\boldsymbol{A}) = n - k$. Recall that for homogeneous equation systems $\boldsymbol{A}\boldsymbol{x} = \boldsymbol{0}$, the solution was a vector subspace, which we can also think of as a special affine space with support point $\boldsymbol{x}_0 = \boldsymbol{0}$. ♢

2.8.2 Affine Mappings

Similar to linear mappings between vector spaces, which we discussed in Section 2.7, we can define affine mappings between two affine spaces. Linear and affine mappings are closely related. Therefore, many properties that we already know from linear mappings, e.g., that the composition of linear mappings is a linear mapping, also hold for affine mappings.

Definition 2.26 (Affine Mapping)**.** For two vector spaces V, W, a linear mapping $\Phi : V \to W$, and $\boldsymbol{a} \in W$, the mapping

$$\phi : V \to W \tag{2.132}$$
$$\boldsymbol{x} \mapsto \boldsymbol{a} + \Phi(\boldsymbol{x}) \tag{2.133}$$

affine mapping
translation vector

is an *affine mapping* from V to W. The vector $\boldsymbol{a}$ is called the *translation vector* of ϕ.

- Every affine mapping $\phi : V \to W$ is also the composition of a linear mapping $\Phi : V \to W$ and a translation $\tau : W \to W$ in W, such that $\phi = \tau \circ \Phi$. The mappings Φ and τ are uniquely determined.
- The composition $\phi' \circ \phi$ of affine mappings $\phi : V \to W$, $\phi' : W \to X$ is affine.
- Affine mappings keep the geometric structure invariant. They also preserve the dimension and parallelism.

2.9 Further Reading

There are many resources for learning linear algebra, including the textbooks by Strang (2003), Golan (2007), Axler (2015), and Liesen and Mehrmann (2015). There are also several online resources that we mentioned in the introduction to this chapter. We only covered Gaussian elimination here, but there are many other approaches for solving systems of linear equations, and we refer to numerical linear algebra textbooks by Stoer and Burlirsch (2002), Golub and Van Loan (2012), and Horn and Johnson (2013) for an in-depth discussion.

In this book, we distinguish between the topics of linear algebra (e.g., vectors, matrices, linear independence, basis) and topics related to the geometry of a vector space. In Chapter 3, we will introduce the inner product, which induces a norm. These concepts allow us to define angles, lengths, and distances, which we will use for orthogonal projections. Projections turn out to be key in many machine learning algorithms, such as linear regression and principal component analysis, both of which we will cover in Chapters 9 and 10, respectively.

Exercises

2.1 We consider $(\mathbb{R}\backslash\{-1\}, \star)$, where

$$a \star b := ab + a + b, \qquad a, b \in \mathbb{R}\backslash\{-1\} \tag{2.134}$$

a. Show that $(\mathbb{R}\backslash\{-1\}, \star)$ is an Abelian group.

b. Solve

$$3 \star x \star x = 15$$

in the Abelian group $(\mathbb{R}\backslash\{-1\}, \star)$, where $\star$ is defined in (2.134).

2.2 Let n be in $\mathbb{N}\backslash\{0\}$. Let k, x be in $\mathbb{Z}$. We define the congruence class $\bar{k}$ of the integer k as the set

$$\begin{aligned}\bar{k} &= \{x \in \mathbb{Z} \mid x - k = 0 \pmod n\} \\ &= \{x \in \mathbb{Z} \mid (\exists a \in \mathbb{Z})\colon (x - k = n \cdot a)\}\,.\end{aligned}$$

We now define $\mathbb{Z}/n\mathbb{Z}$ (sometimes written $\mathbb{Z}_n$) as the set of all congruence classes modulo n. Euclidean division implies that this set is a finite set containing n elements:

$$\mathbb{Z}_n = \{\overline{0}, \overline{1}, \ldots, \overline{n-1}\}$$

For all $\overline{a}, \overline{b} \in \mathbb{Z}_n$, we define

$$\overline{a} \oplus \overline{b} := \overline{a+b}$$

a. Show that $(\mathbb{Z}_n, \oplus)$ is a group. Is it Abelian?

b. We now define another operation $\otimes$ for all $\overline{a}$ and $\overline{b}$ in $\mathbb{Z}_n$ as

$$\overline{a} \otimes \overline{b} = \overline{a \times b}\,, \tag{2.135}$$

where $a \times b$ represents the usual multiplication in $\mathbb{Z}$.
Let $n = 5$. Draw the times table of the elements of $\mathbb{Z}_5\backslash\{\overline{0}\}$ under $\otimes$, i.e., calculate the products $\overline{a} \otimes \overline{b}$ for all $\overline{a}$ and $\overline{b}$ in $\mathbb{Z}_5\backslash\{\overline{0}\}$.
Hence, show that $\mathbb{Z}_5\backslash\{\overline{0}\}$ is closed under $\otimes$ and possesses a neutral element for $\otimes$. Display the inverse of all elements in $\mathbb{Z}_5\backslash\{\overline{0}\}$ under $\otimes$. Conclude that $(\mathbb{Z}_5\backslash\{\overline{0}\}, \otimes)$ is an Abelian group.

c. Show that $(\mathbb{Z}_8\backslash\{\overline{0}\}, \otimes)$ is not a group.

d. We recall that the Bézout theorem states that two integers a and b are relatively prime (i.e., $gcd(a, b) = 1$) if and only if there exist two integers u and v such that $au + bv = 1$. Show that $(\mathbb{Z}_n\backslash\{\overline{0}\}, \otimes)$ is a group if and only if $n \in \mathbb{N}\backslash\{0\}$ is prime.

2.3 Consider the set $\mathcal{G}$ of 3×3 matrices defined as follows:

$$\mathcal{G} = \left\{ \begin{bmatrix} 1 & x & z \\ 0 & 1 & y \\ 0 & 0 & 1 \end{bmatrix} \in \mathbb{R}^{3\times 3} \,\middle|\, x, y, z \in \mathbb{R} \right\} \tag{2.136}$$

We define $\cdot$ as the standard matrix multiplication.
Is $(\mathcal{G}, \cdot)$ a group? If yes, is it Abelian? Justify your answer.

2.4 Compute the following matrix products, if possible:

a.

$$\begin{bmatrix}1 & 2\\ 4 & 5\\ 7 & 8\end{bmatrix}\begin{bmatrix}1 & 1 & 0\\ 0 & 1 & 1\\ 1 & 0 & 1\end{bmatrix}$$

b.

$$\begin{bmatrix}1 & 2 & 3\\ 4 & 5 & 6\\ 7 & 8 & 9\end{bmatrix}\begin{bmatrix}1 & 1 & 0\\ 0 & 1 & 1\\ 1 & 0 & 1\end{bmatrix}$$

c.

$$\begin{bmatrix}1 & 1 & 0\\ 0 & 1 & 1\\ 1 & 0 & 1\end{bmatrix}\begin{bmatrix}1 & 2 & 3\\ 4 & 5 & 6\\ 7 & 8 & 9\end{bmatrix}$$

d.

$$\begin{bmatrix}1 & 2 & 1 & 2\\ 4 & 1 & -1 & -4\end{bmatrix}\begin{bmatrix}0 & 3\\ 1 & -1\\ 2 & 1\\ 5 & 2\end{bmatrix}$$

e.

$$\begin{bmatrix}0 & 3\\ 1 & -1\\ 2 & 1\\ 5 & 2\end{bmatrix}\begin{bmatrix}1 & 2 & 1 & 2\\ 4 & 1 & -1 & -4\end{bmatrix}$$

2.5 Find the set $\mathcal{S}$ of all solutions in $\boldsymbol{x}$ of the following inhomogeneous linear systems $\boldsymbol{A}\boldsymbol{x} = \boldsymbol{b}$, where $\boldsymbol{A}$ and $\boldsymbol{b}$ are defined as follows:

a.

$$\boldsymbol{A} = \begin{bmatrix}1 & 1 & -1 & -1\\ 2 & 5 & -7 & -5\\ 2 & -1 & 1 & 3\\ 5 & 2 & -4 & 2\end{bmatrix}, \quad \boldsymbol{b} = \begin{bmatrix}1\\ -2\\ 4\\ 6\end{bmatrix}$$

b.

$$\boldsymbol{A} = \begin{bmatrix}1 & -1 & 0 & 0 & 1\\ 1 & 1 & 0 & -3 & 0\\ 2 & -1 & 0 & 1 & -1\\ -1 & 2 & 0 & -2 & -1\end{bmatrix}, \quad \boldsymbol{b} = \begin{bmatrix}3\\ 6\\ 5\\ -1\end{bmatrix}$$

2.6 Using Gaussian elimination, find all solutions of the inhomogeneous equation system $\boldsymbol{A}\boldsymbol{x} = \boldsymbol{b}$ with

$$\boldsymbol{A} = \begin{bmatrix}0 & 1 & 0 & 0 & 1 & 0\\ 0 & 0 & 0 & 1 & 1 & 0\\ 0 & 1 & 0 & 0 & 0 & 1\end{bmatrix}, \quad \boldsymbol{b} = \begin{bmatrix}2\\ -1\\ 1\end{bmatrix}$$

2.7 Find all solutions in $\boldsymbol{x} = \begin{bmatrix} x_1 \\ x_2 \\ x_3 \end{bmatrix} \in \mathbb{R}^3$ of the equation system $\boldsymbol{A}\boldsymbol{x} = 12\boldsymbol{x}$, where

$$\boldsymbol{A} = \begin{bmatrix} 6 & 4 & 3 \\ 6 & 0 & 9 \\ 0 & 8 & 0 \end{bmatrix}$$

and $\sum_{i=1}^{3} x_i = 1$.

2.8 Determine the inverses of the following matrices if possible:

a.

$$\boldsymbol{A} = \begin{bmatrix} 2 & 3 & 4 \\ 3 & 4 & 5 \\ 4 & 5 & 6 \end{bmatrix}$$

b.

$$\boldsymbol{A} = \begin{bmatrix} 1 & 0 & 1 & 0 \\ 0 & 1 & 1 & 0 \\ 1 & 1 & 0 & 1 \\ 1 & 1 & 1 & 0 \end{bmatrix}$$

2.9 Which of the following sets are subspaces of $\mathbb{R}^3$?

a. $A = \{(\lambda, \lambda + \mu^3, \lambda - \mu^3) \mid \lambda, \mu \in \mathbb{R}\}$
b. $B = \{(\lambda^2, -\lambda^2, 0) \mid \lambda \in \mathbb{R}\}$
c. Let γ be in $\mathbb{R}$.
$C = \{(\xi_1, \xi_2, \xi_3) \in \mathbb{R}^3 \mid \xi_1 - 2\xi_2 + 3\xi_3 = \gamma\}$
d. $D = \{(\xi_1, \xi_2, \xi_3) \in \mathbb{R}^3 \mid \xi_2 \in \mathbb{Z}\}$

2.10 Are the following sets of vectors linearly independent?

a.

$$\boldsymbol{x}_1 = \begin{bmatrix} 2 \\ -1 \\ 3 \end{bmatrix}, \quad \boldsymbol{x}_2 = \begin{bmatrix} 1 \\ 1 \\ -2 \end{bmatrix}, \quad \boldsymbol{x}_3 = \begin{bmatrix} 3 \\ -3 \\ 8 \end{bmatrix}$$

b.

$$\boldsymbol{x}_1 = \begin{bmatrix} 1 \\ 2 \\ 1 \\ 0 \\ 0 \end{bmatrix}, \quad \boldsymbol{x}_2 = \begin{bmatrix} 1 \\ 1 \\ 0 \\ 1 \\ 1 \end{bmatrix}, \quad \boldsymbol{x}_3 = \begin{bmatrix} 1 \\ 0 \\ 0 \\ 1 \\ 1 \end{bmatrix}$$

2.11 Write

$$\boldsymbol{y} = \begin{bmatrix} 1 \\ -2 \\ 5 \end{bmatrix}$$

as linear combination of

$$\boldsymbol{x}_1 = \begin{bmatrix} 1 \\ 1 \\ 1 \end{bmatrix}, \quad \boldsymbol{x}_2 = \begin{bmatrix} 1 \\ 2 \\ 3 \end{bmatrix}, \quad \boldsymbol{x}_3 = \begin{bmatrix} 2 \\ -1 \\ 1 \end{bmatrix}$$

2.12 Consider two subspaces of $\mathbb{R}^4$:

$$U_1 = \text{span}[\begin{bmatrix} 1 \\ 1 \\ -3 \\ 1 \end{bmatrix}, \begin{bmatrix} 2 \\ -1 \\ 0 \\ -1 \end{bmatrix}, \begin{bmatrix} -1 \\ 1 \\ -1 \\ 1 \end{bmatrix}], \quad U_2 = \text{span}[\begin{bmatrix} -1 \\ -2 \\ 2 \\ 1 \end{bmatrix}, \begin{bmatrix} 2 \\ -2 \\ 0 \\ 0 \end{bmatrix}, \begin{bmatrix} -3 \\ 6 \\ -2 \\ -1 \end{bmatrix}].$$

Determine a basis of $U_1 \cap U_2$.

2.13 Consider two subspaces U_1 and U_2, where U_1 is the solution space of the homogeneous equation system $\boldsymbol{A}_1\boldsymbol{x} = \boldsymbol{0}$ and U_2 is the solution space of the homogeneous equation system $\boldsymbol{A}_2\boldsymbol{x} = \boldsymbol{0}$ with

$$\boldsymbol{A}_1 = \begin{bmatrix} 1 & 0 & 1 \\ 1 & -2 & -1 \\ 2 & 1 & 3 \\ 1 & 0 & 1 \end{bmatrix}, \quad \boldsymbol{A}_2 = \begin{bmatrix} 3 & -3 & 0 \\ 1 & 2 & 3 \\ 7 & -5 & 2 \\ 3 & -1 & 2 \end{bmatrix}.$$

a. Determine the dimension of U_1, U_2.
b. Determine bases of U_1 and U_2.
c. Determine a basis of $U_1 \cap U_2$.

2.14 Consider two subspaces U_1 and U_2, where U_1 is spanned by the columns of $\boldsymbol{A}_1$ and U_2 is spanned by the columns of $\boldsymbol{A}_2$ with

$$\boldsymbol{A}_1 = \begin{bmatrix} 1 & 0 & 1 \\ 1 & -2 & -1 \\ 2 & 1 & 3 \\ 1 & 0 & 1 \end{bmatrix}, \quad \boldsymbol{A}_2 = \begin{bmatrix} 3 & -3 & 0 \\ 1 & 2 & 3 \\ 7 & -5 & 2 \\ 3 & -1 & 2 \end{bmatrix}.$$

a. Determine the dimension of U_1, U_2
b. Determine bases of U_1 and U_2
c. Determine a basis of $U_1 \cap U_2$

2.15 Let $F = \{(x, y, z) \in \mathbb{R}^3 \mid x + y - z = 0\}$ and $G = \{(a - b, a + b, a - 3b) \mid a, b \in \mathbb{R}\}$.

a. Show that F and G are subspaces of $\mathbb{R}^3$.
b. Calculate $F \cap G$ without resorting to any basis vector.
c. Find one basis for F and one for G, calculate $F \cap G$ using the basis vectors previously found and check your result with the previous question.

2.16 Are the following mappings linear?

a. Let $a, b \in \mathbb{R}$.

$$\Phi : L^1([a, b]) \to \mathbb{R}$$
$$f \mapsto \Phi(f) = \int_a^b f(x)dx\,,$$

where $L^1([a, b])$ denotes the set of integrable functions on $[a, b]$.

b.

$$\Phi : C^1 \to C^0$$
$$f \mapsto \Phi(f) = f'\,,$$

where for $k \geqslant 1$, C^k denotes the set of k times continuously differentiable functions, and C^0 denotes the set of continuous functions.

c.

$$\begin{aligned}\Phi : \mathbb{R} &\to \mathbb{R}\\ x &\mapsto \Phi(x) = \cos(x)\end{aligned}$$

d.

$$\begin{aligned}\Phi : \mathbb{R}^3 &\to \mathbb{R}^2\\ \boldsymbol{x} &\mapsto \begin{bmatrix}1 & 2 & 3\\ 1 & 4 & 3\end{bmatrix}\boldsymbol{x}\end{aligned}$$

e. Let θ be in $[0, 2\pi[$.

$$\begin{aligned}\Phi : \mathbb{R}^2 &\to \mathbb{R}^2\\ \boldsymbol{x} &\mapsto \begin{bmatrix}\cos(\theta) & \sin(\theta)\\ -\sin(\theta) & \cos(\theta)\end{bmatrix}\boldsymbol{x}\end{aligned}$$

2.17 Consider the linear mapping

$$\begin{aligned}\Phi : \mathbb{R}^3 &\to \mathbb{R}^4\\ \Phi\left(\begin{bmatrix}x_1\\ x_2\\ x_3\end{bmatrix}\right) &= \begin{bmatrix}3x_1 + 2x_2 + x_3\\ x_1 + x_2 + x_3\\ x_1 - 3x_2\\ 2x_1 + 3x_2 + x_3\end{bmatrix}\end{aligned}$$

- Find the transformation matrix $\boldsymbol{A}_\Phi$.
- Determine $\text{rk}(\boldsymbol{A}_\Phi)$.
- Compute the kernel and image of Φ. What are $\dim(\ker(\Phi))$ and $\dim(\text{Im}(\Phi))$?

2.18 Let E be a vector space. Let f and g be two automorphisms on E such that $f \circ g = \text{id}_E$ (i.e., $f \circ g$ is the identity mapping id_E). Show that $\ker(f) = \ker(g \circ f)$, $\text{Im}(g) = \text{Im}(g \circ f)$ and that $\ker(f) \cap \text{Im}(g) = \{\boldsymbol{0}_E\}$.

2.19 Consider an endomorphism $\Phi : \mathbb{R}^3 \to \mathbb{R}^3$ whose transformation matrix (with respect to the standard basis in $\mathbb{R}^3$) is

$$\boldsymbol{A}_\Phi = \begin{bmatrix}1 & 1 & 0\\ 1 & -1 & 0\\ 1 & 1 & 1\end{bmatrix}.$$

a. Determine $\ker(\Phi)$ and $\text{Im}(\Phi)$.

b. Determine the transformation matrix $\tilde{\boldsymbol{A}}_\Phi$ with respect to the basis

$$B = \left(\begin{bmatrix}1\\ 1\\ 1\end{bmatrix}, \begin{bmatrix}1\\ 2\\ 1\end{bmatrix}, \begin{bmatrix}1\\ 0\\ 0\end{bmatrix}\right),$$

i.e., perform a basis change toward the new basis B.

2.20 Let us consider $\boldsymbol{b}_1, \boldsymbol{b}_2, \boldsymbol{b}_1', \boldsymbol{b}_2'$, four vectors of $\mathbb{R}^2$ expressed in the standard basis of $\mathbb{R}^2$ as

$$\boldsymbol{b}_1 = \begin{bmatrix}2\\ 1\end{bmatrix}, \quad \boldsymbol{b}_2 = \begin{bmatrix}-1\\ -1\end{bmatrix}, \quad \boldsymbol{b}_1' = \begin{bmatrix}2\\ -2\end{bmatrix}, \quad \boldsymbol{b}_2' = \begin{bmatrix}1\\ 1\end{bmatrix}$$

and let us define two ordered bases $B = (\boldsymbol{b}_1, \boldsymbol{b}_2)$ and $B' = (\boldsymbol{b}_1', \boldsymbol{b}_2')$ of $\mathbb{R}^2$.

a. Show that B and B' are two bases of $\mathbb{R}^2$ and draw those basis vectors.

b. Compute the matrix $\boldsymbol{P}_1$ that performs a basis change from B' to B.

c. We consider $\boldsymbol{c}_1, \boldsymbol{c}_2, \boldsymbol{c}_3$, three vectors of $\mathbb{R}^3$ defined in the standard basis of $\mathbb{R}$ as

$$\boldsymbol{c}_1 = \begin{bmatrix} 1 \\ 2 \\ -1 \end{bmatrix}, \quad \boldsymbol{c}_2 = \begin{bmatrix} 0 \\ -1 \\ 2 \end{bmatrix}, \quad \boldsymbol{c}_3 = \begin{bmatrix} 1 \\ 0 \\ -1 \end{bmatrix}$$

and we define $C = (\boldsymbol{c}_1, \boldsymbol{c}_2, \boldsymbol{c}_3)$.

1. Show that C is a basis of $\mathbb{R}^3$, e.g., by using determinants (see Section 4.1).
2. Let us call $C' = (\boldsymbol{c}'_1, \boldsymbol{c}'_2, \boldsymbol{c}'_3)$ the standard basis of $\mathbb{R}^3$. Determine the matrix $\boldsymbol{P}_2$ that performs the basis change from C to C'.

d. We consider a homomorphism $\Phi : \mathbb{R}^2 \longrightarrow \mathbb{R}^3$, such that

$$\begin{aligned} \Phi(\boldsymbol{b}_1 + \boldsymbol{b}_2) &= \boldsymbol{c}_2 + \boldsymbol{c}_3 \\ \Phi(\boldsymbol{b}_1 - \boldsymbol{b}_2) &= 2\boldsymbol{c}_1 - \boldsymbol{c}_2 + 3\boldsymbol{c}_3 \end{aligned}$$

where $B = (\boldsymbol{b}_1, \boldsymbol{b}_2)$ and $C = (\boldsymbol{c}_1, \boldsymbol{c}_2, \boldsymbol{c}_3)$ are ordered bases of $\mathbb{R}^2$ and $\mathbb{R}^3$, respectively.

Determine the transformation matrix $\boldsymbol{A}_\Phi$ of Φ with respect to the ordered bases B and C.

e. Determine $\boldsymbol{A}'$, the transformation matrix of Φ with respect to the bases $\boldsymbol{B}'$ and $\boldsymbol{C}'$.

f. Let us consider the vector $\boldsymbol{x} \in \mathbb{R}^2$ whose coordinates in B' are $[2, 3]^\top$. In other words, $\boldsymbol{x} = 2\boldsymbol{b}'_1 + 3\boldsymbol{b}'_2$.

1. Calculate the coordinates of $\boldsymbol{x}$ in B.
2. Based on that, compute the coordinates of $\Phi(\boldsymbol{x})$ expressed in C.
3. Then write $\Phi(\boldsymbol{x})$ in terms of $\boldsymbol{c}'_1, \boldsymbol{c}'_2, \boldsymbol{c}'_3$.
4. Use the representation of $\boldsymbol{x}$ in B' and the matrix $\boldsymbol{A}'$ to find this result directly.

3

Analytic Geometry

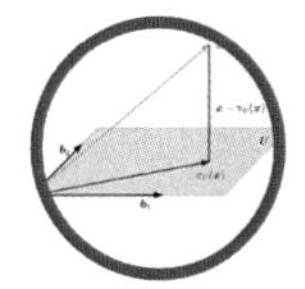

In Chapter 2, we studied vectors, vector spaces, and linear mappings at a general but abstract level. In this chapter, we will add some geometric interpretation and intuition to all of these concepts. In particular, we will look at geometric vectors and compute their lengths and distances or angles between two vectors. To be able to do this, we equip the vector space with an inner product that induces the geometry of the vector space. Inner products and their corresponding norms and metrics capture the intuitive notions of similarity and distances, which we use to develop the support vector machine in Chapter 12. We will then use the concepts of lengths and angles between vectors to discuss orthogonal projections, which will play a central role when we discuss principal component analysis in Chapter 10 and regression via maximum likelihood estimation in Chapter 9. Figure 3.1 gives an overview of how concepts in this chapter are related and how they are connected to other chapters of the book.

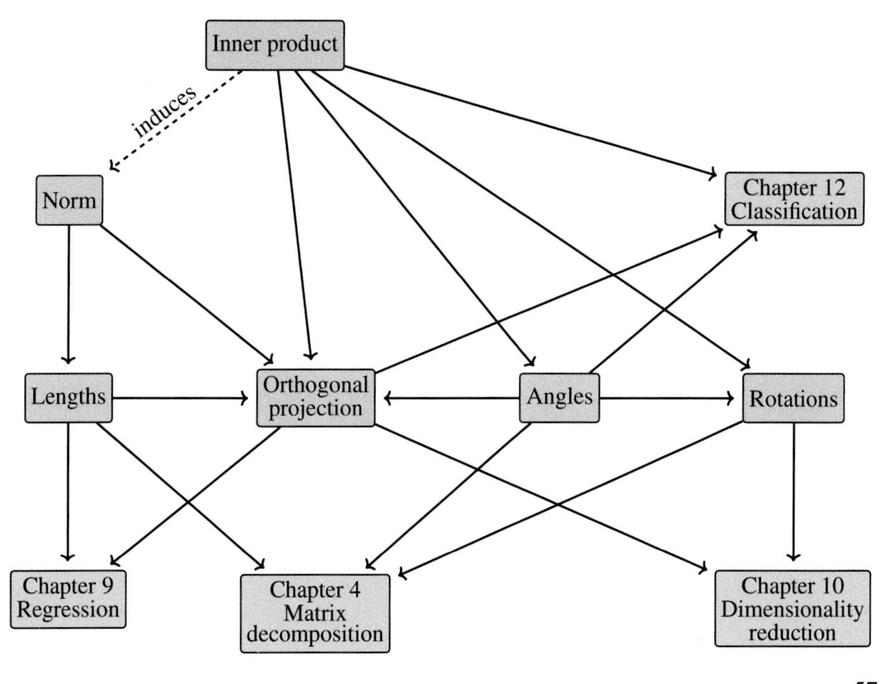

Figure 3.1 A mind map of the concepts introduced in this chapter, along with when they are used in other parts of the book.

3.1 Norms

When we think of geometric vectors, i.e., directed line segments that start at the origin, then intuitively the length of a vector is the distance of the "end" of this directed line segment from the origin. In the following, we will discuss the notion of the length of vectors using the concept of a norm.

norm

Definition 3.1 (Norm). A *norm* on a vector space V is a function

$$\|\cdot\| : V \to \mathbb{R}, \tag{3.1}$$

$$\boldsymbol{x} \mapsto \|\boldsymbol{x}\|, \tag{3.2}$$

length

which assigns each vector $\boldsymbol{x}$ its *length* $\|\boldsymbol{x}\| \in \mathbb{R}$, such that for all $\lambda \in \mathbb{R}$ and $\boldsymbol{x}, \boldsymbol{y} \in V$ the following hold:

absolutely homogeneous

triangle inequality

positive definite

- *Absolutely homogeneous:* $\|\lambda\boldsymbol{x}\| = |\lambda|\|\boldsymbol{x}\|$
- *Triangle inequality:* $\|\boldsymbol{x} + \boldsymbol{y}\| \leqslant \|\boldsymbol{x}\| + \|\boldsymbol{y}\|$
- *Positive definite:* $\|\boldsymbol{x}\| \geqslant 0$ and $\|\boldsymbol{x}\| = 0 \iff \boldsymbol{x} = \boldsymbol{0}$

Figure 3.2 Triangle inequality.

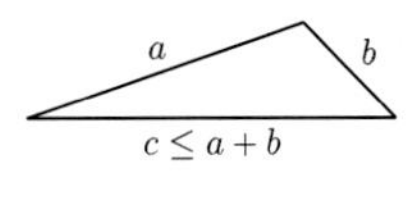

In geometric terms, the triangle inequality states that for any triangle, the sum of the lengths of any two sides must be greater than or equal to the length of the remaining side; see Figure 3.2 for an illustration. Definition 3.1 is in terms of a general vector space V (Section 2.4), but in this book we will only consider a finite-dimensional vector space $\mathbb{R}^n$. Recall that for a vector $\boldsymbol{x} \in \mathbb{R}^n$ we denote the elements of the vector using a subscript, that is, x_i is the ith element of the vector $\boldsymbol{x}$.

Example 3.1 (Manhattan Norm)

Manhattan norm

The *Manhattan norm* on $\mathbb{R}^n$ is defined for $\boldsymbol{x} \in \mathbb{R}^n$ as

$$\|\boldsymbol{x}\|_1 := \sum_{i=1}^{n} |x_i|, \tag{3.3}$$

ℓ_1 norm

where $|\cdot|$ is the absolute value. The left panel of Figure 3.3 shows all vectors $\boldsymbol{x} \in \mathbb{R}^2$ with $\|\boldsymbol{x}\|_1 = 1$. The Manhattan norm is also called *ℓ_1 norm.*

Example 3.2 (Euclidean Norm)

Euclidean norm

The *Euclidean norm* of $\boldsymbol{x} \in \mathbb{R}^n$ is defined as

$$\|\boldsymbol{x}\|_2 := \sqrt{\sum_{i=1}^{n} x_i^2} = \sqrt{\boldsymbol{x}^\top \boldsymbol{x}} \tag{3.4}$$

Figure 3.3 For different norms, the red lines indicate the set of vectors with norm 1. Left: Manhattan norm; Right: Euclidean distance.

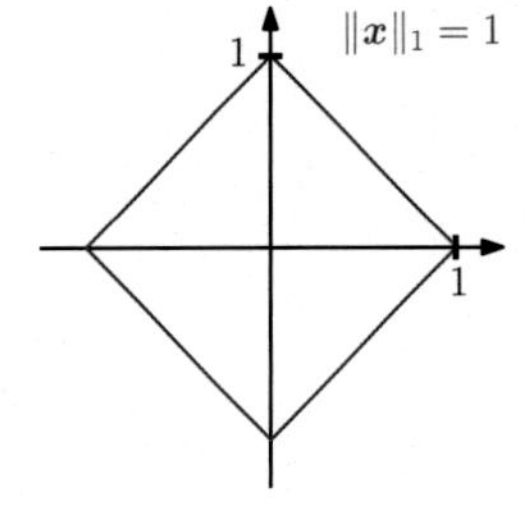

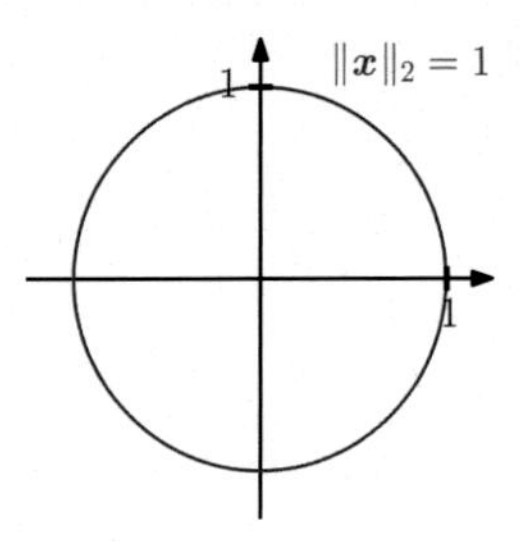

and computes the *Euclidean distance* of $\boldsymbol{x}$ from the origin. The right panel of Figure 3.3 shows all vectors $\boldsymbol{x} \in \mathbb{R}^2$ with $\|\boldsymbol{x}\|_2 = 1$. The Euclidean norm is also called ℓ_2 *norm*.

Euclidean distance

ℓ_2 norm

Remark. Throughout this book, we will use the Euclidean norm (3.4) by default if not stated otherwise. ◇

3.2 Inner Products

Inner products allow for the introduction of intuitive geometrical concepts, such as the length of a vector and the angle or distance between two vectors. A major purpose of inner products is to determine whether vectors are orthogonal to each other.

3.2.1 Dot Product

We may already be familiar with a particular type of inner product, the *scalar product/dot product* in $\mathbb{R}^n$, which is given by

scalar product

dot product

$$\boldsymbol{x}^\top \boldsymbol{y} = \sum_{i=1}^{n} x_i y_i \,. \tag{3.5}$$

We will refer to this particular inner product as the dot product in this book. However, inner products are more general concepts with specific properties, which we will now introduce.

3.2.2 General Inner Products

Recall the linear mapping from Section 2.7, where we can rearrange the mapping with respect to addition and multiplication with a scalar. A *bilinear mapping* Ω is a mapping with two arguments, and it is linear in each argument, i.e., when we look at a vector space V then it holds that for all $\boldsymbol{x}, \boldsymbol{y}, \boldsymbol{z} \in V,\ \lambda, \psi \in \mathbb{R}$ that

bilinear mapping

$$\Omega(\lambda \boldsymbol{x} + \psi \boldsymbol{y}, \boldsymbol{z}) = \lambda \Omega(\boldsymbol{x}, \boldsymbol{z}) + \psi \Omega(\boldsymbol{y}, \boldsymbol{z}) \tag{3.6}$$

$$\Omega(\boldsymbol{x}, \lambda \boldsymbol{y} + \psi \boldsymbol{z}) = \lambda \Omega(\boldsymbol{x}, \boldsymbol{y}) + \psi \Omega(\boldsymbol{x}, \boldsymbol{z}) \,. \tag{3.7}$$

Here, (3.6) asserts that Ω is linear in the first argument, and (3.7) asserts that Ω is linear in the second argument (see also (2.87)).

Definition 3.2. Let V be a vector space and $\Omega : V \times V \to \mathbb{R}$ be a bilinear mapping that takes two vectors and maps them onto a real number. Then

- Ω is called *symmetric* if $\Omega(\boldsymbol{x}, \boldsymbol{y}) = \Omega(\boldsymbol{y}, \boldsymbol{x})$ for all $\boldsymbol{x}, \boldsymbol{y} \in V$, i.e., the order of the arguments does not matter.
- Ω is called *positive definite* if

symmetric

positive definite

$$\forall \boldsymbol{x} \in V \backslash \{\boldsymbol{0}\} : \Omega(\boldsymbol{x}, \boldsymbol{x}) > 0 \,, \quad \Omega(\boldsymbol{0}, \boldsymbol{0}) = 0 \,. \tag{3.8}$$

Definition 3.3. Let V be a vector space and $\Omega : V \times V \to \mathbb{R}$ be a bilinear mapping that takes two vectors and maps them onto a real number. Then

- A positive definite, symmetric bilinear mapping $\Omega : V \times V \to \mathbb{R}$ is called an *inner product* on V. We typically write $\langle \boldsymbol{x}, \boldsymbol{y} \rangle$ instead of $\Omega(\boldsymbol{x}, \boldsymbol{y})$.
- The pair $(V, \langle \cdot, \cdot \rangle)$ is called an *inner product space* or (real) *vector space with inner product*. If we use the dot product defined in (3.5), we call $(V, \langle \cdot, \cdot \rangle)$ a *Euclidean vector space*.

inner product
inner product space
vector space with inner product
Euclidean vector space

We will refer to these spaces as inner product spaces in this book.

Example 3.3 (Inner Product That Is Not the Dot Product)
Consider $V = \mathbb{R}^2$. If we define

$$\langle \boldsymbol{x}, \boldsymbol{y} \rangle := x_1 y_1 - (x_1 y_2 + x_2 y_1) + 2 x_2 y_2 \tag{3.9}$$

then $\langle \cdot, \cdot \rangle$ is an inner product but different from the dot product. The proof will be an exercise.

3.2.3 Symmetric, Positive Definite Matrices

Symmetric, positive definite matrices play an important role in machine learning, and they are defined via the inner product. In Section 4.3, we will return to symmetric, positive definite matrices in the context of matrix decompositions. The idea of symmetric positive semidefinite matrices is key in the definition of kernels (Section 12.4).

Consider an n-dimensional vector space V with an inner product $\langle \cdot, \cdot \rangle : V \times V \to \mathbb{R}$ (see Definition 3.3) and an ordered basis $B = (\boldsymbol{b}_1, \dots, \boldsymbol{b}_n)$ of V. Recall from Section 2.6.1 that any vectors $\boldsymbol{x}, \boldsymbol{y} \in V$ can be written as linear combinations of the basis vectors so that $\boldsymbol{x} = \sum_{i=1}^{n} \psi_i \boldsymbol{b}_i \in V$ and $\boldsymbol{y} = \sum_{j=1}^{n} \lambda_j \boldsymbol{b}_j \in V$ for suitable $\psi_i, \lambda_j \in \mathbb{R}$. Due to the bilinearity of the inner product, it holds for all $\boldsymbol{x}, \boldsymbol{y} \in V$ that

$$\langle \boldsymbol{x}, \boldsymbol{y} \rangle = \left\langle \sum_{i=1}^{n} \psi_i \boldsymbol{b}_i, \sum_{j=1}^{n} \lambda_j \boldsymbol{b}_j \right\rangle = \sum_{i=1}^{n} \sum_{j=1}^{n} \psi_i \langle \boldsymbol{b}_i, \boldsymbol{b}_j \rangle \lambda_j = \hat{\boldsymbol{x}}^\top \boldsymbol{A} \hat{\boldsymbol{y}}\,, \tag{3.10}$$

where $A_{ij} := \langle \boldsymbol{b}_i, \boldsymbol{b}_j \rangle$ and $\hat{\boldsymbol{x}}, \hat{\boldsymbol{y}}$ are the coordinates of $\boldsymbol{x}$ and $\boldsymbol{y}$ with respect to the basis B. This implies that the inner product $\langle \cdot, \cdot \rangle$ is uniquely determined through $\boldsymbol{A}$. The symmetry of the inner product also means that $\boldsymbol{A}$ is symmetric. Furthermore, the positive definiteness of the inner product implies that

$$\forall \boldsymbol{x} \in V \backslash \{\boldsymbol{0}\} : \boldsymbol{x}^\top \boldsymbol{A} \boldsymbol{x} > 0\,. \tag{3.11}$$

Definition 3.4 (Symmetric, Positive Definite Matrix)**.** A symmetric matrix $\boldsymbol{A} \in \mathbb{R}^{n \times n}$ that satisfies (3.11) is called *symmetric, positive definite*, or just *positive definite*. If only $\geqslant$ holds in (3.11), then $\boldsymbol{A}$ is called *symmetric, positive semidefinite*.

symmetric, positive definite
positive definite
symmetric, positive semidefinite

Example 3.4 (Symmetric, Positive Definite Matrices)
Consider the matrices

$$\boldsymbol{A}_1 = \begin{bmatrix} 9 & 6 \\ 6 & 5 \end{bmatrix}, \quad \boldsymbol{A}_2 = \begin{bmatrix} 9 & 6 \\ 6 & 3 \end{bmatrix}. \tag{3.12}$$

A_1 is positive definite because it is symmetric and

$$\boldsymbol{x}^\top \boldsymbol{A}_1 \boldsymbol{x} = \begin{bmatrix} x_1 & x_2 \end{bmatrix} \begin{bmatrix} 9 & 6 \\ 6 & 5 \end{bmatrix} \begin{bmatrix} x_1 \\ x_2 \end{bmatrix} \tag{3.13a}$$

$$= 9x_1^2 + 12x_1x_2 + 5x_2^2 = (3x_1 + 2x_2)^2 + x_2^2 > 0 \tag{3.13b}$$

for all $\boldsymbol{x} \in V \backslash \{\boldsymbol{0}\}$. In contrast, $\boldsymbol{A}_2$ is symmetric but not positive definite because $\boldsymbol{x}^\top \boldsymbol{A}_2 \boldsymbol{x} = 9x_1^2 + 12x_1x_2 + 3x_2^2 = (3x_1 + 2x_2)^2 - x_2^2$ can be less than 0, e.g., for $\boldsymbol{x} = [2, -3]^\top$.

If $\boldsymbol{A} \in \mathbb{R}^{n \times n}$ is symmetric, positive definite, then

$$\langle \boldsymbol{x}, \boldsymbol{y} \rangle = \hat{\boldsymbol{x}}^\top \boldsymbol{A} \hat{\boldsymbol{y}} \tag{3.14}$$

defines an inner product with respect to an ordered basis B, where $\hat{\boldsymbol{x}}$ and $\hat{\boldsymbol{y}}$ are the coordinate representations of $\boldsymbol{x}, \boldsymbol{y} \in V$ with respect to B.

Theorem 3.5. *For a real-valued, finite-dimensional vector space V and an ordered basis B of V, it holds that $\langle \cdot, \cdot \rangle : V \times V \to \mathbb{R}$ is an inner product if and only if there exists a symmetric, positive definite matrix $\boldsymbol{A} \in \mathbb{R}^{n \times n}$ with*

$$\langle \boldsymbol{x}, \boldsymbol{y} \rangle = \hat{\boldsymbol{x}}^\top \boldsymbol{A} \hat{\boldsymbol{y}} \,. \tag{3.15}$$

The following properties hold if $\boldsymbol{A} \in \mathbb{R}^{n \times n}$ is symmetric and positive definite:

- The null space (kernel) of $\boldsymbol{A}$ consists only of $\boldsymbol{0}$ because $\boldsymbol{x}^\top \boldsymbol{A} \boldsymbol{x} > 0$ for all $\boldsymbol{x} \neq \boldsymbol{0}$. This implies that $\boldsymbol{A}\boldsymbol{x} \neq \boldsymbol{0}$ if $\boldsymbol{x} \neq \boldsymbol{0}$.
- The diagonal elements a_{ii} of $\boldsymbol{A}$ are positive because $a_{ii} = \boldsymbol{e}_i^\top \boldsymbol{A} \boldsymbol{e}_i > 0$, where $\boldsymbol{e}_i$ is the ith vector of the standard basis in $\mathbb{R}^n$.

3.3 Lengths and Distances

In Section 3.1, we already discussed norms that we can use to compute the length of a vector. Inner products and norms are closely related in the sense that any inner product induces a norm

Inner products induce norms.

$$\|\boldsymbol{x}\| := \sqrt{\langle \boldsymbol{x}, \boldsymbol{x} \rangle} \tag{3.16}$$

in a natural way, such that we can compute lengths of vectors using the inner product. However, not every norm is induced by an inner product. The Manhattan norm (3.3) is an example of a norm without a corresponding inner product. In the following, we will focus on norms that are induced by inner products and introduce geometric concepts, such as lengths, distances, and angles.

Remark (Cauchy–Schwarz Inequality). For an inner product vector space $(V, \langle \cdot, \cdot \rangle)$, the induced norm $\| \cdot \|$ satisfies the *Cauchy–Schwarz inequality*

Cauchy–Schwarz inequality

$$|\langle \boldsymbol{x}, \boldsymbol{y} \rangle| \leqslant \|\boldsymbol{x}\| \|\boldsymbol{y}\| \,. \tag{3.17}$$

Example 3.5 (Lengths of Vectors Using Inner Products)
In geometry, we are often interested in lengths of vectors. We can now use an inner product to compute them using (3.16). Let us take $\boldsymbol{x} = [1, 1]^\top \in \mathbb{R}^2$. If we use the dot product as the inner product, with (3.16) we obtain

$$\|\boldsymbol{x}\| = \sqrt{\boldsymbol{x}^\top \boldsymbol{x}} = \sqrt{1^2 + 1^2} = \sqrt{2} \tag{3.18}$$

as the length of $\boldsymbol{x}$. Let us now choose a different inner product:

$$\langle \boldsymbol{x}, \boldsymbol{y} \rangle := \boldsymbol{x}^\top \begin{bmatrix} 1 & -\frac{1}{2} \\ -\frac{1}{2} & 1 \end{bmatrix} \boldsymbol{y} = x_1 y_1 - \frac{1}{2}(x_1 y_2 + x_2 y_1) + x_2 y_2 \,. \tag{3.19}$$

If we compute the norm of a vector, then this inner product returns smaller values than the dot product if x_1 and x_2 have the same sign (and $x_1 x_2 > 0$); otherwise, it returns greater values than the dot product. With this inner product, we obtain

$$\langle \boldsymbol{x}, \boldsymbol{x} \rangle = x_1^2 - x_1 x_2 + x_2^2 = 1 - 1 + 1 = 1 \implies \|\boldsymbol{x}\| = \sqrt{1} = 1 \,, \tag{3.20}$$

such that $\boldsymbol{x}$ is "shorter" with this inner product than with the dot product.

Definition 3.6 (Distance and Metric)**.** Consider an inner product space $(V, \langle \cdot, \cdot \rangle)$. Then

$$d(\boldsymbol{x}, \boldsymbol{y}) := \|\boldsymbol{x} - \boldsymbol{y}\| = \sqrt{\langle \boldsymbol{x} - \boldsymbol{y}, \boldsymbol{x} - \boldsymbol{y} \rangle} \tag{3.21}$$

distance
Euclidean distance

is called the *distance* between $\boldsymbol{x}$ and $\boldsymbol{y}$ for $\boldsymbol{x}, \boldsymbol{y} \in V$. If we use the dot product as the inner product, then the distance is called *Euclidean distance*. The mapping

$$d : V \times V \to \mathbb{R} \tag{3.22}$$

$$(\boldsymbol{x}, \boldsymbol{y}) \mapsto d(\boldsymbol{x}, \boldsymbol{y}) \tag{3.23}$$

metric

is called a *metric*.

Remark. Similar to the length of a vector, the distance between vectors does not require an inner product: a norm is sufficient. If we have a norm induced by an inner product, the distance may vary depending on the choice of the inner product. ◇

A metric d satisfies the following:

positive definite
symmetric
triangle inequality

1. d is *positive definite*, i.e., $d(\boldsymbol{x}, \boldsymbol{y}) \geqslant 0$ for all $\boldsymbol{x}, \boldsymbol{y} \in V$ and $d(\boldsymbol{x}, \boldsymbol{y}) = 0 \iff \boldsymbol{x} = \boldsymbol{y}$.
2. d is *symmetric*, i.e., $d(\boldsymbol{x}, \boldsymbol{y}) = d(\boldsymbol{y}, \boldsymbol{x})$ for all $\boldsymbol{x}, \boldsymbol{y} \in V$.
3. *Triangle inequality:* $d(\boldsymbol{x}, \boldsymbol{z}) \leqslant d(\boldsymbol{x}, \boldsymbol{y}) + d(\boldsymbol{y}, \boldsymbol{z})$ for all $\boldsymbol{x}, \boldsymbol{y}, \boldsymbol{z} \in V$.

Remark. At first glance, the lists of properties of inner products and metrics look very similar. However, by comparing Definition 3.3 with Definition 3.6 we observe that $\langle \boldsymbol{x}, \boldsymbol{y} \rangle$ and $d(\boldsymbol{x}, \boldsymbol{y})$ behave in opposite directions. Very similar $\boldsymbol{x}$ and $\boldsymbol{y}$ will result in a large value for the inner product and a small value for the metric. ◇

3.4 Angles and Orthogonality

Figure 3.4 When restricted to $[0, \pi]$, then $f(\omega) = \cos(\omega)$ returns a unique number in the interval $[-1, 1]$.

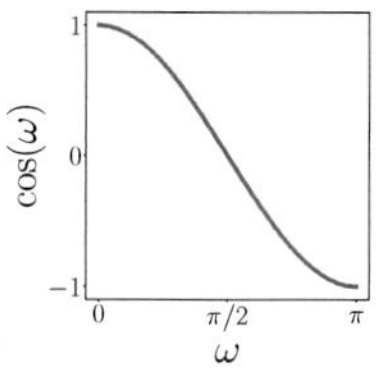

In addition to enabling the definition of lengths of vectors, as well as the distance between two vectors, inner products also capture the geometry of a vector space by defining the angle ω between two vectors. We use the Cauchy–Schwarz inequality (3.17) to define angles ω in inner product spaces between two vectors $\boldsymbol{x}, \boldsymbol{y}$, and this notion coincides with our intuition in $\mathbb{R}^2$ and $\mathbb{R}^3$. Assume that $\boldsymbol{x} \neq \boldsymbol{0}, \boldsymbol{y} \neq \boldsymbol{0}$. Then

$$-1 \leqslant \frac{\langle \boldsymbol{x}, \boldsymbol{y} \rangle}{\|\boldsymbol{x}\| \, \|\boldsymbol{y}\|} \leqslant 1 \, . \tag{3.24}$$

Therefore, there exists a unique $\omega \in [0, \pi]$, illustrated in Figure 3.4, with

$$\cos \omega = \frac{\langle \boldsymbol{x}, \boldsymbol{y} \rangle}{\|\boldsymbol{x}\| \, \|\boldsymbol{y}\|} \, . \tag{3.25}$$

angle

The number ω is the *angle* between the vectors $\boldsymbol{x}$ and $\boldsymbol{y}$. Intuitively, the angle between two vectors tells us how similar their orientations are. For example, using the dot product, the angle between $\boldsymbol{x}$ and $\boldsymbol{y} = 4\boldsymbol{x}$, i.e., $\boldsymbol{y}$ is a scaled version of $\boldsymbol{x}$, is 0: Their orientation is the same.

Example 3.6 (Angle between Vectors)

Let us compute the angle between $\boldsymbol{x} = [1, 1]^\top \in \mathbb{R}^2$ and $\boldsymbol{y} = [1, 2]^\top \in \mathbb{R}^2$; see Figure 3.5, where we use the dot product as the inner product. Then we get

$$\cos \omega = \frac{\langle \boldsymbol{x}, \boldsymbol{y} \rangle}{\sqrt{\langle \boldsymbol{x}, \boldsymbol{x} \rangle \langle \boldsymbol{y}, \boldsymbol{y} \rangle}} = \frac{\boldsymbol{x}^\top \boldsymbol{y}}{\sqrt{\boldsymbol{x}^\top \boldsymbol{x} \boldsymbol{y}^\top \boldsymbol{y}}} = \frac{3}{\sqrt{10}} \, , \tag{3.26}$$

and the angle between the two vectors is $\arccos(\frac{3}{\sqrt{10}}) \approx 0.32 \, \text{rad}$, which corresponds to about $18°$.

Figure 3.5 The angle ω between two vectors $\boldsymbol{x}, \boldsymbol{y}$ is computed using the inner product.

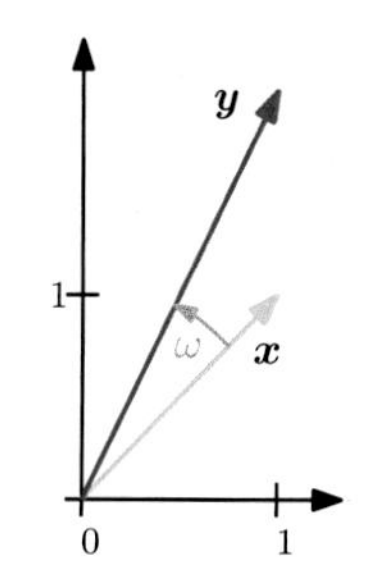

A key feature of the inner product is that it also allows us to characterize vectors that are orthogonal.

orthogonal

orthonormal

Definition 3.7 (Orthogonality)**.** Two vectors $\boldsymbol{x}$ and $\boldsymbol{y}$ are *orthogonal* if and only if $\langle \boldsymbol{x}, \boldsymbol{y} \rangle = 0$, and we write $\boldsymbol{x} \perp \boldsymbol{y}$. If additionally $\|\boldsymbol{x}\| = 1 = \|\boldsymbol{y}\|$, i.e., the vectors are unit vectors, then $\boldsymbol{x}$ and $\boldsymbol{y}$ are *orthonormal*.

An implication of this definition is that the $\boldsymbol{0}$-vector is orthogonal to every vector in the vector space.

Remark. Orthogonality is the generalization of the concept of perpendicularity to bilinear forms that do not have to be the dot product. In our context, geometrically, we can think of orthogonal vectors as having a right angle with respect to a specific inner product. ♢

Example 3.7 (Orthogonal Vectors)

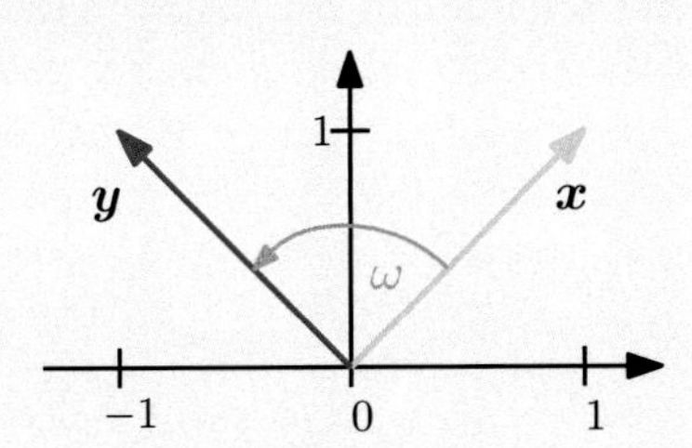

Figure 3.6 The angle ω between two vectors $\boldsymbol{x}, \boldsymbol{y}$ can change depending on the inner product.

Consider two vectors $\boldsymbol{x} = [1, 1]^\top, \boldsymbol{y} = [-1, 1]^\top \in \mathbb{R}^2$; see Figure 3.6. We are interested in determining the angle ω between them using two different inner products. Using the dot product as the inner product yields an angle ω between $\boldsymbol{x}$ and $\boldsymbol{y}$ of $90°$, such that $\boldsymbol{x} \perp \boldsymbol{y}$. However, if we choose the inner product

$$\langle \boldsymbol{x}, \boldsymbol{y} \rangle = \boldsymbol{x}^\top \begin{bmatrix} 2 & 0 \\ 0 & 1 \end{bmatrix} \boldsymbol{y}, \tag{3.27}$$

we get that the angle ω between $\boldsymbol{x}$ and $\boldsymbol{y}$ is given by

$$\cos \omega = \frac{\langle \boldsymbol{x}, \boldsymbol{y} \rangle}{\|\boldsymbol{x}\| \|\boldsymbol{y}\|} = -\frac{1}{3} \implies \omega \approx 1.91 \text{ rad} \approx 109.5°, \tag{3.28}$$

and $\boldsymbol{x}$ and $\boldsymbol{y}$ are not orthogonal. Therefore, vectors that are orthogonal with respect to one inner product do not have to be orthogonal with respect to a different inner product.

orthogonal matrix

Definition 3.8 (Orthogonal Matrix). A square matrix $\boldsymbol{A} \in \mathbb{R}^{n \times n}$ is an *orthogonal matrix* if and only if its columns are orthonormal so that

$$\boldsymbol{A}\boldsymbol{A}^\top = \boldsymbol{I} = \boldsymbol{A}^\top \boldsymbol{A}, \tag{3.29}$$

which implies that

$$\boldsymbol{A}^{-1} = \boldsymbol{A}^\top, \tag{3.30}$$

i.e., the inverse is obtained by simply transposing the matrix.

It is convention to call these matrices "orthogonal" but a more precise description would be 'orthonormal'. Transformations with orthogonal matrices preserve distances and angles.

Transformations by orthogonal matrices are special because the length of a vector $\boldsymbol{x}$ is not changed when transforming it using an orthogonal matrix $\boldsymbol{A}$. For the dot product, we obtain

$$\|\boldsymbol{A}\boldsymbol{x}\|^2 = (\boldsymbol{A}\boldsymbol{x})^\top (\boldsymbol{A}\boldsymbol{x}) = \boldsymbol{x}^\top \boldsymbol{A}^\top \boldsymbol{A}\boldsymbol{x} = \boldsymbol{x}^\top \boldsymbol{I}\boldsymbol{x} = \boldsymbol{x}^\top \boldsymbol{x} = \|\boldsymbol{x}\|^2. \tag{3.31}$$

Moreover, the angle between any two vectors $\boldsymbol{x}, \boldsymbol{y}$, as measured by their inner product, is also unchanged when transforming both of them using an orthogonal matrix $\boldsymbol{A}$. Assuming the dot product as the inner product, the angle of the images $\boldsymbol{A}\boldsymbol{x}$ and $\boldsymbol{A}\boldsymbol{y}$ is given as

$$\cos \omega = \frac{(\boldsymbol{A}\boldsymbol{x})^\top (\boldsymbol{A}\boldsymbol{y})}{\|\boldsymbol{A}\boldsymbol{x}\| \, \|\boldsymbol{A}\boldsymbol{y}\|} = \frac{\boldsymbol{x}^\top \boldsymbol{A}^\top \boldsymbol{A}\boldsymbol{y}}{\sqrt{\boldsymbol{x}^\top \boldsymbol{A}^\top \boldsymbol{A}\boldsymbol{x}\boldsymbol{y}^\top \boldsymbol{A}^\top \boldsymbol{A}\boldsymbol{y}}} = \frac{\boldsymbol{x}^\top \boldsymbol{y}}{\|\boldsymbol{x}\| \, \|\boldsymbol{y}\|}, \tag{3.32}$$

which gives exactly the angle between $\boldsymbol{x}$ and $\boldsymbol{y}$. This means that orthogonal matrices $\boldsymbol{A}$ with $\boldsymbol{A}^\top = \boldsymbol{A}^{-1}$ preserve both angles and distances. It turns out that orthogonal matrices define transformations that are rotations (with the possibility of flips). In Section 3.9, we will discuss more details about rotations.

3.5 Orthonormal Basis

In Section 2.6.1, we characterized properties of basis vectors and found that in an n-dimensional vector space, we need n basis vectors, i.e., n vectors that are linearly independent. In Sections 3.3 and 3.4, we used inner products to compute the length of vectors and the angle between vectors. In the following, we will discuss the special case where the basis vectors are orthogonal to each other and where the length of each basis vector is 1. We will call this basis then an orthonormal basis.

Let us introduce this more formally.

Definition 3.9 (Orthonormal Basis)**.** Consider an n-dimensional vector space V and a basis $\{\boldsymbol{b}_1, \dots, \boldsymbol{b}_n\}$ of V. If

$$\langle \boldsymbol{b}_i, \boldsymbol{b}_j \rangle = 0 \quad \text{for } i \neq j \tag{3.33}$$

$$\langle \boldsymbol{b}_i, \boldsymbol{b}_i \rangle = 1 \tag{3.34}$$

for all $i, j = 1, \dots, n$ then the basis is called an *orthonormal basis* (*ONB*). If only (3.33) is satisfied, then the basis is called an *orthogonal basis*. Note that (3.34) implies that every basis vector has length/norm 1.

orthonormal basis
ONB
orthogonal basis

Recall from Section 2.6.1 that we can use Gaussian elimination to find a basis for a vector space spanned by a set of vectors. Assume we are given a set $\{\tilde{\boldsymbol{b}}_1, \dots, \tilde{\boldsymbol{b}}_n\}$ of nonorthogonal and unnormalized basis vectors. We concatenate them into a matrix $\tilde{\boldsymbol{B}} = [\tilde{\boldsymbol{b}}_1, \dots, \tilde{\boldsymbol{b}}_n]$ and apply Gaussian elimination to the augmented matrix (Section 2.3.2) $[\tilde{\boldsymbol{B}}\tilde{\boldsymbol{B}}^\top | \tilde{\boldsymbol{B}}]$ to obtain an orthonormal basis. This constructive way to iteratively build an orthonormal basis $\{\boldsymbol{b}_1, \dots, \boldsymbol{b}_n\}$ is called the *Gram–Schmidt process* (Strang, 2003).

Example 3.8 (Orthonormal Basis)

The canonical/standard basis for a Euclidean vector space $\mathbb{R}^n$ is an orthonormal basis, where the inner product is the dot product of vectors.

In $\mathbb{R}^2$, the vectors

$$\boldsymbol{b}_1 = \frac{1}{\sqrt{2}}\begin{bmatrix}1\\1\end{bmatrix}, \quad \boldsymbol{b}_2 = \frac{1}{\sqrt{2}}\begin{bmatrix}1\\-1\end{bmatrix} \tag{3.35}$$

form an orthonormal basis since $\boldsymbol{b}_1^\top \boldsymbol{b}_2 = 0$ and $\|\boldsymbol{b}_1\| = 1 = \|\boldsymbol{b}_2\|$.

We will exploit the concept of an orthonormal basis in Chapters 10 and 12 when we discuss support vector machines and principal component analysis.

3.6 Orthogonal Complement

Having defined orthogonality, we will now look at vector spaces that are orthogonal to each other. This will play an important role in Chapter 10, when we discuss linear dimensionality reduction from a geometric perspective.

Consider a D-dimensional vector space V and an M-dimensional subspace $U \subseteq V$. Then its *orthogonal complement* $U^\perp$ is a $(D-M)$-dimensional subspace

orthogonal complement

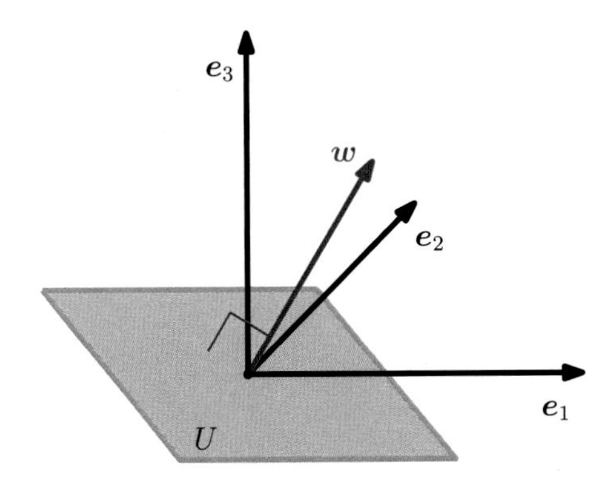

Figure 3.7 A plane U in a three-dimensional vector space can be described by its normal vector, which spans its orthogonal complement $U^\perp$.

of V and contains all vectors in V that are orthogonal to every vector in U. Furthermore, $U \cap U^\perp = \{\mathbf{0}\}$ so that any vector $\boldsymbol{x} \in V$ can be uniquely decomposed into

$$\boldsymbol{x} = \sum_{m=1}^{M} \lambda_m \boldsymbol{b}_m + \sum_{j=1}^{D-M} \psi_j \boldsymbol{b}_j^\perp, \quad \lambda_m,\ \psi_j \in \mathbb{R}, \tag{3.36}$$

where $(\boldsymbol{b}_1, \dots, \boldsymbol{b}_M)$ is a basis of U and $(\boldsymbol{b}_1^\perp, \dots, \boldsymbol{b}_{D-M}^\perp)$ is a basis of $U^\perp$.

Therefore, the orthogonal complement can also be used to describe a plane U (two-dimensional subspace) in a three-dimensional vector space. More specifically, the vector $\boldsymbol{w}$ with $\|\boldsymbol{w}\| = 1$, which is orthogonal to the plane U, is the basis vector of $U^\perp$. Figure 3.7 illustrates this setting. All vectors that are orthogonal to $\boldsymbol{w}$ must (by construction) lie in the plane U. The vector $\boldsymbol{w}$ is called the *normal vector* of U.

normal vector

Generally, orthogonal complements can be used to describe hyperplanes in n-dimensional vector and affine spaces.

3.7 Inner Product of Functions

Thus far, we looked at properties of inner products to compute lengths, angles, and distances. We focused on inner products of finite-dimensional vectors. In the following, we will look at an example of inner products of a different type of vectors: inner products of functions.

The inner products we discussed so far were defined for vectors with a finite number of entries. We can think of a vector $\boldsymbol{x} \in \mathbb{R}^n$ as function with n function values. The concept of an inner product can be generalized to vectors with an infinite number of entries (countably infinite) and also continuous-valued functions (uncountably infinite). Then the sum over individual components of vectors – see Equation (3.5), for example – turns into an integral.

An inner product of two functions $u : \mathbb{R} \to \mathbb{R}$ and $v : \mathbb{R} \to \mathbb{R}$ can be defined as the definite integral

$$\langle u, v \rangle := \int_a^b u(x)v(x)dx \tag{3.37}$$

for lower and upper limits $a, b < \infty$, respectively. As with our usual inner product, we can define norms and orthogonality by looking at the inner product. If (3.37) evaluates to 0, the functions u and v are orthogonal. To make the preceding inner product mathematically precise, we need to take care of measures and the definition of integrals, leading to the definition of a Hilbert space. Furthermore, unlike inner products on finite-dimensional vectors, inner products on functions may diverge (have infinite value). All this requires diving into some more intricate details of real and functional analysis, which we do not cover in this book.

Example 3.9 (Inner Product of Functions)
If we choose $u = \sin(x)$ and $v = \cos(x)$, the integrand $f(x) = u(x)v(x)$ of (3.37) is shown in Figure 3.8. We see that this function is odd, i.e., $f(-x) = -f(x)$. Therefore, the integral with limits $a = -\pi, b = \pi$ of this product evaluates to 0. Therefore, sin and cos are orthogonal functions.

Figure 3.8 $f(x) = \sin(x)\cos(x)$.

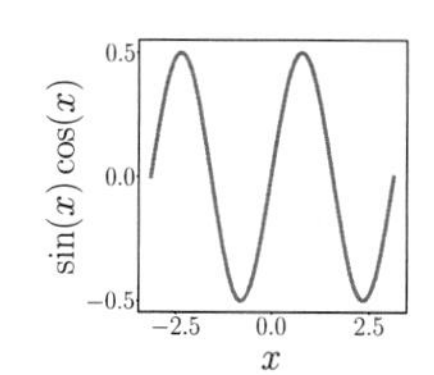

Remark. It also holds that the collection of functions

$$\{1, \cos(x), \cos(2x), \cos(3x), \dots\} \tag{3.38}$$

is orthogonal if we integrate from $-\pi$ to π, i.e., any pair of functions are orthogonal to each other. The collection of functions in (3.38) spans a large subspace of the functions that are even and periodic on $[-\pi, \pi)$, and projecting functions onto this subspace is the fundamental idea behind Fourier series. ◇

In Section 6.4.6, we will have a look at a second type of unconventional inner products: the inner product of random variables.

3.8 Orthogonal Projections

Projections are an important class of linear transformations (besides rotations and reflections) and play an important role in graphics, coding theory, statistics, and machine learning. In machine learning, we often deal with data that is high-dimensional. High-dimensional data is often hard to analyze or visualize. However, high-dimensional data quite often possess the property that only a few dimensions contain most information, and most other dimensions are not essential to describe key properties of the data. When we compress or visualize high-dimensional data, we will lose information. To minimize this compression loss, we ideally find the most informative dimensions in the data. As discussed in Chapter 1, data can be represented as vectors, and in this chapter, we will discuss some of the fundamental tools for data compression. More specifically, we can project the original high-dimensional data onto a lower-dimensional feature space and work in this lower-dimensional space to learn more about the dataset and extract relevant patterns. For example, machine learning algorithms, such as principal component analysis (PCA) by Pearson (1901) and Hotelling (1933) and deep neural networks (e.g., deep auto-encoders (Deng et al., 2010)), heavily exploit the idea of dimensionality reduction. In the following, we will focus on orthogonal projections, which we will use in Chapter 10 for linear dimensionality reduction and in Chapter 12 for classification. Even linear regression, which we discuss in Chapter 9, can be interpreted using orthogonal projections. For a given lower-dimensional subspace, orthogonal projections of high-dimensional data retain as much information as possible and minimize the difference/error

"Feature" is a common expression for data representation.

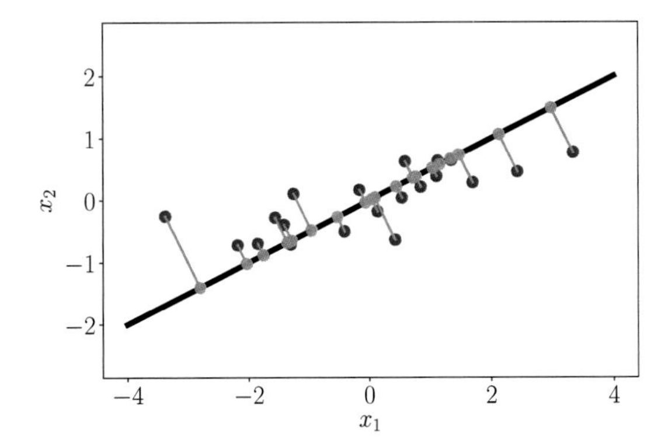

Figure 3.9 Orthogonal projection (orange dots) of a two-dimensional dataset (blue dots) onto a one-dimensional subspace (straight line).

between the original data and the corresponding projection. An illustration of such an orthogonal projection is given in Figure 3.9. Before we detail how to obtain these projections, let us define what a projection actually is.

projection

Definition 3.10 (Projection)**.** Let V be a vector space and $U \subseteq V$ a subspace of V. A linear mapping $\pi : V \to U$ is called a *projection* if $\pi^2 = \pi \circ \pi = \pi$.

projection matrix

Since linear mappings can be expressed by transformation matrices (see Section 2.7), the preceding definition applies equally to a special kind of transformation matrices, the *projection matrices* $\boldsymbol{P}_\pi$, which exhibit the property that $\boldsymbol{P}_\pi^2 = \boldsymbol{P}_\pi$.

line

In the following, we will derive orthogonal projections of vectors in the inner product space $(\mathbb{R}^n, \langle\cdot,\cdot\rangle)$ onto subspaces. We will start with one-dimensional subspaces, which are also called *lines*. If not mentioned otherwise, we assume the dot product $\langle \boldsymbol{x}, \boldsymbol{y}\rangle = \boldsymbol{x}^\top \boldsymbol{y}$ as the inner product.

3.8.1 Projection onto One-Dimensional Subspaces (Lines)

Assume we are given a line (one-dimensional subspace) through the origin with basis vector $\boldsymbol{b} \in \mathbb{R}^n$. The line is a one-dimensional subspace $U \subseteq \mathbb{R}^n$ spanned by $\boldsymbol{b}$. When we project $\boldsymbol{x} \in \mathbb{R}^n$ onto U, we seek the vector $\pi_U(\boldsymbol{x}) \in U$ that is closest to $\boldsymbol{x}$. Using geometric arguments, let us characterize some properties of the projection $\pi_U(\boldsymbol{x})$ (Figure 3.10(a) serves as an illustration):

- The projection $\pi_U(\boldsymbol{x})$ is closest to $\boldsymbol{x}$, where "closest" implies that the distance $\|\boldsymbol{x}-\pi_U(\boldsymbol{x})\|$ is minimal. It follows that the segment $\pi_U(\boldsymbol{x})-\boldsymbol{x}$ from $\pi_U(\boldsymbol{x})$ to $\boldsymbol{x}$ is orthogonal to U, and therefore the basis vector $\boldsymbol{b}$ of U. The orthogonality condition yields $\langle \pi_U(\boldsymbol{x}) - \boldsymbol{x}, \boldsymbol{b}\rangle = 0$ since angles between vectors are defined via the inner product.
- The projection $\pi_U(\boldsymbol{x})$ of $\boldsymbol{x}$ onto U must be an element of U, and therefore a multiple of the basis vector $\boldsymbol{b}$ that spans U. Hence, $\pi_U(\boldsymbol{x}) = \lambda\boldsymbol{b}$, for some $\lambda \in \mathbb{R}$.

λ is then the coordinate of $\pi_U(\boldsymbol{x})$ with respect to $\boldsymbol{b}$.

In the following three steps, we determine the coordinate λ, the projection $\pi_U(\boldsymbol{x}) \in U$, and the projection matrix $\boldsymbol{P}_\pi$ that maps any $\boldsymbol{x} \in \mathbb{R}^n$ onto U:

1. Finding the coordinate λ. The orthogonality condition yields

$$\langle \boldsymbol{x} - \pi_U(\boldsymbol{x}), \boldsymbol{b}\rangle = 0 \overset{\pi_U(\boldsymbol{x})=\lambda\boldsymbol{b}}{\iff} \langle \boldsymbol{x} - \lambda\boldsymbol{b}, \boldsymbol{b}\rangle = 0\,. \tag{3.39}$$

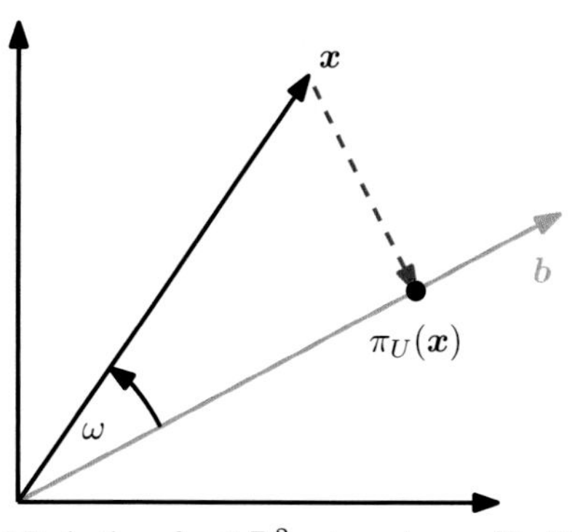

(a) Projection of $\boldsymbol{x} \in \mathbb{R}^2$ onto a subspace U with basis vector $\boldsymbol{b}$

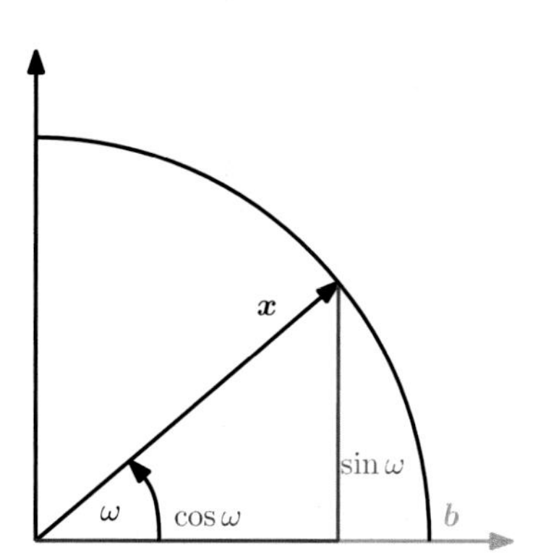

(b) Projection of a two-dimensional vector $\boldsymbol{x}$ with $\|\boldsymbol{x}\| = 1$ onto a one-dimensional subspace spanned by $\boldsymbol{b}$

Figure 3.10 Examples of projections onto one-dimensional subspaces.

We can now exploit the bilinearity of the inner product and arrive at

$$\langle \boldsymbol{x}, \boldsymbol{b} \rangle - \lambda \langle \boldsymbol{b}, \boldsymbol{b} \rangle = 0 \iff \lambda = \frac{\langle \boldsymbol{x}, \boldsymbol{b} \rangle}{\langle \boldsymbol{b}, \boldsymbol{b} \rangle} = \frac{\langle \boldsymbol{b}, \boldsymbol{x} \rangle}{\|\boldsymbol{b}\|^2}. \tag{3.40}$$

With a general inner product, we get $\lambda = \langle \boldsymbol{x}, \boldsymbol{b} \rangle$ if $\|\boldsymbol{b}\| = 1$.

In the last step, we exploited the fact that inner products are symmetric. If we choose $\langle \cdot, \cdot \rangle$ to be the dot product, we obtain

$$\lambda = \frac{\boldsymbol{b}^\top \boldsymbol{x}}{\boldsymbol{b}^\top \boldsymbol{b}} = \frac{\boldsymbol{b}^\top \boldsymbol{x}}{\|\boldsymbol{b}\|^2}. \tag{3.41}$$

If $\|\boldsymbol{b}\| = 1$, then the coordinate λ of the projection is given by $\boldsymbol{b}^\top \boldsymbol{x}$.

2. Finding the projection point $\pi_U(\boldsymbol{x}) \in U$. Since $\pi_U(\boldsymbol{x}) = \lambda \boldsymbol{b}$, we immediately obtain with (3.40) that

$$\pi_U(\boldsymbol{x}) = \lambda \boldsymbol{b} = \frac{\langle \boldsymbol{x}, \boldsymbol{b} \rangle}{\|\boldsymbol{b}\|^2} \boldsymbol{b} = \frac{\boldsymbol{b}^\top \boldsymbol{x}}{\|\boldsymbol{b}\|^2} \boldsymbol{b}, \tag{3.42}$$

where the last equality holds for the dot product only. We can also compute the length of $\pi_U(\boldsymbol{x})$ by means of Definition 3.1 as

$$\|\pi_U(\boldsymbol{x})\| = \|\lambda \boldsymbol{b}\| = |\lambda| \, \|\boldsymbol{b}\|. \tag{3.43}$$

Hence, our projection is of length $|\lambda|$ times the length of $\boldsymbol{b}$. This also adds the intuition that λ is the coordinate of $\pi_U(\boldsymbol{x})$ with respect to the basis vector $\boldsymbol{b}$ that spans our one-dimensional subspace U.

If we use the dot product as an inner product, we get

$$\|\pi_U(\boldsymbol{x})\| \overset{(3.42)}{=} \frac{|\boldsymbol{b}^\top \boldsymbol{x}|}{\|\boldsymbol{b}\|^2} \|\boldsymbol{b}\| \overset{(3.25)}{=} |\cos\omega| \, \|\boldsymbol{x}\| \, \|\boldsymbol{b}\| \frac{\|\boldsymbol{b}\|}{\|\boldsymbol{b}\|^2} = |\cos\omega| \, \|\boldsymbol{x}\|. \tag{3.44}$$

Here, ω is the angle between $\boldsymbol{x}$ and $\boldsymbol{b}$. This equation should be familiar from trigonometry: If $\|\boldsymbol{x}\| = 1$, then $\boldsymbol{x}$ lies on the unit circle. It follows that the projection onto the horizontal axis spanned by $\boldsymbol{b}$ is exactly $\cos\omega$, and the length of the corresponding vector $\pi_U(\boldsymbol{x}) = |\cos\omega|$. An illustration is given in Figure 3.10(b).

The horizontal axis is a one-dimensional subspace.

3. Finding the projection matrix $\boldsymbol{P}_\pi$. We know that a projection is a linear mapping (see Definition 3.10). Therefore, there exists a projection matrix $\boldsymbol{P}_\pi$, such that $\pi_U(\boldsymbol{x}) = \boldsymbol{P}_\pi \boldsymbol{x}$. With the dot product as inner product and

$$\pi_U(\boldsymbol{x}) = \lambda \boldsymbol{b} = \boldsymbol{b}\lambda = \boldsymbol{b}\frac{\boldsymbol{b}^\top \boldsymbol{x}}{\|\boldsymbol{b}\|^2} = \frac{\boldsymbol{b}\boldsymbol{b}^\top}{\|\boldsymbol{b}\|^2}\boldsymbol{x}\,, \tag{3.45}$$

we immediately see that

$$\boldsymbol{P}_\pi = \frac{\boldsymbol{b}\boldsymbol{b}^\top}{\|\boldsymbol{b}\|^2}\,. \tag{3.46}$$

Projection matrices are always symmetric.

Note that $\boldsymbol{b}\boldsymbol{b}^\top$ (and, consequently, $\boldsymbol{P}_\pi$) is a symmetric matrix (of rank 1), and $\|\boldsymbol{b}\|^2 = \langle \boldsymbol{b}, \boldsymbol{b}\rangle$ is a scalar.

The projection matrix $\boldsymbol{P}_\pi$ projects any vector $\boldsymbol{x} \in \mathbb{R}^n$ onto the line through the origin with direction $\boldsymbol{b}$ (equivalently, the subspace U spanned by $\boldsymbol{b}$).

Remark. The projection $\pi_U(\boldsymbol{x}) \in \mathbb{R}^n$ is still an n-dimensional vector and not a scalar. However, we no longer require n coordinates to represent the projection, but only a single one if we want to express it with respect to the basis vector $\boldsymbol{b}$ that spans the subspace U: λ. ◇

Example 3.10 (Projection onto a Line)

Find the projection matrix $\boldsymbol{P}_\pi$ onto the line through the origin spanned by $\boldsymbol{b} = \begin{bmatrix}1 & 2 & 2\end{bmatrix}^\top$. $\boldsymbol{b}$ is a direction and a basis of the one-dimensional subspace (line through origin).

With (3.46), we obtain

$$\boldsymbol{P}_\pi = \frac{\boldsymbol{b}\boldsymbol{b}^\top}{\boldsymbol{b}^\top\boldsymbol{b}} = \frac{1}{9}\begin{bmatrix}1\\2\\2\end{bmatrix}\begin{bmatrix}1 & 2 & 2\end{bmatrix} = \frac{1}{9}\begin{bmatrix}1 & 2 & 2\\2 & 4 & 4\\2 & 4 & 4\end{bmatrix}. \tag{3.47}$$

Let us now choose a particular $\boldsymbol{x}$ and see whether it lies in the subspace spanned by $\boldsymbol{b}$. For $\boldsymbol{x} = \begin{bmatrix}1 & 1 & 1\end{bmatrix}^\top$, the projection is

$$\pi_U(\boldsymbol{x}) = \boldsymbol{P}_\pi\boldsymbol{x} = \frac{1}{9}\begin{bmatrix}1 & 2 & 2\\2 & 4 & 4\\2 & 4 & 4\end{bmatrix}\begin{bmatrix}1\\1\\1\end{bmatrix} = \frac{1}{9}\begin{bmatrix}5\\10\\10\end{bmatrix} \in \mathrm{span}[\begin{bmatrix}1\\2\\2\end{bmatrix}]. \tag{3.48}$$

Note that the application of $\boldsymbol{P}_\pi$ to $\pi_U(\boldsymbol{x})$ does not change anything, i.e., $\boldsymbol{P}_\pi\pi_U(\boldsymbol{x}) = \pi_U(\boldsymbol{x})$. This is expected because according to Definition 3.10, we know that a projection matrix $\boldsymbol{P}_\pi$ satisfies $\boldsymbol{P}_\pi^2\boldsymbol{x} = \boldsymbol{P}_\pi\boldsymbol{x}$ for all $\boldsymbol{x}$.

Remark. With the results from Chapter 4, we can show that $\pi_U(\boldsymbol{x})$ is an eigenvector of $\boldsymbol{P}_\pi$, and the corresponding eigenvalue is 1. ◇

If U is given by a set of spanning vectors, which are not a basis, make sure you determine a basis $\boldsymbol{b}_1, \dots, \boldsymbol{b}_m$ before proceeding.

3.8.2 Projection onto General Subspaces

In the following, we look at orthogonal projections of vectors $\boldsymbol{x} \in \mathbb{R}^n$ onto lower-dimensional subspaces $U \subseteq \mathbb{R}^n$ with $\dim(U) = m \geqslant 1$. An illustration is given in Figure 3.11.

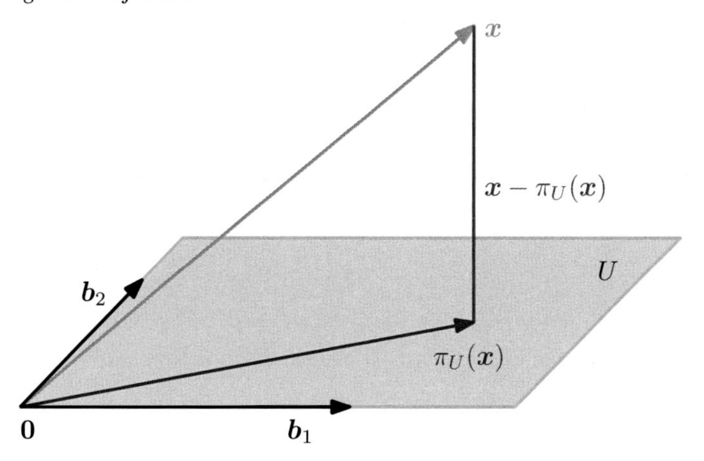

Figure 3.11 Projection onto a two-dimensional subspace U with basis b_1, b_2. The projection $\pi_U(x)$ of $x \in \mathbb{R}^3$ onto U can be expressed as a linear combination of b_1, b_2 and the displacement vector $x - \pi_U(x)$ is orthogonal to both b_1 and b_2.

Assume that $(b_1, \ldots, b_m)$ is an ordered basis of U. Any projection $\pi_U(x)$ onto U is necessarily an element of U. Therefore, they can be represented as linear combinations of the basis vectors $b_1, \ldots, b_m$ of U, such that $\pi_U(x) = \sum_{i=1}^{m} \lambda_i b_i$.

The basis vectors form the columns of $B \in \mathbb{R}^{n \times m}$, where $B = [b_1, \ldots, b_m]$.

As in the 1D case, we follow a three-step procedure to find the projection $\pi_U(x)$ and the projection matrix P_π:

1. Find the coordinates $\lambda_1, \ldots, \lambda_m$ of the projection (with respect to the basis of U), such that the linear combination

$$\pi_U(x) = \sum_{i=1}^{m} \lambda_i b_i = B\lambda, \tag{3.49}$$

$$B = [b_1, \ldots, b_m] \in \mathbb{R}^{n \times m}, \quad \lambda = [\lambda_1, \ldots, \lambda_m]^\top \in \mathbb{R}^m, \tag{3.50}$$

is closest to $x \in \mathbb{R}^n$. As in the 1D case, "closest" means "minimum distance," which implies that the vector connecting $\pi_U(x) \in U$ and $x \in \mathbb{R}^n$ must be orthogonal to all basis vectors of U. Therefore, we obtain m simultaneous conditions (assuming the dot product as the inner product)

$$\langle b_1, x - \pi_U(x) \rangle = b_1^\top (x - \pi_U(x)) = 0 \tag{3.51}$$

$$\vdots$$

$$\langle b_m, x - \pi_U(x) \rangle = b_m^\top (x - \pi_U(x)) = 0 \tag{3.52}$$

which, with $\pi_U(x) = B\lambda$, can be written as

$$b_1^\top (x - B\lambda) = 0 \tag{3.53}$$

$$\vdots$$

$$b_m^\top (x - B\lambda) = 0 \tag{3.54}$$

such that we obtain a homogeneous linear equation system

$$\begin{bmatrix} b_1^\top \\ \vdots \\ b_m^\top \end{bmatrix} \begin{bmatrix} x - B\lambda \end{bmatrix} = \mathbf{0} \iff B^\top (x - B\lambda) = \mathbf{0} \tag{3.55}$$

$$\iff B^\top B\lambda = B^\top x\,. \tag{3.56}$$

normal equation

The last expression is called *normal equation*. Since $\boldsymbol{b}_1, \ldots, \boldsymbol{b}_m$ are a basis of U and, therefore linearly independent, $\boldsymbol{B}^\top \boldsymbol{B} \in \mathbb{R}^{m \times m}$ is regular and can be inverted. This allows us to solve for the coefficients/coordinates

$$\boldsymbol{\lambda} = (\boldsymbol{B}^\top \boldsymbol{B})^{-1} \boldsymbol{B}^\top \boldsymbol{x} \,. \tag{3.57}$$

pseudo-inverse

The matrix $(\boldsymbol{B}^\top \boldsymbol{B})^{-1} \boldsymbol{B}^\top$ is also called the *pseudo-inverse* of $\boldsymbol{B}$, which can be computed for nonsquare matrices $\boldsymbol{B}$. It only requires that $\boldsymbol{B}^\top \boldsymbol{B}$ is positive definite, which is the case if $\boldsymbol{B}$ is full rank. In practical applications (e.g., linear regression), we often add a "jitter term" $\epsilon \boldsymbol{I}$ to $\boldsymbol{B}^\top \boldsymbol{B}$ to guarantee increased numerical stability and positive definiteness. This "ridge" can be rigorously derived using Bayesian inference. See Chapter 9 for details.

2. Find the projection $\pi_U(\boldsymbol{x}) \in U$. We already established that $\pi_U(\boldsymbol{x}) = \boldsymbol{B\lambda}$. Therefore, with (3.57)

$$\pi_U(\boldsymbol{x}) = \boldsymbol{B}(\boldsymbol{B}^\top \boldsymbol{B})^{-1} \boldsymbol{B}^\top \boldsymbol{x} \,. \tag{3.58}$$

3. Find the projection matrix $\boldsymbol{P}_\pi$. From (3.58), we can immediately see that the projection matrix that solves $\boldsymbol{P}_\pi \boldsymbol{x} = \pi_U(\boldsymbol{x})$ must be

$$\boldsymbol{P}_\pi = \boldsymbol{B}(\boldsymbol{B}^\top \boldsymbol{B})^{-1} \boldsymbol{B}^\top \,. \tag{3.59}$$

Remark. The solution for projecting onto general subspaces includes the 1D case as a special case: If $\dim(U) = 1$, then $\boldsymbol{B}^\top \boldsymbol{B} \in \mathbb{R}$ is a scalar and we can rewrite the projection matrix in (3.59) $\boldsymbol{P}_\pi = \boldsymbol{B}(\boldsymbol{B}^\top \boldsymbol{B})^{-1} \boldsymbol{B}^\top$ as $\boldsymbol{P}_\pi = \frac{\boldsymbol{B}\boldsymbol{B}^\top}{\boldsymbol{B}^\top \boldsymbol{B}}$, which is exactly the projection matrix in (3.46). ♢

Example 3.11 (Projection onto a Two-Dimensional Subspace)

For a subspace $U = \text{span}[\begin{bmatrix}1\\1\\1\end{bmatrix}, \begin{bmatrix}0\\1\\2\end{bmatrix}] \subseteq \mathbb{R}^3$ and $\boldsymbol{x} = \begin{bmatrix}6\\0\\0\end{bmatrix} \in \mathbb{R}^3$, find the coordinates $\boldsymbol{\lambda}$ of $\boldsymbol{x}$ in terms of the subspace U, the projection point $\pi_U(\boldsymbol{x})$, and the projection matrix $\boldsymbol{P}_\pi$.

First, we see that the generating set of U is a basis (linear independence) and write the basis vectors of U into a matrix $\boldsymbol{B} = \begin{bmatrix}1 & 0\\1 & 1\\1 & 2\end{bmatrix}$.

Second, we compute the matrix $\boldsymbol{B}^\top \boldsymbol{B}$ and the vector $\boldsymbol{B}^\top \boldsymbol{x}$ as

$$\boldsymbol{B}^\top \boldsymbol{B} = \begin{bmatrix}1 & 1 & 1\\0 & 1 & 2\end{bmatrix} \begin{bmatrix}1 & 0\\1 & 1\\1 & 2\end{bmatrix} = \begin{bmatrix}3 & 3\\3 & 5\end{bmatrix}, \quad \boldsymbol{B}^\top \boldsymbol{x} = \begin{bmatrix}1 & 1 & 1\\0 & 1 & 2\end{bmatrix} \begin{bmatrix}6\\0\\0\end{bmatrix} = \begin{bmatrix}6\\0\end{bmatrix}. \tag{3.60}$$

Third, we solve the normal equation $\boldsymbol{B}^\top \boldsymbol{B\lambda} = \boldsymbol{B}^\top \boldsymbol{x}$ to find $\boldsymbol{\lambda}$:

$$\begin{bmatrix}3 & 3\\3 & 5\end{bmatrix} \begin{bmatrix}\lambda_1\\\lambda_2\end{bmatrix} = \begin{bmatrix}6\\0\end{bmatrix} \iff \boldsymbol{\lambda} = \begin{bmatrix}5\\-3\end{bmatrix}. \tag{3.61}$$

Fourth, the projection $\pi_U(\boldsymbol{x})$ of $\boldsymbol{x}$ onto U, i.e., into the column space of $\boldsymbol{B}$, can be directly computed via

$$\pi_U(\boldsymbol{x}) = \boldsymbol{B}\boldsymbol{\lambda} = \begin{bmatrix} 5 \\ 2 \\ -1 \end{bmatrix}. \tag{3.62}$$

The corresponding *projection error* is the norm of the difference vector between the original vector and its projection onto U, i.e.,

projection error

The projection error is also called the *reconstruction error*.

$$\|\boldsymbol{x} - \pi_U(\boldsymbol{x})\| = \left\| \begin{bmatrix} 1 & -2 & 1 \end{bmatrix}^\top \right\| = \sqrt{6}. \tag{3.63}$$

Fifth, the projection matrix (for any $\boldsymbol{x} \in \mathbb{R}^3$) is given by

$$\boldsymbol{P}_\pi = \boldsymbol{B}(\boldsymbol{B}^\top\boldsymbol{B})^{-1}\boldsymbol{B}^\top = \frac{1}{6}\begin{bmatrix} 5 & 2 & -1 \\ 2 & 2 & 2 \\ -1 & 2 & 5 \end{bmatrix}. \tag{3.64}$$

To verify the results, we can (a) check whether the displacement vector $\pi_U(\boldsymbol{x}) - \boldsymbol{x}$ is orthogonal to all basis vectors of U, and (b) verify that $\boldsymbol{P}_\pi = \boldsymbol{P}_\pi^2$ (see Definition 3.10).

Remark. The projections $\pi_U(\boldsymbol{x})$ are still vectors in $\mathbb{R}^n$ although they lie in an m-dimensional subspace $U \subseteq \mathbb{R}^n$. However, to represent a projected vector we only need the m coordinates $\lambda_1, \ldots, \lambda_m$ with respect to the basis vectors $\boldsymbol{b}_1, \ldots, \boldsymbol{b}_m$ of U. ◇

Remark. In vector spaces with general inner products, we have to pay attention when computing angles and distances, which are defined by means of the inner product. ◇

We can find approximate solutions to unsolvable linear equation systems using projections.

Projections allow us to look at situations where we have a linear system $\boldsymbol{A}\boldsymbol{x} = \boldsymbol{b}$ without a solution. Recall that this means that $\boldsymbol{b}$ does not lie in the span of $\boldsymbol{A}$, i.e., the vector $\boldsymbol{b}$ does not lie in the subspace spanned by the columns of $\boldsymbol{A}$. Given that the linear equation cannot be solved exactly, we can find an *approximate solution*. The idea is to find the vector in the subspace spanned by the columns of $\boldsymbol{A}$ that is closest to $\boldsymbol{b}$, i.e., we compute the orthogonal projection of $\boldsymbol{b}$ onto the subspace spanned by the columns of $\boldsymbol{A}$. This problem arises often in practice, and the solution is called the *least-squares solution* (assuming the dot product as the inner product) of an overdetermined system. This is discussed further in Section 9.4. Using reconstruction errors (3.63) is one possible approach to derive principal component analysis (Section 10.3).

least-squares solution

Remark. We just looked at projections of vectors $\boldsymbol{x}$ onto a subspace U with basis vectors $\{\boldsymbol{b}_1, \ldots, \boldsymbol{b}_k\}$. If this basis is an ONB, i.e., (3.33) and (3.34) are satisfied, the projection equation (3.58) simplifies greatly to

$$\pi_U(\boldsymbol{x}) = \boldsymbol{B}\boldsymbol{B}^\top\boldsymbol{x} \tag{3.65}$$

since $\boldsymbol{B}^\top \boldsymbol{B} = \boldsymbol{I}$ with coordinates

$$\boldsymbol{\lambda} = \boldsymbol{B}^\top \boldsymbol{x} . \tag{3.66}$$

This means that we no longer have to compute the inverse from (3.58), which saves computation time. ◇

3.8.3 Gram–Schmidt Orthogonalization

Projections are at the core of the Gram–Schmidt method that allows us to constructively transform any basis $(\boldsymbol{b}_1, \ldots, \boldsymbol{b}_n)$ of an n-dimensional vector space V into an orthogonal/orthonormal basis $(\boldsymbol{u}_1, \ldots, \boldsymbol{u}_n)$ of V. This basis always exists (Liesen and Mehrmann, 2015) and $\text{span}[\boldsymbol{b}_1, \ldots, \boldsymbol{b}_n] = \text{span}[\boldsymbol{u}_1, \ldots, \boldsymbol{u}_n]$. The *Gram–Schmidt orthogonalization* method iteratively constructs an orthogonal basis $(\boldsymbol{u}_1, \ldots, \boldsymbol{u}_n)$ from any basis $(\boldsymbol{b}_1, \ldots, \boldsymbol{b}_n)$ of V as follows:

Gram–Schmidt orthogonalization

$$\boldsymbol{u}_1 := \boldsymbol{b}_1 \tag{3.67}$$

$$\boldsymbol{u}_k := \boldsymbol{b}_k - \pi_{\text{span}[\boldsymbol{u}_1, \ldots, \boldsymbol{u}_{k-1}]}(\boldsymbol{b}_k) , \quad k = 2, \ldots, n . \tag{3.68}$$

In (3.68), the kth basis vector $\boldsymbol{b}_k$ is projected onto the subspace spanned by the first $k-1$ constructed orthogonal vectors $\boldsymbol{u}_1, \ldots, \boldsymbol{u}_{k-1}$; see Section 3.8.2. This projection is then subtracted from $\boldsymbol{b}_k$ and yields a vector $\boldsymbol{u}_k$ that is orthogonal to the $(k-1)$-dimensional subspace spanned by $\boldsymbol{u}_1, \ldots, \boldsymbol{u}_{k-1}$. Repeating this procedure for all n basis vectors $\boldsymbol{b}_1, \ldots, \boldsymbol{b}_n$ yields an orthogonal basis $(\boldsymbol{u}_1, \ldots, \boldsymbol{u}_n)$ of V. If we normalize the $\boldsymbol{u}_k$, we obtain an ONB where $\|\boldsymbol{u}_k\| = 1$ for $k = 1, \ldots, n$.

Example 3.12 (Gram–Schmidt Orthogonalization)

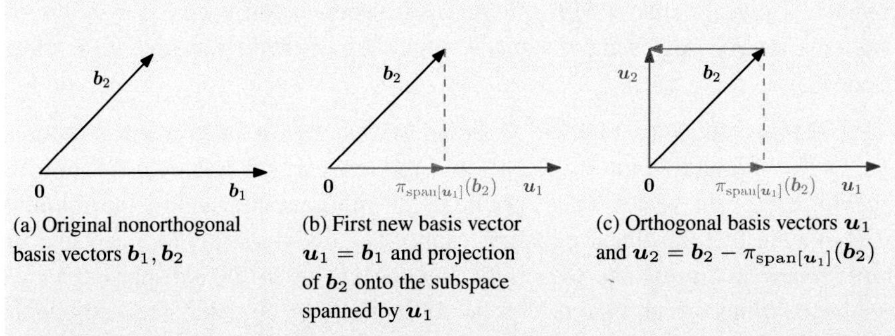

(a) Original nonorthogonal basis vectors $\boldsymbol{b}_1, \boldsymbol{b}_2$

(b) First new basis vector $\boldsymbol{u}_1 = \boldsymbol{b}_1$ and projection of $\boldsymbol{b}_2$ onto the subspace spanned by $\boldsymbol{u}_1$

(c) Orthogonal basis vectors $\boldsymbol{u}_1$ and $\boldsymbol{u}_2 = \boldsymbol{b}_2 - \pi_{\text{span}[\boldsymbol{u}_1]}(\boldsymbol{b}_2)$

Figure 3.12 Gram–Schmidt orthogonalization. (a) non-orthogonal basis $(\boldsymbol{b}_1, \boldsymbol{b}_2)$ of $\mathbb{R}^2$; (b) first constructed basis vector $\boldsymbol{u}_1$ and orthogonal projection of $\boldsymbol{b}_2$ onto $\text{span}[\boldsymbol{u}_1]$; (c) orthogonal basis $(\boldsymbol{u}_1, \boldsymbol{u}_2)$ of $\mathbb{R}^2$.

Consider a basis $(\boldsymbol{b}_1, \boldsymbol{b}_2)$ of $\mathbb{R}^2$, where

$$\boldsymbol{b}_1 = \begin{bmatrix} 2 \\ 0 \end{bmatrix} , \quad \boldsymbol{b}_2 = \begin{bmatrix} 1 \\ 1 \end{bmatrix} ; \tag{3.69}$$

see also Figure 3.12(a). Using the Gram–Schmidt method, we construct an orthogonal basis $(\boldsymbol{u}_1, \boldsymbol{u}_2)$ of $\mathbb{R}^2$ as follows (assuming the dot product as the inner product):

$$\boldsymbol{u}_1 := \boldsymbol{b}_1 = \begin{bmatrix} 2 \\ 0 \end{bmatrix} , \tag{3.70}$$

$$u_2 := b_2 - \pi_{\text{span}[u_1]}(b_2) \overset{(3.45)}{=} b_2 - \frac{u_1 u_1^\top}{\|u_1\|^2} b_2 = \begin{bmatrix}1\\1\end{bmatrix} - \begin{bmatrix}1 & 0\\0 & 0\end{bmatrix}\begin{bmatrix}1\\1\end{bmatrix} = \begin{bmatrix}0\\1\end{bmatrix}. \tag{3.71}$$

These steps are illustrated in Figure 3.12(b) and (c). We immediately see that u_1 and u_2 are orthogonal, i.e., $u_1^\top u_2 = 0$.

3.8.4 Projection onto Affine Subspaces

Thus far, we discussed how to project a vector onto a lower-dimensional subspace U. In the following, we provide a solution to projecting a vector onto an affine subspace.

Consider the setting in Figure 3.13(a). We are given an affine space $L = x_0 + U$, where b_1, b_2 are basis vectors of U. To determine the orthogonal projection $\pi_L(x)$ of x onto L, we transform the problem into a problem that we know how to solve: the projection onto a vector subspace. In order to get there, we subtract the support point x_0 from x and from L, so that $L - x_0 = U$ is exactly the vector subspace U. We can now use the orthogonal projections onto a subspace we discussed in Section 3.8.2 and obtain the projection $\pi_U(x - x_0)$, which is illustrated in Figure 3.13(b). This projection can now be translated back into L by adding x_0, such that we obtain the orthogonal projection onto an affine space L as

$$\pi_L(x) = x_0 + \pi_U(x - x_0), \tag{3.72}$$

where $\pi_U(\cdot)$ is the orthogonal projection onto the subspace U, i.e., the direction space of L; see Figure 3.13(c).

From Figure 3.13, it is also evident that the distance of x from the affine space L is identical to the distance of $x - x_0$ from U, i.e.,

$$d(x, L) = \|x - \pi_L(x)\| = \|x - (x_0 + \pi_U(x - x_0))\| \tag{3.73a}$$

$$= d(x - x_0, \pi_U(x - x_0)). \tag{3.73b}$$

We will use projections onto an affine subspace to derive the concept of a separating hyperplane in Section 12.1.

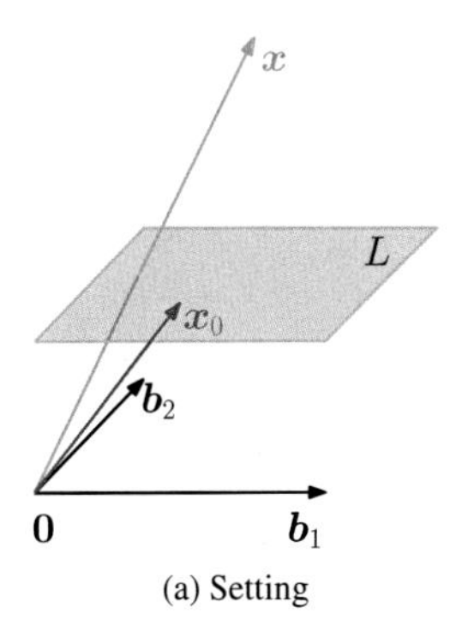

(a) Setting

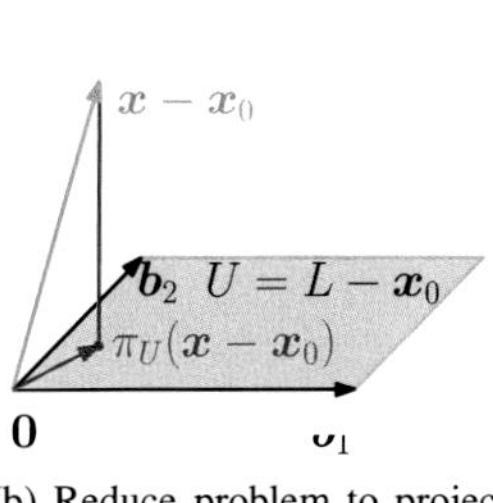

(b) Reduce problem to projection π_U onto vector subspace

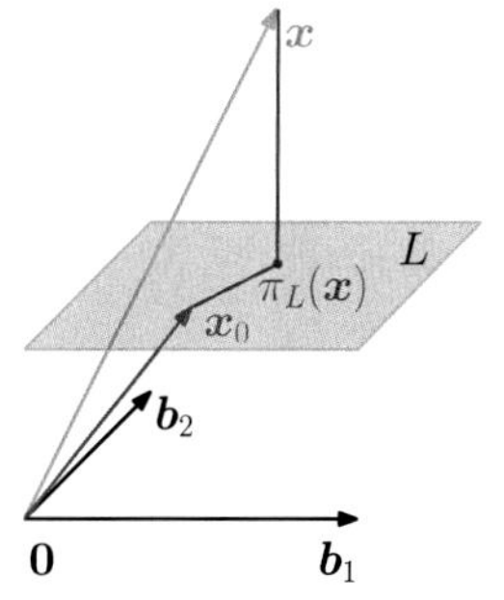

(c) Add support point back in to get affine projection π_L

Figure 3.13 Projection onto an affine space. (a) original setting; (b) setting shifted by $-x_0$ so that $x - x_0$ can be projected onto the direction space U; (c) projection is translated back to $x_0 + \pi_U(x - x_0)$, which gives the final orthogonal projection $\pi_L(x)$.

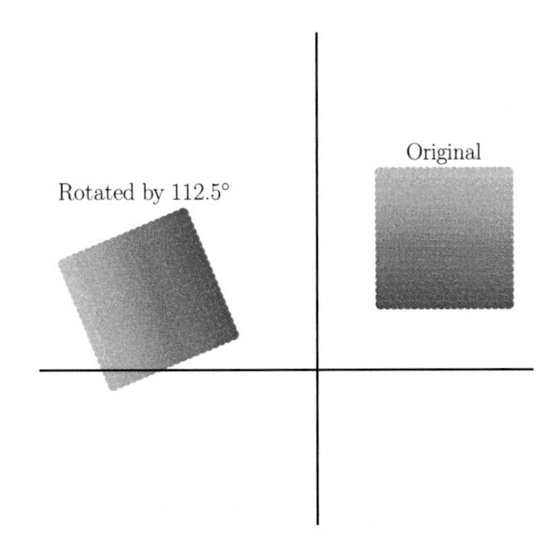

Figure 3.14 A rotation rotates objects in a plane about the origin. If the rotation angle is positive, we rotate counterclockwise.

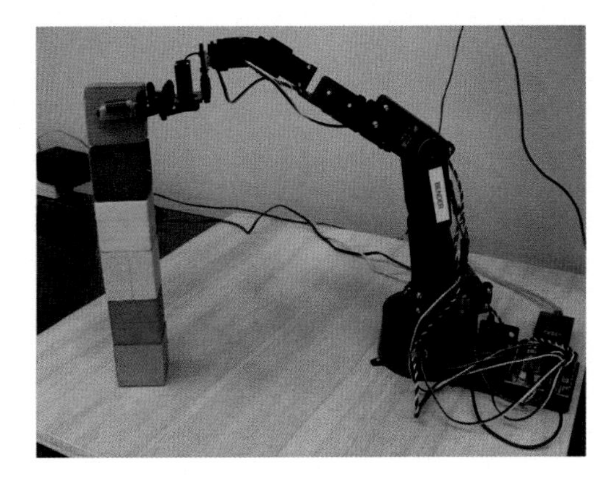

Figure 3.15 The robotic arm needs to rotate its joints in order to pick up objects or to place them correctly. Figure taken from Deisenroth et al. (2015).

3.9 Rotations

Length and angle preservation, as discussed in Section 3.4, are the two characteristics of linear mappings with orthogonal transformation matrices. In the following, we will have a closer look at specific orthogonal transformation matrices, which describe rotations.

rotation

A *rotation* is a linear mapping (more specifically, an automorphism of a Euclidean vector space) that rotates a plane by an angle θ about the origin, i.e., the origin is a fixed point. For a positive angle $\theta > 0$, by common convention, we rotate in a counterclockwise direction. An example is shown in Figure 3.14, where the transformation matrix is

$$\boldsymbol{R} = \begin{bmatrix} -0.38 & -0.92 \\ 0.92 & -0.38 \end{bmatrix} . \tag{3.74}$$

Important application areas of rotations include computer graphics and robotics. For example, in robotics, it is often important to know how to rotate the joints of a robotic arm in order to pick up or place an object; see Figure 3.15.

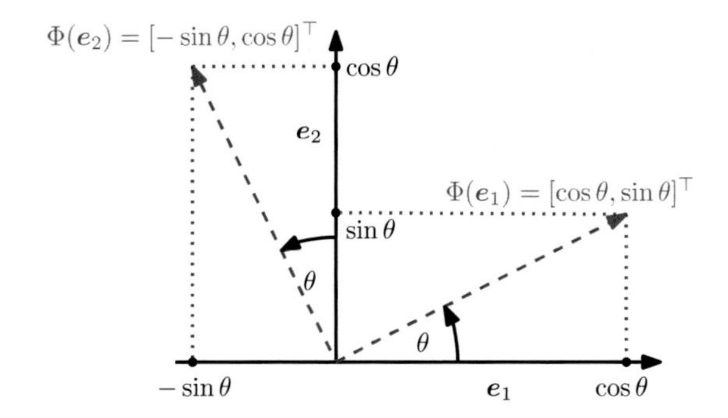

Figure 3.16 Rotation of the standard basis in $\mathbb{R}^2$ by an angle θ.

3.9.1 Rotations in $\mathbb{R}^2$

Consider the standard basis $\left\{ e_1 = \begin{bmatrix} 1 \\ 0 \end{bmatrix}, \; e_2 = \begin{bmatrix} 0 \\ 1 \end{bmatrix} \right\}$ of $\mathbb{R}^2$, which defines the standard coordinate system in $\mathbb{R}^2$. We aim to rotate this coordinate system by an angle θ as illustrated in Figure 3.16. Note that the rotated vectors are still linearly independent and therefore are a basis of $\mathbb{R}^2$. This means that the rotation performs a basis change.

Rotations Φ are linear mappings so that we can express them by a *rotation matrix* $\boldsymbol{R}(\theta)$. Trigonometry (see Figure 3.16) allows us to determine the coordinates of the rotated axes (the image of Φ) with respect to the standard basis in $\mathbb{R}^2$. We obtain

rotation matrix

$$\Phi(e_1) = \begin{bmatrix} \cos\theta \\ \sin\theta \end{bmatrix}, \quad \Phi(e_2) = \begin{bmatrix} -\sin\theta \\ \cos\theta \end{bmatrix}. \tag{3.75}$$

Therefore, the rotation matrix that performs the basis change into the rotated coordinates $\boldsymbol{R}(\theta)$ is given as

$$\boldsymbol{R}(\theta) = \begin{bmatrix} \Phi(e_1) & \Phi(e_2) \end{bmatrix} = \begin{bmatrix} \cos\theta & -\sin\theta \\ \sin\theta & \cos\theta \end{bmatrix}. \tag{3.76}$$

3.9.2 Rotations in $\mathbb{R}^3$

In contrast to the $\mathbb{R}^2$ case, in $\mathbb{R}^3$ we can rotate any two-dimensional plane about a one-dimensional axis. The easiest way to specify the general rotation matrix is to specify how the images of the standard basis e_1, e_2, e_3 are supposed to be rotated, and making sure these images $\boldsymbol{R}e_1, \boldsymbol{R}e_2, \boldsymbol{R}e_3$ are orthonormal to each other. We can then obtain a general rotation matrix $\boldsymbol{R}$ by combining the images of the standard basis.

To have a meaningful rotation angle, we have to define what "counterclockwise" means when we operate in more than two dimensions. We use the convention that a "counterclockwise" (planar) rotation about an axis refers to a rotation about an axis when we look at the axis "head on, from the end toward the origin." In $\mathbb{R}^3$, there are therefore three (planar) rotations about the three standard basis vectors (see Figure 3.17):

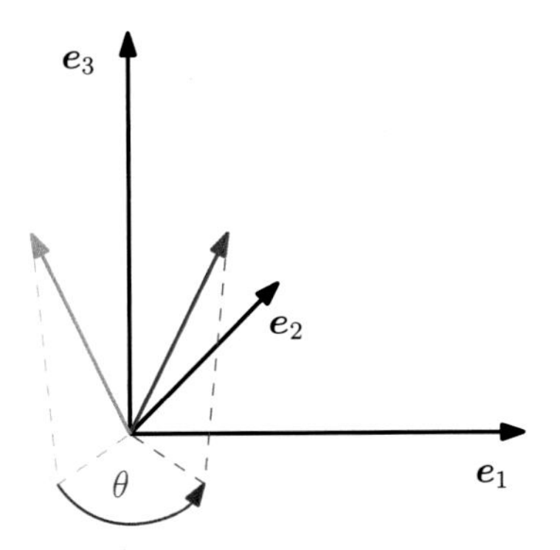

Figure 3.17 Rotation of a vector (gray) in $\mathbb{R}^3$ by an angle θ about the e_3-axis. The rotated vector is shown in blue.

- Rotation about the e_1-axis

$$R_1(\theta) = \begin{bmatrix}\Phi(e_1) & \Phi(e_2) & \Phi(e_3)\end{bmatrix} = \begin{bmatrix}1 & 0 & 0 \\ 0 & \cos\theta & -\sin\theta \\ 0 & \sin\theta & \cos\theta\end{bmatrix}. \tag{3.77}$$

Here, the e_1 coordinate is fixed, and the counterclockwise rotation is performed in the e_2e_3 plane.

- Rotation about the e_2-axis

$$R_2(\theta) = \begin{bmatrix}\cos\theta & 0 & \sin\theta \\ 0 & 1 & 0 \\ -\sin\theta & 0 & \cos\theta\end{bmatrix}. \tag{3.78}$$

If we rotate the e_1e_3 plane about the e_2 axis, we need to look at the e_2 axis from its "tip" toward the origin.

- Rotation about the e_3-axis

$$R_3(\theta) = \begin{bmatrix}\cos\theta & -\sin\theta & 0 \\ \sin\theta & \cos\theta & 0 \\ 0 & 0 & 1\end{bmatrix}. \tag{3.79}$$

Figure 3.17 illustrates this.

3.9.3 Rotations in n Dimensions

The generalization of rotations from 2D and 3D to n-dimensional Euclidean vector spaces can be intuitively described as fixing $n-2$ dimensions and restrict the rotation to a two-dimensional plane in the n-dimensional space. As in the three-dimensional case, we can rotate any plane (two-dimensional subspace of $\mathbb{R}^n$).

Definition 3.11 (Givens Rotation). Let V be an n-dimensional Euclidean vector space and $\Phi : V \to V$ an automorphism with transformation matrix

$$R_{ij}(\theta) := \begin{bmatrix}\boldsymbol{I}_{i-1} & \boldsymbol{0} & \cdots & \cdots & \boldsymbol{0} \\ \boldsymbol{0} & \cos\theta & \boldsymbol{0} & -\sin\theta & \boldsymbol{0} \\ \boldsymbol{0} & \boldsymbol{0} & \boldsymbol{I}_{j-i-1} & \boldsymbol{0} & \boldsymbol{0} \\ \boldsymbol{0} & \sin\theta & \boldsymbol{0} & \cos\theta & \boldsymbol{0} \\ \boldsymbol{0} & \cdots & \cdots & \boldsymbol{0} & \boldsymbol{I}_{n-j}\end{bmatrix} \in \mathbb{R}^{n\times n}, \tag{3.80}$$

for $1 \leqslant i < j \leqslant n$ and $\theta \in \mathbb{R}$. Then $\boldsymbol{R}_{ij}(\theta)$ is called a *Givens rotation*. Essentially, $\boldsymbol{R}_{ij}(\theta)$ is the identity matrix $\boldsymbol{I}_n$ with

Givens rotation

$$r_{ii} = \cos\theta\,, \quad r_{ij} = -\sin\theta\,, \quad r_{ji} = \sin\theta\,, \quad r_{jj} = \cos\theta\,. \tag{3.81}$$

In two dimensions (i.e., $n = 2$), we obtain (3.76) as a special case.

3.9.4 Properties of Rotations

Rotations exhibit a number of useful properties, which can be derived by considering them as orthogonal matrices (Definition 3.8):

- Rotations preserve distances, i.e., $\|\boldsymbol{x} - \boldsymbol{y}\| = \|\boldsymbol{R}_\theta(\boldsymbol{x}) - \boldsymbol{R}_\theta(\boldsymbol{y})\|$. In other words, rotations leave the distance between any two points unchanged after the transformation.
- Rotations preserve angles, i.e., the angle between $\boldsymbol{R}_\theta\boldsymbol{x}$ and $\boldsymbol{R}_\theta\boldsymbol{y}$ equals the angle between $\boldsymbol{x}$ and $\boldsymbol{y}$.
- Rotations in three (or more) dimensions are generally not commutative. Therefore, the order in which rotations are applied is important, even if they rotate about the same point. Only in two dimensions vector rotations are commutative, such that $\boldsymbol{R}(\phi)\boldsymbol{R}(\theta) = \boldsymbol{R}(\theta)\boldsymbol{R}(\phi)$ for all $\phi, \theta \in [0, 2\pi)$. They form an Abelian group (with multiplication) only if they rotate about the same point (e.g., the origin).

3.10 Further Reading

In this chapter, we gave a brief overview of some of the important concepts of analytic geometry, which we will use in later chapters of the book. For a broader and more in-depth overview of some the concepts we presented, we refer to the following excellent books: Axler (2015) and Boyd and Vandenberghe (2018).

Inner products allow us to determine specific bases of vector (sub)spaces, where each vector is orthogonal to all others (orthogonal bases) using the Gram–Schmidt method. These bases are important in optimization and numerical algorithms for solving linear equation systems. For instance, Krylov subspace methods, such as conjugate gradients or the generalized minimal residual method (GMRES), minimize residual errors that are orthogonal to each other (Stoer and Burlirsch, 2002).

In machine learning, inner products are important in the context of kernel methods (Schölkopf and Smola, 2002). Kernel methods exploit the fact that many linear algorithms can be expressed purely by inner product computations. Then, the "kernel trick" allows us to compute these inner products implicitly in a (potentially infinite-dimensional) feature space, without even knowing this feature space explicitly. This allowed the "nonlinearization" of many algorithms used in machine learning, such as kernel-PCA (Schölkopf et al., 1997) for dimensionality reduction. Gaussian processes (Rasmussen and Williams, 2006) also fall into the category of kernel methods and are the current state of the art in probabilistic regression (fitting curves to data points). The idea of kernels is explored further in Chapter 12.

Projections are often used in computer graphics, e.g., to generate shadows. In optimization, orthogonal projections are often used to (iteratively) minimize residual errors. This also has applications in machine learning, e.g., in linear regression where we want to find a (linear) function that minimizes the residual errors, i.e., the lengths of the orthogonal projections of the data onto the linear function (Bishop, 2006). We will investigate this further in Chapter 9. PCA (Pearson, 1901; Hotelling, 1933) also uses projections to reduce the dimensionality of high-dimensional data. We will discuss this in more detail in Chapter 10.

Exercises

3.1 Show that $\langle\cdot,\cdot\rangle$ defined for all $\boldsymbol{x} = [x_1, x_2]^\top \in \mathbb{R}^2$ and $\boldsymbol{y} = [y_1, y_2]^\top \in \mathbb{R}^2$ by

$$\langle \boldsymbol{x}, \boldsymbol{y} \rangle := x_1y_1 - (x_1y_2 + x_2y_1) + 2(x_2y_2)$$

is an inner product.

3.2 Consider $\mathbb{R}^2$ with $\langle\cdot,\cdot\rangle$ defined for all $\boldsymbol{x}$ and $\boldsymbol{y}$ in $\mathbb{R}^2$ as

$$\langle \boldsymbol{x}, \boldsymbol{y} \rangle := \boldsymbol{x}^\top \underbrace{\begin{bmatrix} 2 & 0 \\ 1 & 2 \end{bmatrix}}_{=:\boldsymbol{A}} \boldsymbol{y}\,.$$

Is $\langle\cdot,\cdot\rangle$ an inner product?

3.3 Compute the distance between

$$\boldsymbol{x} = \begin{bmatrix} 1 \\ 2 \\ 3 \end{bmatrix}, \quad \boldsymbol{y} = \begin{bmatrix} -1 \\ -1 \\ 0 \end{bmatrix}$$

using

a. $\langle \boldsymbol{x}, \boldsymbol{y} \rangle := \boldsymbol{x}^\top \boldsymbol{y}$

b. $\langle \boldsymbol{x}, \boldsymbol{y} \rangle := \boldsymbol{x}^\top \boldsymbol{A}\boldsymbol{y}\,, \quad \boldsymbol{A} := \begin{bmatrix} 2 & 1 & 0 \\ 1 & 3 & -1 \\ 0 & -1 & 2 \end{bmatrix}$

3.4 Compute the angle between

$$\boldsymbol{x} = \begin{bmatrix} 1 \\ 2 \end{bmatrix}, \quad \boldsymbol{y} = \begin{bmatrix} -1 \\ -1 \end{bmatrix}$$

using

a. $\langle \boldsymbol{x}, \boldsymbol{y} \rangle := \boldsymbol{x}^\top \boldsymbol{y}$

b. $\langle \boldsymbol{x}, \boldsymbol{y} \rangle := \boldsymbol{x}^\top \boldsymbol{B}\boldsymbol{y}\,, \quad \boldsymbol{B} := \begin{bmatrix} 2 & 1 \\ 1 & 3 \end{bmatrix}$

3.5 Consider the Euclidean vector space $\mathbb{R}^5$ with the dot product. A subspace $U \subseteq \mathbb{R}^5$ and $\boldsymbol{x} \in \mathbb{R}^5$ are given by

$$U = \operatorname{span}[\begin{bmatrix} 0 \\ -1 \\ 2 \\ 0 \\ 2 \end{bmatrix}, \begin{bmatrix} 1 \\ -3 \\ 1 \\ -1 \\ 2 \end{bmatrix}, \begin{bmatrix} -3 \\ 4 \\ 1 \\ 2 \\ 1 \end{bmatrix}, \begin{bmatrix} -1 \\ -3 \\ 5 \\ 0 \\ 7 \end{bmatrix}], \quad \boldsymbol{x} = \begin{bmatrix} -1 \\ -9 \\ -1 \\ 4 \\ 1 \end{bmatrix}$$

a. Determine the orthogonal projection $\pi_U(\boldsymbol{x})$ of $\boldsymbol{x}$ onto U.
b. Determine the distance $d(\boldsymbol{x}, U)$.

3.6 Consider $\mathbb{R}^3$ with the inner product

$$\langle \boldsymbol{x}, \boldsymbol{y} \rangle := \boldsymbol{x}^\top \begin{bmatrix} 2 & 1 & 0 \\ 1 & 2 & -1 \\ 0 & -1 & 2 \end{bmatrix} \boldsymbol{y} .$$

Furthermore, we define $\boldsymbol{e}_1, \boldsymbol{e}_2, \boldsymbol{e}_3$ as the standard/canonical basis in $\mathbb{R}^3$.

a. Determine the orthogonal projection $\pi_U(\boldsymbol{e}_2)$ of $\boldsymbol{e}_2$ onto

$$U = \text{span}[\boldsymbol{e}_1, \boldsymbol{e}_3] .$$

Hint: Orthogonality is defined through the inner product.
b. Compute the distance $d(\boldsymbol{e}_2, U)$.
c. Draw the scenario: standard basis vectors and $\pi_U(\boldsymbol{e}_2)$.

3.7 Let V be a vector space π, an endomorphism of V.

a. Prove that π is a projection if and only if $\text{id}_V - \pi$ is a projection, where id_V is the identity endomorphism on V.
b. Assume now that π is a projection. Calculate $\text{Im}(\text{id}_V - \pi)$ and $\ker(\text{id}_V - \pi)$ as a function of $\text{Im}(\pi)$ and $\ker(\pi)$.

3.8 Using the Gram–Schmidt method, turn the basis $B = (\boldsymbol{b}_1, \boldsymbol{b}_2)$ of a two-dimensional subspace $U \subseteq \mathbb{R}^3$ into an ONB $C = (\boldsymbol{c}_1, \boldsymbol{c}_2)$ of U, where

$$\boldsymbol{b}_1 := \begin{bmatrix} 1 \\ 1 \\ 1 \end{bmatrix}, \quad \boldsymbol{b}_2 := \begin{bmatrix} -1 \\ 2 \\ 0 \end{bmatrix} .$$

3.9 Let $n \in \mathbb{N}^*$ and let $x_1, \ldots, x_n > 0$ be n positive real numbers so that $x_1 + \cdots + x_n = 1$. Use the Cauchy–Schwarz inequality and show that

a. $\sum_{i=1}^n x_i^2 \geqslant \frac{1}{n}$
b. $\sum_{i=1}^n \frac{1}{x_i} \geqslant n^2$

Hint: Think about the dot product on $\mathbb{R}^n$. Then choose specific vectors $\boldsymbol{x}, \boldsymbol{y} \in \mathbb{R}^n$ and apply the Cauchy–Schwarz inequality.

3.10 Rotate the vectors

$$\boldsymbol{x}_1 := \begin{bmatrix} 2 \\ 3 \end{bmatrix}, \quad \boldsymbol{x}_2 := \begin{bmatrix} 0 \\ -1 \end{bmatrix}$$

by 30°.

4

Matrix Decompositions

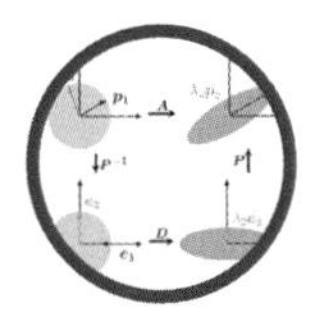

In Chapters 2 and 3, we studied ways to manipulate and measure vectors, projections of vectors, and linear mappings. Mappings and transformations of vectors can be conveniently described as operations performed by matrices. Moreover, data is often represented in matrix form as well, e.g., where the rows of the matrix represent different people and the columns describe different features of the people, such as weight, height, and socio-economic status. In this chapter, we present three aspects of matrices: how to summarize matrices, how matrices can be decomposed, and how these decompositions can be used for matrix approximations.

We first consider methods that allow us to describe matrices with just a few numbers that characterize the overall properties of matrices. We will do this in the sections on determinants (Section 4.1) and eigenvalues (Section 4.2) for the important special case of square matrices. These characteristic numbers have important mathematical consequences and allow us to quickly grasp what useful properties a matrix has. From here we will proceed to matrix decomposition methods: An analogy for matrix decomposition is the factoring of numbers, such as the factoring of 21 into prime numbers $7 \cdot 3$. For this reason matrix decomposition is also often referred to as *matrix factorization*. Matrix decompositions are used to describe a matrix by means of a different representation using factors of interpretable matrices.

matrix factorization

We will first cover a square-root-like operation for symmetric, positive definite matrices, the Cholesky decomposition (Section 4.3). From here we will look at two related methods for factorizing matrices into canonical forms. The first one is known as matrix diagonalization (Section 4.4), which allows us to represent the linear mapping using a diagonal transformation matrix if we choose an appropriate basis. The second method, singular value decomposition (Section 4.5), extends this factorization to nonsquare matrices, and it is considered one of the fundamental concepts in linear algebra. These decompositions are helpful, as matrices representing numerical data are often very large and hard to analyze. We conclude the chapter with a systematic overview of the types of matrices and the characteristic properties that distinguish them in the form of a matrix taxonomy (Section 4.7).

The methods that we cover in this chapter will become important in both subsequent mathematical chapters, such as Chapter 6, but also in applied chapters, such as dimensionality reduction in Chapter 10 or density estimation in Chapter 11. This chapter's overall structure is depicted in the mind map of Figure 4.1.

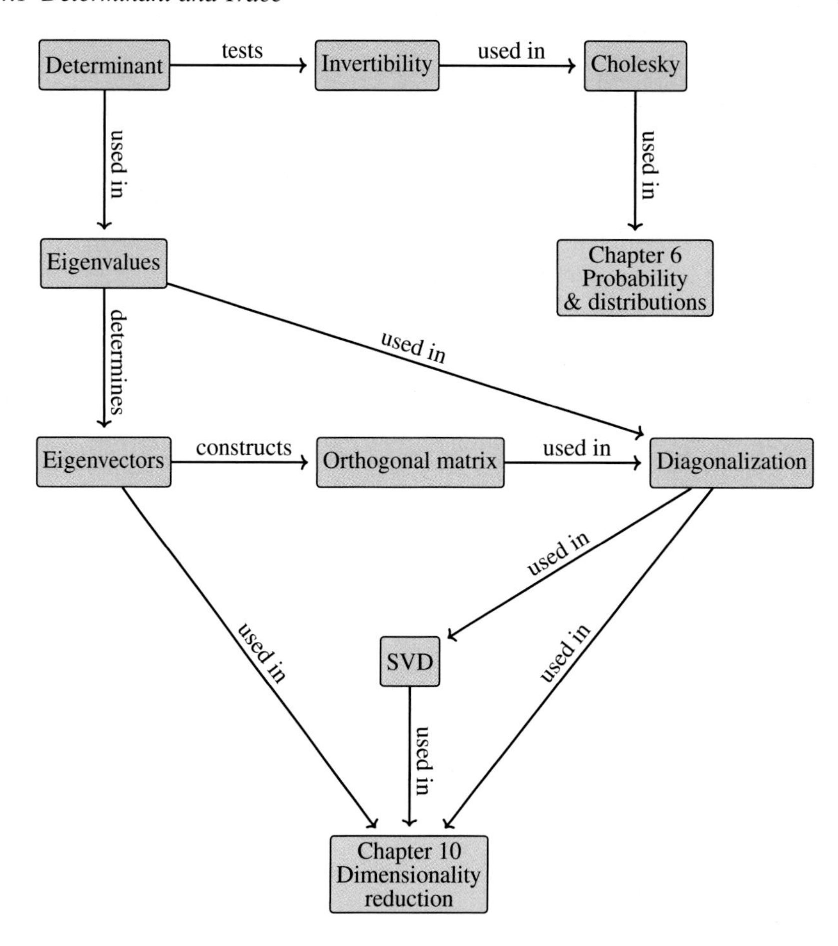

Figure 4.1 A mind map of the concepts introduced in this chapter, along with where they are used in other parts of the book.

4.1 Determinant and Trace

The determinant notation $|A|$ must not be confused with the absolute value.

Determinants are important concepts in linear algebra. A determinant is a mathematical object in the analysis and solution of systems of linear equations. Determinants are only defined for square matrices $\boldsymbol{A} \in \mathbb{R}^{n \times n}$, i.e., matrices with the same number of rows and columns. In this book, we write the determinant as $\det(\boldsymbol{A})$ or sometimes as $|\boldsymbol{A}|$ so that

$$\det(\boldsymbol{A}) = \begin{vmatrix} a_{11} & a_{12} & \dots & a_{1n} \\ a_{21} & a_{22} & \dots & a_{2n} \\ \vdots & & \ddots & \vdots \\ a_{n1} & a_{n2} & \dots & a_{nn} \end{vmatrix} . \tag{4.1}$$

determinant

The *determinant* of a square matrix $\boldsymbol{A} \in \mathbb{R}^{n \times n}$ is a function that maps $\boldsymbol{A}$ onto a real number. Before providing a definition of the determinant for general $n \times n$ matrices, let us have a look at some motivating examples and define determinants for some special matrices.

Example 4.1 (Testing for Matrix Invertibility)
Let us begin with exploring if a square matrix $\boldsymbol{A}$ is invertible (see Section 2.2.2). For the smallest cases, we already know when a matrix is invertible. If $\boldsymbol{A}$ is a 1×1 matrix, i.e., it is a scalar number, then $\boldsymbol{A} = a \implies \boldsymbol{A}^{-1} = \frac{1}{a}$. Thus $a \frac{1}{a} = 1$ holds, if and only if $a \neq 0$.

For 2×2 matrices, by the definition of the inverse (Definition 2.3), we know that $\boldsymbol{A}\boldsymbol{A}^{-1} = \boldsymbol{I}$. Then, with (2.24), the inverse of $\boldsymbol{A}$ is

$$\boldsymbol{A}^{-1} = \frac{1}{a_{11}a_{22} - a_{12}a_{21}} \begin{bmatrix} a_{22} & -a_{12} \\ -a_{21} & a_{11} \end{bmatrix}. \tag{4.2}$$

Hence, $\boldsymbol{A}$ is invertible if and only if

$$a_{11}a_{22} - a_{12}a_{21} \neq 0. \tag{4.3}$$

This quantity is the determinant of $\boldsymbol{A} \in \mathbb{R}^{2\times 2}$, i.e.,

$$\det(\boldsymbol{A}) = \begin{vmatrix} a_{11} & a_{12} \\ a_{21} & a_{22} \end{vmatrix} = a_{11}a_{22} - a_{12}a_{21}. \tag{4.4}$$

Example 4.1 points already at the relationship between determinants and the existence of inverse matrices. The next theorem states the same result for $n \times n$ matrices.

Theorem 4.1. *For any square matrix $\boldsymbol{A} \in \mathbb{R}^{n\times n}$ it holds that $\boldsymbol{A}$ is invertible if and only if* $\det(\boldsymbol{A}) \neq 0$.

We have explicit (closed-form) expressions for determinants of small matrices in terms of the elements of the matrix. For $n = 1$,

$$\det(\boldsymbol{A}) = \det(a_{11}) = a_{11}. \tag{4.5}$$

For $n = 2$,

$$\det(\boldsymbol{A}) = \begin{vmatrix} a_{11} & a_{12} \\ a_{21} & a_{22} \end{vmatrix} = a_{11}a_{22} - a_{12}a_{21}, \tag{4.6}$$

which we have observed in the preceding example.

For $n = 3$ (known as Sarrus' rule),

$$\begin{vmatrix} a_{11} & a_{12} & a_{13} \\ a_{21} & a_{22} & a_{23} \\ a_{31} & a_{32} & a_{33} \end{vmatrix} = a_{11}a_{22}a_{33} + a_{21}a_{32}a_{13} + a_{31}a_{12}a_{23} \tag{4.7}$$
$$- a_{31}a_{22}a_{13} - a_{11}a_{32}a_{23} - a_{21}a_{12}a_{33}.$$

For a memory aid of the product terms in Sarrus' rule, try tracing the elements of the triple products in the matrix.

upper-triangular matrix
lower-triangular matrix

We call a square matrix $\boldsymbol{T}$ an *upper-triangular matrix* if $T_{ij} = 0$ for $i > j$, i.e., the matrix is zero below its diagonal. Analogously, we define a *lower-triangular matrix* as a matrix with zeros above its diagonal. For a triangular matrix $\boldsymbol{T} \in \mathbb{R}^{n\times n}$, the determinant is the product of the diagonal elements, i.e.,

$$\det(\boldsymbol{T}) = \prod_{i=1}^{n} T_{ii}. \tag{4.8}$$

Example 4.2 (Determinants as Measures of Volume)
The notion of a determinant is natural when we consider it as a mapping from a set of n vectors spanning an object in $\mathbb{R}^n$. It turns out that the determinant $\det(\boldsymbol{A})$ is the signed volume of an n-dimensional parallelepiped formed by columns of the matrix $\boldsymbol{A}$.

The determinant is the signed volume of the parallelepiped formed by the columns of the matrix.

For $n = 2$, the columns of the matrix form a parallelogram; see Figure 4.2. As the angle between vectors gets smaller, the area of a parallelogram shrinks, too. Consider two vectors $\boldsymbol{b}, \boldsymbol{g}$ that form the columns of a matrix $\boldsymbol{A} = [\boldsymbol{b}, \boldsymbol{g}]$. Then the absolute value of the determinant of $\boldsymbol{A}$ is the area of the parallelogram with vertices $\boldsymbol{0}, \boldsymbol{b}, \boldsymbol{g}, \boldsymbol{b}+\boldsymbol{g}$. In particular, if $\boldsymbol{b}, \boldsymbol{g}$ are linearly dependent so that $\boldsymbol{b} = \lambda \boldsymbol{g}$ for some $\lambda \in \mathbb{R}$, they no longer form a two-dimensional parallelogram. Therefore, the corresponding area is 0. On the contrary, if $\boldsymbol{b}, \boldsymbol{g}$ are linearly independent and are multiples of the canonical basis vectors $\boldsymbol{e}_1, \boldsymbol{e}_2$, then they can be written as $\boldsymbol{b} = \begin{bmatrix} b \\ 0 \end{bmatrix}$ and $\boldsymbol{g} = \begin{bmatrix} 0 \\ g \end{bmatrix}$, and the determinant is $\begin{vmatrix} b & 0 \\ 0 & g \end{vmatrix} = bg - 0 = bg$.

Figure 4.2 The area of the parallelogram (shaded region) spanned by the vectors $\boldsymbol{b}$ and $\boldsymbol{g}$ is $|\det([\boldsymbol{b}, \boldsymbol{g}])|$.

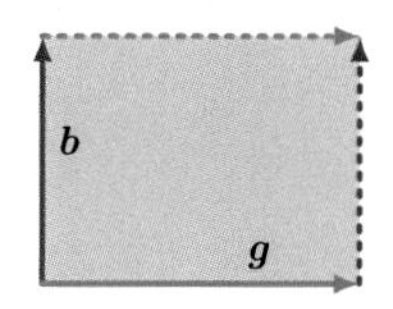

The sign of the determinant indicates the orientation of the spanning vectors $\boldsymbol{b}, \boldsymbol{g}$ with respect to the standard basis $(\boldsymbol{e}_1, \boldsymbol{e}_2)$. In our figure, flipping the order to $\boldsymbol{g}, \boldsymbol{b}$ swaps the columns of $\boldsymbol{A}$ and reverses the orientation of the shaded area. This becomes the familiar formula: area $=$ height $\times$ length. This intuition extends to higher dimensions. In $\mathbb{R}^3$, we consider three vectors $\boldsymbol{r}, \boldsymbol{b}, \boldsymbol{g} \in \mathbb{R}^3$ spanning the edges of a parallelepiped, i.e., a solid with faces that are parallel parallelograms (see Figure 4.3). The absolute value of the determinant of the 3×3 matrix $[\boldsymbol{r},\ \boldsymbol{b},\ \boldsymbol{g}]$ is the volume of the solid. Thus the determinant acts as a function that measures the signed volume formed by column vectors composed in a matrix.

Figure 4.3 The volume of the parallelepiped (shaded volume) spanned by vectors $\boldsymbol{r}, \boldsymbol{b}, \boldsymbol{g}$ is $|\det([\boldsymbol{r},\ \boldsymbol{b},\ \boldsymbol{g}])|$.

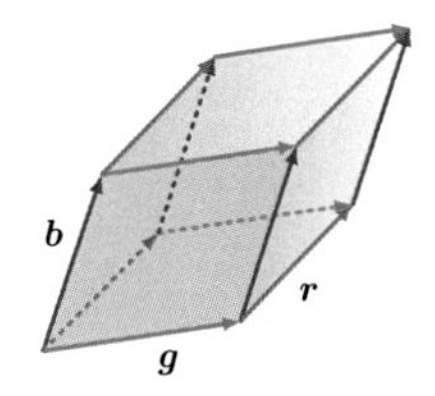

The sign of the determinant indicates the orientation of the spanning vectors.

Consider the three linearly independent vectors $\boldsymbol{r}, \boldsymbol{g}, \boldsymbol{b} \in \mathbb{R}^3$ given as

$$\boldsymbol{r} = \begin{bmatrix} 2 \\ 0 \\ -8 \end{bmatrix}, \quad \boldsymbol{g} = \begin{bmatrix} 6 \\ 1 \\ 0 \end{bmatrix}, \quad \boldsymbol{b} = \begin{bmatrix} 1 \\ 4 \\ -1 \end{bmatrix}. \tag{4.9}$$

Writing these vectors as the columns of a matrix

$$\boldsymbol{A} = [\boldsymbol{r},\ \boldsymbol{g},\ \boldsymbol{b}] = \begin{bmatrix} 2 & 6 & 1 \\ 0 & 1 & 4 \\ -8 & 0 & -1 \end{bmatrix} \tag{4.10}$$

allows us to compute the desired volume as

$$V = |\det(\boldsymbol{A})| = 186\,. \tag{4.11}$$

Computing the determinant of an $n \times n$ matrix requires a general algorithm to solve the cases for $n > 3$, which we are going to explore in the following. Theorem 4.2 below reduces the problem of computing the determinant of an $n \times n$ matrix to computing the determinant of $(n - 1) \times (n - 1)$ matrices. By recursively applying the Laplace expansion (Theorem 4.2), we can therefore compute determinants of $n \times n$ matrices by ultimately computing determinants of 2×2 matrices.

Laplace expansion

Theorem 4.2 (Laplace Expansion)**.** *Consider a matrix* $\boldsymbol{A} \in \mathbb{R}^{n\times n}$*. Then, for all* $j = 1, \ldots, n$*:*

$\det(\boldsymbol{A}_{k,j})$ is called a *minor* and $(-1)^{k+j}\det(\boldsymbol{A}_{k,j})$ a *cofactor*.

1. *Expansion along column* j

$$\det(\boldsymbol{A}) = \sum_{k=1}^{n}(-1)^{k+j}a_{kj}\det(\boldsymbol{A}_{k,j}). \tag{4.12}$$

2. *Expansion along row* j

$$\det(\boldsymbol{A}) = \sum_{k=1}^{n}(-1)^{k+j}a_{jk}\det(\boldsymbol{A}_{j,k}). \tag{4.13}$$

Here $\boldsymbol{A}_{k,j} \in \mathbb{R}^{(n-1)\times(n-1)}$ *is the submatrix of* $\boldsymbol{A}$ *that we obtain when deleting row* k *and column* j*.*

Example 4.3 (Laplace Expansion)

Let us compute the determinant of

$$\boldsymbol{A} = \begin{bmatrix} 1 & 2 & 3 \\ 3 & 1 & 2 \\ 0 & 0 & 1 \end{bmatrix} \tag{4.14}$$

using the Laplace expansion along the first row. Applying (4.13) yields

$$\begin{aligned} \begin{vmatrix} 1 & 2 & 3 \\ 3 & 1 & 2 \\ 0 & 0 & 1 \end{vmatrix} &= (-1)^{1+1}\cdot 1 \begin{vmatrix} 1 & 2 \\ 0 & 1 \end{vmatrix} \\ &+ (-1)^{1+2}\cdot 2 \begin{vmatrix} 3 & 2 \\ 0 & 1 \end{vmatrix} + (-1)^{1+3}\cdot 3 \begin{vmatrix} 3 & 1 \\ 0 & 0 \end{vmatrix}. \end{aligned} \tag{4.15}$$

We use (4.6) to compute the determinants of all 2×2 matrices and obtain

$$\det(\boldsymbol{A}) = 1(1-0) - 2(3-0) + 3(0-0) = -5. \tag{4.16}$$

For completeness we can compare this result to computing the determinant using Sarrus' rule (4.7):

$$\det(\boldsymbol{A}) = 1\cdot1\cdot1 + 3\cdot0\cdot3 + 0\cdot2\cdot2 - 0\cdot1\cdot3 - 1\cdot0\cdot2 - 3\cdot2\cdot1 = 1-6 = -5. \tag{4.17}$$

For $\boldsymbol{A} \in \mathbb{R}^{n\times n}$, the determinant exhibits the following properties:

- The determinant of a matrix product is the product of the corresponding determinants, $\det(\boldsymbol{AB}) = \det(\boldsymbol{A})\det(\boldsymbol{B})$.
- Determinants are invariant to transposition, i.e., $\det(\boldsymbol{A}) = \det(\boldsymbol{A}^\top)$.
- If $\boldsymbol{A}$ is regular (invertible), then $\det(\boldsymbol{A}^{-1}) = \frac{1}{\det(\boldsymbol{A})}$.
- Similar matrices (Definition 2.22) possess the same determinant. Therefore, for a linear mapping $\Phi : V \to V$ all transformation matrices $\boldsymbol{A}_\Phi$ of Φ have

the same determinant. Thus, the determinant is invariant to the choice of basis of a linear mapping.
- Adding a multiple of a column/row to another one does not change $\det(\boldsymbol{A})$.
- Multiplication of a column/row with $\lambda \in \mathbb{R}$ scales $\det(\boldsymbol{A})$ by λ. In particular, $\det(\lambda \boldsymbol{A}) = \lambda^n \det(\boldsymbol{A})$.
- Swapping two rows/columns changes the sign of $\det(\boldsymbol{A})$.

Because of the last three properties, we can use Gaussian elimination (see Section 2.1) to compute $\det(\boldsymbol{A})$ by bringing $\boldsymbol{A}$ into row-echelon form. We can stop Gaussian elimination when we have $\boldsymbol{A}$ in a triangular form where the elements below the diagonal are all 0. Recall from (4.8) that the determinant of a triangular matrix is the product of the diagonal elements.

Theorem 4.3. *A square matrix* $\boldsymbol{A} \in \mathbb{R}^{n \times n}$ *has* $\det(\boldsymbol{A}) \neq 0$ *if and only if* $\mathrm{rk}(\boldsymbol{A}) = n$. *In other words,* $\boldsymbol{A}$ *is invertible if and only if it is full rank.*

When mathematics was mainly performed by hand, the determinant calculation was considered an essential way to analyze matrix invertibility. However, contemporary approaches in machine learning use direct numerical methods that superseded the explicit calculation of the determinant. For example, in Chapter 2, we learned that inverse matrices can be computed by Gaussian elimination. Gaussian elimination can thus be used to compute the determinant of a matrix.

Determinants will play an important theoretical role for the following sections, especially when we learn about eigenvalues and eigenvectors (Section 4.2) through the characteristic polynomial.

Definition 4.4. The *trace* of a square matrix $\boldsymbol{A} \in \mathbb{R}^{n \times n}$ is defined as

trace

$$\mathrm{tr}(\boldsymbol{A}) := \sum_{i=1}^{n} a_{ii} \,, \tag{4.18}$$

i.e., the trace is the sum of the diagonal elements of $\boldsymbol{A}$.

The trace satisfies the following properties:

- $\mathrm{tr}(\boldsymbol{A} + \boldsymbol{B}) = \mathrm{tr}(\boldsymbol{A}) + \mathrm{tr}(\boldsymbol{B})$ for $\boldsymbol{A}, \boldsymbol{B} \in \mathbb{R}^{n \times n}$
- $\mathrm{tr}(\alpha \boldsymbol{A}) = \alpha \mathrm{tr}(\boldsymbol{A}) \,, \alpha \in \mathbb{R}$ for $\boldsymbol{A} \in \mathbb{R}^{n \times n}$
- $\mathrm{tr}(\boldsymbol{I}_n) = n$
- $\mathrm{tr}(\boldsymbol{A}\boldsymbol{B}) = \mathrm{tr}(\boldsymbol{B}\boldsymbol{A})$ for $\boldsymbol{A} \in \mathbb{R}^{n \times k}, \boldsymbol{B} \in \mathbb{R}^{k \times n}$

It can be shown that only one function satisfies these four properties together – the trace (Gohberg et al., 2012).

The properties of the trace of matrix products are more general. Specifically, the trace is invariant under cyclic permutations, i.e.,

The trace is invariant under cyclic permutations.

$$\mathrm{tr}(\boldsymbol{A}\boldsymbol{K}\boldsymbol{L}) = \mathrm{tr}(\boldsymbol{K}\boldsymbol{L}\boldsymbol{A}) \tag{4.19}$$

for matrices $\boldsymbol{A} \in \mathbb{R}^{a \times k}, \boldsymbol{K} \in \mathbb{R}^{k \times l}, \boldsymbol{L} \in \mathbb{R}^{l \times a}$. This property generalizes to products of an arbitrary number of matrices. As a special case of (4.19), it follows that for two vectors $\boldsymbol{x}, \boldsymbol{y} \in \mathbb{R}^n$

$$\text{tr}(\boldsymbol{x}\boldsymbol{y}^\top) = \text{tr}(\boldsymbol{y}^\top\boldsymbol{x}) = \boldsymbol{y}^\top\boldsymbol{x} \in \mathbb{R}\,. \tag{4.20}$$

Given a linear mapping $\Phi : V \to V$, where V is a vector space, we define the trace of this map by using the trace of matrix representation of Φ. For a given basis of V, we can describe Φ by means of the transformation matrix $\boldsymbol{A}$. Then the trace of Φ is the trace of $\boldsymbol{A}$. For a different basis of V, it holds that the corresponding transformation matrix $\boldsymbol{B}$ of Φ can be obtained by a basis change of the form $\boldsymbol{S}^{-1}\boldsymbol{A}\boldsymbol{S}$ for suitable $\boldsymbol{S}$ (see Section 2.7.2). For the corresponding trace of Φ, this means

$$\text{tr}(\boldsymbol{B}) = \text{tr}(\boldsymbol{S}^{-1}\boldsymbol{A}\boldsymbol{S}) \overset{(4.19)}{=} \text{tr}(\boldsymbol{A}\boldsymbol{S}\boldsymbol{S}^{-1}) = \text{tr}(\boldsymbol{A})\,. \tag{4.21}$$

Hence, while matrix representations of linear mappings are basis dependent, the trace of a linear mapping Φ is independent of the basis.

In this section, we covered determinants and traces as functions characterizing a square matrix. Taking together our understanding of determinants and traces, we can now define an important equation describing a matrix $\boldsymbol{A}$ in terms of a polynomial, which we will use extensively in the following sections.

Definition 4.5 (Characteristic Polynomial)**.** For $\lambda \in \mathbb{R}$ and a square matrix $\boldsymbol{A} \in \mathbb{R}^{n\times n}$,

$$p_{\boldsymbol{A}}(\lambda) := \det(\boldsymbol{A} - \lambda\boldsymbol{I}) \tag{4.22a}$$

$$= c_0 + c_1\lambda + c_2\lambda^2 + \cdots + c_{n-1}\lambda^{n-1} + (-1)^n\lambda^n\,, \tag{4.22b}$$

characteristic polynomial

$c_0, \ldots, c_{n-1} \in \mathbb{R}$, is the *characteristic polynomial* of $\boldsymbol{A}$. In particular,

$$c_0 = \det(\boldsymbol{A})\,, \tag{4.23}$$

$$c_{n-1} = (-1)^{n-1}\text{tr}(\boldsymbol{A})\,. \tag{4.24}$$

The characteristic polynomial (4.22a) will allow us to compute eigenvalues and eigenvectors, covered in the next section.

4.2 Eigenvalues and Eigenvectors

We will now get to know a new way to characterize a matrix and its associated linear mapping. Recall from Section 2.7.1 that every linear mapping has a unique transformation matrix given an ordered basis. We can interpret linear mappings and their associated transformation matrices by performing an "eigen" analysis. As we will see, the eigenvalues of a linear mapping will tell us how a special set of vectors, the eigenvectors, is transformed by the linear mapping.

Eigen is a German word meaning "characteristic," "self," or "own."

eigenvalue

eigenvector

Definition 4.6. Let $\boldsymbol{A} \in \mathbb{R}^{n\times n}$ be a square matrix. Then $\lambda \in \mathbb{R}$ is an *eigenvalue* of $\boldsymbol{A}$ and $\boldsymbol{x} \in \mathbb{R}^n\backslash\{\boldsymbol{0}\}$ is the corresponding *eigenvector* of $\boldsymbol{A}$ if

$$\boldsymbol{A}\boldsymbol{x} = \lambda\boldsymbol{x}\,. \tag{4.25}$$

eigenvalue equation

We call (4.25) the *eigenvalue equation.*

Remark. In the linear algebra literature and software, it is often a convention that eigenvalues are sorted in descending order, so that the largest eigenvalue and associated eigenvector are called the first eigenvalue and its associated eigenvector, and the second largest called the second eigenvalue and its associated eigenvector, and so on. However, textbooks and publications may have different or no notion of orderings. We do not want to presume an ordering in this book if not stated explicitly. ◇

The following statements are equivalent:

- λ is an eigenvalue of $\boldsymbol{A} \in \mathbb{R}^{n \times n}$.
- There exists an $\boldsymbol{x} \in \mathbb{R}^n \backslash \{\boldsymbol{0}\}$ with $\boldsymbol{A}\boldsymbol{x} = \lambda\boldsymbol{x}$, or equivalently, $(\boldsymbol{A} - \lambda\boldsymbol{I}_n)\boldsymbol{x} = \boldsymbol{0}$ can be solved nontrivially, i.e., $\boldsymbol{x} \neq \boldsymbol{0}$.
- $\text{rk}(\boldsymbol{A} - \lambda\boldsymbol{I}_n) < n$.
- $\det(\boldsymbol{A} - \lambda\boldsymbol{I}_n) = 0$.

Definition 4.7 (Collinearity and Codirection)**.** Two vectors that point in the same direction are called *codirected*. Two vectors are *collinear* if they point in the same or the opposite direction.

codirected

collinear

Remark (Nonuniqueness of eigenvectors). If $\boldsymbol{x}$ is an eigenvector of $\boldsymbol{A}$ associated with eigenvalue λ, then for any $c \in \mathbb{R}\backslash\{0\}$ it holds that $c\boldsymbol{x}$ is an eigenvector of $\boldsymbol{A}$ with the same eigenvalue since

$$\boldsymbol{A}(c\boldsymbol{x}) = c\boldsymbol{A}\boldsymbol{x} = c\lambda\boldsymbol{x} = \lambda(c\boldsymbol{x}). \tag{4.26}$$

Thus, all vectors that are collinear to $\boldsymbol{x}$ are also eigenvectors of $\boldsymbol{A}$. ◇

Theorem 4.8. *$\lambda \in \mathbb{R}$ is eigenvalue of $\boldsymbol{A} \in \mathbb{R}^{n \times n}$ if and only if λ is a root of the characteristic polynomial $p_{\boldsymbol{A}}(\lambda)$ of $\boldsymbol{A}$.*

Definition 4.9. Let a square matrix $\boldsymbol{A}$ have an eigenvalue λ_i. The *algebraic multiplicity* of λ_i is the number of times the root appears in the characteristic polynomial.

algebraic multiplicity

Definition 4.10 (Eigenspace and Eigenspectrum)**.** For $\boldsymbol{A} \in \mathbb{R}^{n \times n}$, the set of all eigenvectors of $\boldsymbol{A}$ associated with an eigenvalue λ spans a subspace of $\mathbb{R}^n$, which is called the *eigenspace* of $\boldsymbol{A}$ with respect to λ and is denoted by E_λ. The set of all eigenvalues of $\boldsymbol{A}$ is called the *eigenspectrum*, or just *spectrum*, of $\boldsymbol{A}$.

eigenspace

eigenspectrum

spectrum

If λ is an eigenvalue of $\boldsymbol{A} \in \mathbb{R}^{n \times n}$, then the corresponding eigenspace E_λ is the solution space of the homogeneous system of linear equations $(\boldsymbol{A} - \lambda\boldsymbol{I})\boldsymbol{x} = \boldsymbol{0}$. Geometrically, the eigenvector corresponding to a nonzero eigenvalue points in a direction that is stretched by the linear mapping. The eigenvalue is the factor by which it is stretched. If the eigenvalue is negative, the direction of the stretching is flipped.

Example 4.4 (The Case of the Identity Matrix)

The identity matrix $\boldsymbol{I} \in \mathbb{R}^{n \times n}$ has characteristic polynomial $p_{\boldsymbol{I}}(\lambda) = \det(\boldsymbol{I} - \lambda\boldsymbol{I}) = (1 - \lambda\boldsymbol{I})^n = 0$, which has only one eigenvalue $\lambda = 1$ that occurs n times. Moreover, $\boldsymbol{I}\boldsymbol{x} = \lambda\boldsymbol{x} = 1\boldsymbol{x}$ holds for all vectors $\boldsymbol{x} \in \mathbb{R}^n\backslash\{\boldsymbol{0}\}$. Because of this, the sole eigenspace E_1 of the identity matrix spans n dimensions, and all n standard basis vectors of $\mathbb{R}^n$ are eigenvectors of $\boldsymbol{I}$.

Useful properties regarding eigenvalues and eigenvectors include the following:

- A matrix $\boldsymbol{A}$ and its transpose $\boldsymbol{A}^\top$ possess the same eigenvalues, but not necessarily the same eigenvectors.
- The eigenspace E_λ is the null space of $\boldsymbol{A} - \lambda \boldsymbol{I}$ since

$$\boldsymbol{A}\boldsymbol{x} = \lambda \boldsymbol{x} \iff \boldsymbol{A}\boldsymbol{x} - \lambda \boldsymbol{x} = \boldsymbol{0} \tag{4.27a}$$

$$\iff (\boldsymbol{A} - \lambda \boldsymbol{I})\boldsymbol{x} = \boldsymbol{0} \iff \boldsymbol{x} \in \ker(\boldsymbol{A} - \lambda \boldsymbol{I}). \tag{4.27b}$$

- Similar matrices (see Definition 2.22) possess the same eigenvalues. Therefore, a linear mapping Φ has eigenvalues that are independent of the choice of basis of its transformation matrix. This makes eigenvalues, together with the determinant and the trace, key characteristic parameters of a linear mapping as they are all invariant under basis change.
- Symmetric, positive definite matrices always have positive, real eigenvalues.

Example 4.5 (Computing Eigenvalues, Eigenvectors, and Eigenspaces)

Let us find the eigenvalues and eigenvectors of the 2×2 matrix

$$\boldsymbol{A} = \begin{bmatrix} 4 & 2 \\ 1 & 3 \end{bmatrix}. \tag{4.28}$$

Step 1: Characteristic Polynomial. From our definition of the eigenvector $\boldsymbol{x} \neq \boldsymbol{0}$ and eigenvalue λ of $\boldsymbol{A}$, there will be a vector such that $\boldsymbol{A}\boldsymbol{x} = \lambda \boldsymbol{x}$, i.e., $(\boldsymbol{A} - \lambda \boldsymbol{I})\boldsymbol{x} = \boldsymbol{0}$. Since $\boldsymbol{x} \neq \boldsymbol{0}$, this requires that the kernel (null space) of $\boldsymbol{A} - \lambda \boldsymbol{I}$ contains more elements than just $\boldsymbol{0}$. This means that $\boldsymbol{A} - \lambda \boldsymbol{I}$ is not invertible and therefore $\det(\boldsymbol{A} - \lambda \boldsymbol{I}) = 0$. Hence, we need to compute the roots of the characteristic polynomial (4.22a) to find the eigenvalues.

Step 2: Eigenvalues. The characteristic polynomial is

$$p_{\boldsymbol{A}}(\lambda) = \det(\boldsymbol{A} - \lambda \boldsymbol{I}) \tag{4.29a}$$

$$= \det\left(\begin{bmatrix} 4 & 2 \\ 1 & 3 \end{bmatrix} - \begin{bmatrix} \lambda & 0 \\ 0 & \lambda \end{bmatrix}\right) = \begin{vmatrix} 4-\lambda & 2 \\ 1 & 3-\lambda \end{vmatrix} \tag{4.29b}$$

$$= (4-\lambda)(3-\lambda) - 2 \cdot 1. \tag{4.29c}$$

We factorize the characteristic polynomial and obtain

$$p(\lambda) = (4-\lambda)(3-\lambda) - 2 \cdot 1 = 10 - 7\lambda + \lambda^2 = (2-\lambda)(5-\lambda) \tag{4.30}$$

giving the roots $\lambda_1 = 2$ and $\lambda_2 = 5$.

Step 3: Eigenvectors and Eigenspaces. We find the eigenvectors that correspond to these eigenvalues by looking at vectors $\boldsymbol{x}$ such that

$$\begin{bmatrix} 4-\lambda & 2 \\ 1 & 3-\lambda \end{bmatrix} \boldsymbol{x} = \boldsymbol{0}. \tag{4.31}$$

For $\lambda = 5$, we obtain

$$\begin{bmatrix} 4-5 & 2 \\ 1 & 3-5 \end{bmatrix} \begin{bmatrix} x_1 \\ x_2 \end{bmatrix} = \begin{bmatrix} -1 & 2 \\ 1 & -2 \end{bmatrix} \begin{bmatrix} x_1 \\ x_2 \end{bmatrix} = \boldsymbol{0}. \tag{4.32}$$

We solve this homogeneous system and obtain a solution space

$$E_5 = \text{span}[\begin{bmatrix} 2 \\ 1 \end{bmatrix}]. \tag{4.33}$$

This eigenspace is one-dimensional as it possesses a single basis vector.

Analogously, we find the eigenvector for $\lambda = 2$ by solving the homogeneous system of equations

$$\begin{bmatrix} 4-2 & 2 \\ 1 & 3-2 \end{bmatrix} \boldsymbol{x} = \begin{bmatrix} 2 & 2 \\ 1 & 1 \end{bmatrix} \boldsymbol{x} = \boldsymbol{0}. \tag{4.34}$$

This means any vector $\boldsymbol{x} = \begin{bmatrix} x_1 \\ x_2 \end{bmatrix}$, where $x_2 = -x_1$, such as $\begin{bmatrix} 1 \\ -1 \end{bmatrix}$, is an eigenvector with eigenvalue 2. The corresponding eigenspace is given as

$$E_2 = \text{span}[\begin{bmatrix} 1 \\ -1 \end{bmatrix}]. \tag{4.35}$$

The two eigenspaces E_5 and E_2 in Example 4.5 are one-dimensional as they are each spanned by a single vector. However, in other cases we may have multiple identical eigenvalues (see Definition 4.9) and the eigenspace may have more than one dimension.

Definition 4.11. Let λ_i be an eigenvalue of a square matrix $\boldsymbol{A}$. Then the *geometric multiplicity* of λ_i is the number of linearly independent eigenvectors associated with λ_i. In other words, it is the dimensionality of the eigenspace spanned by the eigenvectors associated with λ_i.

geometric multiplicity

Remark. A specific eigenvalue's geometric multiplicity must be at least one because every eigenvalue has at least one associated eigenvector. An eigenvalue's geometric multiplicity cannot exceed its algebraic multiplicity, but it may be lower. ◇

Example 4.6

The matrix $\boldsymbol{A} = \begin{bmatrix} 2 & 1 \\ 0 & 2 \end{bmatrix}$ has two repeated eigenvalues $\lambda_1 = \lambda_2 = 2$ and an algebraic multiplicity of 2. The eigenvalue has, however, only one distinct unit eigenvector $\boldsymbol{x}_1 = \begin{bmatrix} 1 \\ 0 \end{bmatrix}$ and, thus, geometric multiplicity 1.

4.2.1 Graphical Intuition in Two Dimensions

Let us gain some intuition for determinants, eigenvectors, and eigenvalues using different linear mappings. Figure 4.4 depicts five transformation matrices $\boldsymbol{A}_1, \ldots, \boldsymbol{A}_5$ and their impact on a square grid of points, centered at the origin:

- $\boldsymbol{A}_1 = \begin{bmatrix} \frac{1}{2} & 0 \\ 0 & 2 \end{bmatrix}$. The direction of the two eigenvectors correspond to the canonical basis vectors in $\mathbb{R}^2$, i.e., to two cardinal axes. The vertical axis is extended by a factor of 2 (eigenvalue $\lambda_1 = 2$), and the horizontal axis is compressed by factor $\frac{1}{2}$ (eigenvalue $\lambda_2 = \frac{1}{2}$). The mapping is area preserving ($\det(\boldsymbol{A}_1) = 1 = 2 \cdot \frac{1}{2}$).

In geometry, the area-preserving properties of this type of shearing parallel to an axis is also known as Cavalieri's principle of equal areas for parallelograms (Katz, 2004).

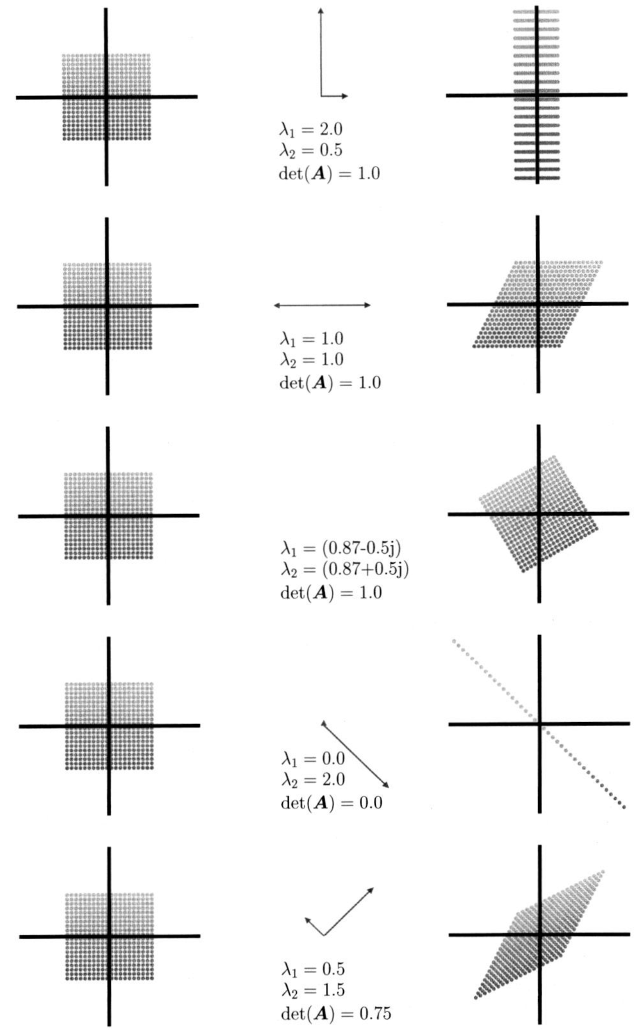

Figure 4.4 Determinants and eigenspaces. Overview of five linear mappings and their associated transformation matrices $\boldsymbol{A}_i \in \mathbb{R}^{2\times 2}$ projecting 400 color-coded points $\boldsymbol{x} \in \mathbb{R}^2$ (left column) onto target points $\boldsymbol{A}_i\boldsymbol{x}$ (right column). The central column depicts the first eigenvector, stretched by its associated eigenvalue λ_1, and the second eigenvector stretched by its eigenvalue λ_2. Each row depicts the effect of one of five transformation matrices $\boldsymbol{A}_i$ with respect to the standard basis.

- $\boldsymbol{A}_2 = \begin{bmatrix} 1 & \frac{1}{2} \\ 0 & 1 \end{bmatrix}$ corresponds to a shearing mapping, i.e., it shears the points along the horizontal axis to the right if they are on the positive half of the vertical axis, and to the left vice versa. This mapping is area preserving ($\det(\boldsymbol{A}_2) = 1$). The eigenvalue $\lambda_1 = 1 = \lambda_2$ is repeated, and the eigenvectors are collinear (drawn here for emphasis in two opposite directions). This indicates that the mapping acts only along one direction (the horizontal axis).
- $\boldsymbol{A}_3 = \begin{bmatrix} \cos(\frac{\pi}{6}) & -\sin(\frac{\pi}{6}) \\ \sin(\frac{\pi}{6}) & \cos(\frac{\pi}{6}) \end{bmatrix} = \frac{1}{2}\begin{bmatrix} \sqrt{3} & -1 \\ 1 & \sqrt{3} \end{bmatrix}$ The matrix $\boldsymbol{A}_3$ rotates the points by $\frac{\pi}{6}$ rad $= 30°$ counter clockwise and has only complex eigenvalues, reflecting that the mapping is a rotation (hence, no eigenvectors are drawn). A rotation has to be volume preserving, and so the determinant is 1. For more details on rotations, we refer to Section 3.9.
- $\boldsymbol{A}_4 = \begin{bmatrix} 1 & -1 \\ -1 & 1 \end{bmatrix}$ represents a mapping in the standard basis that collapses a two-dimensional domain onto one dimension. Since one eigenvalue is 0, the

space in direction of the (blue) eigenvector corresponding to $\lambda_1 = 0$ collapses, while the orthogonal (red) eigenvector stretches space by a factor of $\lambda_2 = 2$. Therefore, the area of the image is 0.

- $\boldsymbol{A}_5 = \begin{bmatrix} 1 & \frac{1}{2} \\ \frac{1}{2} & 1 \end{bmatrix}$ is a shear-and-stretch mapping that shrinks space by 75% since $|\det(\boldsymbol{A}_5)| = \frac{3}{4}$. It stretches space along the (blue) eigenvector of λ_2 by a factor 1.5 and compresses it along the orthogonal (blue) eigenvector by a factor of 0.5.

Example 4.7 (Eigenspectrum of a Biological Neural Network)
Methods to analyze and learn from network data are an essential component of machine learning methods. The key to understanding networks is the connectivity between network nodes, especially if two nodes are connected to each other or not. In data science applications, it is often useful to study the matrix that captures this connectivity data.

We build a connectivity/adjacency matrix $\boldsymbol{A} \in \mathbb{R}^{277 \times 277}$ of the complete neural network of the worm *C.Elegans*. Each row/column represents one of the 277 neurons of this worm's brain. The connectivity matrix $\boldsymbol{A}$ has a value of $a_{ij} = 1$ if neuron i talks to neuron j through a synapse, and $a_{ij} = 0$ otherwise. The connectivity matrix is not symmetric, which implies that eigenvalues may not be real valued. Therefore, we compute a symmetrized version of the connectivity matrix as $\boldsymbol{A}_{sym} := \boldsymbol{A} + \boldsymbol{A}^\top$. This new matrix $\boldsymbol{A}_{sym}$ is shown in Figure 4.5(a) and has a nonzero value a_{ij} if and only if two neurons are connected (white pixels), irrespective of the direction of the connection. In Figure 4.5(b), we show the corresponding eigenspectrum of $\boldsymbol{A}_{sym}$. The horizontal axis shows the index of the eigenvalues, sorted in descending order. The vertical axis shows the corresponding eigenvalue. The S-like shape of this eigenspectrum is typical for many biological neural networks. The underlying mechanism responsible for this is an area of active neuroscience research.

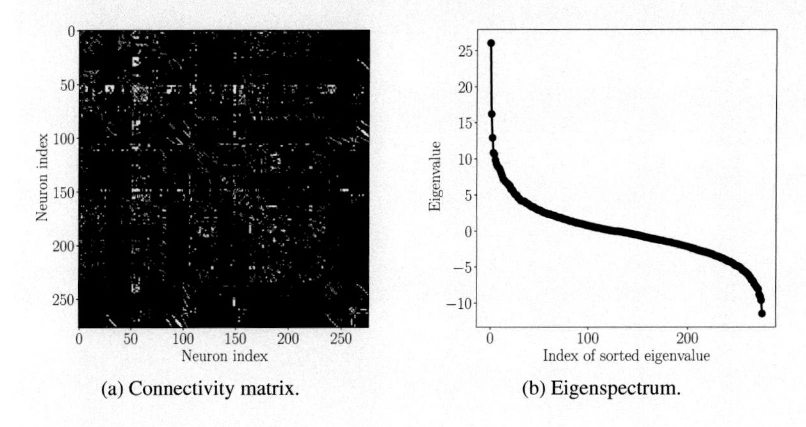

(a) Connectivity matrix. (b) Eigenspectrum.

Figure 4.5 Caenorhabditis elegans neural network (Kaiser and Hilgetag, 2006). (a) Symmetrized connectivity matrix; (b) eigenspectrum.

Theorem 4.12. *The eigenvectors $\boldsymbol{x}_1, \ldots, \boldsymbol{x}_n$ of a matrix $\boldsymbol{A} \in \mathbb{R}^{n \times n}$ with n distinct eigenvalues $\lambda_1, \ldots, \lambda_n$ are linearly independent.*

This theorem states that eigenvectors of a matrix with n distinct eigenvalues form a basis of $\mathbb{R}^n$.

defective

Definition 4.13. A square matrix $\boldsymbol{A} \in \mathbb{R}^{n \times n}$ is *defective* if it possesses fewer than n linearly independent eigenvectors.

A nondefective matrix $\boldsymbol{A} \in \mathbb{R}^{n \times n}$ does not necessarily require n distinct eigenvalues, but it does require that the eigenvectors form a basis of $\mathbb{R}^n$. Looking at the eigenspaces of a defective matrix, it follows that the sum of the dimensions of the eigenspaces is less than n. Specifically, a defective matrix has at least one eigenvalue λ_i with an algebraic multiplicity $m > 1$ a geometric multiplicity of less than m.

Remark. A defective matrix cannot have n distinct eigenvalues, as distinct eigenvalues have linearly independent eigenvectors (Theorem 4.12). ◇

Theorem 4.14. *Given a matrix $\boldsymbol{A} \in \mathbb{R}^{m \times n}$, we can always obtain a symmetric, positive semidefinite matrix $\boldsymbol{S} \in \mathbb{R}^{n \times n}$ by defining*

$$\boldsymbol{S} := \boldsymbol{A}^\top \boldsymbol{A} . \tag{4.36}$$

Remark. If $\text{rk}(\boldsymbol{A}) = n$, then $\boldsymbol{S} := \boldsymbol{A}^\top \boldsymbol{A}$ is symmetric, positive definite. ◇

Understanding why Theorem 4.14 holds is insightful for how we can use symmetrized matrices: Symmetry requires $\boldsymbol{S} = \boldsymbol{S}^\top$, and by inserting (4.36) we obtain $\boldsymbol{S} = \boldsymbol{A}^\top \boldsymbol{A} = \boldsymbol{A}^\top (\boldsymbol{A}^\top)^\top = (\boldsymbol{A}^\top \boldsymbol{A})^\top = \boldsymbol{S}^\top$. Moreover, positive semidefiniteness (Section 3.2.3) requires that $\boldsymbol{x}^\top \boldsymbol{S} \boldsymbol{x} \geqslant 0$ and inserting (4.36) we obtain $\boldsymbol{x}^\top \boldsymbol{S} \boldsymbol{x} = \boldsymbol{x}^\top \boldsymbol{A}^\top \boldsymbol{A} \boldsymbol{x} = (\boldsymbol{x}^\top \boldsymbol{A}^\top)(\boldsymbol{A}\boldsymbol{x}) = (\boldsymbol{A}\boldsymbol{x})^\top (\boldsymbol{A}\boldsymbol{x}) \geqslant 0$, because the dot product computes a sum of squares (which are themselves nonnegative).

spectral theorem

Theorem 4.15 (Spectral Theorem)**.** *If $\boldsymbol{A} \in \mathbb{R}^{n \times n}$ is symmetric, there exists an orthonormal basis of the corresponding vector space V consisting of eigenvectors of $\boldsymbol{A}$, and each eigenvalue is real.*

A direct implication of the spectral theorem is that the eigendecomposition of a symmetric matrix $\boldsymbol{A}$ exists (with real eigenvalues), and that we can find an ONB of eigenvectors so that $\boldsymbol{A} = \boldsymbol{P}\boldsymbol{D}\boldsymbol{P}^\top$, where $\boldsymbol{D}$ is diagonal and the columns of $\boldsymbol{P}$ contain the eigenvectors.

Example 4.8

Consider the matrix

$$\boldsymbol{A} = \begin{bmatrix} 3 & 2 & 2 \\ 2 & 3 & 2 \\ 2 & 2 & 3 \end{bmatrix} . \tag{4.37}$$

The characteristic polynomial of $\boldsymbol{A}$ is

$$p_{\boldsymbol{A}}(\lambda) = -(\lambda - 1)^2(\lambda - 7) , \tag{4.38}$$

so that we obtain the eigenvalues $\lambda_1 = 1$ and $\lambda_2 = 7$, where λ_1 is a repeated eigenvalue. Following our standard procedure for computing eigenvectors, we obtain the eigenspaces

$$E_1 = \text{span}[\underbrace{\begin{bmatrix}-1\\1\\0\end{bmatrix}}_{=:\boldsymbol{x}_1}, \underbrace{\begin{bmatrix}-1\\0\\1\end{bmatrix}}_{=:\boldsymbol{x}_2}], \quad E_7 = \text{span}[\underbrace{\begin{bmatrix}1\\1\\1\end{bmatrix}}_{=:\boldsymbol{x}_3}]. \tag{4.39}$$

We see that $\boldsymbol{x}_3$ is orthogonal to both $\boldsymbol{x}_1$ and $\boldsymbol{x}_2$. However, since $\boldsymbol{x}_1^\top \boldsymbol{x}_2 = 1 \neq 0$, they are not orthogonal. The spectral theorem (Theorem 4.15) states that there exists an orthogonal basis, but the one we have is not orthogonal. However, we can construct one.

To construct such a basis, we exploit the fact that $\boldsymbol{x}_1, \boldsymbol{x}_2$ are eigenvectors associated with the same eigenvalue λ. Therefore, for any $\alpha, \beta \in \mathbb{R}$ it holds that

$$\boldsymbol{A}(\alpha\boldsymbol{x}_1 + \beta\boldsymbol{x}_2) = \boldsymbol{A}\boldsymbol{x}_1\alpha + \boldsymbol{A}\boldsymbol{x}_2\beta = \lambda(\alpha\boldsymbol{x}_1 + \beta\boldsymbol{x}_2), \tag{4.40}$$

i.e., any linear combination of $\boldsymbol{x}_1$ and $\boldsymbol{x}_2$ is also an eigenvector of $\boldsymbol{A}$ associated with λ. The Gram–Schmidt algorithm (Section 3.8.3) is a method for iteratively constructing an orthogonal/orthonormal basis from a set of basis vectors using such linear combinations. Therefore, even if $\boldsymbol{x}_1$ and $\boldsymbol{x}_2$ are not orthogonal, we can apply the Gram–Schmidt algorithm and find eigenvectors associated with $\lambda_1 = 1$ that are orthogonal to each other (and to $\boldsymbol{x}_3$). In our example, we will obtain

$$\boldsymbol{x}_1' = \begin{bmatrix}-1\\1\\0\end{bmatrix}, \quad \boldsymbol{x}_2' = \frac{1}{2}\begin{bmatrix}-1\\-1\\2\end{bmatrix}, \tag{4.41}$$

which are orthogonal to each other, orthogonal to $\boldsymbol{x}_3$, and eigenvectors of $\boldsymbol{A}$ associated with $\lambda_1 = 1$.

Before we conclude our considerations of eigenvalues and eigenvectors, it is useful to tie these matrix characteristics together with the concepts of the determinant and the trace.

Theorem 4.16. *The determinant of a matrix $\boldsymbol{A} \in \mathbb{R}^{n\times n}$ is the product of its eigenvalues, i.e.,*

$$\det(\boldsymbol{A}) = \prod_{i=1}^{n} \lambda_i, \tag{4.42}$$

where λ_i are (possibly repeated) eigenvalues of $\boldsymbol{A}$.

Theorem 4.17. *The trace of a matrix $\boldsymbol{A} \in \mathbb{R}^{n\times n}$ is the sum of its eigenvalues, i.e.,*

$$tr(\boldsymbol{A}) = \sum_{i=1}^{n} \lambda_i, \tag{4.43}$$

where λ_i are (possibly repeated) eigenvalues of $\boldsymbol{A}$.

Let us provide a geometric intuition of these two theorems. Consider a matrix $\boldsymbol{A} \in \mathbb{R}^{2\times 2}$ that possesses two linearly independent eigenvectors $\boldsymbol{x}_1, \boldsymbol{x}_2$. For this

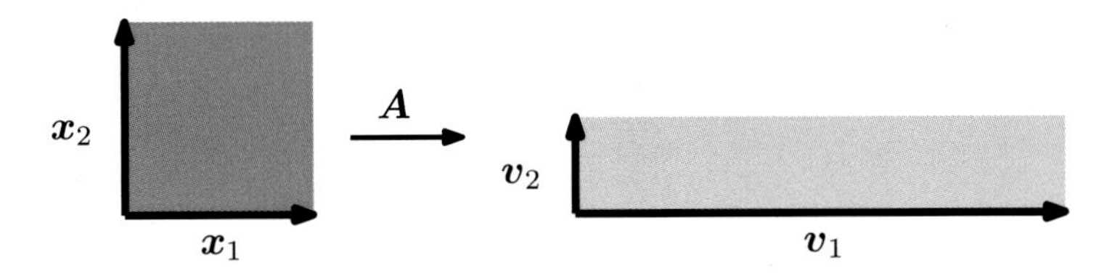

Figure 4.6 Geometric interpretation of eigenvalues. The eigenvectors of $\boldsymbol{A}$ get stretched by the corresponding eigenvalues. The area of the unit square changes by $|\lambda_1\lambda_2|$, the circumference changes by a factor $2(|\lambda_1| + |\lambda_2|)$.

example, we assume $(\boldsymbol{x}_1, \boldsymbol{x}_2)$ are an ONB of $\mathbb{R}^2$ so that they are orthogonal and the area of the square they span is 1; see Figure 4.6. From Section 4.1, we know that the determinant computes the change of area of unit square under the transformation $\boldsymbol{A}$. In this example, we can compute the change of area explicitly: Mapping the eigenvectors using $\boldsymbol{A}$ gives us vectors $\boldsymbol{v}_1 = \boldsymbol{A}\boldsymbol{x}_1 = \lambda_1\boldsymbol{x}_1$ and $\boldsymbol{v}_2 = \boldsymbol{A}\boldsymbol{x}_2 = \lambda_2\boldsymbol{x}_2$, i.e., the new vectors $\boldsymbol{v}_i$ are scaled versions of the eigenvectors $\boldsymbol{x}_i$, and the scaling factors are the corresponding eigenvalues λ_i. $\boldsymbol{v}_1, \boldsymbol{v}_2$ are still orthogonal, and the area of the rectangle they span is $|\lambda_1\lambda_2|$.

Given that $\boldsymbol{x}_1, \boldsymbol{x}_2$ (in our example) are orthonormal, we can directly compute the circumference of the unit square as $2(1+1)$. Mapping the eigenvectors using $\boldsymbol{A}$ creates a rectangle whose circumference is $2(|\lambda_1| + |\lambda_2|)$. Therefore, the sum of the absolute values of the eigenvalues tells us how the circumference of the unit square changes under the transformation matrix $\boldsymbol{A}$.

Example 4.9 (Google's PageRank – Webpages as Eigenvectors)
Google uses the eigenvector corresponding to the maximal eigenvalue of a matrix $\boldsymbol{A}$ to determine the rank of a page for search. The idea for the PageRank algorithm, developed at Stanford University by Larry Page and Sergey Brin in 1996, was that the importance of any web page can be approximated by the importance of pages that link to it. For this, they write down all web sites as a huge directed graph that shows which page links to which. PageRank computes the weight (importance) $x_i \geqslant 0$ of a web site a_i by counting the number of pages pointing to a_i. Moreover, PageRank takes into account the importance of the web sites that link to a_i. The navigation behavior of a user is then modeled by a transition matrix $\boldsymbol{A}$ of this graph that tells us with what (click) probability somebody will end up on a different web site. The matrix $\boldsymbol{A}$ has the property that for any initial rank/importance vector $\boldsymbol{x}$ of a web site the sequence $\boldsymbol{x}, \boldsymbol{A}\boldsymbol{x}, \boldsymbol{A}^2\boldsymbol{x}, \ldots$ converges to a vector $\boldsymbol{x}^*$. This vector is called the *PageRank* and satisfies $\boldsymbol{A}\boldsymbol{x}^* = \boldsymbol{x}^*$, i.e., it is an eigenvector (with corresponding eigenvalue 1) of $\boldsymbol{A}$. After normalizing $\boldsymbol{x}^*$, such that $\|\boldsymbol{x}^*\| = 1$, we can interpret the entries as probabilities. More details and different perspectives on PageRank can be found in the original technical report (Page et al., 1999).

PageRank

4.3 Cholesky Decomposition

There are many ways to factorize special types of matrices that we encounter often in machine learning. In the positive real numbers, we have the square-root operation that gives us a decomposition of the number into identical components, e.g., $9 = 3 \cdot 3$. For matrices, we need to be careful that we compute a square-root-like operation on positive quantities. For symmetric, positive definite matrices (see Section 3.2.3), we can choose from a number of square-root equivalent operations. The *Cholesky decomposition/Cholesky factorization* provides a square-root equivalent operation on symmetric, positive definite matrices that is useful in practice.

Cholesky decomposition

Cholesky factorization

Theorem 4.18 (Cholesky Decomposition). *A symmetric, positive definite matrix* $\boldsymbol{A}$ *can be factorized into a product* $\boldsymbol{A} = \boldsymbol{L}\boldsymbol{L}^\top$*, where* $\boldsymbol{L}$ *is a lower-triangular matrix with positive diagonal elements:*

$$\begin{bmatrix} a_{11} & \cdots & a_{1n} \\ \vdots & \ddots & \vdots \\ a_{n1} & \cdots & a_{nn} \end{bmatrix} = \begin{bmatrix} l_{11} & \cdots & 0 \\ \vdots & \ddots & \vdots \\ l_{n1} & \cdots & l_{nn} \end{bmatrix} \begin{bmatrix} l_{11} & \cdots & l_{n1} \\ \vdots & \ddots & \vdots \\ 0 & \cdots & l_{nn} \end{bmatrix} . \tag{4.44}$$

$\boldsymbol{L}$ *is called the Cholesky factor of* $\boldsymbol{A}$*, and* $\boldsymbol{L}$ *is unique.*

Cholesky factor

Example 4.10 (Cholesky Factorization)

Consider a symmetric, positive definite matrix $\boldsymbol{A} \in \mathbb{R}^{3\times 3}$. We are interested in finding its Cholesky factorization $\boldsymbol{A} = \boldsymbol{L}\boldsymbol{L}^\top$, i.e.,

$$\boldsymbol{A} = \begin{bmatrix} a_{11} & a_{21} & a_{31} \\ a_{21} & a_{22} & a_{32} \\ a_{31} & a_{32} & a_{33} \end{bmatrix} = \boldsymbol{L}\boldsymbol{L}^\top = \begin{bmatrix} l_{11} & 0 & 0 \\ l_{21} & l_{22} & 0 \\ l_{31} & l_{32} & l_{33} \end{bmatrix} \begin{bmatrix} l_{11} & l_{21} & l_{31} \\ 0 & l_{22} & l_{32} \\ 0 & 0 & l_{33} \end{bmatrix} . \tag{4.45}$$

Multiplying out the right-hand side yields

$$\boldsymbol{A} = \begin{bmatrix} l_{11}^2 & l_{21}l_{11} & l_{31}l_{11} \\ l_{21}l_{11} & l_{21}^2 + l_{22}^2 & l_{31}l_{21} + l_{32}l_{22} \\ l_{31}l_{11} & l_{31}l_{21} + l_{32}l_{22} & l_{31}^2 + l_{32}^2 + l_{33}^2 \end{bmatrix} . \tag{4.46}$$

Comparing the left-hand side of (4.45) and the right-hand side of (4.46) shows that there is a simple pattern in the diagonal elements l_{ii}:

$$l_{11} = \sqrt{a_{11}}\,, \quad l_{22} = \sqrt{a_{22} - l_{21}^2}\,, \quad l_{33} = \sqrt{a_{33} - (l_{31}^2 + l_{32}^2)}\,. \tag{4.47}$$

Similarly for the elements below the diagonal (l_{ij}, where $i > j$), there is also a repeating pattern:

$$l_{21} = \frac{1}{l_{11}} a_{21}\,, \quad l_{31} = \frac{1}{l_{11}} a_{31}\,, \quad l_{32} = \frac{1}{l_{22}} (a_{32} - l_{31}l_{21})\,. \tag{4.48}$$

Thus, we constructed the Cholesky decomposition for any symmetric, positive definite 3×3 matrix. The key realization is that we can backward calculate what the components l_{ij} for the $\boldsymbol{L}$ should be, given the values a_{ij} for $\boldsymbol{A}$ and previously computed values of l_{ij}.

The Cholesky decomposition is an important tool for the numerical computations underlying machine learning. Here, symmetric positive definite matrices require frequent manipulation, e.g., the covariance matrix of a multivariate Gaussian variable (see Section 6.5) is symmetric, positive definite. The Cholesky factorization of this covariance matrix allows us to generate samples from a Gaussian distribution. It also allows us to perform a linear transformation of random variables, which is heavily exploited when computing gradients in deep stochastic models, such as the variational auto-encoder (Jimenez Rezende et al., 2014; Kingma and Welling, 2014). The Cholesky decomposition also allows us to compute determinants very efficiently. Given the Cholesky decomposition $\boldsymbol{A} = \boldsymbol{L}\boldsymbol{L}^\top$, we know that $\det(\boldsymbol{A}) = \det(\boldsymbol{L})\det(\boldsymbol{L}^\top) = \det(\boldsymbol{L})^2$. Since $\boldsymbol{L}$ is a triangular matrix, the determinant is simply the product of its diagonal entries so that $\det(\boldsymbol{A}) = \prod_i l_{ii}^2$. Thus, many numerical software packages use the Cholesky decomposition to make computations more efficient.

4.4 Eigendecomposition and Diagonalization

diagonal matrix

A *diagonal matrix* is a matrix that has value zero on all off-diagonal elements, i.e., they are of the form

$$\boldsymbol{D} = \begin{bmatrix} c_1 & \cdots & 0 \\ \vdots & \ddots & \vdots \\ 0 & \cdots & c_n \end{bmatrix}. \tag{4.49}$$

They allow fast computation of determinants, powers, and inverses. The determinant is the product of its diagonal entries, a matrix power $\boldsymbol{D}^k$ is given by each diagonal element raised to the power k, and the inverse $\boldsymbol{D}^{-1}$ is the reciprocal of its diagonal elements if all of them are nonzero.

In this section, we will discuss how to transform matrices into diagonal form. This is an important application of the basis change we discussed in Section 2.7.2 and eigenvalues from Section 4.2.

Recall that two matrices $\boldsymbol{A}, \boldsymbol{D}$ are similar (Definition 2.22) if there exists an invertible matrix $\boldsymbol{P}$, such that $\boldsymbol{D} = \boldsymbol{P}^{-1}\boldsymbol{A}\boldsymbol{P}$. More specifically, we will look at matrices $\boldsymbol{A}$ that are similar to diagonal matrices $\boldsymbol{D}$ that contain the eigenvalues of $\boldsymbol{A}$ on the diagonal.

diagonalizable

Definition 4.19 (Diagonalizable). A matrix $\boldsymbol{A} \in \mathbb{R}^{n\times n}$ is *diagonalizable* if it is similar to a diagonal matrix, i.e., if there exists an invertible matrix $\boldsymbol{P} \in \mathbb{R}^{n\times n}$ such that $\boldsymbol{D} = \boldsymbol{P}^{-1}\boldsymbol{A}\boldsymbol{P}$.

In the following, we will see that diagonalizing a matrix $\boldsymbol{A} \in \mathbb{R}^{n\times n}$ is a way of expressing the same linear mapping but in another basis (see Section 2.6.1), which will turn out to be a basis that consists of the eigenvectors of $\boldsymbol{A}$.

Let $\boldsymbol{A} \in \mathbb{R}^{n\times n}$, let $\lambda_1, \dots, \lambda_n$ be a set of scalars, and let $\boldsymbol{p}_1, \dots, \boldsymbol{p}_n$ be a set of vectors in $\mathbb{R}^n$. We define $\boldsymbol{P} := [\boldsymbol{p}_1, \dots, \boldsymbol{p}_n]$ and let $\boldsymbol{D} \in \mathbb{R}^{n\times n}$ be a diagonal matrix with diagonal entries $\lambda_1, \dots, \lambda_n$. Then we can show that

$$\boldsymbol{A}\boldsymbol{P} = \boldsymbol{P}\boldsymbol{D} \tag{4.50}$$

if and only if $\lambda_1, \dots, \lambda_n$ are the eigenvalues of $\boldsymbol{A}$ and $\boldsymbol{p}_1, \dots, \boldsymbol{p}_n$ are corresponding eigenvectors of $\boldsymbol{A}$.

We can see that this statement holds because

$$AP = A[p_1, \ldots, p_n] = [Ap_1, \ldots, Ap_n], \tag{4.51}$$

$$PD = [p_1, \ldots, p_n] \begin{bmatrix} \lambda_1 & & 0 \\ & \ddots & \\ 0 & & \lambda_n \end{bmatrix} = [\lambda_1 p_1, \ldots, \lambda_n p_n]. \tag{4.52}$$

Thus, (4.50) implies that

$$Ap_1 = \lambda_1 p_1 \tag{4.53}$$

$$\vdots$$

$$Ap_n = \lambda_n p_n. \tag{4.54}$$

Therefore, the columns of P must be eigenvectors of A.

Our definition of diagonalization requires that $P \in \mathbb{R}^{n \times n}$ is invertible, i.e., P has full rank (Theorem 4.3). This requires us to have n linearly independent eigenvectors $p_1, \ldots, p_n$, i.e., the p_i form a basis of $\mathbb{R}^n$.

Theorem 4.20 (Eigendecomposition). *A square matrix $A \in \mathbb{R}^{n \times n}$ can be factored into*

$$A = PDP^{-1}, \tag{4.55}$$

where $P \in \mathbb{R}^{n \times n}$ and D is a diagonal matrix whose diagonal entries are the eigenvalues of A, if and only if the eigenvectors of A form a basis of $\mathbb{R}^n$.

Theorem 4.20 implies that only nondefective matrices can be diagonalized and that the columns of P are the n eigenvectors of A. For symmetric matrices, we can obtain even stronger outcomes for the eigenvalue decomposition.

Theorem 4.21. *A symmetric matrix $S \in \mathbb{R}^{n \times n}$ can always be diagonalized.*

Theorem 4.21 follows directly from the spectral theorem 4.15. Moreover, the spectral theorem states that we can find an ONB of eigenvectors of $\mathbb{R}^n$. This makes P an orthogonal matrix so that $D = P^\top AP$.

Remark. The Jordan normal form of a matrix offers a decomposition that works for defective matrices (Lang, 1987) but is beyond the scope of this book. ◊

4.4.1 Geometric Intuition for the Eigendecomposition

We can interpret the eigendecomposition of a matrix as follows (see also Figure 4.7): Let A be the transformation matrix of a linear mapping with respect to the standard basis. P^{-1} performs a basis change from the standard basis into the eigenbasis. This identifies the eigenvectors p_i (red and orange arrows in Figure 4.7) onto the standard basis vectors e_i. Then the diagonal D scales the vectors along these axes by the eigenvalues λ_i. Finally, P transforms these scaled vectors back into the standard/canonical coordinates yielding $\lambda_i p_i$.

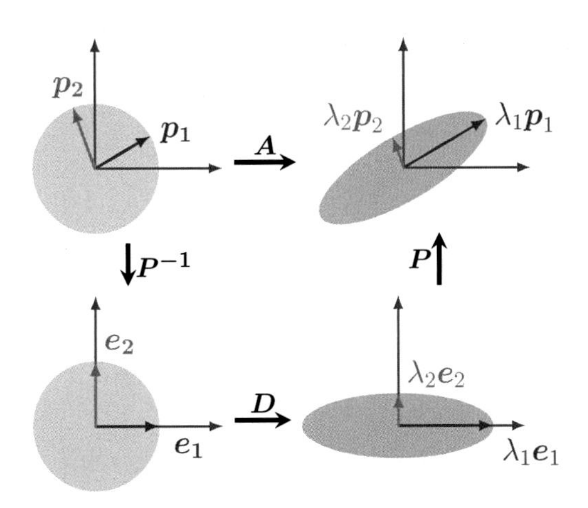

Figure 4.7 Intuition behind the eigendecomposition as sequential transformations. Top-left to bottom-left: $\boldsymbol{P}^{-1}$ performs a basis change (here drawn in $\mathbb{R}^2$ and depicted as a rotationlike operation), mapping the eigenvectors into the standard basis. Bottom-left to bottom-right: $\boldsymbol{D}$ performs a scaling along the remapped orthogonal eigenvectors, depicted here by a circle being stretched to an ellipse. Bottom-right to top-right: $\boldsymbol{P}$ undoes the basis change (depicted as a reverse rotation) and restores the original coordinate frame.

Example 4.11 (Eigendecomposition)

Let us compute the eigendecomposition of $\boldsymbol{A} = \begin{bmatrix} 2 & 1 \\ 1 & 2 \end{bmatrix}$.

Step 1: Compute eigenvalues and eigenvectors. The characteristic polynomial of $\boldsymbol{A}$ is

$$\det(\boldsymbol{A} - \lambda \boldsymbol{I}) = \det\left(\begin{bmatrix} 2-\lambda & 1 \\ 1 & 2-\lambda \end{bmatrix}\right) \tag{4.56a}$$

$$= (2-\lambda)^2 - 1 = \lambda^2 - 4\lambda + 3 = (\lambda - 3)(\lambda - 1) . \tag{4.56b}$$

Therefore, the eigenvalues of $\boldsymbol{A}$ are $\lambda_1 = 1$ and $\lambda_2 = 3$ (the roots of the characteristic polynomial), and the associated (normalized) eigenvectors are obtained via

$$\begin{bmatrix} 2 & 1 \\ 1 & 2 \end{bmatrix} \boldsymbol{p}_1 = 1 \boldsymbol{p}_1 , \quad \begin{bmatrix} 2 & 1 \\ 1 & 2 \end{bmatrix} \boldsymbol{p}_2 = 3 \boldsymbol{p}_2 . \tag{4.57}$$

This yields

$$\boldsymbol{p}_1 = \frac{1}{\sqrt{2}} \begin{bmatrix} 1 \\ -1 \end{bmatrix} , \quad \boldsymbol{p}_2 = \frac{1}{\sqrt{2}} \begin{bmatrix} 1 \\ 1 \end{bmatrix} . \tag{4.58}$$

Step 2: Check for existence. The eigenvectors $\boldsymbol{p}_1, \boldsymbol{p}_2$ form a basis of $\mathbb{R}^2$. Therefore, $\boldsymbol{A}$ can be diagonalized.

Step 3: Construct the matrix $\boldsymbol{P}$ to diagonalize $\boldsymbol{A}$. We collect the eigenvectors of $\boldsymbol{A}$ in $\boldsymbol{P}$ so that

$$\boldsymbol{P} = [\boldsymbol{p}_1, \boldsymbol{p}_2] = \frac{1}{\sqrt{2}} \begin{bmatrix} 1 & 1 \\ -1 & 1 \end{bmatrix} . \tag{4.59}$$

We then obtain

$$\boldsymbol{P}^{-1} \boldsymbol{A} \boldsymbol{P} = \begin{bmatrix} 1 & 0 \\ 0 & 3 \end{bmatrix} = \boldsymbol{D} . \tag{4.60}$$

Equivalently, we get (exploiting that $\boldsymbol{P}^{-1} = \boldsymbol{P}^\top$ since the eigenvectors $\boldsymbol{p}_1$ and $\boldsymbol{p}_2$ in this example form an ONB)

$$\underbrace{\begin{bmatrix} 2 & 1 \\ 1 & 2 \end{bmatrix}}_{\boldsymbol{A}} = \underbrace{\frac{1}{\sqrt{2}} \begin{bmatrix} 1 & 1 \\ -1 & 1 \end{bmatrix}}_{\boldsymbol{P}} \underbrace{\begin{bmatrix} 1 & 0 \\ 0 & 3 \end{bmatrix}}_{\boldsymbol{D}} \underbrace{\frac{1}{\sqrt{2}} \begin{bmatrix} 1 & -1 \\ 1 & 1 \end{bmatrix}}_{\boldsymbol{P}^\top} . \tag{4.61}$$

- Diagonal matrices $\boldsymbol{D}$ can efficiently be raised to a power. Therefore, we can find a matrix power for a matrix $\boldsymbol{A} \in \mathbb{R}^{n \times n}$ via the eigenvalue decomposition (if it exists) so that

$$\boldsymbol{A}^k = (\boldsymbol{P}\boldsymbol{D}\boldsymbol{P}^{-1})^k = \boldsymbol{P}\boldsymbol{D}^k\boldsymbol{P}^{-1}\,. \tag{4.62}$$

  Computing $\boldsymbol{D}^k$ is efficient because we apply this operation individually to any diagonal element.
- Assume that the eigendecomposition $\boldsymbol{A} = \boldsymbol{P}\boldsymbol{D}\boldsymbol{P}^{-1}$ exists. Then,

$$\det(\boldsymbol{A}) = \det(\boldsymbol{P}\boldsymbol{D}\boldsymbol{P}^{-1}) = \det(\boldsymbol{P})\det(\boldsymbol{D})\det(\boldsymbol{P}^{-1}) \tag{4.63a}$$

$$= \det(\boldsymbol{D}) = \prod_i d_{ii} \tag{4.63b}$$

  allows for an efficient computation of the determinant of $\boldsymbol{A}$.

The eigenvalue decomposition requires square matrices. It would be useful to perform a decomposition on general matrices. In the next section, we introduce a more general matrix decomposition technique, the singular value decomposition.

4.5 Singular Value Decomposition

The singular value decomposition (SVD) of a matrix is a central matrix decomposition method in linear algebra. It has been referred to as the "fundamental theorem of linear algebra" (Strang, 1993) because it can be applied to all matrices, not only to square matrices, and it always exists. Moreover, as we will explore in the following, the SVD of a matrix $\boldsymbol{A}$, which represents a linear mapping $\Phi : V \to W$, quantifies the change between the underlying geometry of these two vector spaces. We recommend the work by Kalman (1996) and Roy and Banerjee (2014) for a deeper overview of the mathematics of the SVD.

SVD theorem

Theorem 4.22 (SVD Theorem). *Let $\boldsymbol{A}^{m \times n}$ be a rectangular matrix of rank $r \in [0, \min(m, n)]$. The SVD of $\boldsymbol{A}$ is a decomposition of the form*

SVD
singular value decomposition

$$\underset{m \times n}{\boldsymbol{A}} = \underset{m \times m}{\boldsymbol{U}}\;\underset{m \times n}{\boldsymbol{\Sigma}}\;\underset{n \times n}{\boldsymbol{V}^\top} \tag{4.64}$$

with an orthogonal matrix $\boldsymbol{U} \in \mathbb{R}^{m \times m}$ with column vectors $\boldsymbol{u}_i$, $i = 1, \ldots, m$, and an orthogonal matrix $\boldsymbol{V} \in \mathbb{R}^{n \times n}$ with column vectors $\boldsymbol{v}_j$, $j = 1, \ldots, n$. Moreover, $\boldsymbol{\Sigma}$ is an $m \times n$ matrix with $\Sigma_{ii} = \sigma_i \geqslant 0$ and $\Sigma_{ij} = 0$, $i \neq j$.

singular values
left-singular vectors
right-singular vectors
singular value matrix

The diagonal entries σ_i, $i = 1, \ldots, r$, of $\boldsymbol{\Sigma}$ are called the *singular values*, $\boldsymbol{u}_i$ are called the *left-singular vectors*, and $\boldsymbol{v}_j$ are called the *right-singular vectors*. By convention, the singular values are ordered, i.e., $\sigma_1 \geqslant \sigma_2 \geqslant \sigma_r \geqslant 0$.

The *singular value matrix* $\boldsymbol{\Sigma}$ is unique, but it requires some attention. Observe that the $\boldsymbol{\Sigma} \in \mathbb{R}^{m \times n}$ is rectangular. In particular, $\boldsymbol{\Sigma}$ is of the same size as $\boldsymbol{A}$. This means that $\boldsymbol{\Sigma}$ has a diagonal submatrix that contains the singular values and needs additional zero padding. Specifically, if $m > n$, then the matrix $\boldsymbol{\Sigma}$ has

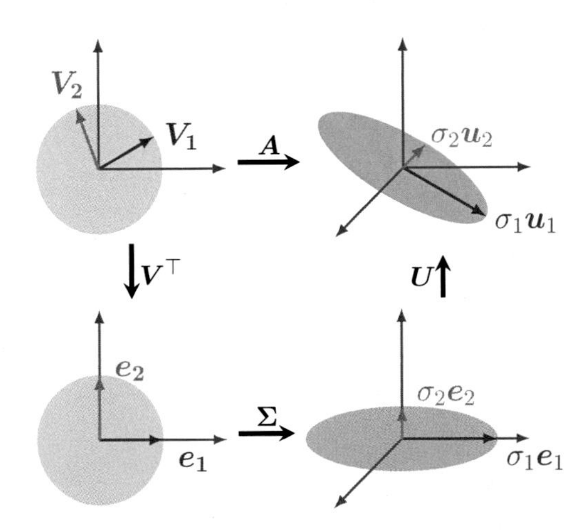

Figure 4.8 Intuition behind the SVD of a matrix $A \in \mathbb{R}^{3\times 2}$ as sequential transformations. Top-left to bottom-left: $V^\top$ performs a basis change in $\mathbb{R}^2$. Bottom-left to bottom-right: Σ scales and maps from $\mathbb{R}^2$ to $\mathbb{R}^3$. The ellipse in the bottom-right lives in $\mathbb{R}^3$. The third dimension is orthogonal to the surface of the elliptical disk. Bottom-right to top-right: U performs a basis change within $\mathbb{R}^3$.

diagonal structure up to row n and then consists of $\mathbf{0}^\top$ row vectors from $n+1$ to m below so that

$$\Sigma = \begin{bmatrix} \sigma_1 & 0 & 0 \\ 0 & \ddots & 0 \\ 0 & 0 & \sigma_n \\ 0 & \dots & 0 \\ \vdots & & \vdots \\ 0 & \dots & 0 \end{bmatrix}. \tag{4.65}$$

If $m < n$, the matrix Σ has a diagonal structure up to column m and columns that consist of $\mathbf{0}$ from $m+1$ to n:

$$\Sigma = \begin{bmatrix} \sigma_1 & 0 & 0 & 0 & \dots & 0 \\ 0 & \ddots & 0 & 0 & & 0 \\ 0 & 0 & \sigma_m & 0 & \dots & 0 \end{bmatrix}. \tag{4.66}$$

Remark. The SVD exists for any matrix $A \in \mathbb{R}^{m\times n}$. ◇

4.5.1 Geometric Intuitions for the SVD

The SVD offers geometric intuitions to describe a transformation matrix A. In the following, we will discuss the SVD as sequential linear transformations performed on the bases. In Example 4.12, we will then apply transformation matrices of the SVD to a set of vectors in $\mathbb{R}^2$, which allows us to visualize the effect of each transformation more clearly.

The SVD of a matrix can be interpreted as a decomposition of a corresponding linear mapping (recall Section 2.7.1) $\Phi : \mathbb{R}^n \to \mathbb{R}^m$ into three operations; see Figure 4.8. The SVD intuition follows superficially a similar structure to our eigendecomposition intuition; see Figure 4.7: Broadly speaking, the SVD performs a basis change via $V^\top$) followed by a scaling and augmentation (or reduction) in dimensionality via the singular value matrix Σ. Finally, it performs a second basis change via $\boldsymbol{U}$. The SVD entails a number of important details and caveats, which is why we will review our intuition in more detail.

Assume we are given a transformation matrix of a linear mapping $\Phi : \mathbb{R}^n \to \mathbb{R}^m$ with respect to the standard bases B and C of $\mathbb{R}^n$ and $\mathbb{R}^m$, respectively. Moreover, assume a second basis $\tilde{B}$ of $\mathbb{R}^n$ and $\tilde{C}$ of $\mathbb{R}^m$. Then

It is useful to revise basis changes (Section 2.7.2), orthogonal matrices (Definition 3.8), and orthonormal bases (Section 3.5).

1. The matrix $\boldsymbol{V}$ performs a basis change in the domain $\mathbb{R}^n$ from $\tilde{B}$ (represented by the red and orange vectors $\boldsymbol{v}_1$ and $\boldsymbol{v}_2$ in the top-left of Figure 4.8) to the standard basis B. $\boldsymbol{V}^\top = \boldsymbol{V}^{-1}$ performs a basis change from B to $\tilde{B}$. The red and orange vectors are now aligned with the canonical basis in the bottom-left of Figure 4.8.
2. Having changed the coordinate system to $\tilde{B}$, $\boldsymbol{\Sigma}$ scales the new coordinates by the singular values σ_i (and adds or deletes dimensions), i.e., $\boldsymbol{\Sigma}$ is the transformation matrix of Φ with respect to $\tilde{B}$ and $\tilde{C}$, represented by the red and orange vectors being stretched and lying in the $\boldsymbol{e}_1$-$\boldsymbol{e}_2$ plane, which is now embedded in a third dimension in the bottom-right of Figure 4.8.
3. $\boldsymbol{U}$ performs a basis change in the codomain $\mathbb{R}^m$ from $\tilde{C}$ into the canonical basis of $\mathbb{R}^m$, represented by a rotation of the red and orange vectors out of the $\boldsymbol{e}_1$-$\boldsymbol{e}_2$ plane. This is shown in the top-right of Figure 4.8.

The SVD expresses a change of basis in both the domain and codomain. This is in contrast with the eigendecomposition that operates within the same vector space, where the same basis change is applied and then undone. What makes the SVD special is that these two different bases are simultaneously linked by the singular value matrix $\boldsymbol{\Sigma}$.

Example 4.12 (Vectors and the SVD)

Consider a mapping of a square grid of vectors $\mathcal{X} \in \mathbb{R}^2$ that fit in a box of size 2×2 centered at the origin. Using the standard basis, we map these vectors using

$$\boldsymbol{A} = \begin{bmatrix} 1 & -0.8 \\ 0 & 1 \\ 1 & 0 \end{bmatrix} = \boldsymbol{U\Sigma V}^\top \tag{4.67a}$$

$$= \begin{bmatrix} -0.79 & 0 & -0.62 \\ 0.38 & -0.78 & -0.49 \\ -0.48 & -0.62 & 0.62 \end{bmatrix} \begin{bmatrix} 1.62 & 0 \\ 0 & 1.0 \\ 0 & 0 \end{bmatrix} \begin{bmatrix} -0.78 & 0.62 \\ -0.62 & -0.78 \end{bmatrix} . \tag{4.67b}$$

We start with a set of vectors $\mathcal{X}$ (colored dots; see the top-left panel of Figure 4.9) arranged in a grid. We then apply $\boldsymbol{V}^\top \in \mathbb{R}^{2\times 2}$, which rotates $\mathcal{X}$. The rotated vectors are shown in the bottom-left panel of Figure 4.9. We now map these vectors using the singular value matrix $\boldsymbol{\Sigma}$ to the codomain $\mathbb{R}^3$ (see the bottom-right panel in Figure 4.9). Note that all vectors lie in the x_1-x_2 plane. The third coordinate is always 0. The vectors in the x_1-x_2 plane have been stretched by the singular values.

The direct mapping of the vectors $\mathcal{X}$ by $\boldsymbol{A}$ to the codomain $\mathbb{R}^3$ equals the transformation of $\mathcal{X}$ by $\boldsymbol{U\Sigma V}^\top$, where $\boldsymbol{U}$ performs a rotation within the codomain $\mathbb{R}^3$ so that the mapped vectors are no longer restricted to the x_1-x_2 plane; they still are on a plane as shown in the top-right panel of Figure 4.9.

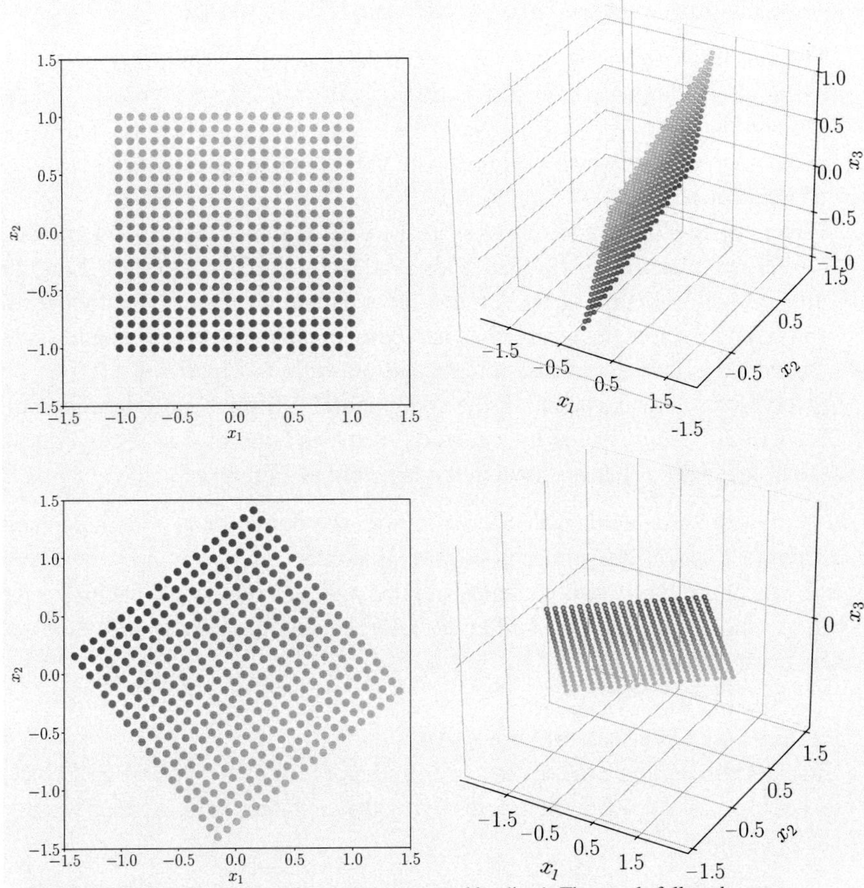

Figure 4.9 SVD and mapping of vectors (represented by discs). The panels follow the same counter clockwise structure of Figure 4.8.

4.5.2 Construction of the SVD

We will next discuss why the SVD exists and show how to compute it in detail. The SVD of a general matrix shares some similarities with the eigendecomposition of a square matrix.

Remark. Compare the eigendecomposition of an SPD matrix

$$\boldsymbol{S} = \boldsymbol{S}^{\top} = \boldsymbol{P}\boldsymbol{D}\boldsymbol{P}^{\top} \tag{4.68}$$

with the corresponding SVD

$$\boldsymbol{S} = \boldsymbol{U}\boldsymbol{\Sigma}\boldsymbol{V}^{\top} . \tag{4.69}$$

If we set

$$\boldsymbol{U} = \boldsymbol{P} = \boldsymbol{V} , \quad \boldsymbol{D} = \boldsymbol{\Sigma} , \tag{4.70}$$

we see that the SVD of SPD matrices is their eigendecomposition. ♢

In the following, we will explore why Theorem 4.22 holds and how the SVD is constructed. Computing the SVD of $\boldsymbol{A} \in \mathbb{R}^{m \times n}$ is equivalent to finding two sets of orthonormal bases $U = (\boldsymbol{u}_1, \dots, \boldsymbol{u}_m)$ and $V = (\boldsymbol{v}_1, \dots, \boldsymbol{v}_n)$ of the codomain $\mathbb{R}^m$ and the domain $\mathbb{R}^n$, respectively. From these ordered bases, we will construct the matrices $\boldsymbol{U}$ and $\boldsymbol{V}$.

Our plan is to start with constructing the orthonormal set of right-singular vectors $\boldsymbol{v}_1, \dots, \boldsymbol{v}_n \in \mathbb{R}^n$. We then construct the orthonormal set of left-singular vectors $\boldsymbol{u}_1, \dots, \boldsymbol{u}_m \in \mathbb{R}^m$. Thereafter, we will link the two and require that the orthogonality of the $\boldsymbol{v}_i$ is preserved under the transformation of $\boldsymbol{A}$. This is important because we know that the images $\boldsymbol{A}\boldsymbol{v}_i$ form a set of orthogonal vectors. We will then normalize these images by scalar factors, which will turn out to be the singular values.

Let us begin with constructing the right-singular vectors. The spectral theorem (Theorem 4.15) tells us that a symmetric matrix possesses an ONB of eigenvectors, which also means it can be diagonalized. Moreover, from Theorem 4.14 we can always construct a symmetric, positive semi-definite matrix $\boldsymbol{A}^\top \boldsymbol{A} \in \mathbb{R}^{n \times n}$ from any rectangular matrix $\boldsymbol{A} \in \mathbb{R}^{m \times n}$. Thus, we can always diagonalize $\boldsymbol{A}^\top \boldsymbol{A}$ and obtain

$$\boldsymbol{A}^\top \boldsymbol{A} = \boldsymbol{P}\boldsymbol{D}\boldsymbol{P}^\top = \boldsymbol{P} \begin{bmatrix} \lambda_1 & \cdots & 0 \\ \vdots & \ddots & \vdots \\ 0 & \cdots & \lambda_n \end{bmatrix} \boldsymbol{P}^\top , \tag{4.71}$$

where $\boldsymbol{P}$ is an orthogonal matrix, which is composed of the orthonormal eigenbasis. The $\lambda_i \geqslant 0$ are the eigenvalues of $\boldsymbol{A}^\top \boldsymbol{A}$. Let us assume the SVD of $\boldsymbol{A}$ exists and inject (4.64) into (4.71). This yields

$$\boldsymbol{A}^\top \boldsymbol{A} = (\boldsymbol{U}\boldsymbol{\Sigma}\boldsymbol{V}^\top)^\top (\boldsymbol{U}\boldsymbol{\Sigma}\boldsymbol{V}^\top) = \boldsymbol{V}\boldsymbol{\Sigma}^\top \boldsymbol{U}^\top \boldsymbol{U}\boldsymbol{\Sigma}\boldsymbol{V}^\top , \tag{4.72}$$

where $\boldsymbol{U}, \boldsymbol{V}$ are orthogonal matrices. Therefore, with $\boldsymbol{U}^\top \boldsymbol{U} = \boldsymbol{I}$ we obtain

$$\boldsymbol{A}^\top \boldsymbol{A} = \boldsymbol{V}\boldsymbol{\Sigma}^\top \boldsymbol{\Sigma}\boldsymbol{V}^\top = \boldsymbol{V} \begin{bmatrix} \sigma_1^2 & 0 & 0 \\ 0 & \ddots & 0 \\ 0 & 0 & \sigma_n^2 \end{bmatrix} \boldsymbol{V}^\top . \tag{4.73}$$

Comparing now (4.71) and (4.73), we identify

$$\boldsymbol{V}^\top = \boldsymbol{P}^\top , \tag{4.74}$$

$$\sigma_i^2 = \lambda_i . \tag{4.75}$$

Therefore, the eigenvectors of $\boldsymbol{A}^\top \boldsymbol{A}$ that compose $\boldsymbol{P}$ are the right-singular vectors $\boldsymbol{V}$ of $\boldsymbol{A}$ (see (4.74)). The eigenvalues of $\boldsymbol{A}^\top \boldsymbol{A}$ are the squared singular values of $\boldsymbol{\Sigma}$ (see (4.75)).

To obtain the left-singular vectors $\boldsymbol{U}$, we follow a similar procedure. We start by computing the SVD of the symmetric matrix $\boldsymbol{A}\boldsymbol{A}^\top \in \mathbb{R}^{m \times m}$ (instead of the previous $\boldsymbol{A}^\top \boldsymbol{A} \in \mathbb{R}^{n \times n}$). The SVD of $\boldsymbol{A}$ yields

$$\boldsymbol{A}\boldsymbol{A}^\top = (\boldsymbol{U}\boldsymbol{\Sigma}\boldsymbol{V}^\top)(\boldsymbol{U}\boldsymbol{\Sigma}\boldsymbol{V}^\top)^\top = \boldsymbol{U}\boldsymbol{\Sigma}\boldsymbol{V}^\top \boldsymbol{V}\boldsymbol{\Sigma}^\top \boldsymbol{U}^\top \tag{4.76a}$$

$$= \boldsymbol{U} \begin{bmatrix} \sigma_1^2 & 0 & 0 \\ 0 & \ddots & 0 \\ 0 & 0 & \sigma_m^2 \end{bmatrix} \boldsymbol{U}^\top . \tag{4.76b}$$

The spectral theorem tells us that $\boldsymbol{A}\boldsymbol{A}^\top = \boldsymbol{S}\boldsymbol{D}\boldsymbol{S}^\top$ can be diagonalized and we can find an ONB of eigenvectors of $\boldsymbol{A}\boldsymbol{A}^\top$, which are collected in $\boldsymbol{S}$. The orthonormal eigenvectors of $\boldsymbol{A}\boldsymbol{A}^\top$ are the left-singular vectors $\boldsymbol{U}$ and form an orthonormal basis set in the codomain of the SVD.

This leaves the question of the structure of the matrix $\boldsymbol{\Sigma}$. Since $\boldsymbol{A}\boldsymbol{A}^\top$ and $\boldsymbol{A}^\top\boldsymbol{A}$ have the same nonzero eigenvalues (see page 90) the nonzero entries of the $\boldsymbol{\Sigma}$ matrices in the SVD for both cases have to be the same.

The last step is to link up all the parts we touched upon so far. We have an orthonormal set of right-singular vectors in $\boldsymbol{V}$. To finish the construction of the SVD, we connect them with the orthonormal vectors $\boldsymbol{U}$. To reach this goal, we use the fact the images of the $\boldsymbol{v}_i$ under $\boldsymbol{A}$ have to be orthogonal, too. We can show this by using the results from Section 3.4. We require that the inner product between $\boldsymbol{A}\boldsymbol{v}_i$ and $\boldsymbol{A}\boldsymbol{v}_j$ must be 0 for $i \neq j$. For any two orthogonal eigenvectors $\boldsymbol{v}_i, \boldsymbol{v}_j$, $i \neq j$, it holds that

$$(\boldsymbol{A}\boldsymbol{v}_i)^\top(\boldsymbol{A}\boldsymbol{v}_j) = \boldsymbol{v}_i^\top(\boldsymbol{A}^\top\boldsymbol{A})\boldsymbol{v}_j = \boldsymbol{v}_i^\top(\lambda_j\boldsymbol{v}_j) = \lambda_j\boldsymbol{v}_i^\top\boldsymbol{v}_j = 0\,. \tag{4.77}$$

For the case $m \geqslant r$, it holds that $\{\boldsymbol{A}\boldsymbol{v}_1, \ldots, \boldsymbol{A}\boldsymbol{v}_r\}$ is a basis of an r-dimensional subspace of $\mathbb{R}^m$.

To complete the SVD construction, we need left-singular vectors that are ortho*normal*: We normalize the images of the right-singular vectors $\boldsymbol{A}\boldsymbol{v}_i$ and obtain

$$\boldsymbol{u}_i := \frac{\boldsymbol{A}\boldsymbol{v}_i}{\|\boldsymbol{A}\boldsymbol{v}_i\|} = \frac{1}{\sqrt{\lambda_i}}\boldsymbol{A}\boldsymbol{v}_i = \frac{1}{\sigma_i}\boldsymbol{A}\boldsymbol{v}_i\,, \tag{4.78}$$

where the last equality was obtained from (4.75) and (4.76b), showing us that the eigenvalues of $\boldsymbol{A}\boldsymbol{A}^\top$ are such that $\sigma_i^2 = \lambda_i$.

Therefore, the eigenvectors of $\boldsymbol{A}^\top\boldsymbol{A}$, which we know are the right-singular vectors $\boldsymbol{v}_i$, and their normalized images under $\boldsymbol{A}$, the left-singular vectors $\boldsymbol{u}_i$, form two self-consistent ONBs that are connected through the singular value matrix $\boldsymbol{\Sigma}$.

singular value equation

Let us rearrange (4.78) to obtain the *singular value equation*

$$\boldsymbol{A}\boldsymbol{v}_i = \sigma_i\boldsymbol{u}_i\,, \quad i = 1, \ldots, r\,. \tag{4.79}$$

This equation closely resembles the eigenvalue equation (4.25), but the vectors on the left- and the right-hand sides are not the same.

For $n > m$, (4.79) holds only for $i \leqslant m$ and (4.79) says nothing about the $\boldsymbol{u}_i$ for $i > m$. However, we know by construction that they are orthonormal. Conversely, for $m > n$, (4.79) holds only for $i \leqslant n$. For $i > n$, we have $\boldsymbol{A}\boldsymbol{v}_i = \boldsymbol{0}$ and we still know that the $\boldsymbol{v}_i$ form an orthonormal set. This means that the SVD also supplies an orthonormal basis of the kernel (null space) of $\boldsymbol{A}$, the set of vectors $\boldsymbol{x}$ with $\boldsymbol{A}\boldsymbol{x} = \boldsymbol{0}$ (see Section 2.7.3).

Moreover, concatenating the $\boldsymbol{v}_i$ as the columns of $\boldsymbol{V}$ and the $\boldsymbol{u}_i$ as the columns of $\boldsymbol{U}$ yields

$$\boldsymbol{A}\boldsymbol{V} = \boldsymbol{U}\boldsymbol{\Sigma}\,, \tag{4.80}$$

where $\boldsymbol{\Sigma}$ has the same dimensions as $\boldsymbol{A}$ and a diagonal structure for rows $1, \ldots, r$. Hence, right-multiplying with $\boldsymbol{V}^\top$ yields $\boldsymbol{A} = \boldsymbol{U}\boldsymbol{\Sigma}\boldsymbol{V}^\top$, which is the SVD of $\boldsymbol{A}$.

Example 4.13 (Computing the SVD)
Let us find the singular value decomposition of

$$\boldsymbol{A} = \begin{bmatrix} 1 & 0 & 1 \\ -2 & 1 & 0 \end{bmatrix}. \tag{4.81}$$

The SVD requires us to compute the right-singular vectors $\boldsymbol{v}_j$, the singular values σ_k, and the left-singular vectors $\boldsymbol{u}_i$.

Step 1: Right-singular vectors as the eigenbasis of $\boldsymbol{A}^\top \boldsymbol{A}$.
We start by computing

$$\boldsymbol{A}^\top \boldsymbol{A} = \begin{bmatrix} 1 & -2 \\ 0 & 1 \\ 1 & 0 \end{bmatrix} \begin{bmatrix} 1 & 0 & 1 \\ -2 & 1 & 0 \end{bmatrix} = \begin{bmatrix} 5 & -2 & 1 \\ -2 & 1 & 0 \\ 1 & 0 & 1 \end{bmatrix}. \tag{4.82}$$

We compute the singular values and right-singular vectors $\boldsymbol{v}_j$ through the eigenvalue decomposition of $\boldsymbol{A}^\top \boldsymbol{A}$, which is given as

$$\boldsymbol{A}^\top \boldsymbol{A} = \begin{bmatrix} \frac{5}{\sqrt{30}} & 0 & \frac{-1}{\sqrt{6}} \\ \frac{-2}{\sqrt{30}} & \frac{1}{\sqrt{5}} & \frac{-2}{\sqrt{6}} \\ \frac{1}{\sqrt{30}} & \frac{2}{\sqrt{5}} & \frac{1}{\sqrt{6}} \end{bmatrix} \begin{bmatrix} 6 & 0 & 0 \\ 0 & 1 & 0 \\ 0 & 0 & 0 \end{bmatrix} \begin{bmatrix} \frac{5}{\sqrt{30}} & \frac{-2}{\sqrt{30}} & \frac{1}{\sqrt{30}} \\ 0 & \frac{1}{\sqrt{5}} & \frac{2}{\sqrt{5}} \\ \frac{-1}{\sqrt{6}} & \frac{-2}{\sqrt{6}} & \frac{1}{\sqrt{6}} \end{bmatrix} = \boldsymbol{P}\boldsymbol{D}\boldsymbol{P}^\top, \tag{4.83}$$

and we obtain the right-singular vectors as the columns of $\boldsymbol{P}$ so that

$$\boldsymbol{V} = \boldsymbol{P} = \begin{bmatrix} \frac{5}{\sqrt{30}} & 0 & \frac{-1}{\sqrt{6}} \\ \frac{-2}{\sqrt{30}} & \frac{1}{\sqrt{5}} & \frac{-2}{\sqrt{6}} \\ \frac{1}{\sqrt{30}} & \frac{2}{\sqrt{5}} & \frac{1}{\sqrt{6}} \end{bmatrix}. \tag{4.84}$$

Step 2: Singular-value matrix.
As the singular values σ_i are the square roots of the eigenvalues of $\boldsymbol{A}^\top \boldsymbol{A}$ we obtain them straight from $\boldsymbol{D}$. Since $\text{rk}(\boldsymbol{A}) = 2$, there are only two nonzero singular values: $\sigma_1 = \sqrt{6}$ and $\sigma_2 = 1$. The singular value matrix must be the same size as $\boldsymbol{A}$, and we obtain

$$\boldsymbol{\Sigma} = \begin{bmatrix} \sqrt{6} & 0 & 0 \\ 0 & 1 & 0 \end{bmatrix}. \tag{4.85}$$

Step 3: Left-singular vectors as the normalized image of the right-singular vectors.
We find the left-singular vectors by computing the image of the right-singular vectors under $\boldsymbol{A}$ and normalizing them by dividing them by their corresponding singular value. We obtain

$$\boldsymbol{u}_1 = \frac{1}{\sigma_1}\boldsymbol{A}\boldsymbol{v}_1 = \frac{1}{\sqrt{6}} \begin{bmatrix} 1 & 0 & 1 \\ -2 & 1 & 0 \end{bmatrix} \begin{bmatrix} \frac{5}{\sqrt{30}} \\ \frac{-2}{\sqrt{30}} \\ \frac{1}{\sqrt{30}} \end{bmatrix} = \begin{bmatrix} \frac{1}{\sqrt{5}} \\ -\frac{2}{\sqrt{5}} \end{bmatrix}, \tag{4.86}$$

$$\boldsymbol{u}_2 = \frac{1}{\sigma_2}\boldsymbol{A}\boldsymbol{v}_2 = \frac{1}{1} \begin{bmatrix} 1 & 0 & 1 \\ -2 & 1 & 0 \end{bmatrix} \begin{bmatrix} 0 \\ \frac{1}{\sqrt{5}} \\ \frac{2}{\sqrt{5}} \end{bmatrix} = \begin{bmatrix} \frac{2}{\sqrt{5}} \\ \frac{1}{\sqrt{5}} \end{bmatrix}, \tag{4.87}$$

$$U = [u_1, u_2] = \frac{1}{\sqrt{5}} \begin{bmatrix} 1 & 2 \\ -2 & 1 \end{bmatrix} . \tag{4.88}$$

Note that on a computer the approach illustrated here has poor numerical behavior, and the SVD of A is normally computed without resorting to the eigenvalue decomposition of $A^\top A$.

4.5.3 Eigenvalue Decomposition vs. Singular Value Decomposition

Let us consider the eigendecomposition $A = PDP^{-1}$ and the SVD $A = U\Sigma V^\top$ and review the core elements of the past sections.

- The SVD always exists for any matrix $\mathbb{R}^{m\times n}$. The eigendecomposition is only defined for square matrices $\mathbb{R}^{n\times n}$ and only exists if we can find a basis of eigenvectors of $\mathbb{R}^n$.
- The vectors in the eigendecomposition matrix P are not necessarily orthogonal, i.e., the change of basis is not a simple rotation and scaling. On the other hand, the vectors in the matrices U and V in the SVD are orthonormal, so they do represent rotations.
- Both the eigendecomposition and the SVD are compositions of three linear mappings:

  1. Change of basis in the domain
  2. Independent scaling of each new basis vector and mapping from domain to codomain
  3. Change of basis in the codomain

  A key difference between the eigendecomposition and the SVD is that in the SVD, domain and codomain can be vector spaces of different dimensions.
- In the SVD, the left- and right-singular vector matrices U and V are generally not inverse of each other (they perform basis changes in different vector spaces). In the eigendecomposition, the basis change matrices P and P^{-1} are inverses of each other.
- In the SVD, the entries in the diagonal matrix Σ are all real and non-negative, which is not generally true for the diagonal matrix in the eigendecomposition.
- The SVD and the eigendecomposition are closely related through their projections

  - The left-singular vectors of A are eigenvectors of $AA^\top$
  - The right-singular vectors of A are eigenvectors of $A^\top A$.
  - The nonzero singular values of A are the square roots of the nonzero eigenvalues of $AA^\top$ and are equal to the nonzero eigenvalues of $A^\top A$.

- For symmetric matrices $A \in \mathbb{R}^{n\times n}$, the eigenvalue decomposition and the SVD are one and the same, which follows from the spectral theorem 4.15.

Example 4.14 (Finding Structure in Movie Ratings and Consumers)
Let us add a practical interpretation of the SVD by analyzing data on people and their preferred movies. Consider three viewers (Ali, Beatrix, Chandra) rating four different movies (*Star Wars*, *Blade Runner*, *Amelie*, *Delicatessen*). Their ratings are values between 0 (worst) and 5 (best) and encoded in a data matrix $\boldsymbol{A} \in \mathbb{R}^{4\times 3}$ as shown in Figure 4.10. Each row represents a movie and each column a user. Thus, the column vectors of movie ratings, one for each viewer, are $\boldsymbol{x}_{\text{Ali}}, \boldsymbol{x}_{\text{Beatrix}}, \boldsymbol{x}_{\text{Chandra}}$.

Factoring $\boldsymbol{A}$ using the SVD offers us a way to capture the relationships of how people rate movies, and especially if there is a structure linking which people like which movies. Applying the SVD to our data matrix $\boldsymbol{A}$ makes a number of assumptions:

1. All viewers rate movies consistently using the same linear mapping.
2. There are no errors or noise in the ratings.
3. We interpret the left-singular vectors $\boldsymbol{u}_i$ as stereotypical movies and the right-singular vectors $\boldsymbol{v}_j$ as stereotypical viewers.

We then make the assumption that any viewer's specific movie preferences can be expressed as a linear combination of the $\boldsymbol{v}_j$. Similarly, any movie's like ability can be expressed as a linear combination of the $\boldsymbol{u}_i$. Therefore, a vector in the domain of the SVD can be interpreted as a viewer in the "space" of stereotypical viewers, and a vector in the codomain of the SVD correspondingly as a movie in the "space" of stereotypical movies. Let us inspect the SVD of our movie-user matrix. The first left-singular vector $\boldsymbol{u}_1$ has large absolute values for the two science fiction movies and a large first singular value (red shading in Figure 4.10). Thus, this groups a type of users with a specific set of movies (science fiction theme). Similarly, the first right-singular $\boldsymbol{v}_1$ shows large absolute values for Ali and Beatrix, who give high ratings to science fiction movies (green shading in Figure 4.10). This suggests that $\boldsymbol{v}_1$ reflects the notion of a science fiction lover.

These two "spaces" are only meaningfully spanned by the respective viewer and movie data if the data itself covers a sufficient diversity of viewers and movies.

$$
\begin{array}{r} \\ \textit{Star Wars} \\ \textit{Blade Runner} \\ \textit{Amelie} \\ \textit{Delicatessen} \end{array}
\begin{array}{c} \text{Ali}\quad\text{Beatrix}\quad\text{Chandra} \\ \begin{bmatrix} 5 & 4 & 1 \\ 5 & 5 & 0 \\ 0 & 0 & 5 \\ 1 & 0 & 4 \end{bmatrix} \end{array}
=
\begin{bmatrix} -0.6710 & 0.0236 & 0.4647 & -0.5774 \\ -0.7197 & 0.2054 & -0.4759 & 0.4619 \\ -0.0939 & -0.7705 & -0.5268 & -0.3464 \\ -0.1515 & -0.6030 & 0.5293 & -0.5774 \end{bmatrix}
$$

$$
\begin{bmatrix} 9.6438 & 0 & 0 \\ 0 & 6.3639 & 0 \\ 0 & 0 & 0.7056 \\ 0 & 0 & 0 \end{bmatrix}
$$

$$
\begin{bmatrix} -0.7367 & -0.6515 & -0.1811 \\ 0.0852 & 0.1762 & -0.9807 \\ 0.6708 & -0.7379 & -0.0743 \end{bmatrix}
$$

Figure 4.10 Movie ratings of three people for four movies and its SVD decomposition.

Similarly, $\boldsymbol{u}_2$, seems to capture a French art house film theme, and $\boldsymbol{v}_2$ indicates that Chandra is close to an idealized lover of such movies. An idealized science fiction lover is a purist and only loves science fiction movies, so a science fiction lover $\boldsymbol{v}_1$ gives a rating of zero to everything but science fiction themed – this logic is implied the diagonal substructure for the singular value matrix $\boldsymbol{\Sigma}$. A specific movie is therefore represented by how it decomposes (linearly) into its stereotypical movies. Likewise, a person would be represented by how they decompose (via linear combination) into movie themes.

It is worth while to briefly discuss SVD terminology and conventions, as there are different versions used in the literature. The mathematics remains invariant to these differences, but these differences can be confusing:

- For convenience in notation and abstraction, we use an SVD notation where the SVD is described as having two square left- and right-singular vector matrices, but a nonsquare singular value matrix. Our definition (4.64) for the SVD is sometimes called the *full SVD*.

full SVD

- Some authors define the SVD a bit differently and focus on square singular matrices. Then, for $\boldsymbol{A} \in \mathbb{R}^{m\times n}$ and $m \geqslant n$,

$$\underset{m\times n}{\boldsymbol{A}} = \underset{m\times n}{\boldsymbol{U}} \underset{n\times n}{\boldsymbol{\Sigma}} \underset{n\times n}{\boldsymbol{V}^\top} . \tag{4.89}$$

reduced SVD

Sometimes this formulation is called the *reduced SVD* (e.g., Datta, 2010) or *the* SVD (e.g., Press et al., 2007). This alternative format changes merely how the matrices are constructed but leaves the mathematical structure of the SVD unchanged. The convenience of this alternative formulation is that $\boldsymbol{\Sigma}$ is diagonal, as in the eigenvalue decomposition.

- In Section 4.6, we will learn about matrix approximation techniques using the SVD, which is also called the *truncated SVD*.

truncated SVD

- It is possible to define the SVD of a rank-r matrix $\boldsymbol{A}$ so that $\boldsymbol{U}$ is an $m \times r$ matrix, $\boldsymbol{\Sigma}$ a diagonal matrix $r \times r$, and $\boldsymbol{V}$ an $r \times n$ matrix. This construction is very similar to our definition, and ensures that the diagonal matrix $\boldsymbol{\Sigma}$ has only nonzero entries along the diagonal. The main convenience of this alternative notation is that $\boldsymbol{\Sigma}$ is diagonal, as in the eigenvalue decomposition.
- A restriction that the SVD for $\boldsymbol{A}$ only applies to $m \times n$ matrices with $m > n$ is practically unnecessary. When $m < n$, the SVD decomposition will yield $\boldsymbol{\Sigma}$ with more zero columns than rows and, consequently, the singular values $\sigma_{m+1}, \ldots, \sigma_n$ are 0.

The SVD is used in a variety of applications in machine learning from least-squares problems in curve fitting to solving systems of linear equations. These applications harness various important properties of the SVD, its relation to the rank of a matrix, and its ability to approximate matrices of a given rank with lower-rank matrices. Substituting a matrix with its SVD has often the advantage of making calculation more robust to numerical rounding errors. As we will explore in the next section, the SVD's ability to approximate matrices with "simpler" matrices in a principled manner opens up machine learning applications

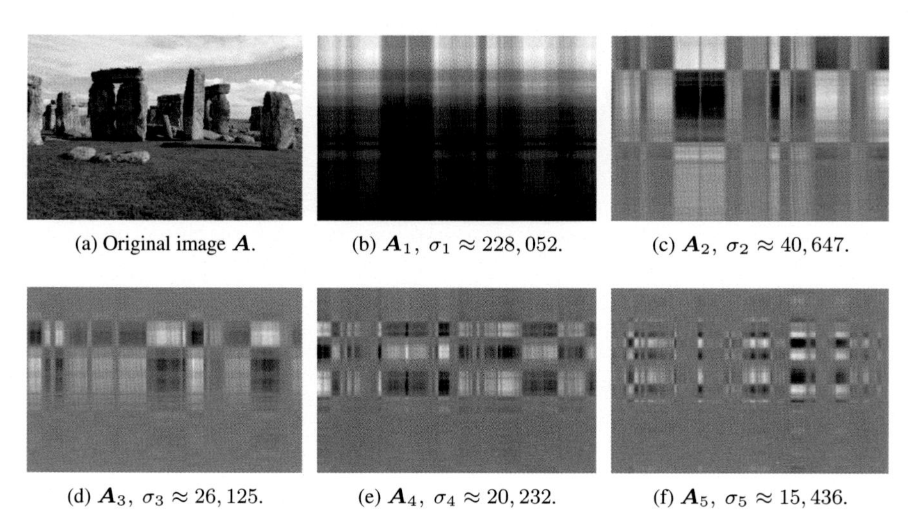

(a) Original image $\boldsymbol{A}$. (b) $\boldsymbol{A}_1,\ \sigma_1 \approx 228{,}052$. (c) $\boldsymbol{A}_2,\ \sigma_2 \approx 40{,}647$. (d) $\boldsymbol{A}_3,\ \sigma_3 \approx 26{,}125$. (e) $\boldsymbol{A}_4,\ \sigma_4 \approx 20{,}232$. (f) $\boldsymbol{A}_5,\ \sigma_5 \approx 15{,}436$.

Figure 4.11 Image processing with the SVD. (a) The original grayscale image is a $1{,}432 \times 1{,}910$ matrix of values between 0 (black) and 1 (white). (b)–(f) Rank-1 matrices $\boldsymbol{A}_1, \ldots, \boldsymbol{A}_5$ and their corresponding singular values $\sigma_1, \ldots, \sigma_5$. The grid like structure of each rank-1 matrix is imposed by the outer product of the left- and right-singular vectors.

ranging from dimensionality reduction and topic modeling to data compression and clustering.

4.6 Matrix Approximation

We considered the SVD as a way to factorize $\boldsymbol{A} = \boldsymbol{U}\boldsymbol{\Sigma}\boldsymbol{V}^\top \in \mathbb{R}^{m \times n}$ into the product of three matrices, where $\boldsymbol{U} \in \mathbb{R}^{m \times m}$ and $\boldsymbol{V} \in \mathbb{R}^{n \times n}$ are orthogonal and $\boldsymbol{\Sigma}$ contains the singular values on its main diagonal. Instead of doing the full SVD factorization, we will now investigate how the SVD allows us to represent a matrix $\boldsymbol{A}$ as a sum of simpler (low-rank) matrices $\boldsymbol{A}_i$, which lends itself to a matrix approximation scheme that is cheaper to compute than the full SVD.

We construct a rank-1 matrix $\boldsymbol{A}_i \in \mathbb{R}^{m \times n}$ as

$$\boldsymbol{A}_i := \boldsymbol{u}_i \boldsymbol{v}_i^\top \,, \tag{4.90}$$

which is formed by the outer product of the ith orthogonal column vector of $\boldsymbol{U}$ and $\boldsymbol{V}$. Figure 4.11 shows an image of Stonehenge, which can be represented by a matrix $\boldsymbol{A} \in \mathbb{R}^{1432 \times 1910}$, and some outer products $\boldsymbol{A}_i$, as defined in (4.90).

A matrix $\boldsymbol{A} \in \mathbb{R}^{m \times n}$ of rank r can be written as a sum of rank-1 matrices $\boldsymbol{A}_i$ so that

$$\boldsymbol{A} = \sum_{i=1}^{r} \sigma_i \boldsymbol{u}_i \boldsymbol{v}_i^\top = \sum_{i=1}^{r} \sigma_i \boldsymbol{A}_i \,, \tag{4.91}$$

where the outer-product matrices $\boldsymbol{A}_i$ are weighted by the ith singular value σ_i. We can see why (4.91) holds: The diagonal structure of the singular value matrix $\boldsymbol{\Sigma}$ multiplies only matching left- and right-singular vectors $\boldsymbol{u}_i \boldsymbol{v}_i^\top$ and scales them by the corresponding singular value σ_i. All terms $\Sigma_{ij} \boldsymbol{u}_i \boldsymbol{v}_j^\top$ vanish for $i \neq j$ because $\boldsymbol{\Sigma}$ is a diagonal matrix. Any terms $i > r$ vanish because the corresponding singular values are 0.

In (4.90), we introduced rank-1 matrices $\boldsymbol{A}_i$. We summed up the r individual rank-1 matrices to obtain a rank-r matrix $\boldsymbol{A}$; see (4.91). If the sum does not run

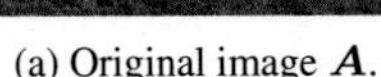
(a) Original image $\boldsymbol{A}$.

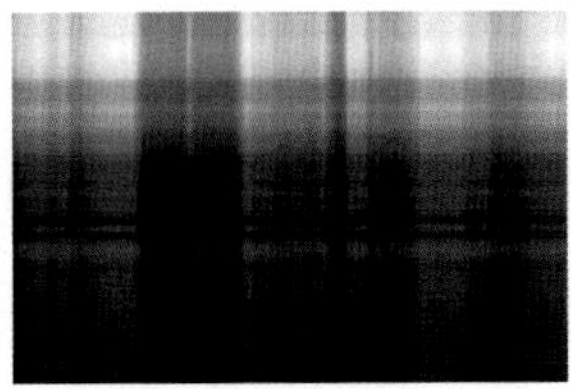
(b) Rank-1 approximation $\widehat{\boldsymbol{A}}(1)$.

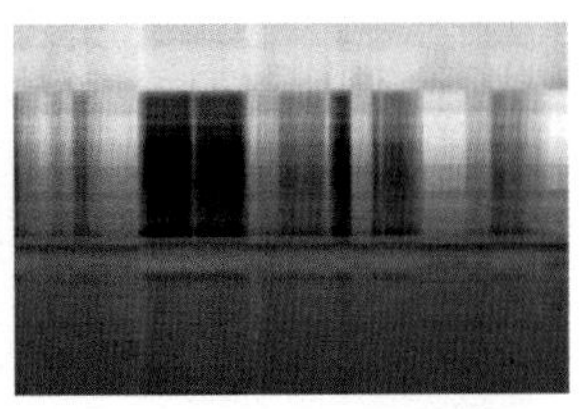
(c) Rank-2 approximation $\widehat{\boldsymbol{A}}(2)$.

(d) Rank-3 approximation $\widehat{\boldsymbol{A}}(3)$.

(e) Rank-4 approximation $\widehat{\boldsymbol{A}}(4)$.

(f) Rank-5 approximation $\widehat{\boldsymbol{A}}(5)$.

Figure 4.12 Image reconstruction with the SVD. (a) Original image. (b)–(f) Image reconstruction using the low-rank approximation of the SVD, where the rank-k approximation is given by $\widehat{\boldsymbol{A}}(k) = \sum_{i=1}^{k} \sigma_i \boldsymbol{A}_i$.

rank-k approximation

over all matrices $\boldsymbol{A}_i$, $i = 1, \ldots, r$, but only up to an intermediate value $k < r$, we obtain a *rank-k approximation*

$$\widehat{\boldsymbol{A}}(k) := \sum_{i=1}^{k} \sigma_i \boldsymbol{u}_i \boldsymbol{v}_i^\top = \sum_{i=1}^{k} \sigma_i \boldsymbol{A}_i \tag{4.92}$$

of $\boldsymbol{A}$ with $\text{rk}(\widehat{\boldsymbol{A}}(k)) = k$. Figure 4.12 shows low-rank approximations $\widehat{\boldsymbol{A}}(k)$ of an original image $\boldsymbol{A}$ of Stonehenge. The shape of the rocks becomes increasingly visible and clearly recognizable in the rank-5 approximation. While the original image requires $1,432 \cdot 1,910 = 2,735,120$ numbers, the rank-5 approximation requires us only to store the five singular values and the five left- and right-singular vectors ($1,432$ and $1,910$-dimensional each) for a total of $5 \cdot (1,432 + 1,910 + 1) = 16,715$ numbers – just above 0.6% of the original.

To measure the difference (error) between $\boldsymbol{A}$ and its rank-k approximation $\widehat{\boldsymbol{A}}(k)$, we need the notion of a norm. In Section 3.1, we already used norms on vectors that measure the length of a vector. By analogy, we can also define norms on matrices.

spectral norm

Definition 4.23 (Spectral Norm of a Matrix). For $\boldsymbol{x} \in \mathbb{R}^n \backslash \{\boldsymbol{0}\}$, the *spectral norm* of a matrix $\boldsymbol{A} \in \mathbb{R}^{m \times n}$ is defined as

$$\|\boldsymbol{A}\|_2 := \max_{\boldsymbol{x}} \frac{\|\boldsymbol{A}\boldsymbol{x}\|_2}{\|\boldsymbol{x}\|_2} . \tag{4.93}$$

We introduce the notation of a subscript in the matrix norm (left-hand side), similar to the Euclidean norm for vectors (right-hand side), which has subscript 2. The spectral norm (4.93) determines how long any vector $\boldsymbol{x}$ can at most become when multiplied by $\boldsymbol{A}$.

Theorem 4.24. *The spectral norm of $\boldsymbol{A}$ is its largest singular value σ_1.*

We leave the proof of this theorem as an exercise.

Theorem 4.25 (Eckart–Young Theorem (Eckart and Young, 1936)). *Consider a matrix $A \in \mathbb{R}^{m\times n}$ of rank r and let $B \in \mathbb{R}^{m\times n}$ be a matrix of rank k. For any $k \leqslant r$ with $\widehat{A}(k) = \sum_{i=1}^{k} \sigma_i u_i v_i^\top$ it holds that*

Eckart–Young theorem

$$\widehat{A}(k) = \text{argmin}_{\text{rk}(B)k} \|A - B\|_2 , \tag{4.94}$$

$$\left\|A - \widehat{A}(k)\right\|_2 = \sigma_{k+1} . \tag{4.95}$$

The Eckart–Young theorem states explicitly how much error we introduce by approximating A using a rank-k approximation. We can interpret the rank-k approximation obtained with the SVD as a projection of the full-rank matrix A onto a lower-dimensional space of rank-at-most-k matrices. Of all possible projections, the SVD minimizes the error (with respect to the spectral norm) between A and any rank-k approximation.

We can retrace some of the steps to understand why (4.95) should hold. We observe that the difference between $A - \widehat{A}(k)$ is a matrix containing the sum of the remaining rank-1 matrices

$$A - \widehat{A}(k) = \sum_{i=k+1}^{r} \sigma_i u_i v_i^\top . \tag{4.96}$$

By Theorem 4.24, we immediately obtain σ_{k+1} as the spectral norm of the difference matrix. Let us have a closer look at (4.94). If we assume that there is another matrix B with $\text{rk}(B) \leqslant k$, such that

$$\|A - B\|_2 < \left\|A - \widehat{A}(k)\right\|_2 , \tag{4.97}$$

then there exists an at least $(n - k)$-dimensional null space $Z \subseteq \mathbb{R}^n$, such that $x \in Z$ implies that $Bx = 0$. Then it follows that

$$\|Ax\|_2 = \|(A - B)x\|_2 , \tag{4.98}$$

and by using a version of the Cauchy–Schwartz inequality (3.17) that encompasses norms of matrices, we obtain

$$\|Ax\|_2 \leqslant \|A - B\|_2 \|x\|_2 < \sigma_{k+1} \|x\|_2 . \tag{4.99}$$

However, there exists a $(k + 1)$-dimensional subspace where $\|Ax\|_2 \geqslant \sigma_{k+1} \|x\|_2$, which is spanned by the right-singular vectors $v_j, j \leqslant k + 1$ of A. Adding up dimensions of these two spaces yields a number greater than n, as there must be a nonzero vector in both spaces. This is a contradiction of the rank-nullity theorem (Theorem 2.24) in Section 2.7.3.

The Eckart–Young theorem implies that we can use SVD to reduce a rank-r matrix A to a rank-k matrix $\widehat{A}$ in a principled, optimal (in the spectral norm sense) manner. We can interpret the approximation of A by a rank-k matrix as a form of lossy compression. Therefore, the low-rank approximation of a matrix appears in many machine learning applications, e.g., image processing, noise filtering, and regularization of ill-posed problems. Furthermore, it plays a key role in dimensionality reduction and principal component analysis, as we will see in Chapter 10.

Example 4.15 (Finding Structure in Movie Ratings and Consumers (continued))

Coming back to our movie-rating example, we can now apply the concept of low-rank approximations to approximate the original data matrix. Recall that our first singular value captures the notion of science fiction theme in movies and science fiction lovers. Thus, by using only the first singular value term in a rank-1 decomposition of the movie-rating matrix, we obtain the predicted ratings

$$\boldsymbol{A}_1 = \boldsymbol{u}_1 \boldsymbol{v}_1^\top = \begin{bmatrix} -0.6710 \\ -0.7197 \\ -0.0939 \\ -0.1515 \end{bmatrix} \begin{bmatrix} -0.7367 & -0.6515 & -0.1811 \end{bmatrix} \tag{4.100a}$$

$$= \begin{bmatrix} 0.4943 & 0.4372 & 0.1215 \\ 0.5302 & 0.4689 & 0.1303 \\ 0.0692 & 0.0612 & 0.0170 \\ 0.1116 & 0.0987 & 0.0274 \end{bmatrix}. \tag{4.100b}$$

This first rank-1 approximation $\boldsymbol{A}_1$ is insightful: It tells us that Ali and Beatrix like science fiction movies, such as *Star Wars* and *Bladerunner* (entries have values > 4), but fails to capture the ratings of the other movies by Chandra. This is not surprising, as Chandra's type of movies is not captured by the first singular value. The second singular value gives us a better rank-1 approximation for those movie-theme lovers:

$$\boldsymbol{A}_2 = \boldsymbol{u}_2 \boldsymbol{v}_2^\top = \begin{bmatrix} 0.0236 \\ 0.2054 \\ -0.7705 \\ -0.6030 \end{bmatrix} \begin{bmatrix} 0.0852 & 0.1762 & -0.9807 \end{bmatrix} \tag{4.101a}$$

$$= \begin{bmatrix} -0.0154 & 0.0042 & -0.0174 \\ -0.1338 & 0.0362 & -0.1516 \\ 0.5019 & -0.1358 & 0.5686 \\ 0.3928 & -0.1063 & 0.445 \end{bmatrix}. \tag{4.101b}$$

In this second rank-1 approximation $\boldsymbol{A}_2$, we capture Chandra's ratings and movie types well, but not the science fiction movies. This leads us to consider the rank-2 approximation $\widehat{\boldsymbol{A}}(2)$, where we combine the first two rank-1 approximations

$$\widehat{\boldsymbol{A}}(2) = \sigma_1 \boldsymbol{A}_1 + \sigma_2 \boldsymbol{A}_2 = \begin{bmatrix} 4.7801 & 4.2419 & 1.0244 \\ 5.2252 & 4.7522 & -0.0250 \\ 0.2493 & -0.2743 & 4.9724 \\ 0.7495 & 0.2756 & 4.0278 \end{bmatrix}. \tag{4.102}$$

$\widehat{A}(2)$ is similar to the original movie ratings table

$$\boldsymbol{A} = \begin{bmatrix} 5 & 4 & 1 \\ 5 & 5 & 0 \\ 0 & 0 & 5 \\ 1 & 0 & 4 \end{bmatrix}, \tag{4.103}$$

and this suggests that we can ignore the contribution of $\boldsymbol{A}_3$. We can interpret this so that in the data table there is no evidence of a third movie-theme/movie-lovers category. This also means that the entire space of movie-themes/movie-lovers in our example is a two-dimensional space spanned by science fiction and French art house movies and lovers.

4.7 Matrix Phylogeny

The word "phylogenetic" describes how we capture the relationships among individuals or groups and derived from the Greek words for "tribe" and "source."

In Chapters 2 and 3, we covered the basics of linear algebra and analytic geometry. In this chapter, we looked at fundamental characteristics of matrices and linear mappings. Figure 4.13 depicts the phylogenetic tree of relationships between different types of matrices (black arrows indicating "is a subset of") and the covered operations we can perform on them (in blue). We consider all *real matrices* $\boldsymbol{A} \in \mathbb{R}^{n\times m}$. For nonsquare matrices (where $n \neq m$), the SVD always exists, as we saw in this chapter. Focusing on *square matrices* $\boldsymbol{A} \in \mathbb{R}^{n\times n}$, the *determinant* informs us whether a square matrix possesses an *inverse matrix*, i.e., whether it belongs to the class of regular, invertible matrices. If the square $n \times n$ matrix possesses n linearly independent eigenvectors, then the matrix is *nondefective* and an *eigendecomposition* exists (Theorem 4.12). We know that repeated eigenvalues may result in defective matrices, which cannot be diagonalized.

Nonsingular and nondefective matrices are not the same. For example, a rotation matrix will be invertible (determinant is nonzero) but not diagonalizable in the real numbers (eigenvalues are not guaranteed to be real numbers).

We dive further into the branch of nondefective square $n \times n$ matrices. $\boldsymbol{A}$ is *normal* if the condition $\boldsymbol{A}^\top\boldsymbol{A} = \boldsymbol{A}\boldsymbol{A}^\top$ holds. Moreover, if the more restrictive condition holds that $\boldsymbol{A}^\top\boldsymbol{A} = \boldsymbol{A}\boldsymbol{A}^\top = \boldsymbol{I}$, then $\boldsymbol{A}$ is called *orthogonal* (see Definition 3.8). The set of orthogonal matrices is a subset of the regular (invertible) matrices and satisfies $\boldsymbol{A}^\top = \boldsymbol{A}^{-1}$.

Normal matrices have a frequently encountered subset, the symmetric matrices $\boldsymbol{S} \in \mathbb{R}^{n\times n}$, which satisfy $\boldsymbol{S} = \boldsymbol{S}^\top$. Symmetric matrices have only real eigenvalues. A subset of the symmetric matrices consists of the positive definite matrices $\boldsymbol{P}$ that satisfy the condition of $\boldsymbol{x}^\top\boldsymbol{P}\boldsymbol{x} > 0$ for all $\boldsymbol{x} \in \mathbb{R}^n\backslash\{\boldsymbol{0}\}$. In this case, a unique *Cholesky decomposition* exists (Theorem 4.18). Positive definite matrices have only positive eigenvalues and are always invertible (i.e., have a nonzero determinant).

Another subset of symmetric matrices consists of the *diagonal matrices* $\boldsymbol{D}$. Diagonal matrices are closed under multiplication and addition, but do not nec-

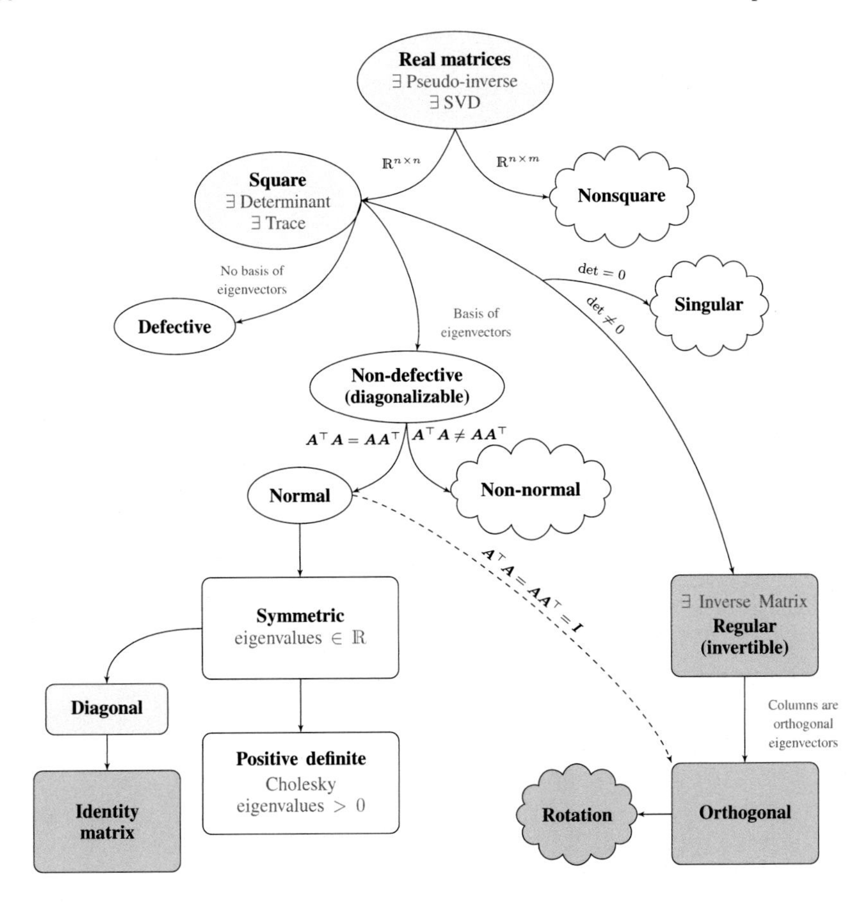

Figure 4.13 A functional phylogeny of matrices encountered in machine learning.

essarily form a group (this is only the case if all diagonal entries are nonzero so that the matrix is invertible). A special diagonal matrix is the identity matrix $\boldsymbol{I}$.

4.8 Further Reading

Most of the content in this chapter establishes underlying mathematics and connects them to methods for studying mappings, many of which are at the heart of machine learning at the level of underpinning software solutions and building blocks for almost all machine learning theory. Matrix characterization using determinants, eigenspectra, and eigenspaces provides fundamental features and conditions for categorizing and analyzing matrices. This extends to all forms of representations of data and mappings involving data, as well as judging the numerical stability of computational operations on such matrices (Press et al., 2007).

Determinants are fundamental tools in order to invert matrices and compute eigenvalues "by hand." However, for almost all but the smallest instances, numerical computation by Gaussian elimination outperforms determinants (Press et al., 2007). Determinants remain nevertheless a powerful theoretical concept, e.g., to gain intuition about the orientation of a basis based on the sign of the determinant. Eigenvectors can be used to perform basis changes to transform data into the coordinates of meaningful orthogonal, feature vectors. Similarly, matrix decomposition methods, such as the Cholesky decomposition, reappear often when we compute or simulate random events (Rubinstein and Kroese, 2016). Therefore, the Cholesky decomposition enables us to compute the *reparametrization trick* where we want to perform continuous differentiation over random variables, e.g., in variational autoencoders (Jimenez Rezende et al., 2014; Kingma and Ba, 2014).

Eigendecomposition is fundamental in enabling us to extract meaningful and interpretable information that characterizes linear mappings. Therefore, the eigendecomposition underlies a general class of machine learning algorithms called *spectral methods* that perform eigendecomposition of a positive-definite kernel. These spectral decomposition methods encompass classical approaches to statistical data analysis, such as the following:

- *Principal component analysis* (PCA (Pearson, 1901), see also Chapter 10), in which a low-dimensional subspace, which explains most of the variability in the data, is sought.
- *Fisher discriminant analysis*, which aims to determine a separating hyperplane for data classification (Mika et al., 1999).
- *Multidimensional scaling* (MDS) (Carroll and Chang, 1970).

principal component analysis

Fisher discriminant analysis

multidimensional scaling

The computational efficiency of these methods typically comes from finding the best rank-k approximation to a symmetric, positive semidefinite matrix. More contemporary examples of spectral methods have different origins, but each of them requires the computation of the eigenvectors and eigenvalues of a positive-definite kernel, such as *Isomap* (Tenenbaum et al., 2000), *Laplacian eigenmaps* (Belkin and Niyogi, 2003), *Hessian eigenmaps* (Donoho and Grimes, 2003), and *spectral clustering* (Shi and Malik, 2000). The core computations of these are generally underpinned by low-rank matrix approximation techniques (Belabbas and Wolfe, 2009) as we encountered here via the SVD.

Isomap

Laplacian eigenmaps

Hessian eigenmaps

spectral clustering

The SVD allows us to discover some of the same kind of information as the eigendecomposition. However, the SVD is more generally applicable to nonsquare matrices and data tables. These matrix factorization methods become relevant whenever we want to identify heterogeneity in data when we want to perform data compression by approximation, e.g., instead of storing $n \times m$ values just storing $(n + m)k$ values, or when we want to perform data pre-processing, e.g., to decorrelate predictor variables of a design matrix (Ormoneit et al., 2001). The SVD operates on matrices, which we can interpret as rectangular arrays with two indices (rows and columns). The extension of a matrixlike structure to higher-dimensional arrays are called tensors. It turns out that the SVD is the special case of a more general family of decompositions that operate on such tensors (Kolda and Bader, 2009). SVD-like operations and low-rank approximations on tensors are, for example, the *Tucker decomposition* (Tucker, 1966) or the *CP decomposition* (Carroll and Chang, 1970).

Tucker decomposition

CP decomposition

The SVD low-rank approximation is frequently used in machine learning for computational efficiency reasons. This is because it reduces the amount of memory and operations with nonzero multiplications we need to perform on potentially very large matrices of data (Trefethen and Bau III, 1997). Moreover, low-rank approximations are used to operate on matrices that may contain missing values as well as for purposes of lossy compression and dimensionality reduction (Moonen and De Moor, 1995; Markovsky, 2011).

Exercises

4.1 Compute the determinant using the Laplace expansion (using the first row) and the Sarrus Rule for

$$\boldsymbol{A} = \begin{bmatrix} 1 & 3 & 5 \\ 2 & 4 & 6 \\ 0 & 2 & 4 \end{bmatrix}.$$

4.2 Compute the following determinant efficiently:

$$\begin{bmatrix} 2 & 0 & 1 & 2 & 0 \\ 2 & -1 & 0 & 1 & 1 \\ 0 & 1 & 2 & 1 & 2 \\ -2 & 0 & 2 & -1 & 2 \\ 2 & 0 & 0 & 1 & 1 \end{bmatrix}.$$

4.3 Compute the eigenspaces of $\begin{bmatrix} 1 & 0 \\ 1 & 1 \end{bmatrix}, \begin{bmatrix} -2 & 2 \\ 2 & 1 \end{bmatrix}$.

4.4 Compute all eigenspaces of

$$\boldsymbol{A} = \begin{bmatrix} 0 & -1 & 1 & 1 \\ -1 & 1 & -2 & 3 \\ 2 & -1 & 0 & 0 \\ 1 & -1 & 1 & 0 \end{bmatrix}.$$

4.5 Diagonalizability of a matrix is unrelated to its invertibility. Determine for the following four matrices whether they are diagonalizable and/or invertible

$$\begin{bmatrix} 1 & 0 \\ 0 & 1 \end{bmatrix}, \quad \begin{bmatrix} 1 & 0 \\ 0 & 0 \end{bmatrix}, \quad \begin{bmatrix} 1 & 1 \\ 0 & 1 \end{bmatrix}, \quad \begin{bmatrix} 0 & 1 \\ 0 & 0 \end{bmatrix}.$$

4.6 Compute the eigenspaces of the following transformation matrices. Are they diagonalizable?

a.

$$\boldsymbol{A} = \begin{bmatrix} 2 & 3 & 0 \\ 1 & 4 & 3 \\ 0 & 0 & 1 \end{bmatrix}$$

b.

$$\boldsymbol{A} = \begin{bmatrix} 1 & 1 & 0 & 0 \\ 0 & 0 & 0 & 0 \\ 0 & 0 & 0 & 0 \\ 0 & 0 & 0 & 0 \end{bmatrix}$$

4.7 Are the following matrices diagonalizable? If yes, determine their diagonal form and a basis with respect to which the transformation matrices are diagonal. If no, give reasons why they are not diagonalizable.

a.

$$A = \begin{bmatrix} 0 & 1 \\ -8 & 4 \end{bmatrix}$$

b.

$$A = \begin{bmatrix} 1 & 1 & 1 \\ 1 & 1 & 1 \\ 1 & 1 & 1 \end{bmatrix}$$

c.

$$A = \begin{bmatrix} 5 & 4 & 2 & 1 \\ 0 & 1 & -1 & -1 \\ -1 & -1 & 3 & 0 \\ 1 & 1 & -1 & 2 \end{bmatrix}$$

d.

$$A = \begin{bmatrix} 5 & -6 & -6 \\ -1 & 4 & 2 \\ 3 & -6 & -4 \end{bmatrix}$$

4.8 Find the SVD of the matrix

$$A = \begin{bmatrix} 3 & 2 & 2 \\ 2 & 3 & -2 \end{bmatrix}.$$

4.9 Find the singular value decomposition of

$$A = \begin{bmatrix} 2 & 2 \\ -1 & 1 \end{bmatrix}.$$

4.10 Find the best rank-1 approximation of

$$A = \begin{bmatrix} 3 & 2 & 2 \\ 2 & 3 & -2 \end{bmatrix}.$$

4.11 Show that for any $A \in \mathbb{R}^{m \times n}$, the matrices $A^\top A$ and $AA^\top$ possess the same nonzero eigenvalues.

4.12 Show that for $x \neq 0$ Theorem 4.24 holds, i.e., show that

$$\max_{x} \frac{\|Ax\|_2}{\|x\|_2} = \sigma_1 ,$$

where σ_1 is the largest singular value of $A \in \mathbb{R}^{m \times n}$.

5

Vector Calculus

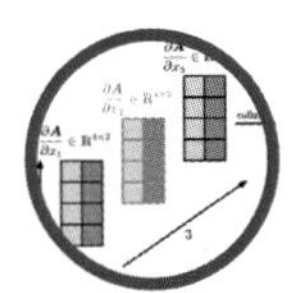

Many algorithms in machine learning optimize an objective function with respect to a set of desired model parameters that control how well a model explains the data: Finding good parameters can be phrased as an optimization problem (see Sections 8.2 and 8.3). Examples include (i) linear regression (see Chapter 9), where we look at curve-fitting problems and optimize linear weight parameters to maximize the likelihood; (ii) neural-network auto-encoders for dimensionality reduction and data compression, where the parameters are the weights and biases of each layer, and where we minimize a reconstruction error by repeated application of the chain rule; and (iii) Gaussian mixture models (see Chapter 11) for modeling data distributions, where we optimize the location and shape parameters of each mixture component to maximize the likelihood of the model. Figure 5.1 illustrates some of these problems, which we typically solve by using optimization algorithms that exploit gradient information (Section 7.1). Figure 5.2 gives an overview of how concepts in this chapter are related and how they are connected to other chapters of the book.

Central to this chapter is the concept of a function. A function f is a quantity that relates two quantities to each other. In this book, these quantities are typically inputs $\boldsymbol{x} \in \mathbb{R}^D$ and targets (function values) $f(\boldsymbol{x})$, which we assume are real-valued if not stated otherwise. Here $\mathbb{R}^D$ is the *domain* of f, and the function values $f(\boldsymbol{x})$ are the *image/codomain* of f. Section 2.7.3 provides much more detailed discussion in the context of linear functions. We often write

domain

image

codomain

$$f : \mathbb{R}^D \to \mathbb{R} \quad (5.1a)$$

$$\boldsymbol{x} \mapsto f(\boldsymbol{x}) \quad (5.1b)$$

Figure 5.1 Vector calculus plays a central role in (a) regression (curve fitting) and (b) density estimation, i.e., modeling data distributions.

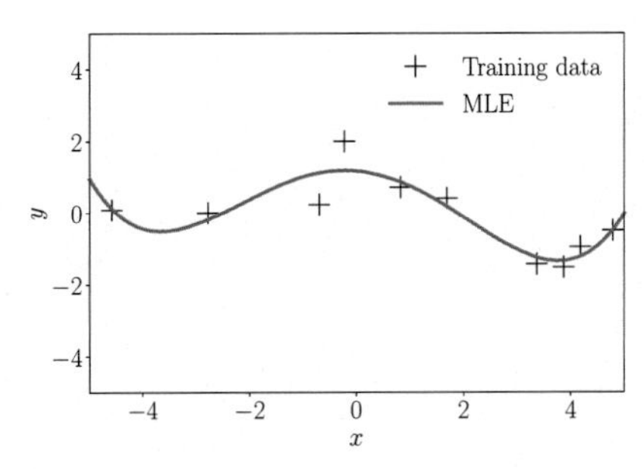

(a) Regression problem: Find parameters, such that the curve explains the observations (crosses) well.

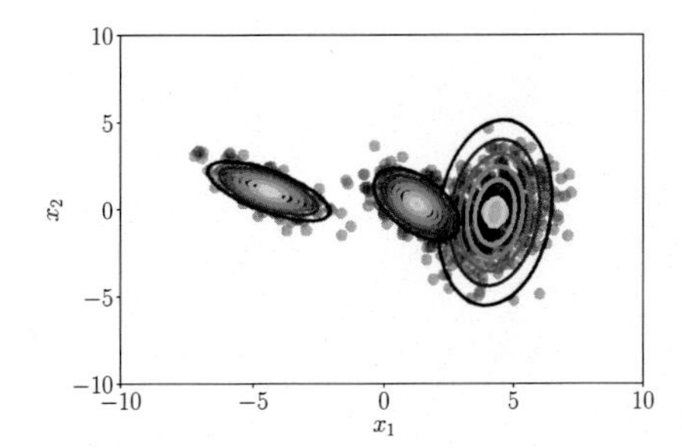

(b) Density estimation with a Gaussian mixture model: Find means and covariances, such that the data (dots) can be explained well.

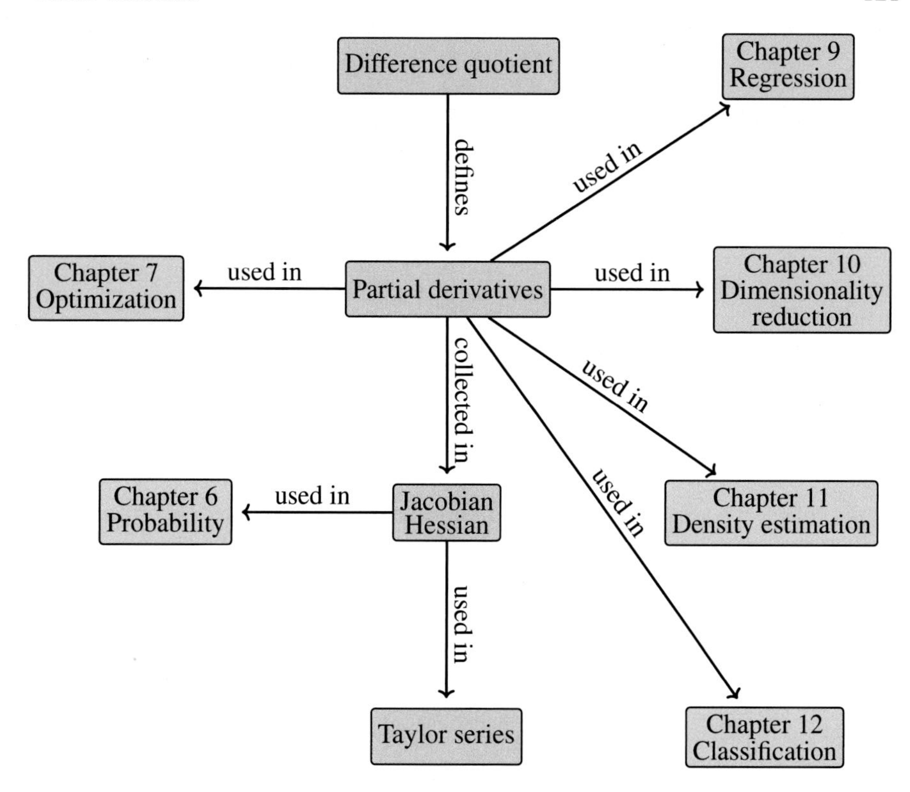

Figure 5.2 A mind map of the concepts introduced in this chapter, along with when they are used in other parts of the book.

to specify a function, where (5.1a) specifies that f is a mapping from $\mathbb{R}^D$ to $\mathbb{R}$ and (5.1b) specifies the explicit assignment of an input $\boldsymbol{x}$ to a function value $f(\boldsymbol{x})$. A function f assigns every input $\boldsymbol{x}$ exactly one function value $f(\boldsymbol{x})$.

Example 5.1

Recall the dot product as a special case of an inner product (Section 3.2). In the previous notation, the function $f(\boldsymbol{x}) = \boldsymbol{x}^\top \boldsymbol{x}$, $\boldsymbol{x} \in \mathbb{R}^2$, would be specified as

$$f : \mathbb{R}^2 \to \mathbb{R} \tag{5.2a}$$

$$\boldsymbol{x} \mapsto x_1^2 + x_2^2 \,. \tag{5.2b}$$

In this chapter, we will discuss how to compute gradients of functions, which is often essential to facilitate learning in machine learning models since the gradient points in the direction of steepest ascent. Therefore, vector calculus is one of the fundamental mathematical tools we need in machine learning. Throughout this book, we assume that functions are differentiable. With some additional technical definitions, which we do not cover here, many of the approaches presented can be extended to subdifferentials (functions that are continuous but not differentiable at certain points). We will look at an extension to the case of functions with constraints in Chapter 7.

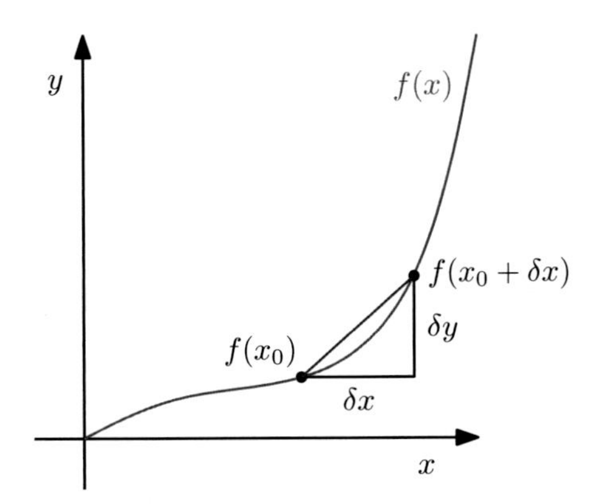

Figure 5.3 The average incline of a function f between x_0 and $x_0 + \delta x$ is the incline of the secant (blue) through $f(x_0)$ and $f(x_0 + \delta x)$ and given by $\delta y/\delta x$.

5.1 Differentiation of Univariate Functions

In the following, we briefly revisit differentiation of a univariate function, which may be familiar from high school mathematics. We start with the difference quotient of a univariate function $y = f(x),\ x, y \in \mathbb{R}$, which we will subsequently use to define derivatives.

difference quotient

Definition 5.1 (Difference Quotient). The *difference quotient*

$$\frac{\delta y}{\delta x} := \frac{f(x + \delta x) - f(x)}{\delta x} \tag{5.3}$$

computes the slope of the secant line through two points on the graph of f. In Figure 5.3, these are the points with x-coordinates x_0 and $x_0 + \delta x$.

The difference quotient can also be considered the average slope of f between x and $x + \delta x$ if we assume f to be a linear function. In the limit for $\delta x \to 0$, we obtain the tangent of f at x, if f is differentiable. The tangent is then the derivative of f at x.

derivative

Definition 5.2 (Derivative). More formally, for $h > 0$ the *derivative* of f at x is defined as the limit

$$\frac{\mathrm{d}f}{\mathrm{d}x} := \lim_{h \to 0} \frac{f(x + h) - f(x)}{h}, \tag{5.4}$$

and the secant in Figure 5.3 becomes a tangent.

The derivative of f points in the direction of steepest ascent of f.

Example 5.2 (Derivative of a Polynomial)
We want to compute the derivative of $f(x) = x^n, n \in \mathbb{N}$. We may already know that the answer will be nx^{n-1}, but we want to derive this result using the definition of the derivative as the limit of the difference quotient.

Using the definition of the derivative in (5.4), we obtain

$$\frac{\mathrm{d}f}{\mathrm{d}x} = \lim_{h \to 0} \frac{f(x + h) - f(x)}{h} \tag{5.5a}$$

$$= \lim_{h \to 0} \frac{(x + h)^n - x^n}{h} \tag{5.5b}$$

$$= \lim_{h\to 0} \frac{\sum_{i=0}^{n} \binom{n}{i} x^{n-i} h^i - x^n}{h}. \tag{5.5c}$$

We see that $x^n = \binom{n}{0} x^{n-0} h^0$. By starting the sum at 1, the x^n-term cancels, and we obtain

$$\frac{\mathrm{d}f}{\mathrm{d}x} = \lim_{h\to 0} \frac{\sum_{i=1}^{n} \binom{n}{i} x^{n-i} h^i}{h} \tag{5.6a}$$

$$= \lim_{h\to 0} \sum_{i=1}^{n} \binom{n}{i} x^{n-i} h^{i-1} \tag{5.6b}$$

$$= \lim_{h\to 0} \binom{n}{1} x^{n-1} + \underbrace{\sum_{i=2}^{n} \binom{n}{i} x^{n-i} h^{i-1}}_{\to 0 \text{ as } h\to 0} \tag{5.6c}$$

$$= \frac{n!}{1!(n-1)!} x^{n-1} = n x^{n-1}. \tag{5.6d}$$

5.1.1 Taylor Series

The Taylor series is a representation of a function f as an infinite sum of terms. These terms are determined using derivatives of f evaluated at x_0.

Definition 5.3 (Taylor Polynomial). The *Taylor polynomial* of degree n of $f : \mathbb{R} \to \mathbb{R}$ at x_0 is defined as

Taylor polynomial

We define $t^0 := 1$ for all $t \in \mathbb{R}$.

$$T_n(x) := \sum_{k=0}^{n} \frac{f^{(k)}(x_0)}{k!} (x - x_0)^k, \tag{5.7}$$

where $f^{(k)}(x_0)$ is the kth derivative of f at x_0 (which we assume exists) and $\frac{f^{(k)}(x_0)}{k!}$ are the coefficients of the polynomial.

Definition 5.4 (Taylor Series). For a smooth function $f \in \mathcal{C}^\infty$, $f : \mathbb{R} \to \mathbb{R}$, the *Taylor series* of f at x_0 is defined as

Taylor series

$$T_\infty(x) = \sum_{k=0}^{\infty} \frac{f^{(k)}(x_0)}{k!} (x - x_0)^k. \tag{5.8}$$

For $x_0 = 0$, we obtain the *Maclaurin series* as a special instance of the Taylor series. If $f(x) = T_\infty(x)$, then f is called *analytic*.

$f \in \mathcal{C}^\infty$ means that f is continuously differentiable infinitely many times.

Maclaurin series

analytic

Remark. In general, a Taylor polynomial of degree n is an approximation of a function, which does not need to be a polynomial. The Taylor polynomial is similar to f in a neighborhood around x_0. However, a Taylor polynomial of degree n is an exact representation of a polynomial f of degree $k \leqslant n$ since all derivatives $f^{(i)}$, $i > k$ vanish. ♢

Example 5.3 (Taylor Polynomial)

We consider the polynomial

$$f(x) = x^4 \tag{5.9}$$

and seek the Taylor polynomial T_6, evaluated at $x_0 = 1$. We start by computing the coefficients $f^{(k)}(1)$ for $k = 0, \ldots, 6$:

$$f(1) = 1 \tag{5.10}$$

$$f'(1) = 4 \tag{5.11}$$

$$f''(1) = 12 \tag{5.12}$$

$$f^{(3)}(1) = 24 \tag{5.13}$$

$$f^{(4)}(1) = 24 \tag{5.14}$$

$$f^{(5)}(1) = 0 \tag{5.15}$$

$$f^{(6)}(1) = 0 \tag{5.16}$$

Therefore, the desired Taylor polynomial is

$$T_6(x) = \sum_{k=0}^{6} \frac{f^{(k)}(x_0)}{k!}(x - x_0)^k \tag{5.17a}$$

$$= 1 + 4(x-1) + 6(x-1)^2 + 4(x-1)^3 + (x-1)^4 + 0\,. \tag{5.17b}$$

Multiplying out and rearranging yields

$$T_6(x) = (1 - 4 + 6 - 4 + 1) + x(4 - 12 + 12 - 4)$$
$$+ x^2(6 - 12 + 6) + x^3(4 - 4) + x^4 \tag{5.18a}$$

$$= x^4 = f(x)\,, \tag{5.18b}$$

i.e., we obtain an exact representation of the original function.

Example 5.4 (Taylor Series)

Consider the function in Figure 5.4 given by

$$f(x) = \sin(x) + \cos(x) \in \mathcal{C}^\infty\,. \tag{5.19}$$

We seek a Taylor series expansion of f at $x_0 = 0$, which is the Maclaurin series expansion of f. We obtain the following derivatives:

$$f(0) = \sin(0) + \cos(0) = 1 \tag{5.20}$$

$$f'(0) = \cos(0) - \sin(0) = 1 \tag{5.21}$$

$$f''(0) = -\sin(0) - \cos(0) = -1 \tag{5.22}$$

$$f^{(3)}(0) = -\cos(0) + \sin(0) = -1 \tag{5.23}$$

$$f^{(4)}(0) = \sin(0) + \cos(0) = f(0) = 1 \tag{5.24}$$

$$\vdots$$

We can see a pattern here: The coefficients in our Taylor series are only ± 1 (since $\sin(0) = 0$), each of which occurs twice before switching to the other one. Furthermore, $f^{(k+4)}(0) = f^{(k)}(0)$.

Therefore, the full Taylor series expansion of f at $x_0 = 0$ is given by

$$
\begin{aligned}
T_\infty(x) &= \sum_{k=0}^{\infty} \frac{f^{(k)}(x_0)}{k!}(x-x_0)^k && \text{(5.25a)}\\
&= 1 + x - \frac{1}{2!}x^2 - \frac{1}{3!}x^3 + \frac{1}{4!}x^4 + \frac{1}{5!}x^5 - \cdots && \text{(5.25b)}\\
&= 1 - \frac{1}{2!}x^2 + \frac{1}{4!}x^4 \mp \cdots + x - \frac{1}{3!}x^3 + \frac{1}{5!}x^5 \mp \cdots && \text{(5.25c)}\\
&= \sum_{k=0}^{\infty}(-1)^k \frac{1}{(2k)!}x^{2k} + \sum_{k=0}^{\infty}(-1)^k \frac{1}{(2k+1)!}x^{2k+1} && \text{(5.25d)}\\
&= \cos(x) + \sin(x)\,, && \text{(5.25e)}
\end{aligned}
$$

where we used the *power series representations*

power series representation

$$\cos(x) = \sum_{k=0}^{\infty}(-1)^k \frac{1}{(2k)!}x^{2k}\,, \tag{5.26}$$

$$\sin(x) = \sum_{k=0}^{\infty}(-1)^k \frac{1}{(2k+1)!}x^{2k+1}\,. \tag{5.27}$$

Figure 5.4 shows the corresponding first Taylor polynomials T_n for $n = 0, 1, 5, 10$.

Remark. A Taylor series is a special case of a power series

$$f(x) = \sum_{k=0}^{\infty} a_k(x-c)^k \tag{5.28}$$

where a_k are coefficients and c is a constant, which has the special form in Definition 5.4. ◇

5.1.2 Differentiation Rules

In the following, we briefly state basic differentiation rules, where we denote the derivative of f by f'.

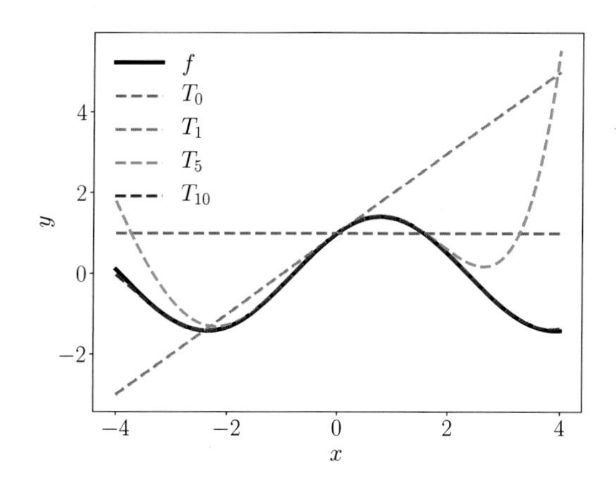

Figure 5.4 Taylor polynomials. The original function $f(x) = \sin(x) + \cos(x)$ (black, solid) is approximated by Taylor polynomials (dashed) around $x_0 = 0$. Higher-order Taylor polynomials approximate the function f better and more globally. T_{10} is already similar to f in $[-4, 4]$.

$$\text{Product rule:} \quad (f(x)g(x))' = f'(x)g(x) + f(x)g'(x) \tag{5.29}$$

$$\text{Quotient rule:} \quad \left(\frac{f(x)}{g(x)}\right)' = \frac{f'(x)g(x) - f(x)g'(x)}{(g(x))^2} \tag{5.30}$$

$$\text{Sum rule:} \quad (f(x) + g(x))' = f'(x) + g'(x) \tag{5.31}$$

$$\text{Chain rule:} \quad \big(g(f(x))\big)' = (g \circ f)'(x) = g'(f(x))f'(x) \tag{5.32}$$

Here, $g \circ f$ denotes function composition $x \mapsto f(x) \mapsto g(f(x))$.

Example 5.5 (Chain Rule)
Let us compute the derivative of the function $h(x) = (2x + 1)^4$ using the chain rule. With

$$h(x) = (2x + 1)^4 = g(f(x))\,, \tag{5.33}$$

$$f(x) = 2x + 1\,, \tag{5.34}$$

$$g(f) = f^4\,, \tag{5.35}$$

we obtain the derivatives of f and g as

$$f'(x) = 2\,, \tag{5.36}$$

$$g'(f) = 4f^3\,, \tag{5.37}$$

such that the derivative of h is given as

$$h'(x) = g'(f)f'(x) = (4f^3) \cdot 2 \overset{(5.34)}{=} 4(2x + 1)^3 \cdot 2 = 8(2x + 1)^3\,, \tag{5.38}$$

where we used the chain rule (5.32) and substituted the definition of f in (5.34) in $g'(f)$.

5.2 Partial Differentiation and Gradients

Differentiation as discussed in Section 5.1 applies to functions f of a scalar variable $x \in \mathbb{R}$. In the following, we consider the general case where the function f depends on one or more variables $\boldsymbol{x} \in \mathbb{R}^n$, e.g., $f(\boldsymbol{x}) = f(x_1, x_2)$. The generalization of the derivative to functions of several variables is the *gradient*.

We find the gradient of the function f with respect to $\boldsymbol{x}$ by *varying one variable at a time* and keeping the others constant. The gradient is then the collection of these *partial derivatives*.

partial derivative

Definition 5.5 (Partial Derivative). For a function $f : \mathbb{R}^n \to \mathbb{R}$, $\boldsymbol{x} \mapsto f(\boldsymbol{x})$, $\boldsymbol{x} \in \mathbb{R}^n$ of n variables $x_1, \dots, x_n$ we define the *partial derivatives* as

$$\begin{aligned} \frac{\partial f}{\partial x_1} &= \lim_{h \to 0} \frac{f(x_1 + h, x_2, \dots, x_n) - f(\boldsymbol{x})}{h} \\ &\vdots \\ \frac{\partial f}{\partial x_n} &= \lim_{h \to 0} \frac{f(x_1, \dots, x_{n-1}, x_n + h) - f(\boldsymbol{x})}{h} \end{aligned} \tag{5.39}$$

and collect them in the row vector

$$\nabla_{\boldsymbol{x}} f = \text{grad} f = \frac{\mathrm{d}f}{\mathrm{d}\boldsymbol{x}} = \begin{bmatrix} \frac{\partial f(\boldsymbol{x})}{\partial x_1} & \frac{\partial f(\boldsymbol{x})}{\partial x_2} & \cdots & \frac{\partial f(\boldsymbol{x})}{\partial x_n} \end{bmatrix} \in \mathbb{R}^{1 \times n}, \tag{5.40}$$

where n is the number of variables and 1 is the dimension of the image/range/codomain of f. Here, we defined the column vector $\boldsymbol{x} = [x_1, \ldots, x_n]^\top \in \mathbb{R}^n$. The row vector in (5.40) is called the *gradient* of f or the *Jacobian* and is the generalization of the derivative from Section 5.1.

gradient

Jacobian

Remark. This definition of the Jacobian is a special case of the general definition of the Jacobian for vector-valued functions as the collection of partial derivatives. We will get back to this in Section 5.3. ◇

Example 5.6 (Partial Derivatives Using the Chain Rule)
For $f(x, y) = (x + 2y^3)^2$, we obtain the partial derivatives

$$\frac{\partial f(x, y)}{\partial x} = 2(x + 2y^3)\frac{\partial}{\partial x}(x + 2y^3) = 2(x + 2y^3), \tag{5.41}$$

$$\frac{\partial f(x, y)}{\partial y} = 2(x + 2y^3)\frac{\partial}{\partial y}(x + 2y^3) = 12(x + 2y^3)y^2, \tag{5.42}$$

where we used the chain rule (5.32) to compute the partial derivatives.

We can use results from scalar differentiation: Each partial derivative is a derivative with respect to a scalar.

Remark (Gradient as a Row Vector). It is not uncommon in the literature to define the gradient vector as a column vector, following the convention that vectors are generally column vectors. The reason why we define the gradient vector as a row vector is twofold: First, we can consistently generalize the gradient to vector-valued functions $f : \mathbb{R}^n \to \mathbb{R}^m$ (then the gradient becomes a matrix). Second, we can immediately apply the multivariate chain rule without paying attention to the dimension of the gradient. We will discuss both points in Section 5.3. ◇

Example 5.7 (Gradient)
For $f(x_1, x_2) = x_1^2 x_2 + x_1 x_2^3 \in \mathbb{R}$, the partial derivatives (i.e., the derivatives of f with respect to x_1 and x_2) are

$$\frac{\partial f(x_1, x_2)}{\partial x_1} = 2x_1 x_2 + x_2^3 \tag{5.43}$$

$$\frac{\partial f(x_1, x_2)}{\partial x_2} = x_1^2 + 3x_1 x_2^2 \tag{5.44}$$

and the gradient is then

$$\frac{\mathrm{d}f}{\mathrm{d}\boldsymbol{x}} = \begin{bmatrix} \frac{\partial f(x_1, x_2)}{\partial x_1} & \frac{\partial f(x_1, x_2)}{\partial x_2} \end{bmatrix} = \begin{bmatrix} 2x_1 x_2 + x_2^3 & x_1^2 + 3x_1 x_2^2 \end{bmatrix} \in \mathbb{R}^{1 \times 2}. \tag{5.45}$$

5.2.1 Basic Rules of Partial Differentiation

Product rule: $(fg)' = f'g + fg'$, Sum rule: $(f+g)' = f' + g'$, Chain rule: $(g(f))' = g'(f)f'$

In the multivariate case, where $\boldsymbol{x} \in \mathbb{R}^n$, the basic differentiation rules that we know from school (e.g., sum rule, product rule, chain rule; see also Section 5.1.2) still apply. However, when we compute derivatives with respect to vectors $\boldsymbol{x} \in \mathbb{R}^n$, we need to pay attention: Our gradients now involve vectors and matrices, and matrix multiplication is not commutative (Section 2.2.1), i.e., the order matters.

Here are the general product rule, sum rule, and chain rule:

$$\text{Product rule:} \quad \frac{\partial}{\partial \boldsymbol{x}}\big(f(\boldsymbol{x})g(\boldsymbol{x})\big) = \frac{\partial f}{\partial \boldsymbol{x}} g(\boldsymbol{x}) + f(\boldsymbol{x})\frac{\partial g}{\partial \boldsymbol{x}} \tag{5.46}$$

$$\text{Sum rule:} \quad \frac{\partial}{\partial \boldsymbol{x}}\big(f(\boldsymbol{x}) + g(\boldsymbol{x})\big) = \frac{\partial f}{\partial \boldsymbol{x}} + \frac{\partial g}{\partial \boldsymbol{x}} \tag{5.47}$$

$$\text{Chain rule:} \quad \frac{\partial}{\partial \boldsymbol{x}}(g \circ f)(\boldsymbol{x}) = \frac{\partial}{\partial \boldsymbol{x}}\big(g(f(\boldsymbol{x}))\big) = \frac{\partial g}{\partial f}\frac{\partial f}{\partial \boldsymbol{x}} \tag{5.48}$$

This is only an intuition, but not mathematically correct since the partial derivative is not a fraction.

Let us have a closer look at the chain rule. The chain rule (5.48) resembles to some degree the rules for matrix multiplication where we said that neighboring dimensions have to match for matrix multiplication to be defined; see Section 2.2.1. If we go from left to right, the chain rule exhibits similar properties: ∂f shows up in the "denominator" of the first factor and in the "numerator" of the second factor. If we multiply the factors together, multiplication is defined, i.e., the dimensions of ∂f match, and ∂f "cancels," such that $\partial g/\partial \boldsymbol{x}$ remains.

5.2.2 Chain Rule

Consider a function $f : \mathbb{R}^2 \to \mathbb{R}$ of two variables x_1, x_2. Furthermore, $x_1(t)$ and $x_2(t)$ are themselves functions of t. To compute the gradient of f with respect to t, we need to apply the chain rule (5.48) for multivariate functions as

$$\frac{\mathrm{d}f}{\mathrm{d}t} = \begin{bmatrix} \frac{\partial f}{\partial x_1} & \frac{\partial f}{\partial x_2} \end{bmatrix} \begin{bmatrix} \frac{\partial x_1(t)}{\partial t} \\ \frac{\partial x_2(t)}{\partial t} \end{bmatrix} = \frac{\partial f}{\partial x_1}\frac{\partial x_1}{\partial t} + \frac{\partial f}{\partial x_2}\frac{\partial x_2}{\partial t}, \tag{5.49}$$

where d denotes the gradient and ∂ partial derivatives.

Example 5.8

Consider $f(x_1, x_2) = x_1^2 + 2x_2$, where $x_1 = \sin t$ and $x_2 = \cos t$, then

$$\frac{\mathrm{d}f}{\mathrm{d}t} = \frac{\partial f}{\partial x_1}\frac{\partial x_1}{\partial t} + \frac{\partial f}{\partial x_2}\frac{\partial x_2}{\partial t} \tag{5.50a}$$

$$= 2\sin t \frac{\partial \sin t}{\partial t} + 2\frac{\partial \cos t}{\partial t} \tag{5.50b}$$

$$= 2\sin t \cos t - 2\sin t = 2\sin t(\cos t - 1) \tag{5.50c}$$

is the corresponding derivative of f with respect to t.

If $f(x_1, x_2)$ is a function of x_1 and x_2, where $x_1(s,t)$ and $x_2(s,t)$ are themselves functions of two variables s and t, the chain rule yields the partial derivatives

$$\frac{\partial f}{\partial s} = \frac{\partial f}{\partial x_1}\frac{\partial x_1}{\partial s} + \frac{\partial f}{\partial x_2}\frac{\partial x_2}{\partial s}, \tag{5.51}$$

$$\frac{\partial f}{\partial t} = \frac{\partial f}{\partial x_1}\frac{\partial x_1}{\partial t} + \frac{\partial f}{\partial x_2}\frac{\partial x_2}{\partial t}, \tag{5.52}$$

and the gradient is obtained by the matrix multiplication

$$\frac{\mathrm{d}f}{\mathrm{d}(s,t)} = \frac{\partial f}{\partial \boldsymbol{x}}\frac{\partial \boldsymbol{x}}{\partial(s,t)} = \underbrace{\begin{bmatrix} \frac{\partial f}{\partial x_1} & \frac{\partial f}{\partial x_2} \end{bmatrix}}_{=\frac{\partial f}{\partial \boldsymbol{x}}} \underbrace{\begin{bmatrix} \frac{\partial x_1}{\partial s} & \frac{\partial x_1}{\partial t} \\ \frac{\partial x_2}{\partial s} & \frac{\partial x_2}{\partial t} \end{bmatrix}}_{=\frac{\partial \boldsymbol{x}}{\partial(s,t)}}. \tag{5.53}$$

This compact way of writing the chain rule as a matrix multiplication only makes sense if the gradient is defined as a row vector. Otherwise, we will need to start transposing gradients for the matrix dimensions to match. This may still be straightforward as long as the gradient is a vector or a matrix; however, when the gradient becomes a tensor (we will discuss this in the following), the transpose is no longer a triviality.

The chain rule can be written as a matrix multiplication.

Remark (Verifying the Correctness of a Gradient Implementation). The definition of the partial derivatives as the limit of the corresponding difference quotient (see (5.39)) can be exploited when numerically checking the correctness of gradients in computer programs: When we compute gradients and implement them, we can use finite differences to numerically test our computation and implementation: We choose the value h to be small (e.g., $h = 10^{-4}$) and compare the finite-difference approximation from (5.39) with our (analytic) implementation of the gradient. If the error is small, our gradient implementation is probably correct. "Small" could mean that $\sqrt{\frac{\sum_i (dh_i - df_i)^2}{\sum_i (dh_i + df_i)^2}} < 10^{-6}$, where dh_i is the finite-difference approximation and df_i is the analytic gradient of f with respect to the ith variable x_i. ◇

gradient checking

5.3 Gradients of Vector-Valued Functions

Thus far, we discussed partial derivatives and gradients of functions $f : \mathbb{R}^n \to \mathbb{R}$ mapping to the real numbers. In the following, we will generalize the concept of the gradient to vector-valued functions (vector fields) $\boldsymbol{f} : \mathbb{R}^n \to \mathbb{R}^m$, where $n \geqslant 1$ and $m > 1$.

For a function $\boldsymbol{f} : \mathbb{R}^n \to \mathbb{R}^m$ and a vector $\boldsymbol{x} = [x_1, \ldots, x_n]^\top \in \mathbb{R}^n$, the corresponding vector of function values is given as

$$\boldsymbol{f}(\boldsymbol{x}) = \begin{bmatrix} f_1(\boldsymbol{x}) \\ \vdots \\ f_m(\boldsymbol{x}) \end{bmatrix} \in \mathbb{R}^m. \tag{5.54}$$

Writing the vector-valued function in this way allows us to view a vector-valued function $\boldsymbol{f} : \mathbb{R}^n \to \mathbb{R}^m$ as a vector of functions $[f_1, \ldots, f_m]^\top$, $f_i : \mathbb{R}^n \to \mathbb{R}$ that map onto $\mathbb{R}$. The differentiation rules for every f_i are exactly the ones we discussed in Section 5.2.

Therefore, the partial derivative of a vector-valued function $\boldsymbol{f}:\mathbb{R}^n\to\mathbb{R}^m$ with respect to $x_i\in\mathbb{R}$, $i=1,\ldots n$, is given as the vector

$$\frac{\partial \boldsymbol{f}}{\partial x_i}=\begin{bmatrix}\frac{\partial f_1}{\partial x_i}\\ \vdots\\ \frac{\partial f_m}{\partial x_i}\end{bmatrix}=\begin{bmatrix}\lim_{h\to 0}\frac{f_1(x_1,\ldots,x_{i-1},x_i+h,x_{i+1},\ldots x_n)-f_1(\boldsymbol{x})}{h}\\ \vdots\\ \lim_{h\to 0}\frac{f_m(x_1,\ldots,x_{i-1},x_i+h,x_{i+1},\ldots x_n)-f_m(\boldsymbol{x})}{h}\end{bmatrix}\in\mathbb{R}^m. \tag{5.55}$$

From (5.40), we know that the gradient of f with respect to a vector is the row vector of the partial derivatives. In (5.55), every partial derivative $\partial f/\partial x_i$ is a column vector. Therefore, we obtain the gradient of $\boldsymbol{f}:\mathbb{R}^n\to\mathbb{R}^m$ with respect to $\boldsymbol{x}\in\mathbb{R}^n$ by collecting these partial derivatives:

$$\frac{\mathrm{d}\boldsymbol{f}(\boldsymbol{x})}{\mathrm{d}\boldsymbol{x}}=\begin{bmatrix}\boxed{\frac{\partial \boldsymbol{f}(\boldsymbol{x})}{\partial x_1}} & \cdots & \boxed{\frac{\partial \boldsymbol{f}(\boldsymbol{x})}{\partial x_n}}\end{bmatrix} \tag{5.56a}$$

$$=\begin{bmatrix}\frac{\partial f_1(\boldsymbol{x})}{\partial x_1} & \cdots & \frac{\partial f_1(\boldsymbol{x})}{\partial x_n}\\ \vdots & & \vdots\\ \frac{\partial f_m(\boldsymbol{x})}{\partial x_1} & \cdots & \frac{\partial f_m(\boldsymbol{x})}{\partial x_n}\end{bmatrix}\in\mathbb{R}^{m\times n}. \tag{5.56b}$$

Jacobian

The gradient of a function $\boldsymbol{f}:\mathbb{R}^n\to\mathbb{R}^m$ is a matrix of size $m\times n$.

Definition 5.6 (Jacobian)**.** The collection of all first-order partial derivatives of a vector-valued function $\boldsymbol{f}:\mathbb{R}^n\to\mathbb{R}^m$ is called the *Jacobian*. The Jacobian $\boldsymbol{J}$ is an $m\times n$ matrix, which we define and arrange as follows:

$$\boldsymbol{J}=\nabla_{\boldsymbol{x}}\boldsymbol{f}=\frac{\mathrm{d}\boldsymbol{f}(\boldsymbol{x})}{\mathrm{d}\boldsymbol{x}}=\begin{bmatrix}\frac{\partial \boldsymbol{f}(\boldsymbol{x})}{\partial x_1} & \cdots & \frac{\partial \boldsymbol{f}(\boldsymbol{x})}{\partial x_n}\end{bmatrix} \tag{5.57}$$

$$=\begin{bmatrix}\frac{\partial f_1(\boldsymbol{x})}{\partial x_1} & \cdots & \frac{\partial f_1(\boldsymbol{x})}{\partial x_n}\\ \vdots & & \vdots\\ \frac{\partial f_m(\boldsymbol{x})}{\partial x_1} & \cdots & \frac{\partial f_m(\boldsymbol{x})}{\partial x_n}\end{bmatrix}, \tag{5.58}$$

$$\boldsymbol{x}=\begin{bmatrix}x_1\\ \vdots\\ x_n\end{bmatrix},\quad J(i,j)=\frac{\partial f_i}{\partial x_j}. \tag{5.59}$$

As a special case of (5.58), a function $f:\mathbb{R}^n\to\mathbb{R}^1$, which maps a vector $\boldsymbol{x}\in\mathbb{R}^n$ onto a scalar (e.g., $f(\boldsymbol{x})=\sum_{i=1}^n x_i$), possesses a Jacobian that is a row vector (matrix of dimension $1\times n$); see (5.40).

numerator layout

denominator layout

Remark. In this book, we use the *numerator layout* of the derivative, i.e., the derivative $\mathrm{d}\boldsymbol{f}/\mathrm{d}\boldsymbol{x}$ of $\boldsymbol{f}\in\mathbb{R}^m$ with respect to $\boldsymbol{x}\in\mathbb{R}^n$ is an $m\times n$ matrix, where the elements of $\boldsymbol{f}$ define the rows and the elements of $\boldsymbol{x}$ define the columns of the corresponding Jacobian; see (5.58). There exists also the *denominator layout*, which is the transpose of the numerator layout. In this book, we will use the numerator layout. ◇

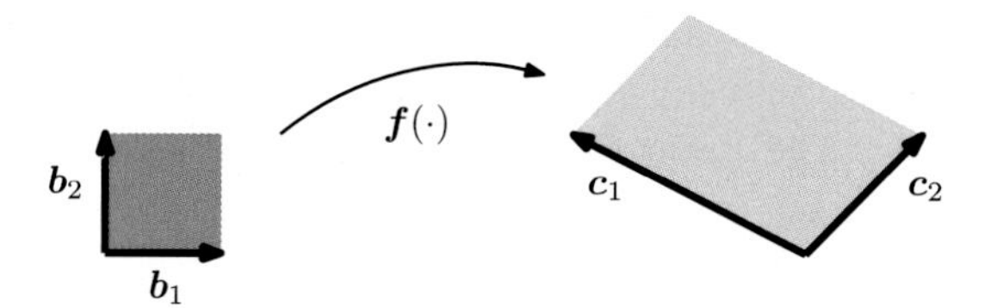

Figure 5.5 The determinant of the Jacobian of $\boldsymbol{f}$ can be used to compute the magnifier between the blue and orange area.

We will see how the Jacobian is used in the change-of-variable method for probability distributions in Section 6.7. The amount of scaling due to the transformation of a variable is provided by the determinant.

In Section 4.1, we saw that the determinant can be used to compute the area of a parallelogram. If we are given two vectors $\boldsymbol{b}_1 = [1, 0]^\top$, $\boldsymbol{b}_2 = [0, 1]^\top$ as the sides of the unit square (blue; see Figure 5.5), the area of this square is

$$\left|\det\left(\begin{bmatrix}1 & 0\\0 & 1\end{bmatrix}\right)\right| = 1\,. \tag{5.60}$$

If we take a parallelogram with the sides $\boldsymbol{c}_1 = [-2, 1]^\top$, $\boldsymbol{c}_2 = [1, 1]^\top$ (orange in Figure 5.5), its area is given as the absolute value of the determinant (see Section 4.1)

$$\left|\det\left(\begin{bmatrix}-2 & 1\\1 & 1\end{bmatrix}\right)\right| = |-3| = 3\,, \tag{5.61}$$

i.e., the area of this is exactly three times the area of the unit square. We can find this scaling factor by finding a mapping that transforms the unit square into the other square. In linear algebra terms, we effectively perform a variable transformation from $(\boldsymbol{b}_1, \boldsymbol{b}_2)$ to $(\boldsymbol{c}_1, \boldsymbol{c}_2)$. In our case, the mapping is linear and the absolute value of the determinant of this mapping gives us exactly the scaling factor we are looking for.

We will describe two approaches to identify this mapping. First, we exploit that the mapping is linear so that we can use the tools from Chapter 2 to identify this mapping. Second, we will find the mapping using partial derivatives using the tools we have been discussing in this chapter.

Approach 1 To get started with the linear algebra approach, we identify both $\{\boldsymbol{b}_1, \boldsymbol{b}_2\}$ and $\{\boldsymbol{c}_1, \boldsymbol{c}_2\}$ as bases of $\mathbb{R}^2$ (see Section 2.6.1 for a recap). What we effectively perform is a change of basis from $(\boldsymbol{b}_1, \boldsymbol{b}_2)$ to $(\boldsymbol{c}_1, \boldsymbol{c}_2)$, and we are looking for the transformation matrix that implements the basis change. Using results from Section 2.7.2, we identify the desired basis change matrix as

$$\boldsymbol{J} = \begin{bmatrix}-2 & 1\\1 & 1\end{bmatrix}, \tag{5.62}$$

such that $\boldsymbol{J}\boldsymbol{b}_1 = \boldsymbol{c}_1$ and $\boldsymbol{J}\boldsymbol{b}_2 = \boldsymbol{c}_2$. The absolute value of the determinant of $\boldsymbol{J}$, which yields the scaling factor we are looking for, is given as $|\det(\boldsymbol{J})| = 3$, i.e., the area of the square spanned by $(\boldsymbol{c}_1, \boldsymbol{c}_2)$ is three times greater than the area spanned by $(\boldsymbol{b}_1, \boldsymbol{b}_2)$.

Approach 2 The linear algebra approach works for linear transformations; for nonlinear transformations (which become relevant in Section 6.7), we follow a more general approach using partial derivatives.

For this approach, we consider a function $\boldsymbol{f} : \mathbb{R}^2 \to \mathbb{R}^2$ that performs a variable transformation. In our example, $\boldsymbol{f}$ maps the coordinate representation of any vector $\boldsymbol{x} \in \mathbb{R}^2$ with respect to $(\boldsymbol{b}_1, \boldsymbol{b}_2)$ onto the coordinate representation $\boldsymbol{y} \in \mathbb{R}^2$ with respect to $(\boldsymbol{c}_1, \boldsymbol{c}_2)$. We want to identify the mapping so that we can compute how an area (or volume) changes when it is being transformed by $\boldsymbol{f}$. For this, we need to find out how $\boldsymbol{f}(\boldsymbol{x})$ changes if we modify $\boldsymbol{x}$ a bit. This question is exactly answered by the Jacobian matrix $\frac{\mathrm{d}\boldsymbol{f}}{\mathrm{d}\boldsymbol{x}} \in \mathbb{R}^{2\times 2}$. Since we can write

$$y_1 = -2x_1 + x_2 \tag{5.63}$$

$$y_2 = x_1 + x_2 , \tag{5.64}$$

we obtain the functional relationship between $\boldsymbol{x}$ and $\boldsymbol{y}$, which allows us to get the partial derivatives

$$\frac{\partial y_1}{\partial x_1} = -2 , \quad \frac{\partial y_1}{\partial x_2} = 1 , \quad \frac{\partial y_2}{\partial x_1} = 1 , \quad \frac{\partial y_2}{\partial x_2} = 1 \tag{5.65}$$

and compose the Jacobian as

$$\boldsymbol{J} = \begin{bmatrix} \frac{\partial y_1}{\partial x_1} & \frac{\partial y_1}{\partial x_2} \\ \frac{\partial y_2}{\partial x_1} & \frac{\partial y_2}{\partial x_2} \end{bmatrix} = \begin{bmatrix} -2 & 1 \\ 1 & 1 \end{bmatrix} . \tag{5.66}$$

Geometrically, the Jacobian determinant gives the magnification/scaling factor when we transform an area or volume.

Jacobian determinant

The Jacobian represents the coordinate transformation we are looking for. It is exact if the coordinate transformation is linear (as in our case), and (5.66) recovers exactly the basis change matrix in (5.62). If the coordinate transformation is nonlinear, the Jacobian approximates this nonlinear transformation locally with a linear one. The absolute value of the *Jacobian determinant* $|\det(\boldsymbol{J})|$ is the factor by which areas or volumes are scaled when coordinates are transformed. Our case yields $|\det(\boldsymbol{J})| = 3$.

The Jacobian determinant and variable transformations will become relevant in Section 6.7 when we transform random variables and probability distributions. These transformations are extremely relevant in machine learning in the context of training deep neural networks using the *reparametrization trick*, also called *infinite perturbation analysis*.

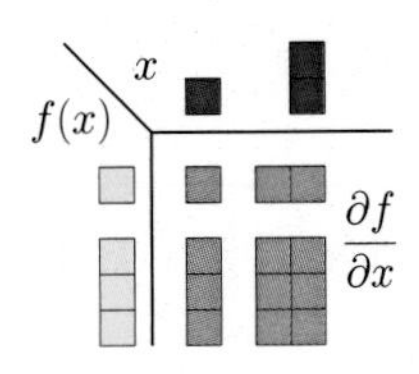

Figure 5.6 Dimensionality of (partial) derivatives.

In this chapter, we encountered derivatives of functions. Figure 5.6 summarizes the dimensions of those derivatives. If $f : \mathbb{R} \to \mathbb{R}$, the gradient is simply a scalar (top-left entry). For $f : \mathbb{R}^D \to \mathbb{R}$, the gradient is a $1 \times D$ row vector (top-right entry). For $\boldsymbol{f} : \mathbb{R} \to \mathbb{R}^E$, the gradient is an $E \times 1$ column vector, and for $\boldsymbol{f} : \mathbb{R}^D \to \mathbb{R}^E$ the gradient is an $E \times D$ matrix.

Example 5.9 (Gradient of a Vector-Valued Function)
We are given

$$\boldsymbol{f}(\boldsymbol{x}) = \boldsymbol{A}\boldsymbol{x}\,, \qquad \boldsymbol{f}(\boldsymbol{x}) \in \mathbb{R}^M, \quad \boldsymbol{A} \in \mathbb{R}^{M\times N}, \quad \boldsymbol{x} \in \mathbb{R}^N\,.$$

To compute the gradient $\mathrm{d}\boldsymbol{f}/\mathrm{d}\boldsymbol{x}$, we first determine the dimension of $\mathrm{d}\boldsymbol{f}/\mathrm{d}\boldsymbol{x}$: Since $\boldsymbol{f} : \mathbb{R}^N \to \mathbb{R}^M$, it follows that $\mathrm{d}\boldsymbol{f}/\mathrm{d}\boldsymbol{x} \in \mathbb{R}^{M\times N}$. Second, to compute the gradient, we determine the partial derivatives of f with respect to every x_j:

$$f_i(\boldsymbol{x}) = \sum_{j=1}^{N} A_{ij} x_j \implies \frac{\partial f_i}{\partial x_j} = A_{ij}\,. \tag{5.67}$$

We collect the partial derivatives in the Jacobian and obtain the gradient

$$\frac{\mathrm{d}\boldsymbol{f}}{\mathrm{d}\boldsymbol{x}} = \begin{bmatrix} \frac{\partial f_1}{\partial x_1} & \cdots & \frac{\partial f_1}{\partial x_N} \\ \vdots & & \vdots \\ \frac{\partial f_M}{\partial x_1} & \cdots & \frac{\partial f_M}{\partial x_N} \end{bmatrix} = \begin{bmatrix} A_{11} & \cdots & A_{1N} \\ \vdots & & \vdots \\ A_{M1} & \cdots & A_{MN} \end{bmatrix} = \boldsymbol{A} \in \mathbb{R}^{M\times N}\,. \tag{5.68}$$

Example 5.10 (Chain Rule)
Consider the function $h : \mathbb{R} \to \mathbb{R}$, $h(t) = (f \circ g)(t)$ with

$$f : \mathbb{R}^2 \to \mathbb{R} \tag{5.69}$$

$$g : \mathbb{R} \to \mathbb{R}^2 \tag{5.70}$$

$$f(\boldsymbol{x}) = \exp(x_1 x_2^2)\,, \tag{5.71}$$

$$\boldsymbol{x} = \begin{bmatrix} x_1 \\ x_2 \end{bmatrix} = g(t) = \begin{bmatrix} t\cos t \\ t\sin t \end{bmatrix} \tag{5.72}$$

and compute the gradient of h with respect to t. Since $f : \mathbb{R}^2 \to \mathbb{R}$ and $g : \mathbb{R} \to \mathbb{R}^2$, we note that

$$\frac{\partial f}{\partial \boldsymbol{x}} \in \mathbb{R}^{1\times 2}\,, \quad \frac{\partial g}{\partial t} \in \mathbb{R}^{2\times 1}\,. \tag{5.73}$$

The desired gradient is computed by applying the chain rule:

$$\frac{\mathrm{d}h}{\mathrm{d}t} = \frac{\partial f}{\partial \boldsymbol{x}}\frac{\partial \boldsymbol{x}}{\partial t} = \begin{bmatrix} \frac{\partial f}{\partial x_1} & \frac{\partial f}{\partial x_2} \end{bmatrix} \begin{bmatrix} \frac{\partial x_1}{\partial t} \\ \frac{\partial x_2}{\partial t} \end{bmatrix} \tag{5.74a}$$

$$= \begin{bmatrix} \exp(x_1 x_2^2) x_2^2 & 2\exp(x_1 x_2^2) x_1 x_2 \end{bmatrix} \begin{bmatrix} \cos t - t\sin t \\ \sin t + t\cos t \end{bmatrix} \tag{5.74b}$$

$$= \exp(x_1 x_2^2)\big(x_2^2(\cos t - t\sin t) + 2x_1 x_2(\sin t + t\cos t)\big)\,, \tag{5.74c}$$

where $x_1 = t\cos t$ and $x_2 = t\sin t$; see (5.72).

We will discuss this model in much more detail in Chapter 9 in the context of linear regression, where we need derivatives of the least-squares loss L with respect to the parameters $\boldsymbol{\theta}$.

Example 5.11 (Gradient of a Least-Squares Loss in a Linear Model)

Let us consider the linear model

$$\boldsymbol{y} = \boldsymbol{\Phi}\boldsymbol{\theta}, \tag{5.75}$$

where $\boldsymbol{\theta} \in \mathbb{R}^D$ is a parameter vector, $\boldsymbol{\Phi} \in \mathbb{R}^{N \times D}$ are input features, and $\boldsymbol{y} \in \mathbb{R}^N$ are the corresponding observations. We define the functions

$$L(\boldsymbol{e}) := \|\boldsymbol{e}\|^2, \tag{5.76}$$

$$\boldsymbol{e}(\boldsymbol{\theta}) := \boldsymbol{y} - \boldsymbol{\Phi}\boldsymbol{\theta}. \tag{5.77}$$

least-squares loss

We seek $\frac{\partial L}{\partial \boldsymbol{\theta}}$, and we will use the chain rule for this purpose. L is called a *least-squares loss* function.

Before we start our calculation, we determine the dimensionality of the gradient as

$$\frac{\partial L}{\partial \boldsymbol{\theta}} \in \mathbb{R}^{1 \times D}. \tag{5.78}$$

The chain rule allows us to compute the gradient as

$$\frac{\partial L}{\partial \boldsymbol{\theta}} = \frac{\partial L}{\partial \boldsymbol{e}} \frac{\partial \boldsymbol{e}}{\partial \boldsymbol{\theta}}, \tag{5.79}$$

```
dLdtheta =
np.einsum(
'n,nd',
dLde,dedtheta)
```

where the dth element is given by

$$\frac{\partial L}{\partial \boldsymbol{\theta}}[1, d] = \sum_{n=1}^{N} \frac{\partial L}{\partial \boldsymbol{e}}[n] \frac{\partial \boldsymbol{e}}{\partial \boldsymbol{\theta}}[n, d]. \tag{5.80}$$

We know that $\|\boldsymbol{e}\|^2 = \boldsymbol{e}^\top \boldsymbol{e}$ (see Section 3.2) and determine

$$\frac{\partial L}{\partial \boldsymbol{e}} = 2\boldsymbol{e}^\top \in \mathbb{R}^{1 \times N}. \tag{5.81}$$

Furthermore, we obtain

$$\frac{\partial \boldsymbol{e}}{\partial \boldsymbol{\theta}} = -\boldsymbol{\Phi} \in \mathbb{R}^{N \times D}, \tag{5.82}$$

such that our desired derivative is

$$\frac{\partial L}{\partial \boldsymbol{\theta}} = -2\boldsymbol{e}^\top \boldsymbol{\Phi} \overset{(5.77)}{=} -\underbrace{2(\boldsymbol{y}^\top - \boldsymbol{\theta}^\top \boldsymbol{\Phi}^\top)}_{1 \times N} \underbrace{\boldsymbol{\Phi}}_{N \times D} \in \mathbb{R}^{1 \times D}. \tag{5.83}$$

Remark. We would have obtained the same result without using the chain rule by immediately looking at the function

$$L_2(\boldsymbol{\theta}) := \|\boldsymbol{y} - \boldsymbol{\Phi}\boldsymbol{\theta}\|^2 = (\boldsymbol{y} - \boldsymbol{\Phi}\boldsymbol{\theta})^\top (\boldsymbol{y} - \boldsymbol{\Phi}\boldsymbol{\theta}). \tag{5.84}$$

This approach is still practical for simple functions like L_2 but becomes impractical for deep function compositions. ◇

5.4 Gradients of Matrices

We will encounter situations where we need to take gradients of matrices with respect to vectors (or other matrices), which results in a multidimensional tensor. We can think of this tensor as a multidimensional array that collects partial derivatives. For example, if we compute the gradient of an $m \times n$ matrix $\boldsymbol{A}$ with respect to a $p \times q$ matrix $\boldsymbol{B}$, the resulting Jacobian would be $(m \times n) \times (p \times q)$, i.e., a four-dimensional tensor $\boldsymbol{J}$, whose entries are given as $J_{ijkl} = \partial A_{ij}/\partial B_{kl}$.

We can think of a tensor as a multidimensional array.

Since matrices represent linear mappings, we can exploit the fact that there is a vector-space isomorphism (linear, invertible mapping) between the space $\mathbb{R}^{m \times n}$ of $m \times n$ matrices and the space $\mathbb{R}^{mn}$ of mn vectors. Therefore, we can reshape our matrices into vectors of lengths mn and pq, respectively. The gradient using these mn vectors results in a Jacobian of size $mn \times pq$. Figure 5.7 visualizes both approaches. In practical applications, it is often desirable to reshape the matrix into a vector and continue working with this Jacobian matrix: The chain rule (5.48) boils down to simple matrix multiplication, whereas in the case of a Jacobian tensor, we will need to pay more attention to what dimensions we need to sum out.

Matrices can be transformed into vectors by stacking the columns of the matrix ("flattening").

Example 5.12 (Gradient of Vectors with Respect to Matrices)
Let us consider the following example, where

$$\boldsymbol{f} = \boldsymbol{A}\boldsymbol{x}, \quad \boldsymbol{f} \in \mathbb{R}^M, \quad \boldsymbol{A} \in \mathbb{R}^{M \times N}, \quad \boldsymbol{x} \in \mathbb{R}^N \tag{5.85}$$

and where we seek the gradient $\mathrm{d}\boldsymbol{f}/\mathrm{d}\boldsymbol{A}$. Let us start again by determining the dimension of the gradient as

$$\frac{\mathrm{d}\boldsymbol{f}}{\mathrm{d}\boldsymbol{A}} \in \mathbb{R}^{M \times (M \times N)}. \tag{5.86}$$

By definition, the gradient is the collection of the partial derivatives:

$$\frac{\mathrm{d}\boldsymbol{f}}{\mathrm{d}\boldsymbol{A}} = \begin{bmatrix} \frac{\partial f_1}{\partial \boldsymbol{A}} \\ \vdots \\ \frac{\partial f_M}{\partial \boldsymbol{A}} \end{bmatrix}, \quad \frac{\partial f_i}{\partial \boldsymbol{A}} \in \mathbb{R}^{1 \times (M \times N)}. \tag{5.87}$$

To compute the partial derivatives, it will be helpful to explicitly write out the matrix vector multiplication:

$$f_i = \sum_{j=1}^{N} A_{ij} x_j, \quad i = 1, \ldots, M, \tag{5.88}$$

and the partial derivatives are then given as

$$\frac{\partial f_i}{\partial A_{iq}} = x_q. \tag{5.89}$$

This allows us to compute the partial derivatives of f_i with respect to a row of $\boldsymbol{A}$, which is given as

$$\frac{\partial f_i}{\partial A_{i,:}} = \boldsymbol{x}^\top \in \mathbb{R}^{1 \times 1 \times N}, \tag{5.90}$$

$$\frac{\partial f_i}{\partial A_{k\neq i,:}} = \mathbf{0}^\top \in \mathbb{R}^{1\times 1\times N}, \tag{5.91}$$

where we have to pay attention to the correct dimensionality. Since f_i maps onto $\mathbb{R}$ and each row of $\boldsymbol{A}$ is of size $1 \times N$, we obtain a $1 \times 1 \times N$-sized tensor as the partial derivative of f_i with respect to a row of $\boldsymbol{A}$.

We stack the partial derivatives (5.91) and get the desired gradient in (5.87) via

$$\frac{\partial f_i}{\partial \boldsymbol{A}} = \begin{bmatrix} \mathbf{0}^\top \\ \vdots \\ \mathbf{0}^\top \\ \boldsymbol{x}^\top \\ \mathbf{0}^\top \\ \vdots \\ \mathbf{0}^\top \end{bmatrix} \in \mathbb{R}^{1\times(M\times N)}. \tag{5.92}$$

Example 5.13 (Gradient of Matrices with Respect to Matrices)
Consider a matrix $\boldsymbol{R} \in \mathbb{R}^{M\times N}$ and $\boldsymbol{f} : \mathbb{R}^{M\times N} \to \mathbb{R}^{N\times N}$ with

$$\boldsymbol{f}(\boldsymbol{R}) = \boldsymbol{R}^\top \boldsymbol{R} =: \boldsymbol{K} \in \mathbb{R}^{N\times N}, \tag{5.93}$$

where we seek the gradient $\mathrm{d}\boldsymbol{K}/\mathrm{d}\boldsymbol{R}$.

To solve this hard problem, let us first write down what we already know: The gradient has the dimensions

$$\frac{\mathrm{d}\boldsymbol{K}}{\mathrm{d}\boldsymbol{R}} \in \mathbb{R}^{(N\times N)\times(M\times N)}, \tag{5.94}$$

which is a tensor. Moreover,

$$\frac{\mathrm{d}K_{pq}}{\mathrm{d}\boldsymbol{R}} \in \mathbb{R}^{1\times M\times N} \tag{5.95}$$

for $p, q = 1, \ldots, N$, where K_{pq} is the (p, q) th entry of $\boldsymbol{K} = \boldsymbol{f}(\boldsymbol{R})$. Denoting the ith column of $\boldsymbol{R}$ by $\boldsymbol{r}_i$, every entry of $\boldsymbol{K}$ is given by the dot product of two columns of $\boldsymbol{R}$, i.e.,

$$K_{pq} = \boldsymbol{r}_p^\top \boldsymbol{r}_q = \sum_{m=1}^{M} R_{mp} R_{mq}. \tag{5.96}$$

When we now compute the partial derivative $\frac{\partial K_{pq}}{\partial R_{ij}}$ we obtain

$$\frac{\partial K_{pq}}{\partial R_{ij}} = \sum_{m=1}^{M} \frac{\partial}{\partial R_{ij}} R_{mp} R_{mq} = \partial_{pqij}, \tag{5.97}$$

$$\partial_{pqij} = \begin{cases} R_{iq} & \text{if } j = p,\ p \neq q \\ R_{ip} & \text{if } j = q,\ p \neq q \\ 2R_{iq} & \text{if } j = p,\ p = q \\ 0 & \text{otherwise} \end{cases}. \tag{5.98}$$

From (5.94), we know that the desired gradient has the dimension $(N \times N) \times (M \times N)$, and every single entry of this tensor is given by ∂_{pqij} in (5.98), where $p, q, j = 1, \ldots, N$ and $i = 1, \ldots, M$.

Figure 5.7 Visualization of gradient computation of a matrix with respect to a vector. We are interested in computing the gradient of $\boldsymbol{A} \in \mathbb{R}^{4\times 2}$ with respect to a vector $\boldsymbol{x} \in \mathbb{R}^3$. We know that gradient $\frac{\mathrm{d}\boldsymbol{A}}{\mathrm{d}\boldsymbol{x}} \in \mathbb{R}^{4\times 2\times 3}$. We follow two equivalent approaches to arrive there: (a) collating partial derivatives into a Jacobian tensor; (b) flattening of the matrix into a vector, computing the Jacobian matrix, reshaping it into a Jacobian tensor.

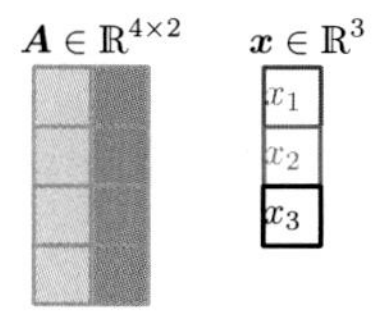

Partial derivatives:

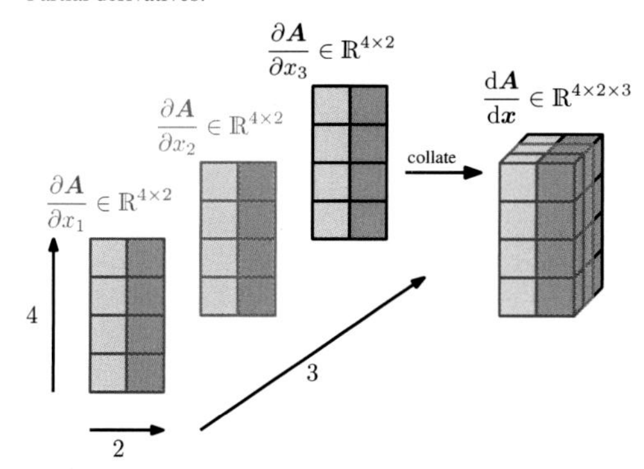

(a) Approach 1: We compute the partial derivative $\frac{\partial \boldsymbol{A}}{\partial x_1}, \frac{\partial \boldsymbol{A}}{\partial x_2}, \frac{\partial \boldsymbol{A}}{\partial x_3}$, each of which is a 4×2 matrix, and collate them in a $4 \times 2 \times 3$ tensor.

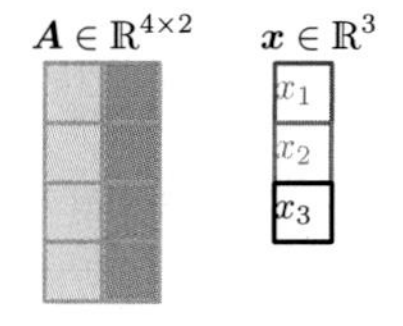

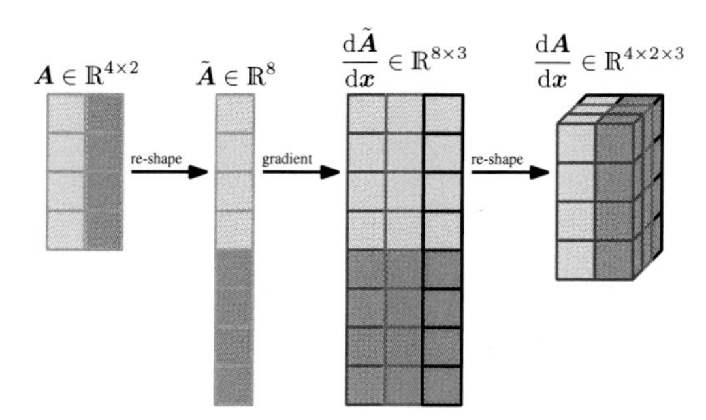

(b) Approach 2: We reshape (flatten) $\boldsymbol{A} \in \mathbb{R}^{4\times 2}$ into a vector $\tilde{\boldsymbol{A}} \in \mathbb{R}^8$. Then we compute the gradient $\frac{\mathrm{d}\tilde{\boldsymbol{A}}}{\mathrm{d}\boldsymbol{x}} \in \mathbb{R}^{8\times 3}$. We obtain the gradient tensor by reshaping this gradient as illustrated as by this figure.

5.5 Useful Identities for Computing Gradients

In the following, we list some useful gradients that are frequently required in a machine learning context (Petersen and Pedersen, 2012). Here we use $\mathrm{tr}(\cdot)$ as the trace (see Definition 4.4), $\det(\cdot)$ as the determinant (see Section 4.1) and $\boldsymbol{f}(\boldsymbol{X})^{-1}$ as the inverse of $\boldsymbol{f}(\boldsymbol{X})$, assuming it exists.

$$\frac{\partial}{\partial \boldsymbol{X}} \boldsymbol{f}(\boldsymbol{X})^\top = \left(\frac{\partial \boldsymbol{f}(\boldsymbol{X})}{\partial \boldsymbol{X}}\right)^\top \tag{5.99}$$

$$\frac{\partial}{\partial \boldsymbol{X}} \mathrm{tr}(\boldsymbol{f}(\boldsymbol{X})) = \mathrm{tr}\left(\frac{\partial \boldsymbol{f}(\boldsymbol{X})}{\partial \boldsymbol{X}}\right) \tag{5.100}$$

$$\frac{\partial}{\partial \boldsymbol{X}} \det(\boldsymbol{f}(\boldsymbol{X})) = \det(\boldsymbol{f}(\boldsymbol{X}))\mathrm{tr}\left(\boldsymbol{f}(\boldsymbol{X})^{-1}\frac{\partial \boldsymbol{f}(\boldsymbol{X})}{\partial \boldsymbol{X}}\right) \tag{5.101}$$

$$\frac{\partial}{\partial \boldsymbol{X}} \boldsymbol{f}(\boldsymbol{X})^{-1} = -\boldsymbol{f}(\boldsymbol{X})^{-1}\frac{\partial \boldsymbol{f}(\boldsymbol{X})}{\partial \boldsymbol{X}}\boldsymbol{f}(\boldsymbol{X})^{-1} \tag{5.102}$$

$$\frac{\partial \boldsymbol{a}^\top \boldsymbol{X}^{-1}\boldsymbol{b}}{\partial \boldsymbol{X}} = -(\boldsymbol{X}^{-1})^\top \boldsymbol{a}\boldsymbol{b}^\top (\boldsymbol{X}^{-1})^\top \tag{5.103}$$

$$\frac{\partial \boldsymbol{x}^\top \boldsymbol{a}}{\partial \boldsymbol{x}} = \boldsymbol{a}^\top \tag{5.104}$$

$$\frac{\partial \boldsymbol{a}^\top \boldsymbol{x}}{\partial \boldsymbol{x}} = \boldsymbol{a}^\top \tag{5.105}$$

$$\frac{\partial \boldsymbol{a}^\top \boldsymbol{X}\boldsymbol{b}}{\partial \boldsymbol{X}} = \boldsymbol{a}\boldsymbol{b}^\top \tag{5.106}$$

$$\frac{\partial \boldsymbol{x}^\top \boldsymbol{B}\boldsymbol{x}}{\partial \boldsymbol{x}} = \boldsymbol{x}^\top(\boldsymbol{B}+\boldsymbol{B}^\top) \tag{5.107}$$

$$\frac{\partial}{\partial \boldsymbol{s}}(\boldsymbol{x}-\boldsymbol{A}\boldsymbol{s})^\top \boldsymbol{W}(\boldsymbol{x}-\boldsymbol{A}\boldsymbol{s}) = -2(\boldsymbol{x}-\boldsymbol{A}\boldsymbol{s})^\top \boldsymbol{W}\boldsymbol{A} \quad \text{for symmetric } \boldsymbol{W} \tag{5.108}$$

Remark. In this book, we only cover traces and transposes of matrices. However, we have seen that derivatives can be higher-dimensional tensors, in which case the usual trace and transpose are not defined. In these cases, the trace of a $D \times D \times E \times F$ tensor would be an $E \times F$-dimensional matrix. This is a special case of a tensor contraction. Similarly, when we "transpose" a tensor, we mean swapping the first two dimensions. Specifically, in (5.99) through (5.102), we require tensor-related computations when we work with multivariate functions $\boldsymbol{f}(\cdot)$ and compute derivatives with respect to matrices (and choose not to vectorize them as discussed in Section 5.4). ◇

5.6 Backpropagation and Automatic Differentiation

A good discussion about backpropagation and the chain rule is available at a blog by Tim Viera at `https://tinyurl.com/ycfm2yrw`.

In many machine learning applications, we find good model parameters by performing gradient descent (Section 7.1), which relies on the fact that we can compute the gradient of a learning objective with respect to the parameters of the model. For a given objective function, we can obtain the gradient with respect to the model parameters using calculus and applying the chain rule; see Section 5.2.2. We already had a taste in Section 5.3 when we looked at the gradient of a squared loss with respect to the parameters of a linear regression model.

Consider the function

$$f(x) = \sqrt{x^2 + \exp(x^2)} + \cos\left(x^2 + \exp(x^2)\right). \tag{5.109}$$

By application of the chain rule, and noting that differentiation is linear, we compute the gradient

$$\begin{aligned}\frac{\mathrm{d}f}{\mathrm{d}x} &= \frac{2x + 2x\exp(x^2)}{2\sqrt{x^2 + \exp(x^2)}} - \sin\left(x^2 + \exp(x^2)\right)\left(2x + 2x\exp(x^2)\right) \\ &= 2x\left(\frac{1}{2\sqrt{x^2 + \exp(x^2)}} - \sin\left(x^2 + \exp(x^2)\right)\right)\left(1 + \exp(x^2)\right).\end{aligned} \tag{5.110}$$

Writing out the gradient in this explicit way is often impractical since it often results in a very lengthy expression for a derivative. In practice, it means that, if we are not careful, the implementation of the gradient could be significantly more expensive than computing the function, which imposes unnecessary overhead. For training deep neural network models, the *backpropagation* algorithm (Kelley, 1960; Bryson, 1961; Dreyfus, 1962; Rumelhart et al., 1986) is an efficient way to compute the gradient of an error function with respect to the parameters of the model.

backpropagation

5.6.1 Gradients in a Deep Network

An area where the chain rule is used to an extreme is that of deep learning, where the function value $\boldsymbol{y}$ is computed as a many-level function composition

$$\boldsymbol{y} = (f_K \circ f_{K-1} \circ \cdots \circ f_1)(\boldsymbol{x}) = f_K(f_{K-1}(\cdots(f_1(\boldsymbol{x}))\cdots)), \tag{5.111}$$

where $\boldsymbol{x}$ are the inputs (e.g., images), $\boldsymbol{y}$ are the observations (e.g., class labels), and every function f_i, $i = 1, \ldots, K$, possesses its own parameters. In neural networks with multiple layers, we have functions $f_i(\boldsymbol{x}_{i-1}) = \sigma(\boldsymbol{A}_{i-1} + \boldsymbol{b}_{i-1})$ in the ith layer. Here $\boldsymbol{x}_{i-1}$ is the output of layer $i-1$ and σ an activation function, such as the logistic sigmoid $\frac{1}{1+e^{-x}}$, tanh or a rectified linear unit (ReLU). In order to train these models, we require the gradient of a loss function L with respect to all model parameters $\boldsymbol{A}_j, \boldsymbol{b}_j$ for $j = 1, \ldots, K$. This also requires us to compute the gradient of L with respect to the inputs of each layer. For example, if we have inputs $\boldsymbol{x}$ and observations $\boldsymbol{y}$ and a network structure defined by

We discuss the case where the activation functions are identical in each layer to unclutter notation.

$$\boldsymbol{f}_0 := \boldsymbol{x} \tag{5.112}$$

$$\boldsymbol{f}_i := \sigma_i(\boldsymbol{A}_{i-1}\boldsymbol{f}_{i-1} + \boldsymbol{b}_{i-1}), \quad i = 1, \ldots, K, \tag{5.113}$$

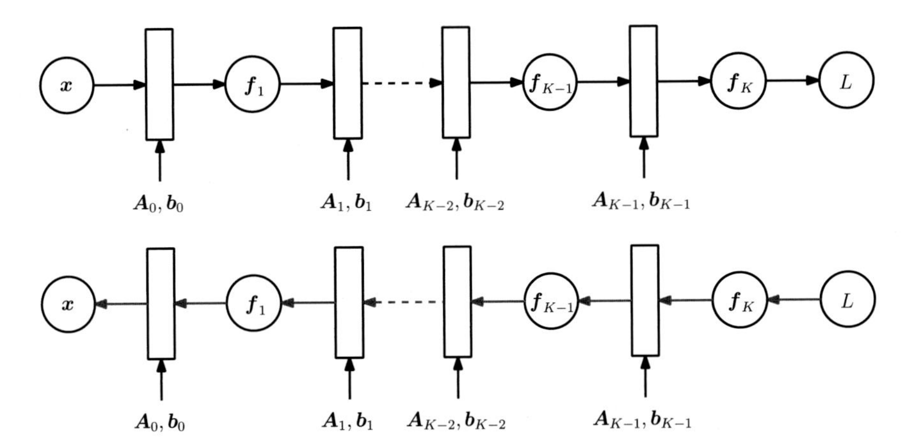

Figure 5.8 Forward pass in a multilayer neural network to compute the loss L as a function of the inputs $\boldsymbol{x}$ and the parameters $\boldsymbol{A}_i$, $\boldsymbol{b}_i$.

Figure 5.9 Backward pass in a multilayer neural network to compute the gradients of the loss function.

(see also Figure 5.8 for a visualization), we may be interested in finding $\boldsymbol{A}_j, \boldsymbol{b}_j$ for $j = 0, \ldots, K-1$, such that the squared loss

$$L(\boldsymbol{\theta}) = \|\boldsymbol{y} - \boldsymbol{f}_K(\boldsymbol{\theta}, \boldsymbol{x})\|^2 \tag{5.114}$$

is minimized, where $\boldsymbol{\theta} = \{\boldsymbol{A}_0, \boldsymbol{b}_0, \ldots, \boldsymbol{A}_{K-1}, \boldsymbol{b}_{K-1}\}$.

To obtain the gradients with respect to the parameter set $\boldsymbol{\theta}$, we require the partial derivatives of L with respect to the parameters $\boldsymbol{\theta}_j = \{\boldsymbol{A}_j, \boldsymbol{b}_j\}$ of each layer $j = 0, \ldots, K-1$. The chain rule allows us to determine the partial derivatives as

A more in-depth discussion about gradients of neural networks can be found in Justin Domke's lecture notes `https://tinyurl.com/yalcxgtv`.

$$\frac{\partial L}{\partial \boldsymbol{\theta}_{K-1}} = \frac{\partial L}{\partial \boldsymbol{f}_K} \frac{\partial \boldsymbol{f}_K}{\partial \boldsymbol{\theta}_{K-1}} \tag{5.115}$$

$$\frac{\partial L}{\partial \boldsymbol{\theta}_{K-2}} = \frac{\partial L}{\partial \boldsymbol{f}_K} \boxed{\frac{\partial \boldsymbol{f}_K}{\partial \boldsymbol{f}_{K-1}} \frac{\partial \boldsymbol{f}_{K-1}}{\partial \boldsymbol{\theta}_{K-2}}} \tag{5.116}$$

$$\frac{\partial L}{\partial \boldsymbol{\theta}_{K-3}} = \frac{\partial L}{\partial \boldsymbol{f}_K} \frac{\partial \boldsymbol{f}_K}{\partial \boldsymbol{f}_{K-1}} \boxed{\frac{\partial \boldsymbol{f}_{K-1}}{\partial \boldsymbol{f}_{K-2}} \frac{\partial \boldsymbol{f}_{K-2}}{\partial \boldsymbol{\theta}_{K-3}}} \tag{5.117}$$

$$\frac{\partial L}{\partial \boldsymbol{\theta}_i} = \frac{\partial L}{\partial \boldsymbol{f}_K} \frac{\partial \boldsymbol{f}_K}{\partial \boldsymbol{f}_{K-1}} \cdots \boxed{\frac{\partial \boldsymbol{f}_{i+2}}{\partial \boldsymbol{f}_{i+1}} \frac{\partial \boldsymbol{f}_{i+1}}{\partial \boldsymbol{\theta}_i}} \tag{5.118}$$

The **orange** terms are partial derivatives of the output of a layer with respect to its inputs, whereas the **blue** terms are partial derivatives of the output of a layer with respect to its parameters. Assuming we have already computed the partial derivatives $\partial L / \partial \boldsymbol{\theta}_{i+1}$, then most of the computation can be reused to compute $\partial L / \partial \boldsymbol{\theta}_i$. The additional terms that we need to compute are indicated by the boxes. Figure 5.9 visualizes that the gradients are passed backward through the network.

5.6.2 Automatic Differentiation

It turns out that backpropagation is a special case of a general technique in numerical analysis called *automatic differentiation*. We can think of automatic

automatic differentiation

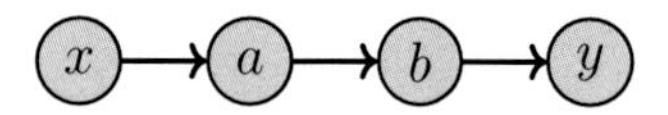

Figure 5.10 Simple graph illustrating the flow of data from x to y via some intermediate variables a, b.

differentation as a set of techniques to numerically (in contrast to symbolically) evaluate the exact (up to machine precision) gradient of a function by working with intermediate variables and applying the chain rule. Automatic differentiation applies a series of elementary arithmetic operations, e.g., addition and multiplication and elementary functions, e.g., $\sin, \cos, \exp, \log$. By applying the chain rule to these operations, the gradient of quite complicated functions can be computed automatically. Automatic differentiation applies to general computer programs and has forward and reverse modes. Baydin et al. (2018) give a great overview of automatic differentiation in machine learning.

Automatic differentiation is different from symbolic differentiation and numerical approximations of the gradient, e.g., by using finite differences.

Figure 5.10 shows a simple graph representing the data flow from inputs x to outputs y via some intermediate variables a, b. If we were to compute the derivative $\mathrm{d}y/\mathrm{d}x$, we would apply the chain rule and obtain

$$\frac{\mathrm{d}y}{\mathrm{d}x} = \frac{\mathrm{d}y}{\mathrm{d}b}\frac{\mathrm{d}b}{\mathrm{d}a}\frac{\mathrm{d}a}{\mathrm{d}x}\,. \tag{5.119}$$

Intuitively, the forward and reverse mode differ in the order of multiplication. Due to the associativity of matrix multiplication, we can choose between

In the general case, we work with Jacobians, which can be vectors, matrices, or tensors.

$$\frac{\mathrm{d}y}{\mathrm{d}x} = \left(\frac{\mathrm{d}y}{\mathrm{d}b}\frac{\mathrm{d}b}{\mathrm{d}a}\right)\frac{\mathrm{d}a}{\mathrm{d}x}\,, \tag{5.120}$$

$$\frac{\mathrm{d}y}{\mathrm{d}x} = \frac{\mathrm{d}y}{\mathrm{d}b}\left(\frac{\mathrm{d}b}{\mathrm{d}a}\frac{\mathrm{d}a}{\mathrm{d}x}\right)\,. \tag{5.121}$$

Equation (5.120) would be the *reverse mode* because gradients are propagated backward through the graph, i.e., reverse to the data flow. Equation (5.121) would be the *forward mode*, where the gradients flow with the data from left to right through the graph.

reverse mode

forward mode

In the following, we will focus on reverse mode automatic differentiation, which is backpropagation. In the context of neural networks, where the input dimensionality is often much higher than the dimensionality of the labels, the reverse mode is computationally significantly cheaper than the forward mode. Let us start with an instructive example.

Example 5.14

Consider the function

$$f(x) = \sqrt{x^2 + \exp(x^2)} + \cos\left(x^2 + \exp(x^2)\right) \tag{5.122}$$

from (5.109). If we were to implement a function f on a computer, we would be able to save some computation by using *intermediate variables*:

intermediate variables

$$a = x^2\,, \tag{5.123}$$

$$b = \exp(a)\,, \tag{5.124}$$

$$c = a + b\,, \tag{5.125}$$

$$d = \sqrt{c}\,, \tag{5.126}$$

$$e = \cos(c)\,, \tag{5.127}$$

$$f = d + e\,. \tag{5.128}$$

This is the same kind of thinking process that occurs when applying the chain rule. Note that the preceding set of equations requires fewer operations than a direct implementation of the function $f(x)$ as defined in (5.109). The corresponding computation graph in Figure 5.11 shows the flow of data and computations required to obtain the function value f.

The set of equations that include intermediate variables can be thought of as a computation graph, a representation that is widely used in implementations of neural network software libraries. We can directly compute the derivatives of the intermediate variables with respect to their corresponding inputs by recalling the definition of the derivative of elementary functions. We obtain the following:

$$\frac{\partial a}{\partial x} = 2x \tag{5.129}$$

$$\frac{\partial b}{\partial a} = \exp(a) \tag{5.130}$$

$$\frac{\partial c}{\partial a} = 1 = \frac{\partial c}{\partial b} \tag{5.131}$$

$$\frac{\partial d}{\partial c} = \frac{1}{2\sqrt{c}} \tag{5.132}$$

$$\frac{\partial e}{\partial c} = -\sin(c) \tag{5.133}$$

$$\frac{\partial f}{\partial d} = 1 = \frac{\partial f}{\partial e}\,. \tag{5.134}$$

By looking at the computation graph in Figure 5.11, we can compute $\partial f/\partial x$ by working backward from the output and obtain

$$\frac{\partial f}{\partial c} = \frac{\partial f}{\partial d}\frac{\partial d}{\partial c} + \frac{\partial f}{\partial e}\frac{\partial e}{\partial c} \tag{5.135}$$

$$\frac{\partial f}{\partial b} = \frac{\partial f}{\partial c}\frac{\partial c}{\partial b} \tag{5.136}$$

$$\frac{\partial f}{\partial a} = \frac{\partial f}{\partial b}\frac{\partial b}{\partial a} + \frac{\partial f}{\partial c}\frac{\partial c}{\partial a} \tag{5.137}$$

$$\frac{\partial f}{\partial x} = \frac{\partial f}{\partial a}\frac{\partial a}{\partial x}\,. \tag{5.138}$$

Note that we implicitly applied the chain rule to obtain $\partial f/\partial x$. By substituting the results of the derivatives of the elementary functions, we get

$$\frac{\partial f}{\partial c} = 1 \cdot \frac{1}{2\sqrt{c}} + 1 \cdot (-\sin(c)) \tag{5.139}$$

$$\frac{\partial f}{\partial b} = \frac{\partial f}{\partial c} \cdot 1 \tag{5.140}$$

$$\frac{\partial f}{\partial a} = \frac{\partial f}{\partial b}\exp(a) + \frac{\partial f}{\partial c} \cdot 1 \tag{5.141}$$

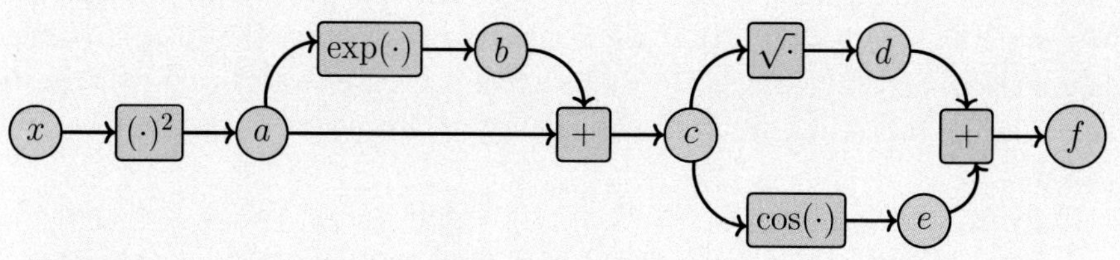

Figure 5.11 Computation graph with inputs x, function values f, and intermediate variables a, b, c, d, e.

$$\frac{\partial f}{\partial x} = \frac{\partial f}{\partial a} \cdot 2x \,. \tag{5.142}$$

By thinking of each of the derivatives above as a variable, we observe that the computation required for calculating the derivative is of similar complexity as the computation of the function itself. This is quite counter-intuitive since the mathematical expression for the derivative $\frac{\partial f}{\partial x}$ (5.110) is significantly more complicated than the mathematical expression of the function $f(x)$ in (5.109).

Automatic differentiation is a formalization of Example 5.14. Let $x_1, \ldots, x_d$ be the input variables to the function, $x_{d+1}, \ldots, x_{D-1}$ be the intermediate variables, and x_D the output variable. Then the computation graph can be expressed as follows:

$$\text{For } i = d+1, \ldots, D: \quad x_i = g_i(x_{\text{Pa}(x_i)}) \,, \tag{5.143}$$

where the $g_i(\cdot)$ are elementary functions and $x_{\text{Pa}(x_i)}$ are the parent nodes of the variable x_i in the graph. Given a function defined in this way, we can use the chain rule to compute the derivative of the function in a step-by-step fashion. Recall that by definition $f = x_D$ and hence

$$\frac{\partial f}{\partial x_D} = 1 \,. \tag{5.144}$$

For other variables x_i, we apply the chain rule

$$\frac{\partial f}{\partial x_i} = \sum_{x_j : x_i \in \text{Pa}(x_j)} \frac{\partial f}{\partial x_j} \frac{\partial x_j}{\partial x_i} = \sum_{x_j : x_i \in \text{Pa}(x_j)} \frac{\partial f}{\partial x_j} \frac{\partial g_j}{\partial x_i} \,, \tag{5.145}$$

where $\text{Pa}(x_j)$ is the set of parent nodes of x_j in the computation graph. Equation (5.143) is the forward propagation of a function, whereas (5.145) is the backpropagation of the gradient through the computation graph. For neural network training, we backpropagate the error of the prediction with respect to the label.

Autodifferentiation in reverse mode requires a parse tree.

The automatic differentiation approach works whenever we have a function that can be expressed as a computation graph, where the elementary functions are differentiable. In fact, the function may not even be a mathematical function but a computer program. However, not all computer programs can be automatically differentiated, e.g., if we cannot find differential elementary functions. Programming structures, such as `for` loops and `if` statements, require more care as well.

5.7 Higher-Order Derivatives

So far, we have discussed gradients, i.e., first-order derivatives. Sometimes, we are interested in derivatives of higher order, e.g., when we want to use Newton's

Method for optimization, which requires second-order derivatives (Nocedal and Wright, 2006). In Section 5.1.1, we discussed the Taylor series to approximate functions using polynomials. In the multivariate case, we can do exactly the same. In the following, we will do exactly this. But let us start with some notation.

Consider a function $f : \mathbb{R}^2 \to \mathbb{R}$ of two variables x, y. We use the following notation for higher-order partial derivatives (and for gradients):

- $\frac{\partial^2 f}{\partial x^2}$ is the second partial derivative of f with respect to x.
- $\frac{\partial^n f}{\partial x^n}$ is the nth partial derivative of f with respect to x.
- $\frac{\partial^2 f}{\partial y \partial x} = \frac{\partial}{\partial y}\left(\frac{\partial f}{\partial x}\right)$ is the partial derivative obtained by first partial differentiating with respect to x and then with respect to y.
- $\frac{\partial^2 f}{\partial x \partial y}$ is the partial derivative obtained by first partial differentiating by y and then x.

Hessian

The *Hessian* is the collection of all second-order partial derivatives.

If $f(x, y)$ is a twice (continuously) differentiable function, then

$$\frac{\partial^2 f}{\partial x \partial y} = \frac{\partial^2 f}{\partial y \partial x}, \tag{5.146}$$

Hessian matrix

i.e., the order of differentiation does not matter, and the corresponding *Hessian matrix*

$$\boldsymbol{H} = \begin{bmatrix} \frac{\partial^2 f}{\partial x^2} & \frac{\partial^2 f}{\partial x \partial y} \\ \frac{\partial^2 f}{\partial x \partial y} & \frac{\partial^2 f}{\partial y^2} \end{bmatrix} \tag{5.147}$$

is symmetric. The Hessian is denoted as $\nabla^2_{x,y} f(x, y)$. Generally, for $\boldsymbol{x} \in \mathbb{R}^n$ and $f : \mathbb{R}^n \to \mathbb{R}$, the Hessian is an $n \times n$ matrix. The Hessian measures the curvature of the function locally around (x, y).

Remark (Hessian of a Vector Field). If $f : \mathbb{R}^n \to \mathbb{R}^m$ is a vector field, the Hessian is an $(m \times n \times n)$-tensor. $\diamondsuit$

5.8 Linearization and Multivariate Taylor Series

The gradient ∇f of a function f is often used for a locally linear approximation of f around $\boldsymbol{x}_0$:

$$f(\boldsymbol{x}) \approx f(\boldsymbol{x}_0) + (\nabla_{\boldsymbol{x}} f)(\boldsymbol{x}_0)(\boldsymbol{x} - \boldsymbol{x}_0). \tag{5.148}$$

Here $(\nabla_{\boldsymbol{x}} f)(\boldsymbol{x}_0)$ is the gradient of f with respect to $\boldsymbol{x}$, evaluated at $\boldsymbol{x}_0$. Figure 5.12 illustrates the linear approximation of a function f at an input x_0. The original function is approximated by a straight line. This approximation is locally accurate, but the farther we move away from x_0, the worse the approximation gets. Equation (5.148) is a special case of a multivariate Taylor series expansion of f at $\boldsymbol{x}_0$, where we consider only the first two terms. We discuss the more general case in the following, which will allow for better approximations.

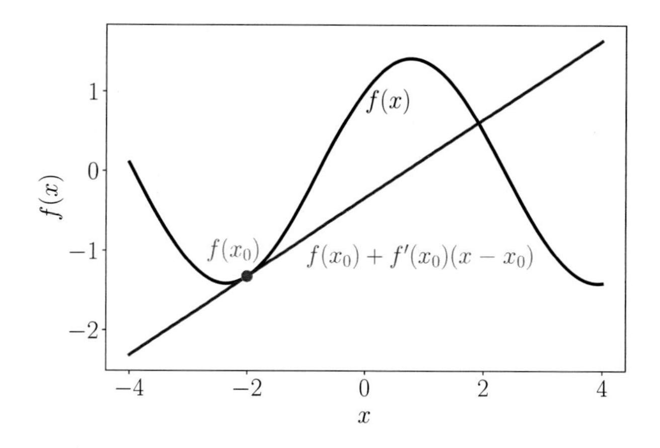

Figure 5.12 Linear approximation of a function. The original function f is linearized at $x_0 = -2$ using a first-order Taylor series expansion.

Definition 5.7 (Multivariate Taylor Series)**.** We consider a function

$$f : \mathbb{R}^D \to \mathbb{R} \tag{5.149}$$

$$\boldsymbol{x} \mapsto f(\boldsymbol{x}), \quad \boldsymbol{x} \in \mathbb{R}^D, \tag{5.150}$$

that is smooth at $\boldsymbol{x}_0$. When we define the difference vector $\boldsymbol{\delta} := \boldsymbol{x} - \boldsymbol{x}_0$, the *multivariate Taylor series* of f at $(\boldsymbol{x}_0)$ is defined as

multivariate Taylor series

$$f(\boldsymbol{x}) = \sum_{k=0}^{\infty} \frac{D_{\boldsymbol{x}}^k f(\boldsymbol{x}_0)}{k!} \boldsymbol{\delta}^k, \tag{5.151}$$

where $D_{\boldsymbol{x}}^k f(\boldsymbol{x}_0)$ is the kth (total) derivative of f with respect to $\boldsymbol{x}$, evaluated at $\boldsymbol{x}_0$.

Definition 5.8 (Taylor Polynomial)**.** The *Taylor polynomial* of degree n of f at $\boldsymbol{x}_0$ contains the first $n+1$ components of the series in (5.151) and is defined as

Taylor polynomial

$$T_n(\boldsymbol{x}) = \sum_{k=0}^{n} \frac{D_{\boldsymbol{x}}^k f(\boldsymbol{x}_0)}{k!} \boldsymbol{\delta}^k. \tag{5.152}$$

In (5.151) and (5.152), we used the slightly sloppy notation of $\boldsymbol{\delta}^k$, which is not defined for vectors $\boldsymbol{x} \in \mathbb{R}^D$, $D > 1$, and $k > 1$. Note that both $D_{\boldsymbol{x}}^k f$ and $\boldsymbol{\delta}^k$ are kth order tensors, i.e., k-dimensional arrays. The kth-order tensor $\boldsymbol{\delta}^k \in \mathbb{R}^{\overbrace{D \times D \times \ldots \times D}^{k \text{ times}}}$ is obtained as a $k$fold outer product, denoted by $\otimes$, of the vector $\boldsymbol{\delta} \in \mathbb{R}^D$. For example,

A vector can be implemented as a one-dimensional array, a matrix as a two-dimensional array.

$$\boldsymbol{\delta}^2 := \boldsymbol{\delta} \otimes \boldsymbol{\delta} = \boldsymbol{\delta}\boldsymbol{\delta}^\top, \quad \boldsymbol{\delta}^2[i,j] = \delta[i]\delta[j] \tag{5.153}$$

$$\boldsymbol{\delta}^3 := \boldsymbol{\delta} \otimes \boldsymbol{\delta} \otimes \boldsymbol{\delta}, \quad \boldsymbol{\delta}^3[i,j,k] = \delta[i]\delta[j]\delta[k]. \tag{5.154}$$

Figure 5.13 visualizes two such outer products. In general, we obtain the terms

$$D_{\boldsymbol{x}}^k f(\boldsymbol{x}_0)\boldsymbol{\delta}^k = \sum_{i_1=1}^{D} \cdots \sum_{i_k=1}^{D} D_{\boldsymbol{x}}^k f(\boldsymbol{x}_0)[i_1, \ldots, i_k]\delta[i_1] \cdots \delta[i_k] \tag{5.155}$$

in the Taylor series, where $D_{\boldsymbol{x}}^k f(\boldsymbol{x}_0)\boldsymbol{\delta}^k$ contains kth-order polynomials.

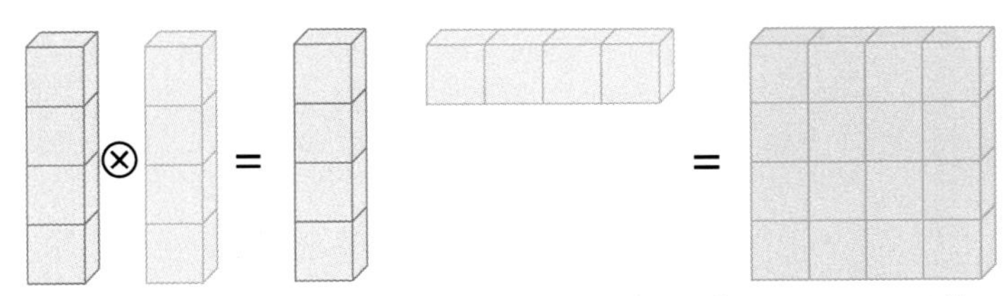

(a) Given a vector $\boldsymbol{\delta} \in \mathbb{R}^4$, we obtain the outer product $\boldsymbol{\delta}^2 := \boldsymbol{\delta} \otimes \boldsymbol{\delta} = \boldsymbol{\delta}\boldsymbol{\delta}^\top \in \mathbb{R}^{4\times 4}$ as a matrix.

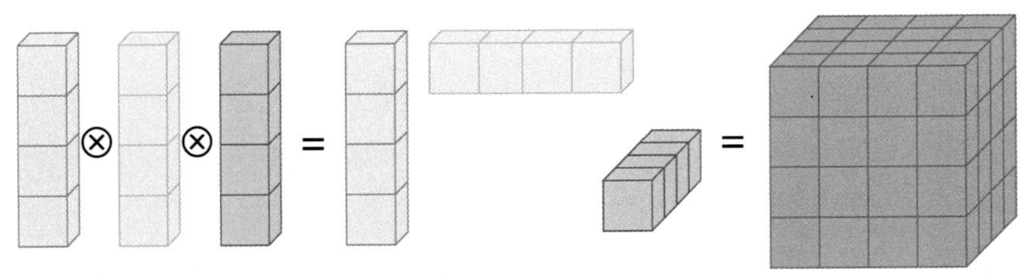

(b) An outer product $\boldsymbol{\delta}^3 := \boldsymbol{\delta} \otimes \boldsymbol{\delta} \otimes \boldsymbol{\delta} \in \mathbb{R}^{4\times 4\times 4}$ results in a third-order tensor ("three-dimensional matrix"), i.e., an array with three indexes.

Figure 5.13 Visualizing outer products. Outer products of vectors increase the dimensionality of the array by 1 per term. (a) The outer product of two vectors results in a matrix; (b) the outer product of three vectors yields a third-order tensor.

Now that we defined the Taylor series for vector fields, let us explicitly write down the first terms $D^k_{\boldsymbol{x}} f(\boldsymbol{x}_0)\boldsymbol{\delta}^k$ of the Taylor series expansion for $k = 0, \ldots, 3$ and $\boldsymbol{\delta} := \boldsymbol{x} - \boldsymbol{x}_0$:

```
np.einsum(
'i,i',Df1,d)
np.einsum(
'ij,i,j',
Df2,d,d)
np.einsum(
'ijk,i,j,k',
Df3,d,d,d)
```

$$k = 0 : D^0_{\boldsymbol{x}} f(\boldsymbol{x}_0)\boldsymbol{\delta}^0 = f(\boldsymbol{x}_0) \in \mathbb{R} \tag{5.156}$$

$$k = 1 : D^1_{\boldsymbol{x}} f(\boldsymbol{x}_0)\boldsymbol{\delta}^1 = \underbrace{\nabla_{\boldsymbol{x}} f(\boldsymbol{x}_0)}_{1\times D} \underbrace{\boldsymbol{\delta}}_{D\times 1} = \sum_{i=1}^{D} \nabla_{\boldsymbol{x}} f(\boldsymbol{x}_0)[i]\delta[i] \in \mathbb{R} \tag{5.157}$$

$$k = 2 : D^2_{\boldsymbol{x}} f(\boldsymbol{x}_0)\boldsymbol{\delta}^2 = \operatorname{tr}\big(\underbrace{\boldsymbol{H}(\boldsymbol{x}_0)}_{D\times D} \underbrace{\boldsymbol{\delta}}_{D\times 1} \underbrace{\boldsymbol{\delta}^\top}_{1\times D}\big) = \boldsymbol{\delta}^\top \boldsymbol{H}(\boldsymbol{x}_0)\boldsymbol{\delta} \tag{5.158}$$

$$= \sum_{i=1}^{D} \sum_{j=1}^{D} H[i,j]\delta[i]\delta[j] \in \mathbb{R} \tag{5.159}$$

$$k = 3 : D^3_{\boldsymbol{x}} f(\boldsymbol{x}_0)\boldsymbol{\delta}^3 = \sum_{i=1}^{D} \sum_{j=1}^{D} \sum_{k=1}^{D} D^3_x f(\boldsymbol{x}_0)[i,j,k]\delta[i]\delta[j]\delta[k] \in \mathbb{R} \tag{5.160}$$

Here, $\boldsymbol{H}(\boldsymbol{x}_0)$ is the Hessian of f evaluated at $\boldsymbol{x}_0$.

Example 5.15 (Taylor Series Expansion of a Function with Two Variables)
Consider the function

$$f(x, y) = x^2 + 2xy + y^3 . \tag{5.161}$$

We want to compute the Taylor series expansion of f at $(x_0, y_0) = (1, 2)$. Before we start, let us discuss what to expect: The function in (5.161) is a polynomial of degree 3. We are looking for a Taylor series expansion, which itself is a linear combination of polynomials. Therefore, we do not expect the Taylor series expansion to contain terms of fourth or higher order to express a third-order polynomial. This means that it should be sufficient to determine the first four terms of (5.151) for an exact alternative representation of (5.161).

To determine the Taylor series expansion, we start with the constant term and the first-order derivatives, which are given by

$$f(1,2) = 13 \tag{5.162}$$

$$\frac{\partial f}{\partial x} = 2x + 2y \implies \frac{\partial f}{\partial x}(1,2) = 6 \tag{5.163}$$

$$\frac{\partial f}{\partial y} = 2x + 3y^2 \implies \frac{\partial f}{\partial y}(1,2) = 14\,. \tag{5.164}$$

Therefore, we obtain

$$D^1_{x,y}f(1,2) = \nabla_{x,y}f(1,2) = \begin{bmatrix}\frac{\partial f}{\partial x}(1,2) & \frac{\partial f}{\partial y}(1,2)\end{bmatrix} = \begin{bmatrix}6 & 14\end{bmatrix} \in \mathbb{R}^{1\times 2} \tag{5.165}$$

such that

$$\frac{D^1_{x,y}f(1,2)}{1!}\boldsymbol{\delta} = \begin{bmatrix}6 & 14\end{bmatrix}\begin{bmatrix}x-1\\ y-2\end{bmatrix} = 6(x-1) + 14(y-2)\,. \tag{5.166}$$

Note that $D^1_{x,y}f(1,2)\boldsymbol{\delta}$ contains only linear terms, i.e., first-order polynomials.

The second-order partial derivatives are given by

$$\frac{\partial^2 f}{\partial x^2} = 2 \implies \frac{\partial^2 f}{\partial x^2}(1,2) = 2 \tag{5.167}$$

$$\frac{\partial^2 f}{\partial y^2} = 6y \implies \frac{\partial^2 f}{\partial y^2}(1,2) = 12 \tag{5.168}$$

$$\frac{\partial^2 f}{\partial y \partial x} = 2 \implies \frac{\partial^2 f}{\partial y \partial x}(1,2) = 2 \tag{5.169}$$

$$\frac{\partial^2 f}{\partial x \partial y} = 2 \implies \frac{\partial^2 f}{\partial x \partial y}(1,2) = 2\,. \tag{5.170}$$

When we collect the second-order partial derivatives, we obtain the Hessian

$$\boldsymbol{H} = \begin{bmatrix}\frac{\partial^2 f}{\partial x^2} & \frac{\partial^2 f}{\partial x \partial y}\\ \frac{\partial^2 f}{\partial y \partial x} & \frac{\partial^2 f}{\partial y^2}\end{bmatrix} = \begin{bmatrix}2 & 2\\ 2 & 6y\end{bmatrix}, \tag{5.171}$$

such that

$$\boldsymbol{H}(1,2) = \begin{bmatrix}2 & 2\\ 2 & 12\end{bmatrix} \in \mathbb{R}^{2\times 2}\,. \tag{5.172}$$

Therefore, the next term of the Taylor series expansion is given by

$$\frac{D^2_{x,y}f(1,2)}{2!}\boldsymbol{\delta}^2 = \frac{1}{2}\boldsymbol{\delta}^\top \boldsymbol{H}(1,2)\boldsymbol{\delta} \tag{5.173a}$$

$$= \frac{1}{2}\begin{bmatrix}x-1 & y-2\end{bmatrix}\begin{bmatrix}2 & 2\\ 2 & 12\end{bmatrix}\begin{bmatrix}x-1\\ y-2\end{bmatrix} \tag{5.173b}$$

$$= (x-1)^2 + 2(x-1)(y-2) + 6(y-2)^2 \,. \tag{5.173c}$$

Here, $D^2_{x,y}f(1,2)\boldsymbol{\delta}^2$ contains only quadratic terms, i.e., second-order polynomials.

The third-order derivatives are obtained as

$$D^3_{x,y}f = \begin{bmatrix}\frac{\partial \boldsymbol{H}}{\partial x} & \frac{\partial \boldsymbol{H}}{\partial y}\end{bmatrix} \in \mathbb{R}^{2\times 2\times 2}\,, \tag{5.174}$$

$$D^3_{x,y}f[:,:,1] = \frac{\partial \boldsymbol{H}}{\partial x} = \begin{bmatrix}\frac{\partial^3 f}{\partial x^3} & \frac{\partial^3 f}{\partial x^2 \partial y}\\ \frac{\partial^3 f}{\partial x \partial y \partial x} & \frac{\partial^3 f}{\partial x \partial y^2}\end{bmatrix}, \tag{5.175}$$

$$D^3_{x,y}f[:,:,2] = \frac{\partial \boldsymbol{H}}{\partial y} = \begin{bmatrix}\frac{\partial^3 f}{\partial y \partial x^2} & \frac{\partial^3 f}{\partial y \partial x \partial y}\\ \frac{\partial^3 f}{\partial y^2 \partial x} & \frac{\partial^3 f}{\partial y^3}\end{bmatrix}. \tag{5.176}$$

Since most second-order partial derivatives in the Hessian in (5.171) are constant, the only nonzero third-order partial derivative is

$$\frac{\partial^3 f}{\partial y^3} = 6 \implies \frac{\partial^3 f}{\partial y^3}(1,2) = 6\,. \tag{5.177}$$

Higher-order derivatives and the mixed derivatives of degree 3 (e.g., $\frac{\partial f^3}{\partial x^2 \partial y}$) vanish, such that

$$D^3_{x,y}f[:,:,1] = \begin{bmatrix}0 & 0\\ 0 & 0\end{bmatrix}, \quad D^3_{x,y}f[:,:,2] = \begin{bmatrix}0 & 0\\ 0 & 6\end{bmatrix} \tag{5.178}$$

and

$$\frac{D^3_{x,y}f(1,2)}{3!}\boldsymbol{\delta}^3 = (y-2)^3\,, \tag{5.179}$$

which collects all cubic terms of the Taylor series. Overall, the (exact) Taylor series expansion of f at $(x_0, y_0) = (1,2)$ is

$$f(x) = f(1,2) + D^1_{x,y}f(1,2)\boldsymbol{\delta} + \frac{D^2_{x,y}f(1,2)}{2!}\boldsymbol{\delta}^2 + \frac{D^3_{x,y}f(1,2)}{3!}\boldsymbol{\delta}^3 \tag{5.180a}$$

$$
\begin{aligned}
&= f(1,2) + \frac{\partial f(1,2)}{\partial x}(x-1) + \frac{\partial f(1,2)}{\partial y}(y-2) \\
&\quad + \frac{1}{2!}\left(\frac{\partial^2 f(1,2)}{\partial x^2}(x-1)^2 + \frac{\partial^2 f(1,2)}{\partial y^2}(y-2)^2\right. \\
&\quad \left. + 2\frac{\partial^2 f(1,2)}{\partial x \partial y}(x-1)(y-2)\right) + \frac{1}{6}\frac{\partial^3 f(1,2)}{\partial y^3}(y-2)^3 \qquad (5.180b)\\
&= 13 + 6(x-1) + 14(y-2) \\
&\quad + (x-1)^2 + 6(y-2)^2 + 2(x-1)(y-2) + (y-2)^3 . \qquad (5.180c)
\end{aligned}
$$

In this case, we obtained an exact Taylor series expansion of the polynomial in (5.161), i.e., the polynomial in (5.180c) is identical to the original polynomial in (5.161). In this particular example, this result is not surprising since the original function was a third-order polynomial, which we expressed through a linear combination of constant terms, first-order, second-order, and third-order polynomials in (5.180c).

5.9 Further Reading

Further details of matrix differentials, along with a short review of the required linear algebra, can be found in Magnus and Neudecker (2007). Automatic differentiation has had a long history, and we refer to Griewank and Walther (2003), Griewank and Walther (2008), and Elliott (2009) and the references therein.

In machine learning (and other disciplines), we often need to compute expectations, i.e., we need to solve integrals of the form

$$
\mathbb{E}_{\boldsymbol{x}}[f(\boldsymbol{x})] = \int f(\boldsymbol{x})p(\boldsymbol{x})d\boldsymbol{x} . \qquad (5.181)
$$

Even if $p(\boldsymbol{x})$ is in a convenient form (e.g., Gaussian), this integral generally cannot be solved analytically. The Taylor series expansion of f is one way of finding an approximate solution: Assuming $p(\boldsymbol{x}) = \mathcal{N}(\boldsymbol{\mu}, \boldsymbol{\Sigma})$ is Gaussian, then the first-order Taylor series expansion around $\boldsymbol{\mu}$ locally linearizes the nonlinear function f. For linear functions, we can compute the mean (and the covariance) exactly if $p(\boldsymbol{x})$ is Gaussian distributed (see Section 6.5). This property is heavily exploited by the *extended Kalman filter* (Maybeck, 1979) for online state estimation in nonlinear dynamical systems (also called "state-space models"). Other deterministic ways to approximate the integral in (5.181) are the *unscented transform* (Julier and Uhlmann, 1997), which does not require any gradients, or the *Laplace approximation* (MacKay, 2003; Bishop, 2006; Murphy, 2012), which uses a second-order Taylor series expansion (requiring the Hessian) for a local Gaussian approximation of $p(\boldsymbol{x})$ around its mode.

extended Kalman filter

unscented transform

Laplace approximation

Exercises

5.1 Compute the derivative $f'(x)$ for

$$f(x) = \log(x^4)\sin(x^3)\,.$$

5.2 Compute the derivative $f'(x)$ of the logistic sigmoid

$$f(x) = \frac{1}{1+\exp(-x)}\,.$$

5.3 Compute the derivative $f'(x)$ of the function

$$f(x) = \exp(-\tfrac{1}{2\sigma^2}(x-\mu)^2)\,,$$

where $\mu,\ \sigma \in \mathbb{R}$ are constants.

5.4 Compute the Taylor polynomials T_n, $n = 0, \dots, 5$ of $f(x) = \sin(x) + \cos(x)$ at $x_0 = 0$.

5.5 Consider the following functions:

$$\begin{aligned} f_1(\boldsymbol{x}) &= \sin(x_1)\cos(x_2)\,, & \boldsymbol{x} &\in \mathbb{R}^2 \\ f_2(\boldsymbol{x}, \boldsymbol{y}) &= \boldsymbol{x}^\top \boldsymbol{y}\,, & \boldsymbol{x}, \boldsymbol{y} &\in \mathbb{R}^n \\ f_3(\boldsymbol{x}) &= \boldsymbol{x}\boldsymbol{x}^\top\,, & \boldsymbol{x} &\in \mathbb{R}^n \end{aligned}$$

a. What are the dimensions of $\frac{\partial f_i}{\partial \boldsymbol{x}}$?

b. Compute the Jacobians.

5.6 Differentiate f with respect to $\boldsymbol{t}$ and g with respect to $\boldsymbol{X}$, where

$$\begin{aligned} f(\boldsymbol{t}) &= \sin(\log(\boldsymbol{t}^\top \boldsymbol{t}))\,, & t &\in \mathbb{R}^D \\ g(\boldsymbol{X}) &= \text{tr}(\boldsymbol{A}\boldsymbol{X}\boldsymbol{B})\,, & \boldsymbol{A} &\in \mathbb{R}^{D\times E}, \boldsymbol{X} \in \mathbb{R}^{E\times F}, \boldsymbol{B} \in \mathbb{R}^{F\times D}\,, \end{aligned}$$

where tr denotes the trace.

5.7 Compute the derivatives $\mathrm{d}f/\mathrm{d}\boldsymbol{x}$ of the following functions by using the chain rule. Provide the dimensions of every single partial derivative. Describe your steps in detail.

a.

$$f(z) = \log(1+z)\,, \quad z = \boldsymbol{x}^\top \boldsymbol{x}\,, \quad \boldsymbol{x} \in \mathbb{R}^D$$

b.

$$f(\boldsymbol{z}) = \sin(\boldsymbol{z})\,, \quad \boldsymbol{z} = \boldsymbol{A}\boldsymbol{x} + \boldsymbol{b}\,, \quad \boldsymbol{A} \in \mathbb{R}^{E\times D}, \boldsymbol{x} \in \mathbb{R}^D, \boldsymbol{b} \in \mathbb{R}^E,$$

where $\sin(\cdot)$ is applied to every element of $\boldsymbol{z}$.

5.8 Compute the derivatives $df/d\boldsymbol{x}$ of the following functions. Describe your steps in detail.

a. Use the chain rule. Provide the dimensions of every single partial derivative.

$$\begin{aligned} f(z) &= \exp(-\tfrac{1}{2}z) \\ z &= g(\boldsymbol{y}) = \boldsymbol{y}^\top \boldsymbol{S}^{-1} \boldsymbol{y} \\ \boldsymbol{y} &= h(\boldsymbol{x}) = \boldsymbol{x} - \boldsymbol{\mu}, \end{aligned}$$

where $\boldsymbol{x}, \boldsymbol{\mu} \in \mathbb{R}^D$, $\boldsymbol{S} \in \mathbb{R}^{D\times D}$.

b.

$$f(\boldsymbol{x}) = \text{tr}(\boldsymbol{x}\boldsymbol{x}^\top + \sigma^2 \boldsymbol{I}), \quad \boldsymbol{x} \in \mathbb{R}^D.$$

Here $\text{tr}(\boldsymbol{A})$ is the trace of $\boldsymbol{A}$, i.e., the sum of the diagonal elements A_{ii}. *Hint: Explicitly write out the outer product.*

c. Use the chain rule. Provide the dimensions of every single partial derivative. You do not need to compute the product of the partial derivatives explicitly.

$$\begin{aligned} \boldsymbol{f} &= \tanh(\boldsymbol{z}) \in \mathbb{R}^M \\ \boldsymbol{z} &= \boldsymbol{A}\boldsymbol{x} + \boldsymbol{b}, \quad \boldsymbol{x} \in \mathbb{R}^N, \boldsymbol{A} \in \mathbb{R}^{M \times N}, \boldsymbol{b} \in \mathbb{R}^M. \end{aligned}$$

Here, $\tanh$ is applied to every component of $\boldsymbol{z}$.

5.9 We define

$$\begin{aligned} g(\boldsymbol{z}, \boldsymbol{\nu}) &:= \log p(\boldsymbol{x}, \boldsymbol{z}) - \log q(\boldsymbol{z}, \boldsymbol{\nu}) \\ \boldsymbol{z} &:= t(\boldsymbol{\epsilon}, \boldsymbol{\nu}) \end{aligned}$$

for differentiable functions p, q, t. By using the chain rule, compute the gradient

$$\frac{\mathrm{d}}{\mathrm{d}\boldsymbol{\nu}} g(\boldsymbol{z}, \boldsymbol{\nu}).$$

6

Probability and Distributions

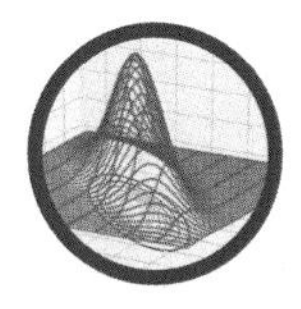

Probability, loosely speaking, concerns the study of uncertainty. Probability can be thought of as the fraction of times an event occurs, or as a degree of belief about an event. We then would like to use this probability to measure the chance of something occurring in an experiment. As mentioned in Chapter 1, we often quantify uncertainty in the data, uncertainty in the machine learning model, and uncertainty in the predictions produced by the model. Quantifying uncertainty requires the idea of a *random variable*, which is a function that maps outcomes of random experiments to a set of properties that we are interested in. Associated with the random variable is a function that measures the probability that a particular outcome (or set of outcomes) will occur; this is called the *probability distribution*.

random variable

probability distribution

Probability distributions are used as a building block for other concepts, such as probabilistic modeling (Section 8.4), graphical models (Section 8.5), and model selection (Section 8.6). In the next section, we present the three concepts that define a probability space (the sample space, the events, and the probability of an event) and how they are related to a fourth concept called the random variable. The presentation is deliberately slightly hand wavy since a rigorous presentation may occlude the intuition behind the concepts. An outline of the concepts presented in this chapter are shown in Figure 6.1.

6.1 Construction of a Probability Space

The theory of probability aims at defining a mathematical structure to describe random outcomes of experiments. For example, when tossing a single coin, we cannot determine the outcome, but by doing a large number of coin tosses, we can observe a regularity in the average outcome. Using this mathematical structure of probability, the goal is to perform automated reasoning, and in this sense, probability generalizes logical reasoning (Jaynes, 2003).

6.1.1 Philosophical Issues

When constructing automated reasoning systems, classical Boolean logic does not allow us to express certain forms of plausible reasoning. Consider the following scenario: We observe that A is false. We find B becomes less plausible, although no conclusion can be drawn from classical logic. We observe that B is

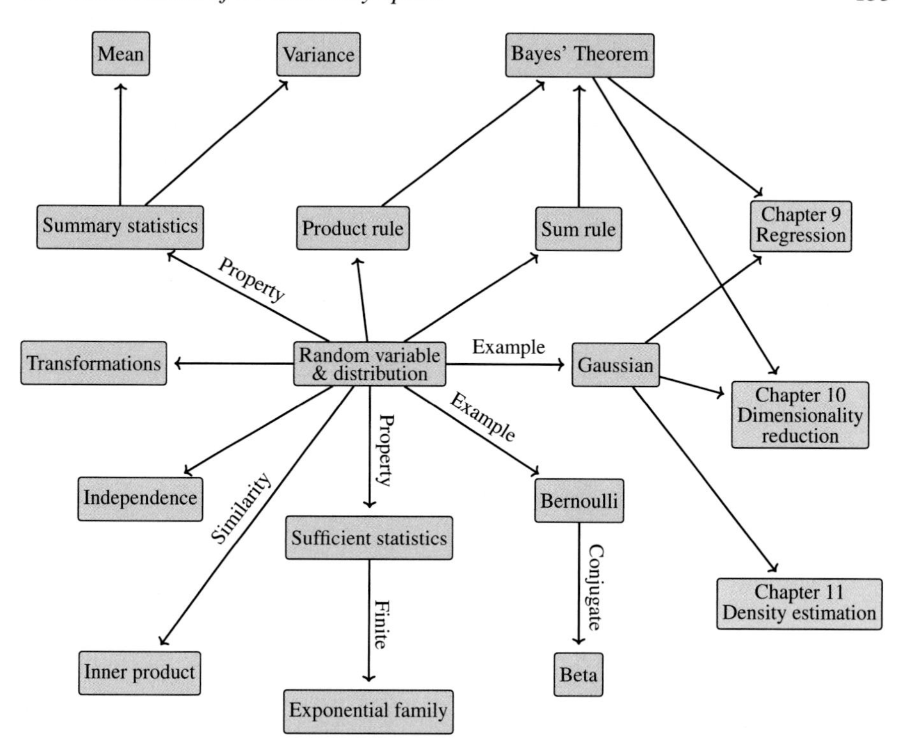

Figure 6.1 A mind map of the concepts related to random variables and probability distributions, as described in this chapter.

true. It seems A becomes more plausible. We use this form of reasoning daily. We are waiting for a friend, and consider three possibilities: H1, she is on time; H2, she has been delayed by traffic and; H3, she has been abducted by aliens. When we observe our friend is late, we must logically rule out H1. We also tend to consider H2 to be more likely, though we are not logically required to do so. Finally, we may consider H3 to be possible, but we continue to consider it quite unlikely. How do we conclude H2 is the most plausible answer? Seen in this way, probability theory can be considered a generalization of Boolean logic. In the context of machine learning, it is often applied in this way to formalize the design of automated reasoning systems. Further arguments about how probability theory is the foundation of reasoning systems can be found in Pearl (1988).

"For plausible reasoning it is necessary to extend the discrete true and false values of truth to continuous plausibilities" (Jaynes, 2003).

The philosophical basis of probability and how it should be somehow related to what we think should be true (in the logical sense) was studied by Cox (Jaynes, 2003). Another way to think about it is that if we are precise about our common sense, we end up constructing probabilities. E. T. Jaynes (1922–1998) identified three mathematical criteria, which must apply to all plausibilities:

1. The degrees of plausibility are represented by real numbers.
2. These numbers must be based on the rules of common sense.
3. The resulting reasoning must be consistent, with the three following meanings of the word "consistent":

   (a) Consistency or noncontradiction: When the same result can be reached through different means, the same plausibility value must be found in all cases.

(b) Honesty: All available data must be taken into account.

(c) Reproducibility: If our state of knowledge about two problems are the same, then we must assign the same degree of plausibility to both of them.

The Cox–Jaynes theorem proves these plausibilities to be sufficient to define the universal mathematical rules that apply to plausibility p, up to transformation by an arbitrary monotonic function. Crucially, these rules *are* the rules of probability.

Remark. In machine learning and statistics, there are two major interpretations of probability: the Bayesian and frequentist interpretations (Bishop, 2006; Efron and Hastie, 2016). The Bayesian interpretation uses probability to specify the degree of uncertainty that the user has about an event. It is sometimes referred to as "subjective probability" or "degree of belief." The frequentist interpretation considers the relative frequencies of events of interest to the total number of events that occurred. The probability of an event is defined as the relative frequency of the event in the limit when one has infinite data. $\diamondsuit$

Some machine learning texts on probabilistic models use lazy notation and jargon, which is confusing. This text is no exception. Multiple distinct concepts are all referred to as "probability distribution," and the reader has to often disentangle the meaning from the context. One trick to help make sense of probability distributions is to check whether we are trying to model something categorical (a discrete random variable) or something continuous (a continuous random variable). The kinds of questions we tackle in machine learning are closely related to whether we are considering categorical or continuous models.

6.1.2 Probability and Random Variables

There are three distinct ideas that are often confused when discussing probabilities. First is the idea of a probability space, which allows us to quantify the idea of a probability. However, we mostly do not work directly with this basic probability space. Instead, we work with random variables (the second idea), which transfers the probability to a more convenient (often numerical) space. The third idea is the idea of a distribution or law associated with a random variable. We will introduce the first two ideas in this section and expand on the third idea in Section 6.2.

Modern probability is based on a set of axioms proposed by Kolmogorov (Grinstead and Snell, 1997; Jaynes, 2003) that introduce the three concepts of sample space, event space, and probability measure. The probability space models a real-world process (referred to as an experiment) with random outcomes.

The sample space Ω

sample space

The *sample space* is the set of all possible outcomes of the experiment, usually denoted by Ω. For example, two successive coin tosses have a sample space of {hh, tt, ht, th}, where "h" denotes "heads" and "t" denotes "tails."

The event space $\mathcal{A}$

The *event space* is the space of potential results of the experiment. A subset A of the sample space Ω is in the event space $\mathcal{A}$ if at the end of the experiment we can observe whether a particular outcome $\omega \in \Omega$ is in A. The event space $\mathcal{A}$ is obtained by considering the collection of subsets of Ω, and for discrete probability distributions (Section 6.2.1) $\mathcal{A}$ is often the power set of Ω.

event space

The probability P

With each event $A \in \mathcal{A}$, we associate a number $P(A)$ that measures the probability or degree of belief that the event will occur. $P(A)$ is called the *probability* of A.

probability

The probability of a single event must lie in the interval $[0, 1]$, and the total probability over all outcomes in the sample space Ω must be 1, i.e., $P(\Omega) = 1$. Given a probability space $(\Omega, \mathcal{A}, P)$, we want to use it to model some real-world phenomenon. In machine learning, we often avoid explicitly referring to the probability space, but instead refer to probabilities on quantities of interest, which we denote by $\mathcal{T}$. In this book, we refer to $\mathcal{T}$ as the *target space* and refer to elements of $\mathcal{T}$ as states. We introduce a function $X : \Omega \to \mathcal{T}$ that takes an element of Ω (an outcome) and returns a particular quantity of interest x, a value in $\mathcal{T}$. This association/mapping from Ω to $\mathcal{T}$ is called a *random variable*. For example, in the case of tossing two coins and counting the number of heads, a random variable X maps to the three possible outcomes: $X(\text{hh}) = 2$, $X(\text{ht}) = 1$, $X(\text{th}) = 1$, and $X(\text{tt}) = 0$. In this particular case, $\mathcal{T} = \{0, 1, 2\}$, and it is the probabilities on elements of $\mathcal{T}$ that we are interested in. For a finite sample space Ω and finite $\mathcal{T}$, the function corresponding to a random variable is essentially a lookup table. For any subset $S \subseteq \mathcal{T}$, we associate $P_X(S) \in [0, 1]$ (the probability) to a particular event occurring corresponding to the random variable X. Example 6.1 provides a concrete illustration of the preceding terminology.

target space

random variable

The name "random variable" is a great source of misunderstanding as it is neither random nor is it a variable. It is a function.

Remark. The aforementioned sample space Ω unfortunately is referred to by different names in different books. Another common name for Ω is "state space" (Jacod and Protter, 2004), but state space is sometimes reserved for referring to states in a dynamical system (Hasselblatt and Katok, 2003). Other names sometimes used to describe Ω are "sample description space," "possibility space," and "event space." ◇

Example 6.1

We assume that the reader is already familiar with computing probabilities of intersections and unions of sets of events. A gentler introduction to probability with many examples can be found in chapter 2 of Walpole et al. (2011).

This toy example is essentially a biased coin flip example.

Consider a statistical experiment where we model a funfair game consisting of drawing two coins from a bag (with replacement). There are coins from USA (denoted as \$) and UK (denoted as £) in the bag, and since we draw two coins from the bag, there are four outcomes in total. The state space or sample space Ω of this experiment is then (\$, \$), (\$, £), (£, \$), (£, £). Let us assume that the composition of the bag of coins is such that a draw returns at random a \$ with probability 0.3.

The event we are interested in is the total number of times the repeated draw returns \$. Let us define a random variable X that maps the sample space Ω to $\mathcal{T}$, which denotes the number of times we draw \$ out of the bag. We can see from the preceding sample space we can get zero \$, one \$, or two \$s, and therefore $\mathcal{T} = \{0, 1, 2\}$. The random variable X (a function or lookup table) can be represented as a table like the following:

$$X((\$, \$)) = 2 \tag{6.1}$$

$$X((\$, £)) = 1 \tag{6.2}$$

$$X((£, \$)) = 1 \tag{6.3}$$

$$X((£, £)) = 0 \,. \tag{6.4}$$

Since we return the first coin we draw before drawing the second, this implies that the two draws are independent of each other, which we will discuss in Section 6.4.5. Note that there are two experimental outcomes, which map to the same event, where only one of the draws returns \$. Therefore, the probability mass function (Section 6.2.1) of X is given by

$$\begin{aligned} P(X = 2) &= P((\$, \$)) \\ &= P(\$) \cdot P(\$) \\ &= 0.3 \cdot 0.3 = 0.09 \end{aligned} \tag{6.5}$$

$$\begin{aligned} P(X = 1) &= P((\$, £) \cup (£, \$)) \\ &= P((\$, £)) + P((£, \$)) \\ &= 0.3 \cdot (1 - 0.3) + (1 - 0.3) \cdot 0.3 = 0.42 \end{aligned} \tag{6.6}$$

$$\begin{aligned} P(X = 0) &= P((£, £)) \\ &= P(£) \cdot P(£) \\ &= (1 - 0.3) \cdot (1 - 0.3) = 0.49 \,. \end{aligned} \tag{6.7}$$

In the preceding calculation, we equated two different concepts, the probability of the output of X and the probability of the samples in Ω. For example, in (6.7) we say $P(X = 0) = P((£, £))$. Consider the random variable $X : \Omega \to \mathcal{T}$ and a subset $S \subseteq \mathcal{T}$ (for example, a single element of $\mathcal{T}$, such as the outcome that one head is obtained when tossing two coins). Let $X^{-1}(S)$ be the preimage of S by X, i.e., the set of elements of Ω that map to S under X; $\{\omega \in \Omega : X(\omega) \in S\}$. One way to understand the transformation of probability from events in Ω via the random variable X is to associate it with the probability of the preimage of S (Jacod and Protter, 2004). For $S \subseteq \mathcal{T}$, we have the notation

$$P_X(S) = P(X \in S) = P(X^{-1}(S)) = P(\{\omega \in \Omega : X(\omega) \in S\}) \,. \tag{6.8}$$

The left-hand side of (6.8) is the probability of the set of possible outcomes (e.g., number of \$ = 1) that we are interested in. Via the random variable X, which maps states to outcomes, we see in the right-hand side of (6.8) that this is the probability of the set of states (in Ω) that have the property (e.g., \$£, £\$). We say that a random variable X is distributed according to a particular probability distribution P_X, which defines the probability mapping between the event and the probability of the outcome of the random variable. In other words, the function P_X or equivalently $P \circ X^{-1}$ is the *law* or *distribution* of random variable X.

law
distribution

Remark. The target space, that is, the range $\mathcal{T}$ of the random variable X, is used to indicate the kind of probability space, i.e., a $\mathcal{T}$ random variable. When $\mathcal{T}$ is finite or countably infinite, this is called a discrete random variable (Section 6.2.1). For continuous random variables (Section 6.2.2), we only consider $\mathcal{T} = \mathbb{R}$ or $\mathcal{T} = \mathbb{R}^D$. ♢

6.1.3 Statistics

Probability theory and statistics are often presented together, but they concern different aspects of uncertainty. One way of contrasting them is by the kinds of problems that are considered. Using probability, we can consider a model of some process, where the underlying uncertainty is captured by random variables, and we use the rules of probability to derive what happens. In statistics, we observe that something has happened and try to figure out the underlying process that explains the observations. In this sense, machine learning is close to statistics in its goals to construct a model that adequately represents the process that generated the data. We can use the rules of probability to obtain a "best-fitting" model for some data.

Another aspect of machine learning systems is that we are interested in generalization error (see Chapter 8). This means that we are actually interested in the performance of our system on instances that we will observe in future, which are not identical to the instances that we have seen so far. This analysis of future performance relies on probability and statistics, most of which is beyond what will be presented in this chapter. The interested reader is encouraged to look at the books by Boucheron et al. (2013) and Shalev-Shwartz and Ben-David (2014). We will see more about statistics in Chapter 8.

6.2 Discrete and Continuous Probabilities

Let us focus our attention on ways to describe the probability of an event as introduced in Section 6.1. Depending on whether the target space is discrete or continuous, the natural way to refer to distributions is different. When the target space $\mathcal{T}$ is discrete, we can specify the probability that a random variable X takes a particular value $x \in \mathcal{T}$, denoted as $P(X = x)$. The expression $P(X = x)$ for a discrete random variable X is known as the *probability mass function*. When the target space $\mathcal{T}$ is continuous, e.g., the real line $\mathbb{R}$, it is more natural to specify the

probability mass function

probability that a random variable X is in an interval, denoted by $P(a \leqslant X \leqslant b)$ for $a < b$. By convention, we specify the probability that a random variable X is less than a particular value x, denoted by $P(X \leqslant x)$. The expression $P(X \leqslant x)$ for a continuous random variable X is known as the *cumulative distribution function*. We will discuss continuous random variables in Section 6.2.2. We will revisit the nomenclature and contrast discrete and continuous random variables in Section 6.2.3.

cumulative distribution function

Remark. We will use the phrase *univariate* distribution to refer to distributions of a single random variable (whose states are denoted by nonbold x). We will refer to distributions of more than one random variable as *multivariate* distributions, and will usually consider a vector of random variables (whose states are denoted by bold $\boldsymbol{x}$). ◇

univariate

multivariate

6.2.1 Discrete Probabilities

When the target space is discrete, we can imagine the probability distribution of multiple random variables as filling out a (multidimensional) array of numbers. Figure 6.2 shows an example. The target space of the joint probability is the Cartesian product of the target spaces of each of the random variables. We define the *joint probability* as the entry of both values jointly

joint probability

$$P(X = x_i, Y = y_j) = \frac{n_{ij}}{N}, \tag{6.9}$$

where n_{ij} is the number of events with state x_i and y_j and N the total number of events. The joint probability is the probability of the intersection of both events, that is, $P(X = x_i, Y = y_j) = P(X = x_i \cap Y = y_j)$. Figure 6.2 illustrates the *probability mass function* (pmf) of a discrete probability distribution. For two random variables X and Y, the probability that $X = x$ and $Y = y$ is (lazily) written as $p(x, y)$ and is called the joint probability. One can think of a probability as a function that takes state x and y and returns a real number, which is the reason we write $p(x, y)$. The *marginal probability* that X takes the value x irrespective of the value of random variable Y is (lazily) written as $p(x)$. We write $X \sim p(x)$ to denote that the random variable X is distributed according to $p(x)$. If we consider only the instances where $X = x$, then the fraction of instances (the *conditional probability*) for which $Y = y$ is written (lazily) as $p(y \mid x)$.

probability mass function

marginal probability

conditional probability

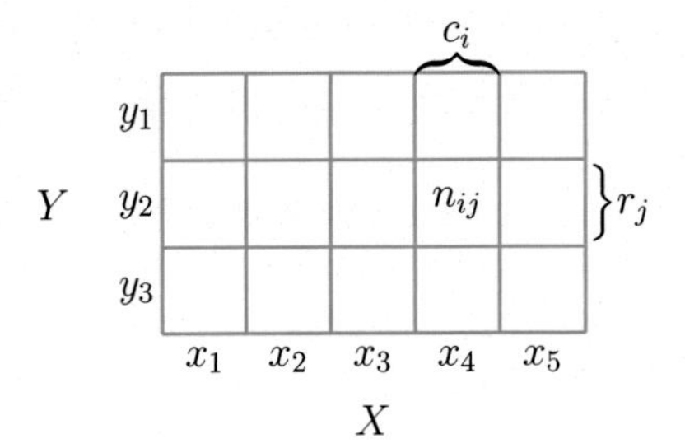

Figure 6.2 Visualization of a discrete bivariate probability mass function, with random variables X and Y. This diagram is adapted from Bishop (2006).

Example 6.2

Consider two random variables X and Y, where X has five possible states and Y has three possible states, as shown in Figure 6.2. We denote by n_{ij} the number of events with state $X = x_i$ and $Y = y_j$, and denote by N the total number of events. The value c_i is the sum of the individual frequencies for the ith column, that is, $c_i = \sum_{j=1}^{3} n_{ij}$. Similarly, the value r_j is the row sum, that is, $r_j = \sum_{i=1}^{5} n_{ij}$. Using these definitions, we can compactly express the distribution of X and Y.

The probability distribution of each random variable, the marginal probability, can be seen as the sum over a row or column

$$P(X = x_i) = \frac{c_i}{N} = \frac{\sum_{j=1}^{3} n_{ij}}{N} \tag{6.10}$$

and

$$P(Y = y_j) = \frac{r_j}{N} = \frac{\sum_{i=1}^{5} n_{ij}}{N}, \tag{6.11}$$

where c_i and r_j are the ith column and jth row of the probability table, respectively. By convention, for discrete random variables with a finite number of events, we assume that probabilties sum up to one, that is,

$$\sum_{i=1}^{5} P(X = x_i) = 1 \quad \text{and} \quad \sum_{j=1}^{3} P(Y = y_j) = 1\,. \tag{6.12}$$

The conditional probability is the fraction of a row or column in a particular cell. For example, the conditional probability of Y given X is

$$P(Y = y_j \,|\, X = x_i) = \frac{n_{ij}}{c_i}, \tag{6.13}$$

and the conditional probability of X given Y is

$$P(X = x_i \,|\, Y = y_j) = \frac{n_{ij}}{r_j}. \tag{6.14}$$

In machine learning, we use discrete probability distributions to model *categorical variables*, i.e., variables that take a finite set of unordered values. They could be categorical features, such as the degree taken at university when used for predicting the salary of a person, or categorical labels, such as letters of the alphabet when doing handwriting recognition. Discrete distributions are also often used to construct probabilistic models that combine a finite number of continuous distributions (Chapter 11).

categorical variable

6.2.2 Continuous Probabilities

We consider real-valued random variables in this section, i.e., we consider target spaces that are intervals of the real line $\mathbb{R}$. In this book, we pretend that we can perform operations on real random variables as if we have discrete probability spaces with finite states. However, this simplification is not precise for two situations: when we repeat something infinitely often, and when we want to draw a point from an interval. The first situation arises when we discuss generalization errors in machine learning (Chapter 8). The second situation arises when we want to discuss continuous distributions, such as the Gaussian (Section 6.5). For our purposes, the lack of precision allows for a briefer introduction to probability.

Remark. In continuous spaces, there are two additional technicalities, which are counterintuitive. First, the set of all subsets (used to define the event space $\mathcal{A}$ in Section 6.1) is not well behaved enough. $\mathcal{A}$ needs to be restricted to behave well under set complements, set intersections, and set unions. Second, the size of a set (which in discrete spaces can be obtained by counting the elements) turns out to be tricky. The size of a set is called its *measure*. For example, the cardinality of discrete sets, the length of an interval in $\mathbb{R}$, and the volume of a region in $\mathbb{R}^d$ are all measures. Sets that behave well under set operations and additionally have a topology are called a *Borel σ-algebra*. Betancourt details a careful construction of probability spaces from set theory without being bogged down in technicalities; see `https://tinyurl.com/yb3t6mfd`. For a more precise construction, we refer to Billingsley (1995) and Jacod and Protter (2004).

measure

Borel σ-algebra

In this book, we consider real-valued random variables with their corresponding Borel σ-algebra. We consider random variables with values in $\mathbb{R}^D$ to be a vector of real-valued random variables. ◇

Definition 6.1 (Probability Density Function)**.** A function $f : \mathbb{R}^D \to \mathbb{R}$ is called a *probability density function* (*pdf*) if

probability density function

pdf

1. $\forall \boldsymbol{x} \in \mathbb{R}^D : f(\boldsymbol{x}) \geqslant 0$
2. Its integral exists and

$$\int_{\mathbb{R}^D} f(\boldsymbol{x}) \mathrm{d}\boldsymbol{x} = 1 \,. \tag{6.15}$$

For probability mass functions (pmf) of discrete random variables, the integral in (6.15) is replaced with a sum (6.12).

Observe that the probability density function is any function f that is nonnegative and integrates to one. We associate a random variable X with this function f by

$$P(a \leqslant X \leqslant b) = \int_a^b f(x) \mathrm{d}x \,, \tag{6.16}$$

where $a, b \in \mathbb{R}$, and $x \in \mathbb{R}$ are outcomes of the continuous random variable X. States $\boldsymbol{x} \in \mathbb{R}^D$ are defined analogously by considering a vector of $x \in \mathbb{R}$. This association (6.16) is called the *law* or *distribution* of the random variable X.

law

$P(X = x)$ is a set of measure zero.

Remark. In contrast to discrete random variables, the probability of a continuous random variable X taking a particular value $P(X = x)$ is zero. This is like trying to specify an interval in (6.16) where $a = b$. ◇

Definition 6.2 (Cumulative Distribution Function)**.** A *cumulative distribution function* (cdf) of a multivariate real-valued random variable X with states $\boldsymbol{x} \in \mathbb{R}^D$ is given by

cumulative distribution function

$$F_X(\boldsymbol{x}) = P(X_1 \leqslant x_1, \ldots, X_D \leqslant x_D), \tag{6.17}$$

where $X = [X_1, \ldots, X_D]^\top$, $\boldsymbol{x} = [x_1, \ldots, x_D]^\top$, and the right-hand side represents the probability that random variable X_i takes the value smaller than or equal to x_i.

There are cdfs that do not have corresponding pdfs.

The cdf can be expressed also as the integral of the probability density function $f(\boldsymbol{x})$ so that

$$F_X(\boldsymbol{x}) = \int_{-\infty}^{x_1} \cdots \int_{-\infty}^{x_D} f(z_1, \ldots, z_D) \mathrm{d}z_1 \cdots \mathrm{d}z_D \,. \tag{6.18}$$

Remark. We reiterate that there are in fact two distinct concepts when talking about distributions. First is the idea of a pdf (denoted by $f(x)$), which is a nonnegative function that sums to one. Second is the law of a random variable X, that is, the association of a random variable X with the pdf $f(x)$. ♢

For most of this book, we will not use the notation $f(x)$ and $F_X(x)$ as we mostly do not need to distinguish between the pdf and cdf. However, we will need to be careful about pdfs and cdfs in Section 6.7.

6.2.3 Contrasting Discrete and Continuous Distributions

Recall from Section 6.1.2 that probabilities are positive and the total probability sums up to one. For discrete random variables (see (6.12)), this implies that the probability of each state must lie in the interval $[0, 1]$. However, for continuous random variables the normalization (see (6.15)) does not imply that the value of the density is less than or equal to 1 for all values. We illustrate this in Figure 6.3 using the *uniform distribution* for both discrete and continuous random variables.

uniform distribution

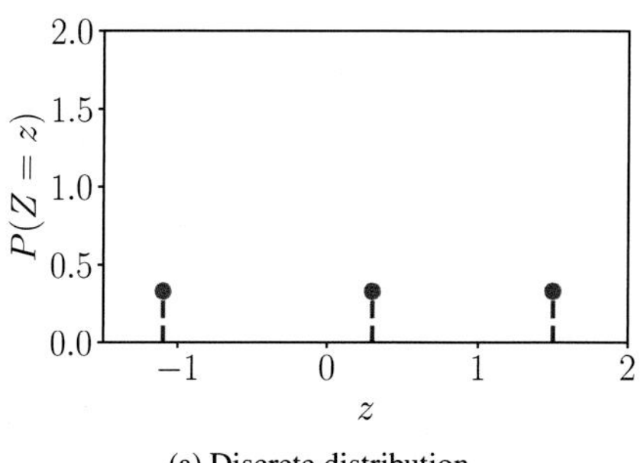

(a) Discrete distribution

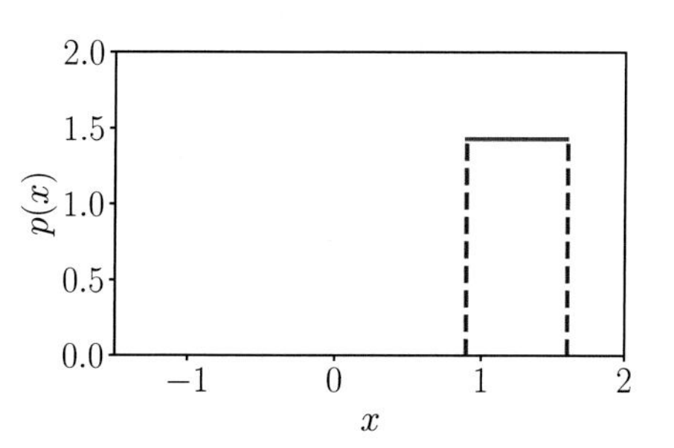

(b) Continuous distribution

Figure 6.3 Examples of (a) discrete and (b) continuous uniform distributions. See Example 6.3 for details of the distributions.

Example 6.3

We consider two examples of the uniform distribution, where each state is equally likely to occur. This example illustrates some differences between discrete and continuous probability distributions.

Let Z be a discrete uniform random variable with three states $\{z = -1.1, z = 0.3, z = 1.5\}$. The probability mass function can be represented as a table of probability values:

The actual values of these states are not meaningful here, and we deliberately chose numbers to drive home the point that we do not want to use (and should ignore) the ordering of the states.

z	-1.1	0.3	1.5
$P(Z = z)$	$\frac{1}{3}$	$\frac{1}{3}$	$\frac{1}{3}$

Alternatively, we can think of this as a graph (Figure 6.4(a)), where we use the fact that the states can be located on the x-axis, and the y-axis represents the probability of a particular state. The y-axis in Figure 6.4(a) is deliberately extended so that is it the same as in Figure 6.4(b).

Let X be a continuous random variable taking values in the range $0.9 \leqslant X \leqslant 1.6$, as represented by Figure 6.4(b). Observe that the height of the density can be greater than 1. However, it needs to hold that

$$\int_{0.9}^{1.6} p(x)\mathrm{d}x = 1\,. \tag{6.19}$$

Remark. There is an additional subtlety with regards to discrete probability distributions. The states $z_1, \dots, z_d$ do not in principle have any structure, i.e., there is usually no way to compare them, for example $z_1 = \text{red}, z_2 = \text{green}, z_3 = \text{blue}$. However, in many machine learning applications discrete states take numerical values, e.g., $z_1 = -1.1, z_2 = 0.3, z_3 = 1.5$, where we could say $z_1 < z_2 < z_3$. Discrete states that assume numerical values are particularly useful because we often consider expected values (Section 6.4.1) of random variables. ♢

Unfortunately, machine learning literature uses notation and nomenclature that hides the distinction between the sample space Ω, the target space $\mathcal{T}$, and the random variable X. For a value x of the set of possible outcomes of the random variable X, i.e., $x \in \mathcal{T}$, $p(x)$ denotes the probability that random variable X has the outcome x. For discrete random variables, this is written as $P(X = x)$, which is known as the probability mass function. The pmf is often referred to as the "distribution." For continuous variables, $p(x)$ is called the probability density function (often referred to as a density). To muddy things even further, the cumulative distribution function $P(X \leqslant x)$ is often also referred to as the "distribution." In this chapter, we will use the notation X to refer to both univariate and multivariate random variables, and denote the states by x and $\boldsymbol{x}$ respectively. We summarize the nomenclature in Table 6.1.

We think of the outcome x as the argument that results in the probability $p(x)$.

Remark. We will be using the expression "probability distribution" not only for discrete probability mass functions but also for continuous probability density

Type	"Point probability"	"Interval probability"
Discrete	$P(X = x)$ Probability mass function	Not applicable
Continuous	$p(x)$ Probability density function	$P(X \leqslant x)$ Cumulative distribution function

Table 6.1 Nomenclature for probability distributions.

functions, although this is technically incorrect. In line with most machine learning literature, we also rely on context to distinguish the different uses of the phrase probability distribution. ♢

6.3 Sum Rule, Product Rule, and Bayes' Theorem

We think of probability theory as an extension to logical reasoning. As we discussed in Section 6.1.1, the rules of probability presented here follow naturally from fulfilling the desiderata (Jaynes, 2003, chapter 2). Probabilistic modeling (Section 8.4) provides a principled foundation for designing machine learning methods. Once we have defined probability distributions (Section 6.2) corresponding to the uncertainties of the data and our problem, it turns out that there are only two fundamental rules, the sum rule and the product rule.

Recall from (6.9) that $p(\boldsymbol{x}, \boldsymbol{y})$ is the joint distribution of the two random variables $\boldsymbol{x}, \boldsymbol{y}$. The distributions $p(\boldsymbol{x})$ and $p(\boldsymbol{y})$ are the corresponding marginal distributions, and $p(\boldsymbol{y} \mid \boldsymbol{x})$ is the conditional distribution of $\boldsymbol{y}$ given $\boldsymbol{x}$. Given the definitions of the marginal and conditional probability for discrete and continuous random variables in Section 6.2, we can now present the two fundamental rules in probability theory.

These two rules arise naturally (Jaynes, 2003) from the requirements we discussed in Section 6.1.1.

The first rule, the *sum rule*, states that

sum rule

$$p(\boldsymbol{x}) = \begin{cases} \sum_{\boldsymbol{y} \in \mathcal{Y}} p(\boldsymbol{x}, \boldsymbol{y}) & \text{if } \boldsymbol{y} \text{ is discrete} \\ \int_{\mathcal{Y}} p(\boldsymbol{x}, \boldsymbol{y}) \mathrm{d}\boldsymbol{y} & \text{if } \boldsymbol{y} \text{ is continuous} \end{cases}, \tag{6.20}$$

where $\mathcal{Y}$ are the states of the target space of random variable Y. This means that we sum out (or integrate out) the set of states $\boldsymbol{y}$ of the random variable Y. The sum rule is also known as the *marginalization property*. The sum rule relates the joint distribution to a marginal distribution. In general, when the joint distribution contains more than two random variables, the sum rule can be applied to any subset of the random variables, resulting in a marginal distribution of potentially more than one random variable. More concretely, if $\boldsymbol{x} = [x_1, \ldots, x_D]^\top$, we obtain the marginal

marginalization property

$$p(x_i) = \int p(x_1, \ldots, x_D) \mathrm{d}\boldsymbol{x}_{\backslash i} \tag{6.21}$$

by repeated application of the sum rule where we integrate/sum out all random variables except x_i, which is indicated by $\backslash i$, which reads "all except i."

Remark. Many of the computational challenges of probabilistic modeling are due to the application of the sum rule. When there are many variables or discrete variables with many states, the sum rule boils down to performing a high-dimensional sum or integral. Performing high-dimensional sums or integrals is generally computationally hard, in the sense that there is no known polynomial-time algorithm to calculate them exactly. ♢

product rule

The second rule, known as the *product rule*, relates the joint distribution to the conditional distribution via

$$p(\boldsymbol{x}, \boldsymbol{y}) = p(\boldsymbol{y} \,|\, \boldsymbol{x}) p(\boldsymbol{x}) \,. \tag{6.22}$$

The product rule can be interpreted as the fact that every joint distribution of two random variables can be factorized (written as a product) of two other distributions. The two factors are the marginal distribution of the first random variable $p(\boldsymbol{x})$, and the conditional distribution of the second random variable given the first $p(\boldsymbol{y} \,|\, \boldsymbol{x})$. Since the ordering of random variables is arbitrary in $p(\boldsymbol{x}, \boldsymbol{y})$, the product rule also implies $p(\boldsymbol{x}, \boldsymbol{y}) = p(\boldsymbol{x} \,|\, \boldsymbol{y}) p(\boldsymbol{y})$. To be precise, (6.22) is expressed in terms of the probability mass functions for discrete random variables. For continuous random variables, the product rule is expressed in terms of the probability density functions (Section 6.2.3).

In machine learning and Bayesian statistics, we are often interested in making inferences of unobserved (latent) random variables given that we have observed other random variables. Let us assume we have some prior knowledge $p(\boldsymbol{x})$ about an unobserved random variable $\boldsymbol{x}$ and some relationship $p(\boldsymbol{y} \,|\, \boldsymbol{x})$ between $\boldsymbol{x}$ and a second random variable $\boldsymbol{y}$, which we can observe. If we observe $\boldsymbol{y}$, we can use Bayes' theorem to draw some conclusions about $\boldsymbol{x}$ given the observed values of $\boldsymbol{y}$. *Bayes' theorem* (also *Bayes' rule* or *Bayes' law*)

Bayes' theorem
Bayes' rule
Bayes' law

$$\underbrace{p(\boldsymbol{x} \,|\, \boldsymbol{y})}_{\text{posterior}} = \frac{\overbrace{p(\boldsymbol{y} \,|\, \boldsymbol{x})}^{\text{likelihood}} \overbrace{p(\boldsymbol{x})}^{\text{prior}}}{\underbrace{p(\boldsymbol{y})}_{\text{evidence}}} \tag{6.23}$$

is a direct consequence of the product rule in (6.20) since

$$p(\boldsymbol{x}, \boldsymbol{y}) = p(\boldsymbol{x} \,|\, \boldsymbol{y}) p(\boldsymbol{y}) \tag{6.24}$$

and

$$p(\boldsymbol{x}, \boldsymbol{y}) = p(\boldsymbol{y} \,|\, \boldsymbol{x}) p(\boldsymbol{x}) \tag{6.25}$$

so that

$$p(\boldsymbol{x} \,|\, \boldsymbol{y}) p(\boldsymbol{y}) = p(\boldsymbol{y} \,|\, \boldsymbol{x}) p(\boldsymbol{x}) \iff p(\boldsymbol{x} \,|\, \boldsymbol{y}) = \frac{p(\boldsymbol{y} \,|\, \boldsymbol{x}) p(\boldsymbol{x})}{p(\boldsymbol{y})} \,. \tag{6.26}$$

prior

In (6.23), $p(\boldsymbol{x})$ is the *prior*, which encapsulates our subjective prior knowledge of the unobserved (latent) variable $\boldsymbol{x}$ before observing any data. We can choose any prior that makes sense to us, but it is critical to ensure that the prior has a nonzero pdf (or pmf) on all plausible $\boldsymbol{x}$, even if they are very rare.

The *likelihood* $p(\boldsymbol{y} \mid \boldsymbol{x})$ describes how $\boldsymbol{x}$ and $\boldsymbol{y}$ are related, and in the case of discrete probability distributions, it is the probability of the data $\boldsymbol{y}$ if we were to know the latent variable $\boldsymbol{x}$. Note that the likelihood is not a distribution in $\boldsymbol{x}$, but only in $\boldsymbol{y}$. We call $p(\boldsymbol{y} \mid \boldsymbol{x})$ either the "likelihood of $\boldsymbol{x}$ (given $\boldsymbol{y}$)" or the "probability of $\boldsymbol{y}$ given $\boldsymbol{x}$" but never the likelihood of $\boldsymbol{y}$ (MacKay, 2003).

likelihood

The likelihood is sometimes also called the "measurement model."

The *posterior* $p(\boldsymbol{x} \mid \boldsymbol{y})$ is the quantity of interest in Bayesian statistics because it expresses exactly what we are interested in, i.e., what we know about $\boldsymbol{x}$ after having observed $\boldsymbol{y}$.

posterior

The quantity

$$p(\boldsymbol{y}) := \int p(\boldsymbol{y} \mid \boldsymbol{x}) p(\boldsymbol{x}) \mathrm{d}\boldsymbol{x} = \mathbb{E}_X[p(\boldsymbol{y} \mid \boldsymbol{x})] \tag{6.27}$$

is the *marginal likelihood/evidence*. The right-hand side of (6.27) uses the expectation operator, which we define in Section 6.4.1. By definition, the marginal likelihood integrates the numerator of (6.23) with respect to the latent variable $\boldsymbol{x}$. Therefore, the marginal likelihood is independent of $\boldsymbol{x}$, and it ensures that the posterior $p(\boldsymbol{x} \mid \boldsymbol{y})$ is normalized. The marginal likelihood can also be interpreted as the expected likelihood where we take the expectation with respect to the prior $p(\boldsymbol{x})$. Beyond normalization of the posterior, the marginal likelihood also plays an important role in Bayesian model selection, as we will discuss in Section 8.6. Due to the integration in (8.44), the evidence is often hard to compute.

marginal likelihood

evidence

Bayes' theorem (6.23) allows us to invert the relationship between $\boldsymbol{x}$ and $\boldsymbol{y}$ given by the likelihood. Therefore, Bayes' theorem is sometimes called the *probabilistic inverse*. We will discuss Bayes' theorem further in Section 8.4.

Bayes' theorem is also called the "probabilistic inverse."

probabilistic inverse

Remark. In Bayesian statistics, the posterior distribution is the quantity of interest as it encapsulates all available information from the prior and the data. Instead of carrying the posterior around, it is possible to focus on some statistic of the posterior, such as the maximum of the posterior, which we will discuss in Section 8.3. However, focusing on some statistic of the posterior leads to loss of information. If we think in a bigger context, then the posterior can be used within a decision-making system, and having the full posterior can be extremely useful and lead to decisions that are robust to disturbances. For example, in the context of model-based reinforcement learning, Deisenroth et al. (2015) show that using the full posterior distribution of plausible transition functions leads to very fast (data/sample efficient) learning, whereas focusing on the maximum of the posterior leads to consistent failures. Therefore, having the full posterior can be very useful for a downstream task. In Chapter 9, we will continue this discussion in the context of linear regression. ◇

6.4 Summary Statistics and Independence

We are often interested in summarizing sets of random variables and comparing pairs of random variables. A statistic of a random variable is a deterministic function of that random variable. The summary statistics of a distribution provide one useful view of how a random variable behaves, and as the name suggests, provide numbers that summarize and characterize the distribution. We describe the mean and the variance, two well-known summary statistics. Then we discuss two ways to compare a pair of random variables: first, how to say that two random variables are independent; and second, how to compute an inner product between them.

6.4.1 Means and Covariances

Mean and (co)variance are often useful to describe properties of probability distributions (expected values and spread). We will see in Section 6.6 that there is a useful family of distributions (called the exponential family), where the statistics of the random variable capture all possible information.

The concept of the expected value is central to machine learning, and the foundational concepts of probability itself can be derived from the expected value (Whittle, 2000).

expected value

Definition 6.3 (Expected Value). The *expected value* of a function $g : \mathbb{R} \to \mathbb{R}$ of a univariate continuous random variable $X \sim p(x)$ is given by

$$\mathbb{E}_X[g(x)] = \int_{\mathcal{X}} g(x)p(x)\mathrm{d}x \,. \tag{6.28}$$

Correspondingly, the expected value of a function g of a discrete random variable $X \sim p(x)$ is given by

$$\mathbb{E}_X[g(x)] = \sum_{x \in \mathcal{X}} g(x)p(x) \,, \tag{6.29}$$

where $\mathcal{X}$ is the set of possible outcomes (the target space) of the random variable X.

In this section, we consider discrete random variables to have numerical outcomes. This can be seen by observing that the function g takes real numbers as inputs.

The expected value of a function of a random variable is sometimes referred to as the law of the unconscious statistician (Casella and Berger, 2002, section 2.2).

Remark. We consider multivariate random variables X as a finite vector of univariate random variables $[X_1, \ldots, X_n]^\top$. For multivariate random variables, we define the expected value element wise

$$\mathbb{E}_X[g(\boldsymbol{x})] = \begin{bmatrix} \mathbb{E}_{X_1}[g(x_1)] \\ \vdots \\ \mathbb{E}_{X_D}[g(x_D)] \end{bmatrix} \in \mathbb{R}^D \,, \tag{6.30}$$

where the subscript $\mathbb{E}_{X_d}$ indicates that we are taking the expected value with respect to the dth element of the vector $\boldsymbol{x}$. ♢

Definition 6.3 defines the meaning of the notation $\mathbb{E}_X$ as the operator indicating that we should take the integral with respect to the probability density (for continuous distributions) or the sum over all states (for discrete distributions). The definition of the mean (Definition 6.4) is a special case of the expected value, obtained by choosing g to be the identity function.

Definition 6.4 (Mean). The *mean* of a random variable X with states $\boldsymbol{x} \in \mathbb{R}^D$ is an average and is defined as

mean

$$\mathbb{E}_X[\boldsymbol{x}] = \begin{bmatrix} \mathbb{E}_{X_1}[x_1] \\ \vdots \\ \mathbb{E}_{X_D}[x_D] \end{bmatrix} \in \mathbb{R}^D, \tag{6.31}$$

where

$$\mathbb{E}_{x_d}[x_d] := \begin{cases} \displaystyle\int_{\mathcal{X}} x_d p(x_d) \mathrm{d}x_d & \text{if } X \text{ is a continuous random variable} \\ \displaystyle\sum_{x_i \in \mathcal{X}} x_i p(x_d = x_i) & \text{if } X \text{ is a discrete random variable} \end{cases} \tag{6.32}$$

for $d = 1, \ldots, D$, where the subscript d indicates the corresponding dimension of $\boldsymbol{x}$. The integral and sum are over the states $\mathcal{X}$ of the target space of the random variable X.

In one dimension, there are two other intuitive notions of "average," which are the *median* and the *mode*. The *median* is the "middle" value if we sort the values, i.e., 50% of the values are greater than the median and 50% are smaller than the median. This idea can be generalized to continuous values by considering the value where the cdf (Definition 6.2) is 0.5. For distributions, which are asymmetric or have long tails, the median provides an estimate of a typical value that is closer to human intuition than the mean value. Furthermore, the median is more robust to outliers than the mean. The generalization of the median to higher dimensions is nontrivial as there is no obvious way to "sort" in more than one dimension (Hallin et al., 2010; Kong and Mizera, 2012). The *mode* is the most frequently occurring value. For a discrete random variable, the mode is defined as the value of x having the highest frequency of occurrence. For a continuous random variable, the mode is defined as a peak in the density $p(\boldsymbol{x})$. A particular density $p(\boldsymbol{x})$ may have more than one mode, and furthermore there may be a very large number of modes in high-dimensional distributions. Therefore, finding all the modes of a distribution can be computationally challenging.

median

mode

Example 6.4

Consider the two-dimensional distribution illustrated in Figure 6.4:

$$p(x) = 0.4\mathcal{N}\left(\boldsymbol{x} \,\middle|\, \begin{bmatrix} 10 \\ 2 \end{bmatrix}, \begin{bmatrix} 1 & 0 \\ 0 & 1 \end{bmatrix}\right) + 0.6\mathcal{N}\left(\boldsymbol{x} \,\middle|\, \begin{bmatrix} 0 \\ 0 \end{bmatrix}, \begin{bmatrix} 8.4 & 2.0 \\ 2.0 & 1.7 \end{bmatrix}\right). \tag{6.33}$$

We will define the Gaussian distribution $\mathcal{N}(\mu, \sigma^2)$ in Section 6.5. Also shown is its corresponding marginal distribution in each dimension. Observe that the distribution is bimodal (has two modes), but one of the marginal distributions is unimodal (has one mode). The horizontal bimodal univariate distribution illustrates that the mean and median can be different from each other. While it is tempting to define the two-dimensional median to be the

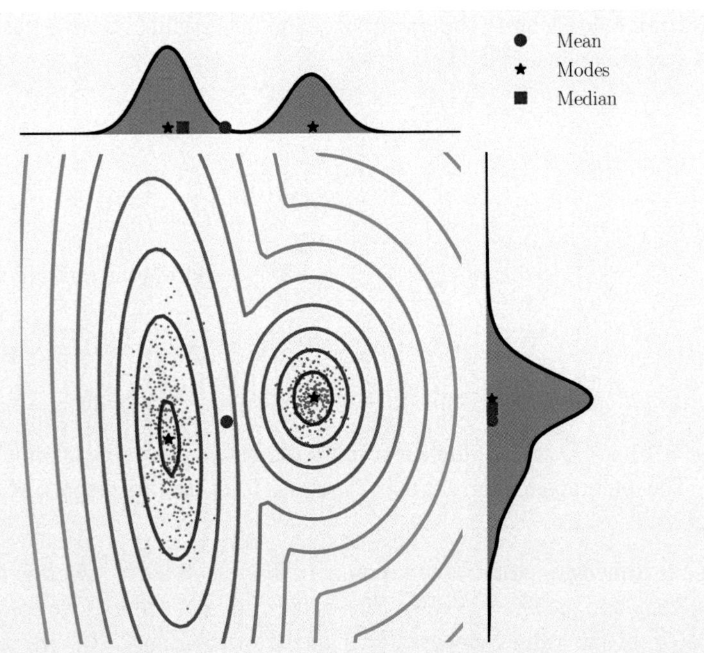

Figure 6.4 Illustration of the mean, mode, and median for a two-dimensional dataset, as well as its marginal densities.

concatenation of the medians in each dimension, the fact that we cannot define an ordering of two-dimensional points makes it difficult. When we say "cannot define an ordering." We mean that there is more than one way to define the relation $<$ so that $\begin{bmatrix}3\\0\end{bmatrix} < \begin{bmatrix}2\\3\end{bmatrix}$.

Remark. The expected value (Definition 6.3) is a linear operator. For example, given a real-valued function $f(\boldsymbol{x}) = ag(\boldsymbol{x}) + bh(\boldsymbol{x})$ where $a, b \in \mathbb{R}$ and $\boldsymbol{x} \in \mathbb{R}^D$, we obtain

$$\mathbb{E}_X[f(\boldsymbol{x})] = \int f(\boldsymbol{x})p(\boldsymbol{x})\mathrm{d}\boldsymbol{x} \tag{6.34a}$$

$$= \int [ag(\boldsymbol{x}) + bh(\boldsymbol{x})]p(\boldsymbol{x})\mathrm{d}\boldsymbol{x} \tag{6.34b}$$

$$= a\int g(\boldsymbol{x})p(\boldsymbol{x})\mathrm{d}x + b\int h(\boldsymbol{x})p(\boldsymbol{x})\mathrm{d}\boldsymbol{x} \tag{6.34c}$$

$$= a\mathbb{E}_X[g(\boldsymbol{x})] + b\mathbb{E}_X[h(\boldsymbol{x})]\,. \tag{6.34d}$$

◇

For two random variables, we may wish to characterize their correspondence to each other. The covariance intuitively represents the notion of how dependent random variables are to one another.

covariance

Definition 6.5 (Covariance (Univariate)). The *covariance* between two univariate random variables $X, Y \in \mathbb{R}$ is given by the expected product of their deviations from their respective means, i.e.,

$$\mathrm{Cov}_{X,Y}[x, y] := \mathbb{E}_{X,Y}\big[(x - \mathbb{E}_X[x])(y - \mathbb{E}_Y[y])\big]\,. \tag{6.35}$$

Remark. When the random variable associated with the expectation or covariance is clear by its arguments, the subscript is often suppressed (for example, $\mathbb{E}_X[x]$ is often written as $\mathbb{E}[x]$). ♢

By using the linearity of expectations, the expression in Definition 6.5 can be rewritten as the expected value of the product minus the product of the expected values, i.e.,

Terminology: The covariance of multivariate random variables $\mathrm{Cov}[x, y]$ is sometimes referred to as cross-covariance, with covariance referring to $\mathrm{Cov}[x, x]$.

$$\mathrm{Cov}[x, y] = \mathbb{E}[xy] - \mathbb{E}[x]\mathbb{E}[y] . \tag{6.36}$$

The covariance of a variable with itself $\mathrm{Cov}[x, x]$ is called the *variance* and is denoted by $\mathbb{V}_X[x]$. The square root of the variance is called the *standard deviation* and is often denoted by $\sigma(x)$. The notion of covariance can be generalized to multivariate random variables.

variance

standard deviation

Definition 6.6 (Covariance (Multivariate))**.** If we consider two multivariate random variables X and Y with states $\boldsymbol{x} \in \mathbb{R}^D$ and $\boldsymbol{y} \in \mathbb{R}^E$ respectively, the *covariance* between X and Y is defined as

covariance

$$\mathrm{Cov}[\boldsymbol{x}, \boldsymbol{y}] = \mathbb{E}[\boldsymbol{x}\boldsymbol{y}^\top] - \mathbb{E}[\boldsymbol{x}]\mathbb{E}[\boldsymbol{y}]^\top = \mathrm{Cov}[\boldsymbol{y}, \boldsymbol{x}]^\top \in \mathbb{R}^{D \times E} . \tag{6.37}$$

Definition 6.6 can be applied with the same multivariate random variable in both arguments, which results in a useful concept that intuitively captures the "spread" of a random variable. For a multivariate random variable, the variance describes the relation between individual dimensions of the random variable.

Definition 6.7 (Variance)**.** The *variance* of a random variable X with states $\boldsymbol{x} \in \mathbb{R}^D$ and a mean vector $\boldsymbol{\mu} \in \mathbb{R}^D$ is defined as

variance

$$\begin{aligned}
\mathbb{V}_X[\boldsymbol{x}] &= \mathrm{Cov}_X[\boldsymbol{x}, \boldsymbol{x}] && (6.38a)\\
&= \mathbb{E}_X[(\boldsymbol{x} - \boldsymbol{\mu})(\boldsymbol{x} - \boldsymbol{\mu})^\top] = \mathbb{E}_X[\boldsymbol{x}\boldsymbol{x}^\top] - \mathbb{E}_X[\boldsymbol{x}]\mathbb{E}_X[\boldsymbol{x}]^\top && (6.38b)\\
&= \begin{bmatrix} \mathrm{Cov}[x_1, x_1] & \mathrm{Cov}[x_1, x_2] & \dots & \mathrm{Cov}[x_1, x_D] \\ \mathrm{Cov}[x_2, x_1] & \mathrm{Cov}[x_2, x_2] & \dots & \mathrm{Cov}[x_2, x_D] \\ \vdots & \vdots & \ddots & \vdots \\ \mathrm{Cov}[x_D, x_1] & \dots & \dots & \mathrm{Cov}[x_D, x_D] \end{bmatrix} . && (6.38c)
\end{aligned}$$

The $D \times D$ matrix in (6.38c) is called the *covariance matrix* of the multivariate random variable X. The covariance matrix is symmetric and positive definite and tells us something about the spread of the data. On its diagonal, the covariance matrix contains the variances of the *marginals*

covariance matrix

marginal

$$p(x_i) = \int p(x_1, \dots, x_D) \mathrm{d}x_{\backslash i} , \tag{6.39}$$

where "$\backslash i$" denotes "all variables but i." The off-diagonal entries are the *cross-covariance* terms $\mathrm{Cov}[x_i, x_j]$ for $i, j = 1, \dots, D, i \neq j$.

cross-covariance

Remark. In this book, we generally assume that covariance matrices are positive definite to enable better intuition. We therefore do not discuss corner cases that result in positive semidefinite (low-rank) covariance matrices. ♢

When we want to compare the covariances between different pairs of random variables, it turns out that the variance of each random variable affects the value of the covariance. The normalized version of covariance is called the correlation.

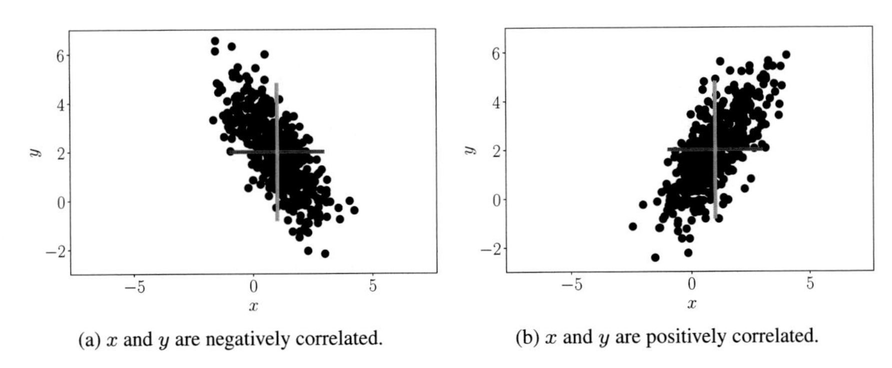

(a) x and y are negatively correlated.

(b) x and y are positively correlated.

Figure 6.5 Two-dimensional datasets with identical means and variances along each axis (colored lines) but with different covariances.

correlation

Definition 6.8 (Correlation). The *correlation* between two random variables X, Y is given by

$$\operatorname{corr}[x, y] = \frac{\operatorname{Cov}[x, y]}{\sqrt{\mathbb{V}[x]\mathbb{V}[y]}} \in [-1, 1] . \tag{6.40}$$

The correlation matrix is the covariance matrix of standardized random variables, $x/\sigma(x)$. In other words, each random variable is divided by its standard deviation (the square root of the variance) in the correlation matrix.

The covariance (and correlation) indicate how two random variables are related; see Figure 6.5. Positive correlation $\operatorname{corr}[x, y]$ means that when x grows, then y is also expected to grow. Negative correlation means that as x increases, then y decreases.

6.4.2 Empirical Means and Covariances

population mean and covariance

The definitions in Section 6.4.1 are often also called the *population mean and covariance*, as it refers to the true statistics for the population. In machine learning, we need to learn from empirical observations of data. Consider a random variable X. There are two conceptual steps to go from population statistics to the realization of empirical statistics. First, we use the fact that we have a finite dataset (of size N) to construct an empirical statistic that is a function of a finite number of identical random variables, $X_1, \ldots, X_N$. Second, we observe the data, that is, we look at the realization $x_1, \ldots, x_N$ of each of the random variables and apply the empirical statistic.

empirical mean
sample mean

Specifically, for the mean (Definition 6.4), given a particular dataset we can obtain an estimate of the mean, which is called the *empirical mean* or *sample mean*. The same holds for the empirical covariance.

empirical mean

Definition 6.9 (Empirical Mean and Covariance). The *empirical mean* vector is the arithmetic average of the observations for each variable, and it is defined as

$$\bar{\boldsymbol{x}} := \frac{1}{N} \sum_{n=1}^{N} \boldsymbol{x}_n , \tag{6.41}$$

where $\boldsymbol{x}_n \in \mathbb{R}^D$.

Similar to the empirical mean, the *empirical covariance* matrix is a $D \times D$ matrix

empirical covariance

$$\boldsymbol{\Sigma} := \frac{1}{N} \sum_{n=1}^{N} (\boldsymbol{x}_n - \bar{\boldsymbol{x}})(\boldsymbol{x}_n - \bar{\boldsymbol{x}})^\top . \tag{6.42}$$

To compute the statistics for a particular dataset, we would use the realizations (observations) $\boldsymbol{x}_1, \ldots, \boldsymbol{x}_N$ and use (6.41) and (6.42). Empirical covariance matrices are symmetric, positive semidefinite (see Section 3.2.3).

Throughout the book, we use the empirical covariance, which is a biased estimate. The unbiased (sometimes called corrected) covariance has the factor $N-1$ in the denominator instead of N.

6.4.3 Three Expressions for the Variance

We now focus on a single random variable X and use the preceding empirical formulas to derive three possible expressions for the variance. The following derivation is the same for the population variance, except that we need to take care of integrals. The standard definition of variance, corresponding to the definition of covariance (Definition 6.5), is the expectation of the squared deviation of a random variable X from its expected value μ, i.e.,

The derivations are exercises at the end of this chapter.

$$\mathbb{V}_X[x] := \mathbb{E}_X[(x-\mu)^2] . \tag{6.43}$$

The expectation in (6.43) and the mean $\mu = \mathbb{E}_X(x)$ are computed using (6.32), depending on whether X is a discrete or continuous random variable. The variance as expressed in (6.43) is the mean of a new random variable $Z := (X-\mu)^2$.

When estimating the variance in (6.43) empirically, we need to resort to a two-pass algorithm: one pass through the data to calculate the mean μ using (6.41), and then a second pass using this estimate $\hat{\mu}$ calculate the variance. It turns out that we can avoid two passes by rearranging the terms. The formula in (6.43) can be converted to the so-called *raw-score formula for variance*:

raw-score formula for variance

$$\mathbb{V}_X[x] = \mathbb{E}_X[x^2] - (\mathbb{E}_X[x])^2 . \tag{6.44}$$

The expression in (6.44) can be remembered as "the mean of the square minus the square of the mean." It can be calculated empirically in one pass through data since we can accumulate x_i (to calculate the mean) and x_i^2 simultaneously, where x_i is the ith observation. Unfortunately, if implemented in this way, it can be numerically unstable. The raw-score version of the variance can be useful in machine learning, e.g., when deriving the bias–variance decomposition (Bishop, 2006).

If the two terms in (6.44) are huge and approximately equal, we may suffer from an unnecessary loss of numerical precision in floating-point arithmetic.

A third way to understand the variance is that it is a sum of pairwise differences between all pairs of observations. Consider a sample $x_1, \ldots, x_N$ of realizations of random variable X, and we compute the squared difference between pairs of x_i and x_j. By expanding the square, we can show that the sum of N^2 pairwise differences is the empirical variance of the observations:

$$\frac{1}{N^2} \sum_{i,j=1}^{N} (x_i - x_j)^2 = 2 \left[\frac{1}{N} \sum_{i=1}^{N} x_i^2 - \left(\frac{1}{N} \sum_{i=1}^{N} x_i \right)^2 \right] . \tag{6.45}$$

We see that (6.45) is twice the raw-score expression (6.44). This means that we can express the sum of pairwise distances (of which there are N^2 of them) as a sum of deviations from the mean (of which there are N). Geometrically, this means that there is an equivalence between the pairwise distances and the distances from the center of the set of points. From a computational perspective, this means that by computing the mean (N terms in the summation), and then computing the variance (again N terms in the summation), we can obtain an expression (left-hand side of (6.45)) that has N^2 terms.

6.4.4 Sums and Transformations of Random Variables

We may want to model a phenomenon that cannot be well explained by textbook distributions (we introduce some in Sections 6.5 and 6.6), and hence may perform simple manipulations of random variables (such as adding two random variables).

Consider two random variables X, Y with states $\boldsymbol{x}, \boldsymbol{y} \in \mathbb{R}^D$. Then:

$$\mathbb{E}[\boldsymbol{x}+\boldsymbol{y}] = \mathbb{E}[\boldsymbol{x}] + \mathbb{E}[\boldsymbol{y}] \tag{6.46}$$

$$\mathbb{E}[\boldsymbol{x}-\boldsymbol{y}] = \mathbb{E}[\boldsymbol{x}] - \mathbb{E}[\boldsymbol{y}] \tag{6.47}$$

$$\mathbb{V}[\boldsymbol{x}+\boldsymbol{y}] = \mathbb{V}[\boldsymbol{x}] + \mathbb{V}[\boldsymbol{y}] + \mathrm{Cov}[\boldsymbol{x},\boldsymbol{y}] + \mathrm{Cov}[\boldsymbol{y},\boldsymbol{x}] \tag{6.48}$$

$$\mathbb{V}[\boldsymbol{x}-\boldsymbol{y}] = \mathbb{V}[\boldsymbol{x}] + \mathbb{V}[\boldsymbol{y}] - \mathrm{Cov}[\boldsymbol{x},\boldsymbol{y}] - \mathrm{Cov}[\boldsymbol{y},\boldsymbol{x}] \,. \tag{6.49}$$

Mean and (co)variance exhibit some useful properties when it comes to affine transformation of random variables. Consider a random variable X with mean $\boldsymbol{\mu}$ and covariance matrix $\boldsymbol{\Sigma}$ and a (deterministic) affine transformation $\boldsymbol{y} = \boldsymbol{A}\boldsymbol{x}+\boldsymbol{b}$ of $\boldsymbol{x}$. Then $\boldsymbol{y}$ is itself a random variable whose mean vector and covariance matrix are given by

$$\mathbb{E}_Y[\boldsymbol{y}] = \mathbb{E}_X[\boldsymbol{A}\boldsymbol{x}+\boldsymbol{b}] = \boldsymbol{A}\mathbb{E}_X[\boldsymbol{x}] + \boldsymbol{b} = \boldsymbol{A}\boldsymbol{\mu} + \boldsymbol{b} \,, \tag{6.50}$$

$$\mathbb{V}_Y[\boldsymbol{y}] = \mathbb{V}_X[\boldsymbol{A}\boldsymbol{x}+\boldsymbol{b}] = \mathbb{V}_X[\boldsymbol{A}\boldsymbol{x}] = \boldsymbol{A}\mathbb{V}_X[\boldsymbol{x}]\boldsymbol{A}^\top = \boldsymbol{A}\boldsymbol{\Sigma}\boldsymbol{A}^\top \,, \tag{6.51}$$

respectively. Furthermore,

This can be shown directly by using the definition of the mean and covariance.

$$\begin{aligned}
\mathrm{Cov}[\boldsymbol{x},\boldsymbol{y}] &= \mathbb{E}[\boldsymbol{x}(\boldsymbol{A}\boldsymbol{x}+\boldsymbol{b})^\top] - \mathbb{E}[\boldsymbol{x}]\mathbb{E}[\boldsymbol{A}\boldsymbol{x}+\boldsymbol{b}]^\top && (6.52\text{a})\\
&= \mathbb{E}[\boldsymbol{x}]\boldsymbol{b}^\top + \mathbb{E}[\boldsymbol{x}\boldsymbol{x}^\top]\boldsymbol{A}^\top - \boldsymbol{\mu}\boldsymbol{b}^\top - \boldsymbol{\mu}\boldsymbol{\mu}^\top\boldsymbol{A}^\top && (6.52\text{b})\\
&= \boldsymbol{\mu}\boldsymbol{b}^\top - \boldsymbol{\mu}\boldsymbol{b}^\top + \left(\mathbb{E}[\boldsymbol{x}\boldsymbol{x}^\top] - \boldsymbol{\mu}\boldsymbol{\mu}^\top\right)\boldsymbol{A}^\top && (6.52\text{c})\\
&\overset{(6.38\text{b})}{=} \boldsymbol{\Sigma}\boldsymbol{A}^\top \,, && (6.52\text{d})
\end{aligned}$$

where $\boldsymbol{\Sigma} = \mathbb{E}[\boldsymbol{x}\boldsymbol{x}^\top] - \boldsymbol{\mu}\boldsymbol{\mu}^\top$ is the covariance of X.

6.4.5 Statistical Independence

statistical independence

Definition 6.10 (Independence). Two random variables X, Y are *statistically independent* if and only if

$$p(\boldsymbol{x},\boldsymbol{y}) = p(\boldsymbol{x})p(\boldsymbol{y}) \,. \tag{6.53}$$

Intuitively, two random variables X and Y are independent if the value of $\boldsymbol{y}$ (once known) does not add any additional information about $\boldsymbol{x}$ (and vice versa). If X, Y are (statistically) independent, then

- $p(\boldsymbol{y} \mid \boldsymbol{x}) = p(\boldsymbol{y})$
- $p(\boldsymbol{x} \mid \boldsymbol{y}) = p(\boldsymbol{x})$
- $\mathbb{V}_{X,Y}[\boldsymbol{x} + \boldsymbol{y}] = \mathbb{V}_X[\boldsymbol{x}] + \mathbb{V}_Y[\boldsymbol{y}]$
- $\mathrm{Cov}_{X,Y}[\boldsymbol{x}, \boldsymbol{y}] = \boldsymbol{0}$

The last point may not hold in converse, i.e., two random variables can have covariance zero but are not statistically independent. To understand why, recall that covariance measures only linear dependence. Therefore, random variables that are nonlinearly dependent could have covariance zero.

Example 6.5
Consider a random variable X with zero mean ($\mathbb{E}_X[x] = 0$) and also $\mathbb{E}_X[x^3] = 0$. Let $y = x^2$ (hence, Y is dependent on X) and consider the covariance (6.36) between X and Y. But this gives

$$\mathrm{Cov}[x, y] = \mathbb{E}[xy] - \mathbb{E}[x]\mathbb{E}[y] = \mathbb{E}[x^3] = 0 \,. \tag{6.54}$$

In machine learning, we often consider problems that can be modeled as *independent and identically distributed* (*i.i.d.*) random variables, $X_1, \ldots, X_N$. For more than two random variables, the word "independent" (Definition 6.10) usually refers to mutually independent random variables, where all subsets are independent (see Pollard (2002, chapter 4) and Jacod and Protter (2004, chapter 3)). The phrase "identically distributed" means that all the random variables are from the same distribution.

independent and identically distributed

i.i.d.

Another concept that is important in machine learning is conditional independence.

Definition 6.11 (Conditional Independence). Two random variables X and Y are *conditionally independent* given Z if and only if

conditionally independent

$$p(\boldsymbol{x}, \boldsymbol{y} \mid \boldsymbol{z}) = p(\boldsymbol{x} \mid \boldsymbol{z}) p(\boldsymbol{y} \mid \boldsymbol{z}) \quad \text{for all} \quad \boldsymbol{z} \in \mathcal{Z} \,, \tag{6.55}$$

where $\mathcal{Z}$ is the set of states of random variable Z. We write $X \perp\!\!\!\perp Y \mid Z$ to denote that X is conditionally independent of Y given Z.

Definition 6.11 requires that the relation in (6.55) must hold true for every value of $\boldsymbol{z}$. The interpretation of (6.55) can be understood as "given knowledge about $\boldsymbol{z}$, the distribution of $\boldsymbol{x}$ and $\boldsymbol{y}$ factorizes." Independence can be cast as a special case of conditional independence if we write $X \perp\!\!\!\perp Y \mid \emptyset$. By using the product rule of probability (6.22), we can expand the left-hand side of (6.55) to obtain

$$p(\boldsymbol{x}, \boldsymbol{y} \mid \boldsymbol{z}) = p(\boldsymbol{x} \mid \boldsymbol{y}, \boldsymbol{z}) p(\boldsymbol{y} \mid \boldsymbol{z}) \,. \tag{6.56}$$

By comparing the right-hand side of (6.55) with (6.56), we see that $p(\boldsymbol{y} \mid \boldsymbol{z})$ appears in both of them so that

$$p(\boldsymbol{x} \mid \boldsymbol{y}, \boldsymbol{z}) = p(\boldsymbol{x} \mid \boldsymbol{z}) \,. \tag{6.57}$$

Equation (6.57) provides an alternative definition of conditional independence, i.e., $X \perp\!\!\!\perp Y \mid Z$. This alternative presentation provides the interpretation "given that we know $\boldsymbol{z}$, knowledge about $\boldsymbol{y}$ does not change our knowledge of $\boldsymbol{x}$."

6.4.6 Inner Products of Random Variables

Recall the definition of inner products from Section 3.2. We can define an inner product between random variables, which we briefly describe in this section. If we have two uncorrelated random variables X, Y, then

$$\mathbb{V}[x + y] = \mathbb{V}[x] + \mathbb{V}[y] \,. \tag{6.58}$$

Since variances are measured in squared units, this looks very much like the Pythagorean theorem for right triangles $c^2 = a^2 + b^2$.

Inner products between multivariate random variables can be treated in a similar fashion

In the following, we see whether we can find a geometric interpretation of the variance relation of uncorrelated random variables in (6.58). Random variables can be considered vectors in a vector space, and we can define inner products to obtain geometric properties of random variables (Eaton, 2007). If we define

$$\langle X, Y \rangle := \mathrm{Cov}[x, y] \tag{6.59}$$

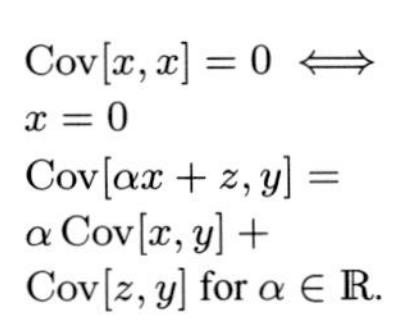
$\mathrm{Cov}[x, x] = 0 \iff x = 0$
$\mathrm{Cov}[\alpha x + z, y] = \alpha\,\mathrm{Cov}[x, y] + \mathrm{Cov}[z, y]$ for $\alpha \in \mathbb{R}$.

for zero mean random variables X and Y, we obtain an inner product. We see that the covariance is symmetric, positive definite, and linear in either argument. The length of a random variable is

$$\|X\| = \sqrt{\mathrm{Cov}[x, x]} = \sqrt{\mathbb{V}[x]} = \sigma[x] \,, \tag{6.60}$$

i.e., its standard deviation. The "longer" the random variable, the more uncertain it is; and a random variable with length 0 is deterministic.

If we look at the angle θ between two random variables X, Y, we get

$$\cos\theta = \frac{\langle X, Y \rangle}{\|X\| \, \|Y\|} = \frac{\mathrm{Cov}[x, y]}{\sqrt{\mathbb{V}[x]\mathbb{V}[y]}} \,, \tag{6.61}$$

which is the correlation (Definition 6.8) between the two random variables. This means that we can think of correlation as the cosine of the angle between two random variables when we consider them geometrically. We know from Definition 3.7 that $X \perp Y \iff \langle X, Y \rangle = 0$. In our case, this means that X and Y are orthogonal if and only if $\mathrm{Cov}[x, y] = 0$, i.e., they are uncorrelated. Figure 6.6 illustrates this relationship.

Remark. While it is tempting to use the Euclidean distance (constructed from the preceding definition of inner products) to compare probability distributions, it is unfortunately not the best way to obtain distances between distributions. Recall that the probability mass (or density) is positive and needs to add up to 1. These constraints mean that distributions live on something called a statistical manifold. The study of this space of probability distributions is called information geometry. Computing distances between distributions are often done using Kullback–Leibler divergence, which is a generalization of distances that account for properties of the statistical manifold. Just like the Euclidean distance is a special case of a metric (Section 3.3), the Kullback–Leibler divergence is a special case of two more general classes of divergences called Bregman divergences and f-divergences. The study of divergences is beyond the scope of this book, and we refer for more details to the recent book by Amari (2016), one of the founders of the field of information geometry. ◇

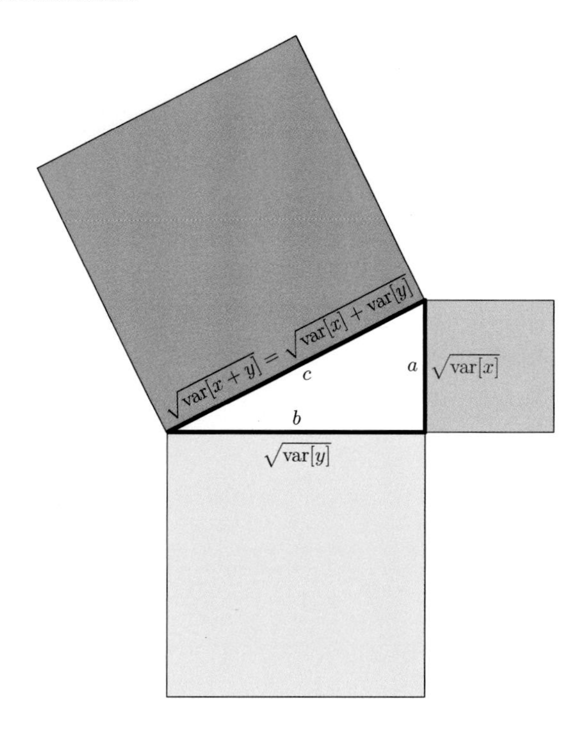

Figure 6.6 Geometry of random variables. If random variables X and Y are uncorrelated, they are orthogonal vectors in a corresponding vector space, and the Pythagorean theorem applies.

6.5 Gaussian Distribution

The Gaussian distribution is the most well-studied probability distribution for continuous-valued random variables. It is also referred to as the *normal distribution*. Its importance originates from the fact that it has many computationally convenient properties, which we will be discussing in the following. In particular, we will use it to define the likelihood and prior for linear regression (Chapter 9), and consider a mixture of Gaussians for density estimation (Chapter 11).

normal distribution

The Gaussian distribution arises naturally when we consider sums of independent and identically distributed random variables. This is known as the central limit theorem (Grinstead and Snell, 1997).

There are many other areas of machine learning that also benefit from using a Gaussian distribution, for example Gaussian processes, variational inference, and reinforcement learning. It is also widely used in other application areas such as signal processing (e.g., Kalman filter), control (e.g., linear quadratic regulator), and statistics (e.g., hypothesis testing).

For a univariate random variable, the Gaussian distribution has a density that is given by

$$p(x \mid \mu, \sigma^2) = \frac{1}{\sqrt{2\pi\sigma^2}} \exp\left(-\frac{(x-\mu)^2}{2\sigma^2}\right). \tag{6.62}$$

The *multivariate Gaussian distribution* is fully characterized by a *mean vector* $\boldsymbol{\mu}$ and a *covariance matrix* $\boldsymbol{\Sigma}$ and defined as

multivariate Gaussian distribution
mean vector
covariance matrix

$$p(\boldsymbol{x} \mid \boldsymbol{\mu}, \boldsymbol{\Sigma}) = (2\pi)^{-\frac{D}{2}} |\boldsymbol{\Sigma}|^{-\frac{1}{2}} \exp\left(-\tfrac{1}{2}(\boldsymbol{x}-\boldsymbol{\mu})^\top \boldsymbol{\Sigma}^{-1}(\boldsymbol{x}-\boldsymbol{\mu})\right), \tag{6.63}$$

where $\boldsymbol{x} \in \mathbb{R}^D$. We write $p(\boldsymbol{x}) = \mathcal{N}(\boldsymbol{x} \mid \boldsymbol{\mu}, \boldsymbol{\Sigma})$ or $X \sim \mathcal{N}(\boldsymbol{\mu}, \boldsymbol{\Sigma})$. Figure 6.7 shows a bivariate Gaussian (mesh), with the corresponding contour plot. Figure 6.8 shows a univariate Gaussian and a bivariate Gaussian with corresponding samples. The special case of the Gaussian with zero mean and identity

Also known as a multivariate normal distribution.

Figure 6.7 Gaussian distribution of two random variables x_1 and x_2.

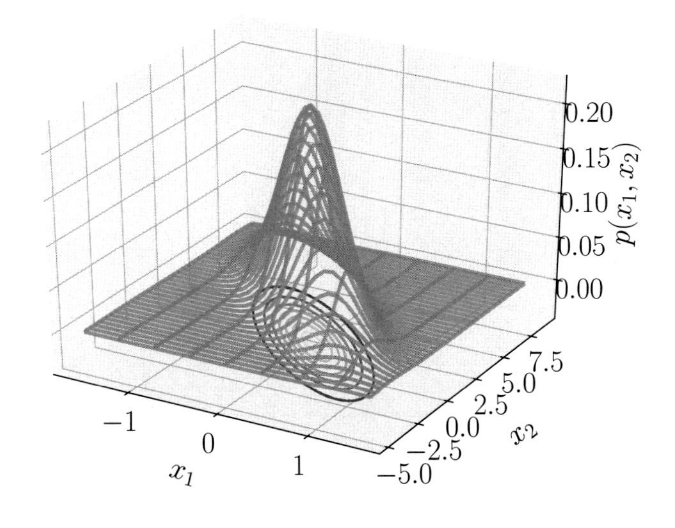

Figure 6.8 Gaussian distributions overlaid with 100 samples: (a) one-dimensional case; (b) two-dimensional case.

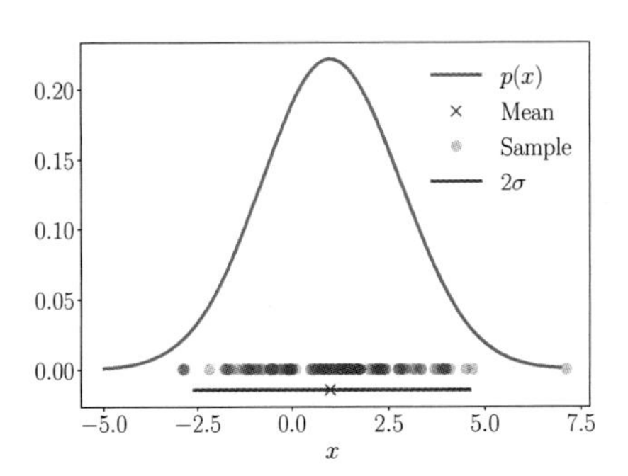

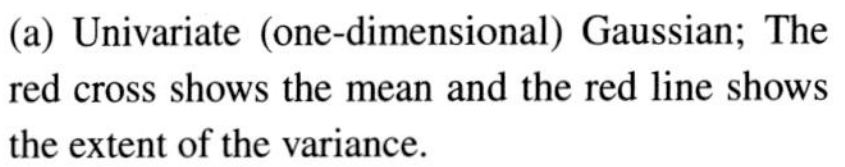

(a) Univariate (one-dimensional) Gaussian; The red cross shows the mean and the red line shows the extent of the variance.

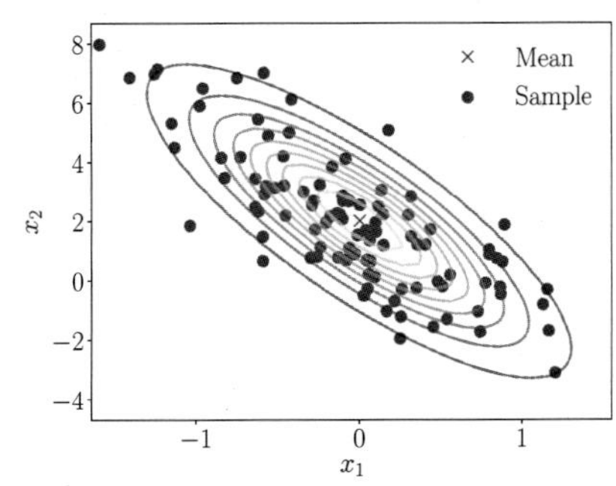

(b) Multivariate (two-dimensional) Gaussian, viewed from top. The red cross shows the mean and the colored lines show the contour lines of the density.

standard normal distribution

covariance, that is, $\boldsymbol{\mu} = \mathbf{0}$ and $\boldsymbol{\Sigma} = \boldsymbol{I}$, is referred to as the *standard normal distribution*.

Gaussians are widely used in statistical estimation and machine learning as they have closed-form expressions for marginal and conditional distributions. In Chapter 9, we use these closed-form expressions extensively for linear regression. A major advantage of modeling with Gaussian random variables is that variable transformations (Section 6.7) are often not needed. Since the Gaussian distribution is fully specified by its mean and covariance, we often can obtain the transformed distribution by applying the transformation to the mean and covariance of the random variable.

6.5.1 Marginals and Conditionals of Gaussians are Gaussians

In the following, we present marginalization and conditioning in the general case of multivariate random variables. If this is confusing at first reading, the reader is advised to consider two univariate random variables instead. Let X and Y be two multivariate random variables that may have different dimensions. To consider the effect of applying the sum rule of probability and the effect of conditioning, we explicitly write the Gaussian distribution in terms of the concatenated states $[\boldsymbol{x}, \boldsymbol{y}]^\top$,

$$p(\boldsymbol{x}, \boldsymbol{y}) = \mathcal{N}\left(\begin{bmatrix}\boldsymbol{\mu}_x \\ \boldsymbol{\mu}_y\end{bmatrix}, \begin{bmatrix}\boldsymbol{\Sigma}_{xx} & \boldsymbol{\Sigma}_{xy} \\ \boldsymbol{\Sigma}_{yx} & \boldsymbol{\Sigma}_{yy}\end{bmatrix}\right). \tag{6.64}$$

where $\boldsymbol{\Sigma}_{xx} = \mathrm{Cov}[\boldsymbol{x}, \boldsymbol{x}]$ and $\boldsymbol{\Sigma}_{yy} = \mathrm{Cov}[\boldsymbol{y}, \boldsymbol{y}]$ are the marginal covariance matrices of $\boldsymbol{x}$ and $\boldsymbol{y}$, respectively, and $\boldsymbol{\Sigma}_{xy} = \mathrm{Cov}[\boldsymbol{x}, \boldsymbol{y}]$ is the cross-covariance matrix between $\boldsymbol{x}$ and $\boldsymbol{y}$.

The conditional distribution $p(\boldsymbol{x} \mid \boldsymbol{y})$ is also Gaussian (illustrated in Figure 6.9(c)) and given by (derived in Section 2.3 of Bishop, 2006)

$$p(\boldsymbol{x} \mid \boldsymbol{y}) = \mathcal{N}\big(\boldsymbol{\mu}_{x \mid y}, \boldsymbol{\Sigma}_{x \mid y}\big) \tag{6.65}$$

$$\boldsymbol{\mu}_{x \mid y} = \boldsymbol{\mu}_x + \boldsymbol{\Sigma}_{xy}\boldsymbol{\Sigma}_{yy}^{-1}(\boldsymbol{y} - \boldsymbol{\mu}_y) \tag{6.66}$$

$$\boldsymbol{\Sigma}_{x \mid y} = \boldsymbol{\Sigma}_{xx} - \boldsymbol{\Sigma}_{xy}\boldsymbol{\Sigma}_{yy}^{-1}\boldsymbol{\Sigma}_{yx}\,. \tag{6.67}$$

Note that in the computation of the mean in (6.66), the $\boldsymbol{y}$-value is an observation and no longer random.

Remark. The conditional Gaussian distribution shows up in many places where we are interested in posterior distributions:

- The Kalman filter (Kalman, 1960), one of the most central algorithms for state estimation in signal processing, does nothing but computing Gaussian conditionals of joint distributions (Deisenroth and Ohlsson, 2011; Särkkä, 2013).
- Gaussian processes (Rasmussen and Williams, 2006), which are a practical implementation of a distribution over functions. In a Gaussian process, we make assumptions of joint Gaussianity of random variables. By (Gaussian) conditioning on observed data, we can determine a posterior distribution over functions.
- Latent linear Gaussian models (Roweis and Ghahramani, 1999; Murphy, 2012), which include probabilistic principal component analysis (PPCA) (Tipping and Bishop, 1999). We will look at PPCA in more detail in Section 10.7.

The marginal distribution $p(\boldsymbol{x})$ of a joint Gaussian distribution $p(\boldsymbol{x}, \boldsymbol{y})$ (see (6.64)) is itself Gaussian and computed by applying the sum rule (6.20) and given by

$$p(\boldsymbol{x}) = \int p(\boldsymbol{x}, \boldsymbol{y})d\boldsymbol{y} = \mathcal{N}\big(\boldsymbol{x} \mid \boldsymbol{\mu}_x, \boldsymbol{\Sigma}_{xx}\big)\,. \tag{6.68}$$

The corresponding result holds for $p(\boldsymbol{y})$, which is obtained by marginalizing with respect to $\boldsymbol{x}$. Intuitively, looking at the joint distribution in (6.64), we ignore (i.e., integrate out) everything we are not interested in. This is illustrated in Figure 6.9(b).

Example 6.6

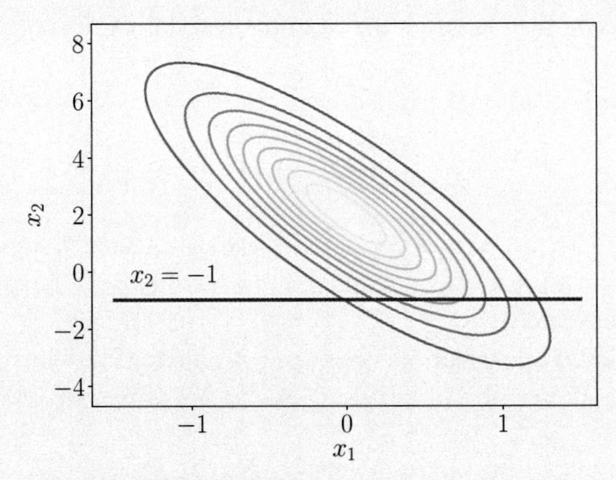

(a) Bivariate Gaussian

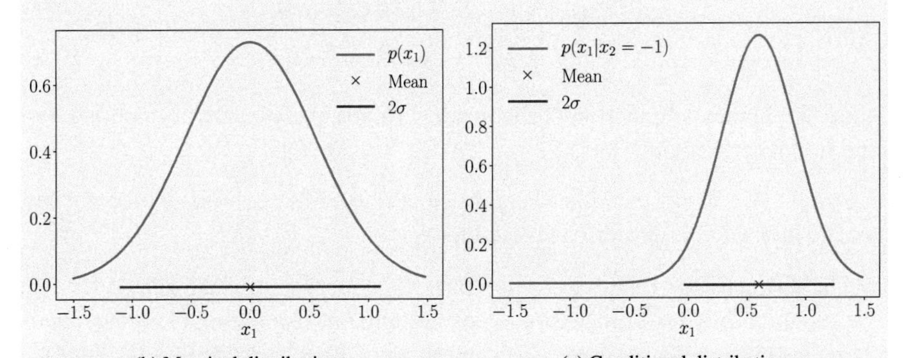

(b) Marginal distribution (c) Conditional distribution

Figure 6.9 (a) Bivariate Gaussian; (b) marginal of a joint Gaussian distribution is Gaussian; (c) the conditional distribution of a Gaussian is also Gaussian.

Consider the bivariate Gaussian distribution (illustrated in Figure 6.9):

$$p(x_1, x_2) = \mathcal{N}\left(\begin{bmatrix}0\\2\end{bmatrix}, \begin{bmatrix}0.3 & -1\\-1 & 5\end{bmatrix}\right). \tag{6.69}$$

We can compute the parameters of the univariate Gaussian, conditioned on $x_2 = -1$, by applying (6.66) and (6.67) to obtain the mean and variance respectively. Numerically, this is

$$\mu_{x_1 \mid x_2=-1} = 0 + (-1) \cdot 0.2 \cdot (-1 - 2) = 0.6 \tag{6.70}$$

and

$$\sigma^2_{x_1 \mid x_2=-1} = 0.3 - (-1) \cdot 0.2 \cdot (-1) = 0.1\,. \tag{6.71}$$

Therefore, the conditional Gaussian is given by

$$p(x_1 \mid x_2 = -1) = \mathcal{N}\big(0.6,\, 0.1\big)\,. \tag{6.72}$$

The marginal distribution $p(x_1)$, in contrast, can be obtained by applying (6.68), which is essentially using the mean and variance of the random variable x_1, giving us

$$p(x_1) = \mathcal{N}\big(0,\, 0.3\big)\,. \tag{6.73}$$

6.5.2 Product of Gaussian Densities

For linear regression (Chapter 9), we need to compute a Gaussian likelihood. Furthermore, we may wish to assume a Gaussian prior (Section 9.3). We apply Bayes' Theorem to compute the posterior, which results in a multiplication of the likelihood and the prior, that is, the multiplication of two Gaussian densities. The *product* of two Gaussians $\mathcal{N}(\boldsymbol{x} \mid \boldsymbol{a}, \boldsymbol{A})\mathcal{N}(\boldsymbol{x} \mid \boldsymbol{b}, \boldsymbol{B})$ is a Gaussian distribution scaled by a $c \in \mathbb{R}$, given by $c\mathcal{N}(\boldsymbol{x} \mid \boldsymbol{c}, \boldsymbol{C})$ with

The derivation is an exercise at the end of this chapter.

$$\boldsymbol{C} = (\boldsymbol{A}^{-1} + \boldsymbol{B}^{-1})^{-1} \tag{6.74}$$

$$\boldsymbol{c} = \boldsymbol{C}(\boldsymbol{A}^{-1}\boldsymbol{a} + \boldsymbol{B}^{-1}\boldsymbol{b}) \tag{6.75}$$

$$c = (2\pi)^{-\frac{D}{2}} |\boldsymbol{A} + \boldsymbol{B}|^{-\frac{1}{2}} \exp\big(-\tfrac{1}{2}(\boldsymbol{a} - \boldsymbol{b})^\top (\boldsymbol{A} + \boldsymbol{B})^{-1}(\boldsymbol{a} - \boldsymbol{b})\big)\,. \tag{6.76}$$

The scaling constant c itself can be written in the form of a Gaussian density either in $\boldsymbol{a}$ or in $\boldsymbol{b}$ with an "inflated" covariance matrix $\boldsymbol{A} + \boldsymbol{B}$, i.e., $c = \mathcal{N}(\boldsymbol{a} \mid \boldsymbol{b}, \boldsymbol{A} + \boldsymbol{B}) = \mathcal{N}(\boldsymbol{b} \mid \boldsymbol{a}, \boldsymbol{A} + \boldsymbol{B})$.

Remark. For notation convenience, we will sometimes use $\mathcal{N}(\boldsymbol{x} \mid \boldsymbol{m}, \boldsymbol{S})$ to describe the functional form of a Gaussian density even if $\boldsymbol{x}$ is not a random variable. We have just done this in the preceding demonstration when we wrote

$$c = \mathcal{N}(\boldsymbol{a} \mid \boldsymbol{b}, \boldsymbol{A} + \boldsymbol{B}) = \mathcal{N}(\boldsymbol{b} \mid \boldsymbol{a}, \boldsymbol{A} + \boldsymbol{B})\,. \tag{6.77}$$

Here, neither $\boldsymbol{a}$ nor $\boldsymbol{b}$ are random variables. However, writing c in this way is more compact than (6.76). ◇

6.5.3 Sums and Linear Transformations

If X, Y are independent Gaussian random variables (i.e., the joint distribution is given as $p(\boldsymbol{x}, \boldsymbol{y}) = p(\boldsymbol{x})p(\boldsymbol{y})$) with $p(\boldsymbol{x}) = \mathcal{N}(\boldsymbol{x} \mid \boldsymbol{\mu}_x, \boldsymbol{\Sigma}_x)$ and $p(\boldsymbol{y}) = \mathcal{N}(\boldsymbol{y} \mid \boldsymbol{\mu}_y, \boldsymbol{\Sigma}_y)$, then $\boldsymbol{x} + \boldsymbol{y}$ is also Gaussian distributed and given by

$$p(\boldsymbol{x} + \boldsymbol{y}) = \mathcal{N}(\boldsymbol{\mu}_x + \boldsymbol{\mu}_y, \boldsymbol{\Sigma}_x + \boldsymbol{\Sigma}_y)\,. \tag{6.78}$$

Knowing that $p(\boldsymbol{x} + \boldsymbol{y})$ is Gaussian, the mean and covariance matrix can be determined immediately using the results from (6.46) through (6.49). This property will be important when we consider i.i.d. Gaussian noise acting on random variables, as is the case for linear regression (Chapter 9).

Example 6.7

Since expectations are linear operations, we can obtain the weighted sum of independent Gaussian random variables

$$p(a\boldsymbol{x} + b\boldsymbol{y}) = \mathcal{N}(a\boldsymbol{\mu}_x + b\boldsymbol{\mu}_y, a^2\boldsymbol{\Sigma}_x + b^2\boldsymbol{\Sigma}_y)\,. \tag{6.79}$$

Remark. A case that will be useful in Chapter 11 is the weighted sum of Gaussian densities. This is different from the weighted sum of Gaussian random variables.

In Theorem 6.12, the random variable x is from a density that is a mixture of two densities $p_1(x)$ and $p_2(x)$, weighted by α. The theorem can be generalized to the multivariate random variable case, since linearity of expectations holds also for multivariate random variables. However, the idea of a squared random variable needs to be replaced by $\boldsymbol{x}\boldsymbol{x}^\top$.

Theorem 6.12. *Consider a mixture of two univariate Gaussian densities*

$$p(x) = \alpha p_1(x) + (1-\alpha)p_2(x)\,, \tag{6.80}$$

where the scalar $0 < \alpha < 1$ *is the mixture weight, and* $p_1(x)$ *and* $p_2(x)$ *are univariate Gaussian densities* (6.62) *with different parameters, i.e.,* $(\mu_1, \sigma_1^2) \neq (\mu_2, \sigma_2^2)$.

Then the mean of the mixture density $p(x)$ *is given by the weighted sum of the means of each random variable*:

$$\mathbb{E}[x] = \alpha\mu_1 + (1-\alpha)\mu_2\,. \tag{6.81}$$

The variance of the mixture density $p(x)$ *is given by*

$$\mathbb{V}[x] = \left[\alpha\sigma_1^2 + (1-\alpha)\sigma_2^2\right] + \left(\left[\alpha\mu_1^2 + (1-\alpha)\mu_2^2\right] - \left[\alpha\mu_1 + (1-\alpha)\mu_2\right]^2\right). \tag{6.82}$$

Proof The mean of the mixture density $p(x)$ is given by the weighted sum of the means of each random variable. We apply the definition of the mean (Definition 6.4), and plug in our mixture (6.80), which yields

$$\begin{aligned}
\mathbb{E}[x] &= \int_{-\infty}^{\infty} x p(x)\mathrm{d}x && (6.83a)\\
&= \int_{-\infty}^{\infty} \alpha x p_1(x) + (1-\alpha) x p_2(x)\mathrm{d}x && (6.83b)\\
&= \alpha\int_{-\infty}^{\infty} x p_1(x)\mathrm{d}x + (1-\alpha)\int_{-\infty}^{\infty} x p_2(x)\mathrm{d}x && (6.83c)\\
&= \alpha\mu_1 + (1-\alpha)\mu_2\,. && (6.83d)
\end{aligned}$$

To compute the variance, we can use the raw-score version of the variance from (6.44), which requires an expression of the expectation of the squared random variable. Here we use the definition of an expectation of a function (the square) of a random variable (Definition 6.3),

$$\begin{aligned}
\mathbb{E}[x^2] &= \int_{-\infty}^{\infty} x^2 p(x)\mathrm{d}x && (6.84a)\\
&= \int_{-\infty}^{\infty} \alpha x^2 p_1(x) + (1-\alpha) x^2 p_2(x)\mathrm{d}x && (6.84b)\\
&= \alpha\int_{-\infty}^{\infty} x^2 p_1(x)\mathrm{d}x + (1-\alpha)\int_{-\infty}^{\infty} x^2 p_2(x)\mathrm{d}x && (6.84c)\\
&= \alpha(\mu_1^2 + \sigma_1^2) + (1-\alpha)(\mu_2^2 + \sigma_2^2)\,, && (6.84d)
\end{aligned}$$

where in the last equality, we again used the raw-score version of the variance (6.44) giving $\sigma^2 = \mathbb{E}[x^2] - \mu^2$. This is rearranged such that the expectation of a squared random variable is the sum of the squared mean and the variance.

Therefore, the variance is given by subtracting (6.83d) from (6.84d),

$$\begin{aligned}\mathbb{V}[x] &= \mathbb{E}[x^2] - (\mathbb{E}[x])^2 &(6.85a)\\ &= \alpha(\mu_1^2 + \sigma_1^2) + (1-\alpha)(\mu_2^2 + \sigma_2^2) - (\alpha\mu_1 + (1-\alpha)\mu_2)^2 &(6.85b)\\ &= \left[\alpha\sigma_1^2 + (1-\alpha)\sigma_2^2\right]\\ &\quad + \left(\left[\alpha\mu_1^2 + (1-\alpha)\mu_2^2\right] - \left[\alpha\mu_1 + (1-\alpha)\mu_2\right]^2\right). &(6.85c)\end{aligned}$$

□

Remark. The preceding derivation holds for any density, but since the Gaussian is fully determined by the mean and variance, the mixture density can be determined in closed form. ◊

For a mixture density, the individual components can be considered to be conditional distributions (conditioned on the component identity). Equation (6.85c) is an example of the conditional variance formula, also known as the *law of total variance*, which generally states that for two random variables X and Y it holds that $\mathbb{V}_X[x] = \mathbb{E}_Y[\mathbb{V}_X[x|y]] + \mathbb{V}_Y[\mathbb{E}_X[x|y]]$, i.e., the (total) variance of X is the expected conditional variance plus the variance of a conditional mean.

law of total variance

We consider in Example 6.17 a bivariate standard Gaussian random variable X and performed a linear transformation $\boldsymbol{A}\boldsymbol{x}$ on it. The outcome is a Gaussian random variable with mean zero and covariance $\boldsymbol{A}\boldsymbol{A}^\top$. Observe that adding a constant vector will change the mean of the distribution, without affecting its variance, that is, the random variable $\boldsymbol{x} + \boldsymbol{\mu}$ is Gaussian with mean $\boldsymbol{\mu}$ and identity covariance. Hence, any linear/affine transformation of a Gaussian random variable is Gaussian distributed.

Any linear/affine transformation of a Gaussian random variable is also Gaussian distributed.

Consider a Gaussian distributed random variable $X \sim \mathcal{N}(\boldsymbol{\mu}, \boldsymbol{\Sigma})$. For a given matrix $\boldsymbol{A}$ of appropriate shape, let Y be a random variable such that $\boldsymbol{y} = \boldsymbol{A}\boldsymbol{x}$ is a transformed version of $\boldsymbol{x}$. We can compute the mean of $\boldsymbol{y}$ by exploiting that the expectation is a linear operator (6.50) as follows:

$$\mathbb{E}[\boldsymbol{y}] = \mathbb{E}[\boldsymbol{A}\boldsymbol{x}] = \boldsymbol{A}\mathbb{E}[\boldsymbol{x}] = \boldsymbol{A}\boldsymbol{\mu}. \tag{6.86}$$

Similarly the variance of $\boldsymbol{y}$ can be found by using (6.51):

$$\mathbb{V}[\boldsymbol{y}] = \mathbb{V}[\boldsymbol{A}\boldsymbol{x}] = \boldsymbol{A}\mathbb{V}[\boldsymbol{x}]\boldsymbol{A}^\top = \boldsymbol{A}\boldsymbol{\Sigma}\boldsymbol{A}^\top. \tag{6.87}$$

This means that the random variable $\boldsymbol{y}$ is distributed according to

$$p(\boldsymbol{y}) = \mathcal{N}\big(\boldsymbol{y} \,|\, \boldsymbol{A}\boldsymbol{\mu},\, \boldsymbol{A}\boldsymbol{\Sigma}\boldsymbol{A}^\top\big). \tag{6.88}$$

Let us now consider the reverse transformation: when we know that a random variable has a mean that is a linear transformation of another random variable. For a given full rank matrix $\boldsymbol{A} \in \mathbb{R}^{M\times N}$, where $M \geqslant N$, let $\boldsymbol{y} \in \mathbb{R}^M$ be a Gaussian random variable with mean $\boldsymbol{A}\boldsymbol{x}$, i.e.,

$$p(\boldsymbol{y}) = \mathcal{N}\big(\boldsymbol{y} \,|\, \boldsymbol{A}\boldsymbol{x},\, \boldsymbol{\Sigma}\big). \tag{6.89}$$

What is the corresponding probability distribution $p(\boldsymbol{x})$? If $\boldsymbol{A}$ is invertible, then we can write $\boldsymbol{x} = \boldsymbol{A}^{-1}\boldsymbol{y}$ and apply the transformation in the previous paragraph. However, in general $\boldsymbol{A}$ is not invertible, and we use an approach similar to that of the pseudo-inverse (3.57). That is, we pre-multiply both sides with $A^\top$ and then invert $A^\top A$, which is symmetric and positive definite, giving us the relation

$$y = Ax \iff (A^\top A)^{-1}A^\top y = x\,. \tag{6.90}$$

Hence, x is a linear transformation of y, and we obtain

$$p(x) = \mathcal{N}\big(x \,|\, (A^\top A)^{-1}A^\top y,\, (A^\top A)^{-1}A^\top \Sigma A(A^\top A)^{-1}\big)\,. \tag{6.91}$$

6.5.4 Sampling from Multivariate Gaussian Distributions

We will not explain the subtleties of random sampling on a computer, and the interested reader is referred to Gentle (2004). In the case of a multivariate Gaussian, this process consists of three stages: first, we need a source of pseudo-random numbers that provide a uniform sample in the interval [0,1]; second, we use a nonlinear transformation such as the Box–Müller transform (Devroye, 1986) to obtain a sample from a univariate Gaussian; and third, we collate a vector of these samples to obtain a sample from a multivariate standard normal $\mathcal{N}(\mathbf{0}, I)$.

For a general multivariate Gaussian, that is, where the mean is nonzero and the covariance is not the identity matrix, we use the properties of linear transformations of a Gaussian random variable. Assume we are interested in generating samples $x_i, i = 1, \ldots, n$, from a multivariate Gaussian distribution with mean μ and covariance matrix Σ. We would like to construct the sample from a sampler that provides samples from the multivariate standard normal $\mathcal{N}(\mathbf{0}, I)$.

To compute the Cholesky factorization of a matrix, it is required that the matrix is symmetric and positive definite (Section 3.2.3). Covariance matrices possess this property.

To obtain samples from a multivariate normal $\mathcal{N}(\mu, \Sigma)$, we can use the properties of a linear transformation of a Gaussian random variable: If $x \sim \mathcal{N}(\mathbf{0}, I)$, then $y = Ax + \mu$, where $AA^\top = \Sigma$ is Gaussian distributed with mean μ and covariance matrix Σ. One convenient choice of A is to use the Cholesky decomposition (Section 4.3) of the covariance matrix $\Sigma = AA^\top$. The Cholesky decomposition has the benefit that A is triangular, leading to efficient computation.

6.6 Conjugacy and the Exponential Family

Many of the probability distributions "with names" that we find in statistics textbooks were discovered to model particular types of phenomena. For example, we have seen the Gaussian distribution in Section 6.5. The distributions are also related to each other in complex ways (Leemis and McQueston, 2008). For a beginner in the field, it can be overwhelming to figure out which distribution to use. In addition, many of these distributions were discovered at a time that statistics and computation were done by pencil and paper. It is natural to ask what are meaningful concepts in the computing age (Efron and Hastie, 2016). In the previous section, we saw that many of the operations required for inference can be conveniently calculated when the distribution is Gaussian. It is worth recalling at this point the desiderata for manipulating probability distributions in the machine learning context:

"Computers" used to be a job description.

1. There is some "closure property" when applying the rules of probability, e.g., Bayes' theorem. By closure, we mean that applying a particular operation returns an object of the same type.
2. As we collect more data, we do not need more parameters to describe the distribution.
3. Since we are interested in learning from data, we want parameter estimation to behave nicely.

It turns out that the class of distributions called the *exponential family* provides the right balance of generality while retaining favorable computation and inference properties. Before we introduce the exponential family, let us see three more members of "named" probability distributions, the Bernoulli (Example 6.8), Binomial (Example 6.9), and Beta (Example 6.10) distributions.

exponential family

Example 6.8

The *Bernoulli distribution* is a distribution for a single binary random variable X with state $x \in \{0, 1\}$. It is governed by a single continuous parameter $\mu \in [0, 1]$ that represents the probability of $X = 1$. The Bernoulli distribution $\text{Ber}(\mu)$ is defined as

Bernoulli distribution

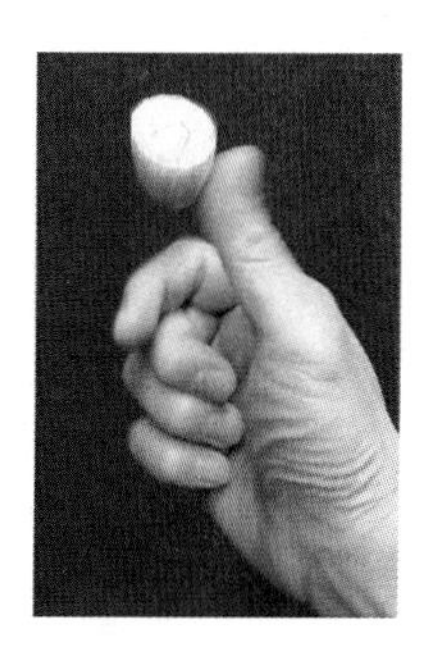

$$p(x \mid \mu) = \mu^x (1 - \mu)^{1-x}, \quad x \in \{0, 1\}, \tag{6.92}$$

$$\mathbb{E}[x] = \mu, \tag{6.93}$$

$$\mathbb{V}[x] = \mu(1 - \mu), \tag{6.94}$$

where $\mathbb{E}[x]$ and $\mathbb{V}[x]$ are the mean and variance of the binary random variable X.

An example where the Bernoulli distribution can be used is when we are interested in modeling the probability of "heads" when flipping a coin.

Remark. The rewriting above of the Bernoulli distribution, where we use Boolean variables as numerical 0 or 1 and express them in the exponents, is a trick that is often used in machine learning textbooks. Another occurence of this is when expressing the Multinomial distribution. ♢

Example 6.9 (Binomial Distribution)

The *Binomial distribution* is a generalization of the Bernoulli distribution to a distribution over integers (illustrated in Figure 6.10). In particular, the Binomial can be used to describe the probability of observing m occurrences of $X = 1$ in a set of N samples from a Bernoulli distribution where $p(X = 1) = \mu \in [0, 1]$. The Binomial distribution $\text{Bin}(N, \mu)$ is defined as

Binomial distribution

$$p(m \mid N, \mu) = \binom{N}{m} \mu^m (1 - \mu)^{N-m}, \tag{6.95}$$

$$\mathbb{E}[m] = N\mu, \tag{6.96}$$

$$\mathbb{V}[m] = N\mu(1 - \mu), \tag{6.97}$$

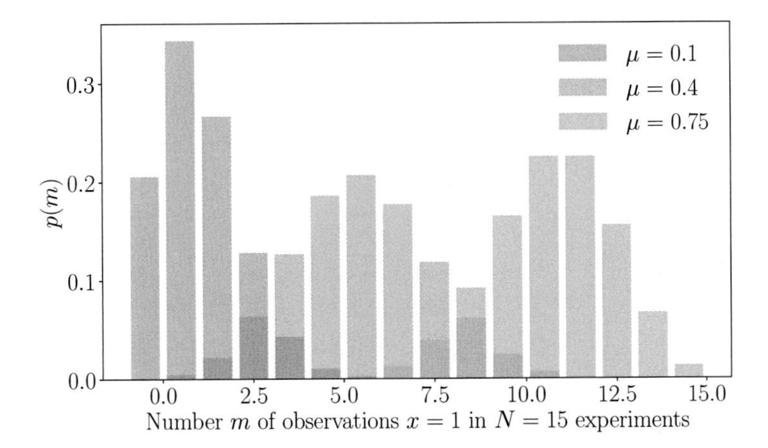

Figure 6.10 Examples of the Binomial distribution for $\mu \in \{0.1, 0.4, 0.75\}$ and $N = 15$.

where $\mathbb{E}[m]$ and $\mathbb{V}[m]$ are the mean and variance of m, respectively. The *Binomial distribution* is a generalization of the Bernoulli distribution to a distribution over integers (illustrated in Figure 6.10).

Binomial distribution

An example where the Binomial could be used is if we want to describe the probability of observing m "heads" in N coin-flip experiments if the probability for observing heads in a single experiment is μ.

Example 6.10 (Beta Distribution)

We may wish to model a continuous random variable on a finite interval. The *Beta distribution* is a distribution over a continuous random variable $\mu \in [0, 1]$, which is often used to represent the probability for some binary event (e.g., the parameter governing the Bernoulli distribution). The Beta distribution $\text{Beta}(\alpha, \beta)$ (illustrated in Figure 6.11) itself is governed by two parameters $\alpha > 0,\ \beta > 0$ and is defined as

Beta distribution

$$p(\mu \mid \alpha, \beta) = \frac{\Gamma(\alpha+\beta)}{\Gamma(\alpha)\Gamma(\beta)} \mu^{\alpha-1}(1-\mu)^{\beta-1} \tag{6.98}$$

$$\mathbb{E}[\mu] = \frac{\alpha}{\alpha+\beta}, \qquad \mathbb{V}[\mu] = \frac{\alpha\beta}{(\alpha+\beta)^2(\alpha+\beta+1)} \tag{6.99}$$

where $\Gamma(\cdot)$ is the Gamma function defined as

$$\Gamma(t) := \int_0^\infty x^{t-1}\exp(-x)dx, \qquad t > 0. \tag{6.100}$$

$$\Gamma(t+1) = t\Gamma(t). \tag{6.101}$$

Note that the fraction of Gamma functions in (6.98) normalizes the Beta distribution.

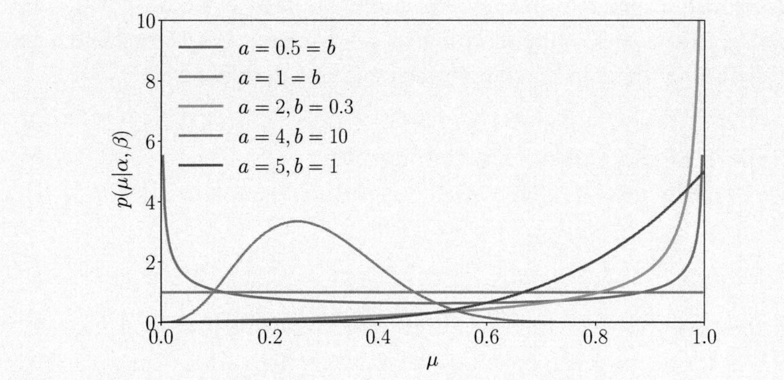

Figure 6.11 Examples of the Beta distribution for different values of α and β.

Intuitively, α moves probability mass toward 1, whereas β moves probability mass toward 0. There are some special cases (Murphy, 2012):

- For $\alpha = 1 = \beta$, we obtain the uniform distribution $\mathcal{U}[0, 1]$.
- For $\alpha, \beta < 1$, we get a bimodal distribution with spikes at 0 and 1.
- For $\alpha, \beta > 1$, the distribution is unimodal.
- For $\alpha, \beta > 1$ and $\alpha = \beta$, the distribution is unimodal, symmetric, and centered in the interval $[0, 1]$, i.e., the mode/mean is at $\frac{1}{2}$.

Remark. There is a whole zoo of distributions with names, and they are related in different ways to each other (Leemis and McQueston, 2008). It is worth keeping in mind that each named distribution is created for a particular reason, but may have other applications. Knowing the reason behind the creation of a particular distribution often allows insight into how to best use it. We introduced the preceding three distributions to be able to illustrate the concepts of conjugacy (Section 6.6.1) and exponential families (Section 6.6.3). ◇

6.6.1 Conjugacy

According to Bayes' theorem (6.23), the posterior is proportional to the product of the prior and the likelihood. The specification of the prior can be tricky for two reasons: First, the prior should encapsulate our knowledge about the problem before we see any data. This is often difficult to describe. Second, it is often not possible to compute the posterior distribution analytically. However, there are some priors that are computationally convenient: *conjugate priors*.

conjugate prior

Definition 6.13 (Conjugate Prior). A prior is *conjugate* for the likelihood function if the posterior is of the same form/type as the prior.

conjugate

Conjugacy is particularly convenient because we can algebraically calculate our posterior distribution by updating the parameters of the prior distribution.

Remark. When considering the geometry of probability distributions, conjugate priors retain the same distance structure as the likelihood (Agarwal and Daumé III, 2010). ◇

To introduce a concrete example of conjugate priors, we describe in Example 6.11 the Binomial distribution (defined on discrete random variables) and the Beta distribution (defined on continuous random variables).

Example 6.11 (Beta–Binomial Conjugacy)
Consider a Binomial random variable $x \sim \text{Bin}(N, \mu)$ where

$$p(x \mid N, \mu) = \binom{N}{x} \mu^x (1-\mu)^{N-x}, \quad x = 0, 1, \ldots, N, \tag{6.102}$$

is the probability of finding x times the outcome "heads" in N coin flips, where μ is the probability of a "head." We place a Beta prior on the parameter μ, that is, $\mu \sim \text{Beta}(\alpha, \beta)$, where

$$p(\mu \mid \alpha, \beta) = \frac{\Gamma(\alpha+\beta)}{\Gamma(\alpha)\Gamma(\beta)} \mu^{\alpha-1}(1-\mu)^{\beta-1}. \tag{6.103}$$

If we now observe some outcome $x = h$, that is, we see h heads in N coin flips, we compute the posterior distribution on μ as

$$p(\mu \mid x = h, N, \alpha, \beta) \propto p(x \mid N, \mu) p(\mu \mid \alpha, \beta) \tag{6.104a}$$
$$\propto \mu^h (1-\mu)^{(N-h)} \mu^{\alpha-1}(1-\mu)^{\beta-1} \tag{6.104b}$$
$$= \mu^{h+\alpha-1}(1-\mu)^{(N-h)+\beta-1} \tag{6.104c}$$
$$\propto \text{Beta}(h+\alpha, N-h+\beta), \tag{6.104d}$$

i.e., the posterior distribution is a Beta distribution as the prior, i.e., the Beta prior is conjugate for the parameter μ in the Binomial likelihood function.

In the following example, we will derive a result that is similar to the Beta–Binomial conjugacy result. Here we will show that the Beta distribution is a conjugate prior for the Bernoulli distribution.

Example 6.12 (Beta–Bernoulli Conjugacy)
Let $x \in \{0, 1\}$ be distributed according to the Bernoulli distribution with parameter $\theta \in [0, 1]$, that is, $p(x = 1 \mid \theta) = \theta$. This can also be expressed as $p(x \mid \theta) = \theta^x (1-\theta)^{1-x}$. Let θ be distributed according to a Beta distribution with parameters α, β, that is, $p(\theta \mid \alpha, \beta) \propto \theta^{\alpha-1}(1-\theta)^{\beta-1}$.

Multiplying the Beta and the Bernoulli distributions, we get

$$p(\theta \mid x, \alpha, \beta) = p(x \mid \theta) p(\theta \mid \alpha, \beta) \tag{6.105a}$$
$$\propto \theta^x (1-\theta)^{1-x} \theta^{\alpha-1}(1-\theta)^{\beta-1} \tag{6.105b}$$
$$= \theta^{\alpha+x-1}(1-\theta)^{\beta+(1-x)-1} \tag{6.105c}$$
$$\propto p(\theta \mid \alpha+x, \beta+(1-x)). \tag{6.105d}$$

The last line is the Beta distribution with parameters $(\alpha + x, \beta + (1-x))$.

Table 6.2 lists examples for conjugate priors for the parameters of some standard likelihoods used in probabilistic modeling. Distributions such as

Multinomial, inverse Gamma, inverse Wishart, and Dirichlet can be found in any statistical text, and are described in Bishop (2006), for example.

The Beta distribution is the conjugate prior for the parameter μ in both the Binomial and the Bernoulli likelihood. For a Gaussian likelihood function, we can place a conjugate Gaussian prior on the mean. The reason why the Gaussian likelihood appears twice in the table is that we need to distinguish the univariate from the multivariate case. In the univariate (scalar) case, the inverse Gamma is the conjugate prior for the variance. In the multivariate case, we use a conjugate inverse Wishart distribution as a prior on the covariance matrix. The Dirichlet distribution is the conjugate prior for the multinomial likelihood function. For further details, we refer to Bishop (2006).

The Gamma prior is conjugate for the precision (inverse variance) in the univariate Gaussian likelihood, and the Wishart prior is conjugate for the precision matrix (inverse covariance matrix) in the multivariate Gaussian likelihood.

6.6.2 Sufficient Statistics

Recall that a statistic of a random variable is a deterministic function of that random variable. For example, if $\boldsymbol{x} = [x_1, \ldots, x_N]^\top$ is a vector of univariate Gaussian random variables, that is, $x_n \sim \mathcal{N}(\mu, \sigma^2)$, then the sample mean $\hat{\mu} = \frac{1}{N}(x_1 + \cdots + x_N)$ is a statistic. Sir Ronald Fisher discovered the notion of *sufficient statistics*: the idea that there are statistics that will contain all available information that can be inferred from data corresponding to the distribution under consideration. In other words, sufficient statistics carry all the information needed to make inference about the population, that is, they are the statistics that are sufficient to represent the distribution.

sufficient statistics

For a set of distributions parametrized by θ, let X be a random variable with distribution $p(x \mid \theta_0)$ given an unknown θ_0. A vector $\phi(x)$ of statistics is called sufficient statistics for θ_0 if they contain all possible information about θ_0. To be more formal about "contain all possible information," this means that the probability of x given θ can be factored into a part that does not depend on θ, and a part that depends on θ only via $\phi(x)$. The Fisher–Neyman factorization theorem formalizes this notion, which we state in Theorem 6.14 without proof.

Theorem 6.14 (Fisher–Neyman). *[Theorem 6.5 in Lehmann and Casella (1998)] Let X have probability density function $p(x \mid \theta)$. Then the statistics $\phi(x)$ are sufficient for θ if and only if $p(x \mid \theta)$ can be written in the form*

Fisher–Neyman theorem

$$p(x \mid \theta) = h(x) g_\theta(\phi(x)), \tag{6.106}$$

where $h(x)$ is a distribution independent of θ and g_θ captures all the dependence on θ via sufficient statistics $\phi(x)$.

If $p(x \mid \theta)$ does not depend on θ, then $\phi(x)$ is trivially a sufficient statistic for any function ϕ. The more interesting case is that $p(x \mid \theta)$ is dependent only on $\phi(x)$ and not x itself. In this case, $\phi(x)$ is a sufficient statistic for θ.

In machine learning, we consider a finite number of samples from a distribution. One could imagine that for simple distributions (such as the Bernoulli in Example 6.8) we only need a small number of samples to estimate the parameters of the distributions. We could also consider the opposite problem: If we have a set of data (a sample from an unknown distribution), which distribution gives the best fit? A natural question to ask is, as we observe more data, do we need more parameters θ to describe the distribution? It turns out that the answer

Table 6.2 Examples of conjugate priors for common likelihood functions.

Likelihood	Conjugate prior	Posterior
Bernoulli	Beta	Beta
Binomial	Beta	Beta
Gaussian	Gaussian/inverse Gamma	Gaussian/inverse Gamma
Gaussian	Gaussian/inverse Wishart	Gaussian/inverse Wishart
Multinomial	Dirichlet	Dirichlet

is yes in general, and this is studied in nonparametric statistics (Wasserman, 2007). A converse question is to consider which class of distributions have finite-dimensional sufficient statistics, that is, the number of parameters needed to describe them does not increase arbitrarily. The answer is exponential family distributions, described in the following section.

6.6.3 Exponential Family

There are three possible levels of abstraction we can have when considering distributions (of discrete or continuous random variables). At level one (the most concrete end of the spectrum), we have a particular named distribution with fixed parameters, for example a univariate Gaussian $\mathcal{N}(0,\, 1)$ with zero mean and unit variance. In machine learning, we often use the second level of abstraction, that is, we fix the parametric form (the univariate Gaussian) and infer the parameters from data. For example, we assume a univariate Gaussian $\mathcal{N}(\mu,\, \sigma^2)$ with unknown mean μ and unknown variance σ^2, and use a maximum likelihood fit to determine the best parameters (μ, σ^2). We will see an example of this when considering linear regression in Chapter 9. A third level of abstraction is to consider families of distributions, and in this book, we consider the exponential family. The univariate Gaussian is an example of a member of the exponential family. Many of the widely used statistical models, including all the "named" models in Table 6.2, are members of the exponential family. They can all be unified into one concept (Brown, 1986).

Remark. A brief historical anecdote: Like many concepts in mathematics and science, exponential families were independently discovered at the same time by different researchers. In the years 1935–1936, Edwin Pitman in Tasmania, Georges Darmois in Paris, and Bernard Koopman in New York independently showed that the exponential families are the only families that enjoy finite-dimensional sufficient statistics under repeated independent sampling (Lehmann and Casella, 1998). ◇

exponential family

An *exponential family* is a family of probability distributions, parameterized by $\boldsymbol{\theta} \in \mathbb{R}^D$, of the form

$$p(\boldsymbol{x} \mid \boldsymbol{\theta}) = h(\boldsymbol{x}) \exp\left(\langle \boldsymbol{\theta}, \boldsymbol{\phi}(\boldsymbol{x}) \rangle - A(\boldsymbol{\theta})\right) , \tag{6.107}$$

where $\boldsymbol{\phi}(\boldsymbol{x})$ is the vector of sufficient statistics. In general, any inner product (Section 3.2) can be used in (6.107), and for concreteness we will use the standard dot product here ($\langle \boldsymbol{\theta}, \boldsymbol{\phi}(\boldsymbol{x}) \rangle = \boldsymbol{\theta}^\top \boldsymbol{\phi}(\boldsymbol{x})$). Note that the form of the exponential family is essentially a particular expression of $g_\theta(\phi(x))$ in the Fisher–Neyman theorem (Theorem 6.14).

The factor $h(\boldsymbol{x})$ can be absorbed into the dot product term by adding another entry ($\log h(\boldsymbol{x})$) to the vector of sufficient statistics $\boldsymbol{\phi}(\boldsymbol{x})$, and constraining the corresponding parameter $\theta_0 = 1$. The term $A(\boldsymbol{\theta})$ is the normalization constant that ensures that the distribution sums up or integrates to one and is called the *log-partition function*. A good intuitive notion of exponential families can be obtained by ignoring these two terms and considering exponential families as distributions of the form

log-partition function

$$p(\boldsymbol{x} \mid \boldsymbol{\theta}) \propto \exp\left(\boldsymbol{\theta}^\top \boldsymbol{\phi}(\boldsymbol{x})\right). \tag{6.108}$$

For this form of parametrization, the parameters $\boldsymbol{\theta}$ are called the *natural parameters*. At first glance, it seems that exponential families are a mundane transformation by adding the exponential function to the result of a dot product. However, there are many implications that allow for convenient modeling and efficient computation based on the fact that we can capture information about data in $\boldsymbol{\phi}(\boldsymbol{x})$.

natural parameters

Example 6.13 (Gaussian as Exponential Family)

Consider the univariate Gaussian distribution $\mathcal{N}(\mu, \sigma^2)$. Let $\phi(x) = \begin{bmatrix} x \\ x^2 \end{bmatrix}$. Then by using the definition of the exponential family,

$$p(x \mid \boldsymbol{\theta}) \propto \exp(\theta_1 x + \theta_2 x^2). \tag{6.109}$$

Setting

$$\boldsymbol{\theta} = \left[\frac{\mu}{\sigma^2}, -\frac{1}{2\sigma^2}\right]^\top \tag{6.110}$$

and substituting into (6.109), we obtain

$$p(x \mid \boldsymbol{\theta}) \propto \exp\left(\frac{\mu x}{\sigma^2} - \frac{x^2}{2\sigma^2}\right) \propto \exp\left(-\frac{1}{2\sigma^2}(x-\mu)^2\right). \tag{6.111}$$

Therefore, the univariate Gaussian distribution is a member of the exponential family with sufficient statistic $\phi(x) = \begin{bmatrix} x \\ x^2 \end{bmatrix}$, and natural parameters given by $\boldsymbol{\theta}$ in (6.110).

Example 6.14 (Bernoulli as Exponential Family)

Recall the Bernoulli distribution from Example 6.8

$$p(x \mid \mu) = \mu^x (1-\mu)^{1-x}, \quad x \in \{0, 1\}. \tag{6.112}$$

This can be written in exponential family form

$$p(x \mid \mu) = \exp\left[\log\left(\mu^x(1-\mu)^{1-x}\right)\right] \tag{6.113a}$$
$$= \exp\left[x \log \mu + (1-x)\log(1-\mu)\right] \tag{6.113b}$$
$$= \exp\left[x \log \mu - x\log(1-\mu) + \log(1-\mu)\right] \tag{6.113c}$$
$$= \exp\left[x \log \tfrac{\mu}{1-\mu} + \log(1-\mu)\right]. \tag{6.113d}$$

The last line (6.113d) can be identified as being in exponential family form (6.107) by observing that

$$h(x) = 1 \tag{6.114}$$
$$\theta = \log \tfrac{\mu}{1-\mu} \tag{6.115}$$
$$\phi(x) = x \tag{6.116}$$
$$A(\theta) = -\log(1-\mu) = \log(1+\exp(\theta)). \tag{6.117}$$

The relationship between θ and μ is invertible so that

$$\mu = \frac{1}{1+\exp(-\theta)}. \tag{6.118}$$

The relation (6.118) is used to obtain the right equality of (6.117).

sigmoid

Remark. The relationship between the original Bernoulli parameter μ and the natural parameter θ is known as the *sigmoid* or logistic function. Observe that $\mu \in (0,1)$ but $\theta \in \mathbb{R}$, and therefore the sigmoid function squeezes a real value into the range $(0,1)$. This property is useful in machine learning, for example it is used in logistic regression (Bishop, 2006, section 4.3.2), as well as as a nonlinear activation functions in neural networks (Goodfellow et al., 2016, chapter 6). ◇

It is often not obvious how to find the parametric form of the conjugate distribution of a particular distribution (for example, those in Table 6.2). Exponential families provide a convenient way to find conjugate pairs of distributions. Consider the random variable X is a member of the exponential family (6.107):

$$p(\boldsymbol{x} \mid \boldsymbol{\theta}) = h(\boldsymbol{x}) \exp\left(\langle \boldsymbol{\theta}, \boldsymbol{\phi}(\boldsymbol{x})\rangle - A(\boldsymbol{\theta})\right). \tag{6.119}$$

Every member of the exponential family has a conjugate prior (Brown, 1986)

$$p(\boldsymbol{\theta} \mid \boldsymbol{\gamma}) = h_c(\boldsymbol{\theta}) \exp\left(\left\langle \begin{bmatrix}\gamma_1\\ \gamma_2\end{bmatrix}, \begin{bmatrix}\boldsymbol{\theta}\\ -A(\boldsymbol{\theta})\end{bmatrix}\right\rangle - A_c(\boldsymbol{\gamma})\right), \tag{6.120}$$

where $\boldsymbol{\gamma} = \begin{bmatrix}\gamma_1\\ \gamma_2\end{bmatrix}$ has dimension $\dim(\boldsymbol{\theta}) + 1$. The sufficient statistics of the conjugate prior are $\begin{bmatrix}\boldsymbol{\theta}\\ -A(\boldsymbol{\theta})\end{bmatrix}$. By using the knowledge of the general form of conjugate priors for exponential families, we can derive functional forms of conjugate priors corresponding to particular distributions.

Example 6.15

Recall the exponential family form of the Bernoulli distribution (6.113d)

$$p(x \mid \mu) = \exp\left[x \log \frac{\mu}{1-\mu} + \log(1-\mu)\right]. \tag{6.121}$$

The canonical conjugate prior has the form

$$p(\mu \mid \alpha, \beta) = \frac{\mu}{1-\mu} \exp\left[\alpha \log \frac{\mu}{1-\mu} + (\beta + \alpha)\log(1-\mu) - A_c(\gamma)\right], \tag{6.122}$$

where we defined $\gamma := [\alpha, \beta + \alpha]^\top$ and $h_c(\mu) := \mu/(1-\mu)$. Equation (6.122) then simplifies to

$$p(\mu \mid \alpha, \beta) = \exp\left[(\alpha - 1)\log \mu + (\beta - 1)\log(1-\mu) - A_c(\alpha, \beta)\right]. \tag{6.123}$$

Putting this in non-exponential family form yields

$$p(\mu \mid \alpha, \beta) \propto \mu^{\alpha-1}(1-\mu)^{\beta-1}, \tag{6.124}$$

which we identify as the Beta distribution (6.98). In example 6.12, we assumed that the Beta distribution is the conjugate prior of the Bernoulli distribution and showed that it was indeed the conjugate prior. In this example, we derived the form of the Beta distribution by looking at the canonical conjugate prior of the Bernoulli distribution in exponential family form.

As mentioned in the previous section, the main motivation for exponential families is that they have finite-dimensional sufficient statistics. Additionally, conjugate distributions are easy to write down, and the conjugate distributions also come from an exponential family. From an inference perspective, maximum likelihood estimation behaves nicely because empirical estimates of sufficient statistics are optimal estimates of the population values of sufficient statistics (recall the mean and covariance of a Gaussian). From an optimization perspective, the log-likelihood function is concave, allowing for efficient optimization approaches to be applied (Chapter 7).

6.7 Change of Variables/Inverse Transform

It may seem that there are very many known distributions, but in reality the set of distributions for which we have names is quite limited. Therefore, it is often useful to understand how transformed random variables are distributed. For example, assuming that X is a random variable distributed according to the univariate normal distribution $\mathcal{N}(0, 1)$, what is the distribution of X^2? Another example, which is quite common in machine learning, is, given that X_1 and X_2 are univariate standard normal, what is the distribution of $\frac{1}{2}(X_1 + X_2)$?

One option to work out the distribution of $\frac{1}{2}(X_1 + X_2)$ is to calculate the mean and variance of X_1 and X_2 and then combine them. As we saw in Section 6.4.4, we can calculate the mean and variance of resulting random variables when we consider affine transformations of random variables. However, we may not be able to obtain the functional form of the distribution under transformations. Furthermore, we may be interested in nonlinear transformations of random variables for which closed-form expressions are not readily available.

Remark (Notation). In this section, we will be explicit about random variables and the values they take. Hence, recall that we use capital letters X, Y to denote random variables and small letters x, y to denote the values in the target space $\mathcal{T}$ that the random variables take. We will explicitly write pmfs of discrete random variables X as $P(X = x)$. For continuous random variables X (Section 6.2.2), the pdf is written as $f(x)$ and the cdf is written as $F_X(x)$. ♢

We will look at two approaches for obtaining distributions of transformations of random variables: a direct approach using the definition of a cumulative distribution function and a change-of-variable approach that uses the chain rule of calculus (Section 5.2.2). The change-of-variable approach is widely used because it provides a "recipe" for attempting to compute the resulting distribution due to a transformation. We will explain the techniques for univariate random variables, and will only briefly provide the results for the general case of multivariate random variables.

Moment generating functions can also be used to study transformations of random variables (Casella and Berger, 2002, chapter 2).

Transformations of discrete random variables can be understood directly. Suppose that there is a discrete random variable X with pmf $P(X = x)$ (Section 6.2.1), and an invertible function $U(x)$. Consider the transformed random variable $Y := U(X)$, with pmf $P(Y = y)$. Then

$$P(Y = y) = P(U(X) = y) \qquad \text{transformation of interest} \tag{6.125a}$$

$$= P(X = U^{-1}(y)) \qquad \text{inverse} \tag{6.125b}$$

where we can observe that $x = U^{-1}(y)$. Therefore, for discrete random variables, transformations directly change the individual events (with the probabilities appropriately transformed).

6.7.1 Distribution Function Technique

The distribution function technique goes back to first principles, and uses the definition of a cdf $F_X(x) = P(X \leqslant x)$ and the fact that its differential is the pdf $f(x)$ (Wasserman, 2004, chapter 2). For a random variable X and a function U, we find the pdf of the random variable $Y := U(X)$ by

1. Finding the cdf:

$$F_Y(y) = P(Y \leqslant y) \tag{6.126}$$

2. Differentiating the cdf $F_Y(y)$ to get the pdf $f(y)$.

$$f(y) = \frac{\mathrm{d}}{\mathrm{d}y} F_Y(y)\,. \tag{6.127}$$

We also need to keep in mind that the domain of the random variable may have changed due to the transformation by U.

Example 6.16

Let X be a continuous random variable with probability density function on $0 \leqslant x \leqslant 1$

$$f(x) = 3x^2 . \tag{6.128}$$

We are interested in finding the pdf of $Y = X^2$.

The function f is an increasing function of x, and therefore the resulting value of y lies in the interval $[0, 1]$. We obtain

$$\begin{aligned}
F_Y(y) &= P(Y \leqslant y) && \text{definition of cdf} && (6.129a)\\
&= P(X^2 \leqslant y) && \text{transformation of interest} && (6.129b)\\
&= P(X \leqslant y^{\frac{1}{2}}) && \text{inverse} && (6.129c)\\
&= F_X(y^{\frac{1}{2}}) && \text{definition of cdf} && (6.129d)\\
&= \int_0^{y^{\frac{1}{2}}} 3t^2 \mathrm{d}t && \text{cdf as a definite integral} && (6.129e)\\
&= \left[t^3\right]_{t=0}^{t=y^{\frac{1}{2}}} && \text{result of integration} && (6.129f)\\
&= y^{\frac{3}{2}} , \quad 0 \leqslant y \leqslant 1 . && && (6.129g)
\end{aligned}$$

Therefore, the cdf of Y is

$$F_Y(y) = y^{\frac{3}{2}} \tag{6.130}$$

for $0 \leqslant y \leqslant 1$. To obtain the pdf, we differentiate the cdf

$$f(y) = \frac{\mathrm{d}}{\mathrm{d}y} F_Y(y) = \frac{3}{2} y^{\frac{1}{2}} \tag{6.131}$$

for $0 \leqslant y \leqslant 1$.

In Example 6.16, we considered a strictly monotonically increasing function $f(x) = 3x^2$. This means that we could compute an inverse function. In general, we require that the function of interest $y = U(x)$ has an inverse $x = U^{-1}(y)$. A useful result can be obtained by considering the cumulative distribution function $F_X(x)$ of a random variable X, and using it as the transformation $U(x)$. This leads to the following theorem.

Functions that have inverses are called bijective functions (Section 2.7).

Theorem 6.15. *[Theorem 2.1.10 in Casella and Berger (2002)] Let X be a continuous random variable with a strictly monotonic cumulative distribution function $F_X(x)$. Then the random variable Y defined as*

$$Y := F_X(x) \tag{6.132}$$

has a uniform distribution.

Theorem 6.15 is known as the *probability integral transform*, and it is used to derive algorithms for sampling from distributions by transforming the result of sampling from a uniform random variable (Bishop, 2006). The algorithm works by first generating a sample from a uniform distribution, then transforming it by

probability integral transform

the inverse cdf (assuming this is available) to obtain a sample from the desired distribution. The probability integral transform is also used for hypothesis testing whether a sample comes from a particular distribution (Lehmann and Romano, 2005). The idea that the output of a cdf gives a uniform distribution also forms the basis of copulas (Nelsen, 2006).

6.7.2 Change of Variables

The distribution function technique in Section 6.7.1 is derived from first principles, based on the definitions of cdfs and using properties of inverses, differentiation, and integration. This argument from first principles relies on two facts:

1. We can transform the cdf of Y into an expression that is a cdf of X.
2. We can differentiate the cdf to obtain the pdf.

Let us break down the reasoning step by step, with the goal of understanding the more general change-of-variables approach in Theorem 6.16.

Change of variables in probability relies on the change-of-variables method in calculus (Tandra, 2014).

Remark. The name "change of variables" comes from the idea of changing the variable of integration when faced with a difficult integral. For univariate functions, we use the substitution rule of integration,

$$\int f(g(x))g'(x)\mathrm{d}x = \int f(u)\mathrm{d}u, \quad \text{where} \quad u = g(x). \tag{6.133}$$

The derivation of this rule is based on the chain rule of calculus (5.32) and by applying twice the fundamental theorem of calculus. The fundamental theorem of calculus formalizes the fact that integration and differentiation are somehow "inverses" of each other. An intuitive understanding of the rule can be obtained by thinking (loosely) about small changes (differentials) to the equation $u = g(x)$, that is, by considering $\Delta u = g'(x)\Delta x$ as a differential of $u = g(x)$. By subsituting $u = g(x)$, the argument inside the integral on the right-hand side of (6.133) becomes $f(g(x))$. By pretending that the term $\mathrm{d}u$ can be approximated by $\mathrm{d}u \approx \Delta u = g'(x)\Delta x$, and that $\mathrm{d}x \approx \Delta x$, we obtain (6.133). ◇

Consider a univariate random variable X and an *invertible* function U, which gives us another random variable $Y = U(X)$. We assume that random variable X has states $x \in [a, b]$. By the definition of the cdf, we have

$$F_Y(y) = P(Y \leqslant y) . \tag{6.134}$$

We are interested in a function U of the random variable

$$P(Y \leqslant y) = P(U(X) \leqslant y) , \tag{6.135}$$

where we assume that the function U is invertible. An invertible function on an interval is either strictly increasing or strictly decreasing. In the case that U is strictly increasing, then its inverse U^{-1} is also strictly increasing. By applying the inverse U^{-1} to the arguments of $P(U(X) \leqslant y)$, we obtain

$$P(U(X) \leqslant y) = P(U^{-1}(U(X)) \leqslant U^{-1}(y)) = P(X \leqslant U^{-1}(y)) . \tag{6.136}$$

The right-most term in (6.136) is an expression of the cdf of X. Recall the definition of the cdf in terms of the pdf

$$P(X \leqslant U^{-1}(y)) = \int_a^{U^{-1}(y)} f(x)\mathrm{d}x\,. \tag{6.137}$$

Now we have an expression of the cdf of Y in terms of x:

$$F_Y(y) = \int_a^{U^{-1}(y)} f(x)\mathrm{d}x\,. \tag{6.138}$$

To obtain the pdf, we differentiate (6.138) with respect to y:

$$f(y) = \frac{\mathrm{d}}{\mathrm{d}y}F_y(y) = \frac{\mathrm{d}}{\mathrm{d}y}\int_a^{U^{-1}(y)} f(x)\mathrm{d}x. \tag{6.139}$$

Note that the integral on the right-hand side is with respect to x, but we need an integral with respect to y because we are differentiating with respect to y. In particular, we use (6.133) to get the substitution

$$\int f(U^{-1}(y))U^{-1\prime}(y)\mathrm{d}y = \int f(x)\mathrm{d}x \quad \text{where} \quad x = U^{-1}(y). \tag{6.140}$$

Using (6.140) on the right-hand side of (6.139) gives us

$$f(y) = \frac{\mathrm{d}}{\mathrm{d}y}\int_a^{U^{-1}(y)} f_x(U^{-1}(y))U^{-1\prime}(y)\mathrm{d}y. \tag{6.141}$$

We then recall that differentiation is a linear operator and we use the subscript x to remind ourselves that $f_x(U^{-1}(y))$ is a function of x and not y. Invoking the fundamental theorem of calculus again gives us

$$f(y) = f_x(U^{-1}(y)) \cdot \left(\frac{\mathrm{d}}{\mathrm{d}y}U^{-1}(y)\right). \tag{6.142}$$

Recall that we assumed that U is a strictly increasing function. For decreasing functions, it turns out that we have a negative sign when we follow the same derivation. We introduce the absolute value of the differential to have the same expression for both increasing and decreasing U:

$$f(y) = f_x(U^{-1}(y)) \cdot \left|\frac{\mathrm{d}}{\mathrm{d}y}U^{-1}(y)\right|. \tag{6.143}$$

This is called the *change-of-variable technique*. The term $\left|\frac{\mathrm{d}}{\mathrm{d}y}U^{-1}(y)\right|$ in (6.143) measures how much a unit volume changes when applying U (see also the definition of the Jacobian in Section 5.3).

change-of-variable technique.

Remark. In comparison to the discrete case in (6.125b), we have an additional factor $\left|\frac{\mathrm{d}}{\mathrm{d}y}U^{-1}(y)\right|$. The continuous case requires more care because $P(Y = y) = 0$ for all y. The probability density function $f(y)$ does not have a description as a probability of an event involving y. ♢

So far in this section, we have been studying univariate change of variables. The case for multivariate random variables is analogous, but complicated by fact that the absolute value cannot be used for multivariate functions. Instead, we use the determinant of the Jacobian matrix. Recall from (5.58) that the Jacobian is a matrix of partial derivatives, and that the existence of a nonzero determinant shows that we can invert the Jacobian. Recall the discussion in Section 4.1 that the determinant arises because our differentials (cubes of volume) are transformed into parallelepipeds by the Jacobian. Let us summarize the preceding discussion in the following theorem, which gives us a recipe for multivariate change of variables.

Theorem 6.16. *[Theorem 17.2 in Billingsley (1995)] Let $f(\boldsymbol{x})$ be the value of the probability density of the multivariate continuous random variable X. If the vector-valued function $\boldsymbol{y} = U(\boldsymbol{x})$ is differentiable and invertible for all values within the domain of $\boldsymbol{x}$, then for corresponding values of $\boldsymbol{y}$, the probability density of $Y = U(X)$ is given by*

$$f(\boldsymbol{y}) = f_{\boldsymbol{x}}(U^{-1}(\boldsymbol{y})) \cdot \left| \det \left(\frac{\partial}{\partial \boldsymbol{y}} U^{-1}(\boldsymbol{y}) \right) \right| . \tag{6.144}$$

The theorem looks intimidating at first glance, but the key point is that a change of variable of a multivariate random variable follows the procedure of the univariate change of variable. First we need to work out the inverse transform, and substitute that into the density of $\boldsymbol{x}$. Then we calculate the determinant of the Jacobian and multiply the result. The following example illustrates the case of a bivariate random variable.

Example 6.17

Consider a bivariate random variable X with states $\boldsymbol{x} = \begin{bmatrix} x_1 \\ x_2 \end{bmatrix}$ and probability density function

$$f\left(\begin{bmatrix} x_1 \\ x_2 \end{bmatrix} \right) = \frac{1}{2\pi} \exp \left(-\frac{1}{2} \begin{bmatrix} x_1 \\ x_2 \end{bmatrix}^\top \begin{bmatrix} x_1 \\ x_2 \end{bmatrix} \right) . \tag{6.145}$$

We use the change-of-variable technique from Theorem 6.16 to derive the effect of a linear transformation (Section 2.7) of the random variable. Consider a matrix $\boldsymbol{A} \in \mathbb{R}^{2\times 2}$ defined as

$$\boldsymbol{A} = \begin{bmatrix} a & b \\ c & d \end{bmatrix} . \tag{6.146}$$

We are interested in finding the probability density function of the transformed bivariate random variable Y with states $\boldsymbol{y} = \boldsymbol{A}\boldsymbol{x}$.

Recall that for change of variables we require the inverse transformation of $\boldsymbol{x}$ as a function of $\boldsymbol{y}$. Since we consider linear transformations, the inverse transformation is given by the matrix inverse (see Section 2.2.2). For 2×2 matrices, we can explicitly write out the formula, given by

$$\begin{bmatrix} x_1 \\ x_2 \end{bmatrix} = \boldsymbol{A}^{-1} \begin{bmatrix} y_1 \\ y_2 \end{bmatrix} = \frac{1}{ad - bc} \begin{bmatrix} d & -b \\ -c & a \end{bmatrix} \begin{bmatrix} y_1 \\ y_2 \end{bmatrix} . \tag{6.147}$$

Observe that $ad - bc$ is the determinant (Section 4.1) of $\boldsymbol{A}$. The corresponding probability density function is given by

$$f(\boldsymbol{x}) = f(\boldsymbol{A}^{-1}\boldsymbol{y}) = \frac{1}{2\pi} \exp\left(-\tfrac{1}{2}\boldsymbol{y}^\top \boldsymbol{A}^{-\top}\boldsymbol{A}^{-1}\boldsymbol{y}\right). \tag{6.148}$$

The partial derivative of a matrix times a vector with respect to the vector is the matrix itself (Section 5.5), and therefore

$$\frac{\partial}{\partial \boldsymbol{y}} \boldsymbol{A}^{-1}\boldsymbol{y} = \boldsymbol{A}^{-1}. \tag{6.149}$$

Recall from Section 4.1 that the determinant of the inverse is the inverse of the determinant so that the determinant of the Jacobian matrix is

$$\det\left(\frac{\partial}{\partial \boldsymbol{y}} \boldsymbol{A}^{-1}\boldsymbol{y}\right) = \frac{1}{ad - bc}. \tag{6.150}$$

We are now able to apply the change-of-variable formula from Theorem 6.16 by multiplying (6.148) with (6.150), which yields

$$\begin{aligned} f(\boldsymbol{y}) &= f(\boldsymbol{x}) \left|\det\left(\frac{\partial}{\partial \boldsymbol{y}} \boldsymbol{A}^{-1}\boldsymbol{y}\right)\right| & (6.151\text{a}) \\ &= \frac{1}{2\pi} \exp\left(-\tfrac{1}{2}\boldsymbol{y}^\top \boldsymbol{A}^{-\top}\boldsymbol{A}^{-1}\boldsymbol{y}\right) |ad - bc|^{-1}. & (6.151\text{b}) \end{aligned}$$

While Example 6.17 is based on a bivariate random variable, which allows us to easily compute the matrix inverse, the preceding relation holds for higher dimensions.

Remark. We saw in Section 6.5 that the density $f(\boldsymbol{x})$ in (6.148) is actually the standard Gaussian distribution, and the transformed density $f(\boldsymbol{y})$ is a bivariate Gaussian with covariance $\boldsymbol{\Sigma} = \boldsymbol{A}\boldsymbol{A}^\top$. ♢

We will use the ideas in this chapter to describe probabilistic modeling in Section 8.4, as well as introduce a graphical language in Section 8.5. We will see direct machine learning applications of these ideas in Chapters 9 and 11.

6.8 Further Reading

This chapter is rather terse at times. Grinstead and Snell (1997) and Walpole et al. (2011) provide more relaxed presentations that are suitable for self-study. Readers interested in more philosophical aspects of probability should consider Hacking (2001), whereas an approach that is more related to software engineering is presented by Downey (2014). An overview of exponential families can be found in Barndorff-Nielsen (2014). We will see more about how to use probability distributions to model machine learning tasks in Chapter 8. Ironically, the recent surge in interest in neural networks has resulted in a broader appreciation of probabilistic models. For example, the idea of normalizing flows (Jimenez Rezende and Mohamed, 2015) relies on change of variables for transforming random variables. An overview of methods for variational inference as applied to neural networks is described in chapters 16 to 20 of the book by Goodfellow et al. (2016).

We sidestepped a large part of the difficulty in continuous random variables by avoiding measure theoretic questions (Billingsley, 1995; Pollard, 2002), and by assuming without construction that we have real numbers, and ways of defining sets on real numbers as well as their appropriate frequency of occurrence. These details do matter, for example, in the specification of conditional probability $p(y \mid x)$ for continuous random variables x, y (Proschan and Presnell, 1998). The lazy notation hides the fact that we want to specify that $X = x$ (which is a set of measure zero). Furthermore, we are interested in the probability density function of y. A more precise notation would have to say $\mathbb{E}_y[f(y) \mid \sigma(x)]$, where we take the expectation over y of a test function f conditioned on the σ-algebra of x. A more technical audience interested in the details of probability theory have many options (Jaynes, 2003; MacKay, 2003; Jacod and Protter, 2004; Grimmett and Welsh, 2014), including some very technical discussions (Shiryayev, 1984; Lehmann and Casella, 1998; Dudley, 2002; Bickel and Doksum, 2006; Çinlar, 2011). An alternative way to approach probability is to start with the concept of expectation, and "work backward" to derive the necessary properties of a probability space (Whittle, 2000). As machine learning allows us to model more intricate distributions on ever more complex types of data, a developer of probabilistic machine learning models would have to understand these more technical aspects. Machine learning texts with a probabilistic modeling focus include the books by MacKay (2003); Bishop (2006); Rasmussen and Williams (2006); Barber (2012); Murphy (2012).

Exercises

6.1 Consider the following bivariate distribution $p(x, y)$ of two discrete random variables X and Y.

Y						
	y_1	0.01	0.02	0.03	0.1	0.1
	y_2	0.05	0.1	0.05	0.07	0.2
	y_3	0.1	0.05	0.03	0.05	0.04
		x_1	x_2	x_3	x_4	x_5
				X		

Compute:

a. The marginal distributions $p(x)$ and $p(y)$.
b. The conditional distributions $p(x|Y = y_1)$ and $p(y|X = x_3)$.

6.2 Consider a mixture of two Gaussian distributions (illustrated in Figure 6.4):

$$0.4\mathcal{N}\left(\begin{bmatrix}10\\2\end{bmatrix}, \begin{bmatrix}1 & 0\\0 & 1\end{bmatrix}\right) + 0.6\mathcal{N}\left(\begin{bmatrix}0\\0\end{bmatrix}, \begin{bmatrix}8.4 & 2.0\\2.0 & 1.7\end{bmatrix}\right).$$

a. Compute the marginal distributions for each dimension.

b. Compute the mean, mode and median for each marginal distribution.

c. Compute the mean and mode for the two-dimensional distribution.

6.3 You have written a computer program that sometimes compiles and sometimes not (code does not change). You decide to model the apparent stochasticity (success vs. no success) x of the compiler using a Bernoulli distribution with parameter μ:

$$p(x \mid \mu) = \mu^x (1-\mu)^{1-x}, \quad x \in \{0, 1\}.$$

Choose a conjugate prior for the Bernoulli likelihood and compute the posterior distribution $p(\mu \mid x_1, \ldots, x_N)$.

6.4 There are two bags. The first bag contains four mangos and two apples; the second bag contains four mangos and four apples.
We also have a biased coin, which shows "heads" with probability 0.6 and "tails" with probability 0.4. If the coin shows "heads". we pick a fruit at random from bag 1; otherwise, we pick a fruit at random from bag 2.
Your friend flips the coin (you cannot see the result), picks a fruit at random from the corresponding bag, and presents you a mango.
What is the probability that the mango was picked from bag 2?
Hint: Use Bayes' theorem.

6.5 Consider the time-series model

$$\begin{aligned} \boldsymbol{x}_{t+1} &= \boldsymbol{A}\boldsymbol{x}_t + \boldsymbol{w}, \quad \boldsymbol{w} \sim \mathcal{N}(\boldsymbol{0}, \boldsymbol{Q}) \\ \boldsymbol{y}_t &= \boldsymbol{C}\boldsymbol{x}_t + \boldsymbol{v}, \quad \boldsymbol{v} \sim \mathcal{N}(\boldsymbol{0}, \boldsymbol{R}), \end{aligned}$$

where $\boldsymbol{w}, \boldsymbol{v}$ are i.i.d. Gaussian noise variables. Further, assume that $p(\boldsymbol{x}_0) = \mathcal{N}(\boldsymbol{\mu}_0, \boldsymbol{\Sigma}_0)$.

a. What is the form of $p(\boldsymbol{x}_0, \boldsymbol{x}_1, \ldots, \boldsymbol{x}_T)$? Justify your answer (you do not have to explicitly compute the joint distribution).

b. Assume that $p(\boldsymbol{x}_t \mid \boldsymbol{y}_1, \ldots, \boldsymbol{y}_t) = \mathcal{N}(\boldsymbol{\mu}_t, \boldsymbol{\Sigma}_t)$.

1. Compute $p(\boldsymbol{x}_{t+1} \mid \boldsymbol{y}_1, \ldots, \boldsymbol{y}_t)$.
2. Compute $p(\boldsymbol{x}_{t+1}, \boldsymbol{y}_{t+1} \mid \boldsymbol{y}_1, \ldots, \boldsymbol{y}_t)$.
3. At time $t+1$, we observe the value $\boldsymbol{y}_{t+1} = \hat{\boldsymbol{y}}$. Compute the conditional distribution $p(\boldsymbol{x}_{t+1} \mid \boldsymbol{y}_1, \ldots, \boldsymbol{y}_{t+1})$.

6.6 Prove the relationship in (6.44), which relates the standard definition of the variance to the raw-score expression for the variance.

6.7 Prove the relationship in (6.45), which relates the pairwise difference between examples in a dataset with the raw-score expression for the variance.

6.8 Express the Bernoulli distribution in the natural parameter form of the exponential family, see (6.107).

6.9 Express the Binomial distribution as an exponential family distribution. Also express the Beta distribution is an exponential family distribution. Show that the product of the Beta and the Binomial distribution is also a member of the exponential family.

6.10 Derive the relationship in Section 6.5.2 in two ways:

a. By completing the square

b. By expressing the Gaussian in its exponential family form

The *product* of two Gaussians $\mathcal{N}\big(\boldsymbol{x}\,|\,\boldsymbol{a},\,\boldsymbol{A}\big)\mathcal{N}\big(\boldsymbol{x}\,|\,\boldsymbol{b},\,\boldsymbol{B}\big)$ is an unnormalized Gaussian distribution $c\mathcal{N}\big(\boldsymbol{x}\,|\,\boldsymbol{c},\,\boldsymbol{C}\big)$ with

$$\boldsymbol{C} = (\boldsymbol{A}^{-1} + \boldsymbol{B}^{-1})^{-1}$$
$$\boldsymbol{c} = \boldsymbol{C}(\boldsymbol{A}^{-1}\boldsymbol{a} + \boldsymbol{B}^{-1}\boldsymbol{b})$$
$$c = (2\pi)^{-\frac{D}{2}}\,|\,\boldsymbol{A}+\boldsymbol{B}\,|^{-\frac{1}{2}}\exp\big(-\tfrac{1}{2}(\boldsymbol{a}-\boldsymbol{b})^{\top}(\boldsymbol{A}+\boldsymbol{B})^{-1}(\boldsymbol{a}-\boldsymbol{b})\big)\,.$$

Note that the normalizing constant c itself can be considered a (normalized) Gaussian distribution either in $\boldsymbol{a}$ or in $\boldsymbol{b}$ with an "inflated" covariance matrix $\boldsymbol{A}+\boldsymbol{B}$, i.e., $c = \mathcal{N}\big(\boldsymbol{a}\,|\,\boldsymbol{b},\,\boldsymbol{A}+\boldsymbol{B}\big) = \mathcal{N}\big(\boldsymbol{b}\,|\,\boldsymbol{a},\,\boldsymbol{A}+\boldsymbol{B}\big)$.

6.11 **Iterated Expectations.**
Consider two random variables x, y with joint distribution $p(x, y)$. Show that

$$\mathbb{E}_X[x] = \mathbb{E}_Y\big[\mathbb{E}_X[x\,|\,y]\big]\,.$$

Here, $\mathbb{E}_X[x\,|\,y]$ denotes the expected value of x under the conditional distribution $p(x\,|\,y)$.

6.12 **Manipulation of Gaussian Random Variables.**
Consider a Gaussian random variable $\boldsymbol{x} \sim \mathcal{N}\big(\boldsymbol{x}\,|\,\boldsymbol{\mu}_x,\,\boldsymbol{\Sigma}_x\big)$, where $\boldsymbol{x} \in \mathbb{R}^D$. Furthermore, we have

$$\boldsymbol{y} = \boldsymbol{A}\boldsymbol{x} + \boldsymbol{b} + \boldsymbol{w}\,,$$

where $\boldsymbol{y} \in \mathbb{R}^E$, $\boldsymbol{A} \in \mathbb{R}^{E\times D}$, $\boldsymbol{b} \in \mathbb{R}^E$, and $\boldsymbol{w} \sim \mathcal{N}\big(\boldsymbol{w}\,|\,\boldsymbol{0},\,\boldsymbol{Q}\big)$ is independent Gaussian noise. "Independent" implies that $\boldsymbol{x}$ and $\boldsymbol{w}$ are independent random variables and that $\boldsymbol{Q}$ is diagonal.

a. Write down the likelihood $p(\boldsymbol{y}\,|\,\boldsymbol{x})$.
b. The distribution $p(\boldsymbol{y}) = \int p(\boldsymbol{y}\,|\,\boldsymbol{x})p(\boldsymbol{x})d\boldsymbol{x}$ is Gaussian. Compute the mean $\boldsymbol{\mu}_y$ and the covariance $\boldsymbol{\Sigma}_y$. Derive your result in detail.
c. The random variable $\boldsymbol{y}$ is being transformed according to the measurement mapping

$$\boldsymbol{z} = \boldsymbol{C}\boldsymbol{y} + \boldsymbol{v}\,,$$

where $\boldsymbol{z} \in \mathbb{R}^F$, $\boldsymbol{C} \in \mathbb{R}^{F\times E}$, and $\boldsymbol{v} \sim \mathcal{N}\big(\boldsymbol{v}\,|\,\boldsymbol{0},\,\boldsymbol{R}\big)$ is independent Gaussian (measurement) noise.

- Write down $p(\boldsymbol{z}\,|\,\boldsymbol{y})$.
- Compute $p(\boldsymbol{z})$, i.e., the mean $\boldsymbol{\mu}_z$ and the covariance $\boldsymbol{\Sigma}_z$. Derive your result in detail.

d. Now, a value $\hat{\boldsymbol{y}}$ is measured. Compute the posterior distribution $p(\boldsymbol{x}\,|\,\hat{\boldsymbol{y}})$.
Hint for solution: This posterior is also Gaussian, i.e., we need to determine only its mean and covariance matrix. Start by explicitly computing the joint Gaussian $p(\boldsymbol{x}, \boldsymbol{y})$. This also requires us to compute the cross-covariances $\mathrm{Cov}_{\boldsymbol{x},\boldsymbol{y}}[\boldsymbol{x}, \boldsymbol{y}]$ and $\mathrm{Cov}_{\boldsymbol{y},\boldsymbol{x}}[\boldsymbol{y}, \boldsymbol{x}]$. Then apply the rules for Gaussian conditioning.

6.13 **Probability Integral Transformation.**
Given a continuous random variable x, with cdf $F_x(x)$, show that the random variable $y = F_x(x)$ is uniformly distributed.

7

Continuous Optimization

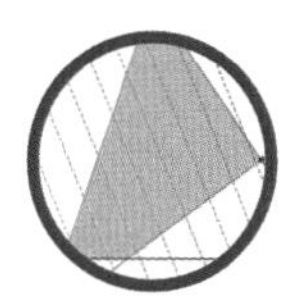

Since machine learning algorithms are implemented on a computer, the mathematical formulations are expressed as numerical optimization methods. This chapter describes the basic numerical methods for training machine learning models. Training a machine learning model often boils down to finding a good set of parameters. The notion of "good" is determined by the objective function or the probabilistic model, which we will see examples of in the second part of this book. Given an objective function, finding the best value is done using optimization algorithms.

Since we consider data and models in $\mathbb{R}^D$, the optimization problems we face are *continuous* optimization problems, as opposed to *combinatorial* optimization problems for discrete variables.

This chapter covers two main branches of continuous optimization (Figure 7.1): unconstrained and constrained optimization. We will assume in this chapter that our objective function is differentiable (see Chapter 5), hence we have access to a gradient at each location in the space to help us find the optimum value. By convention, most objective functions in machine learning are intended to be minimized, that is, the best value is the minimum value. Intuitively finding the best value is like finding the valleys of the objective function, and the gradients point us uphill. The idea is to move downhill (opposite to the gradient) and hope to find the deepest point. For unconstrained optimization, this is the only concept we need, but there are several design choices, which we discuss in Section 7.1. For constrained optimization, we need to introduce other concepts to manage the constraints (Section 7.2). We will also introduce a special class of problems (convex optimization problems in Section 7.3) where we can make statements about reaching the global optimum.

global minimum

Consider the function in Figure 7.2. The function has a *global minimum* around $x = -4.5$, which has the objective function value of approximately -47. Since the function is "smooth," the gradients can be used to help find the minimum by indicating whether we should take a step to the right or left. This assumes that we are in the correct bowl, as there exists another *local minimum* around $x = 0.7$. Recall that we can solve for all the stationary points of a function by calculating its derivative and setting it to zero. For

local minimum

Stationary points are the real roots of the derivative, that is, points that have zero gradient.

$$\ell(x) = x^4 + 7x^3 + 5x^2 - 17x + 3\,, \tag{7.1}$$

we obtain the corresponding gradient as

$$\frac{\mathrm{d}\ell(x)}{\mathrm{d}x} = 4x^3 + 21x^2 + 10x - 17\,. \tag{7.2}$$

Since this is a cubic equation, it has in general three solutions when set to zero. In the example, two of them are minimums and one is a maximum (around

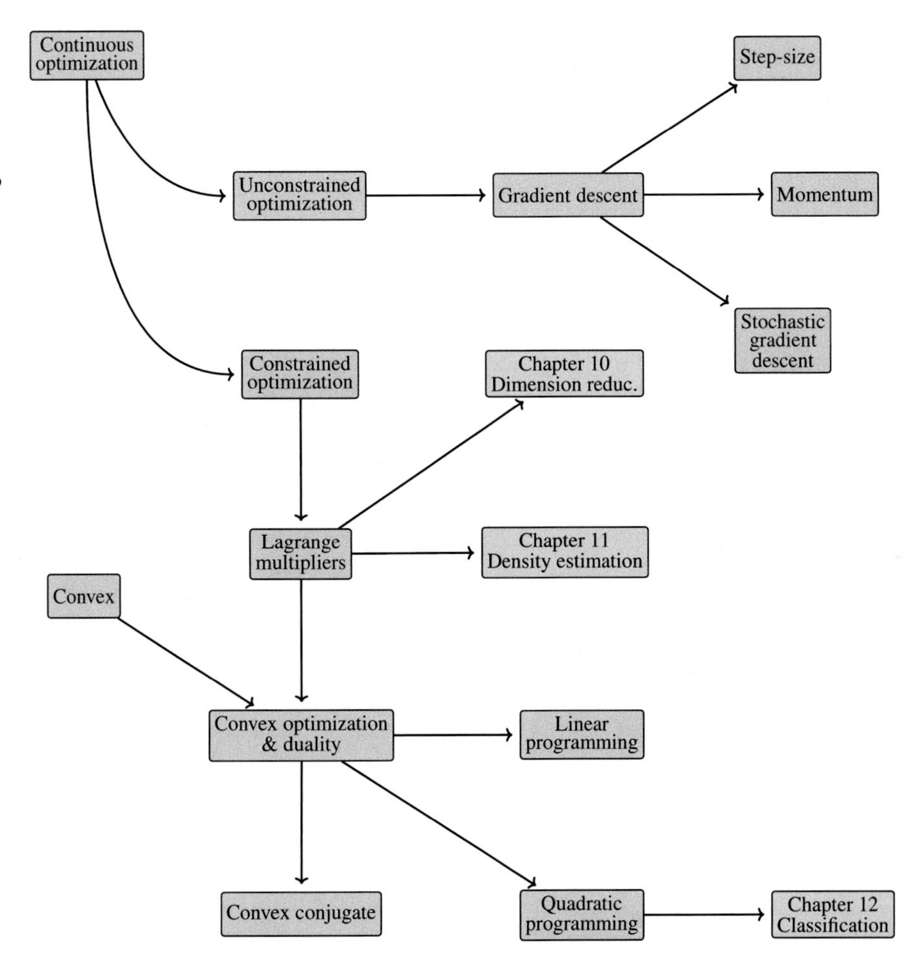

Figure 7.1 A mind map of the concepts related to optimization, as presented in this chapter. There are two main ideas: gradient descent and convex optimization.

$x = -1.4$). To check whether a stationary point is a minimum or maximum, we need to take the derivative a second time and check whether the second derivative is positive or negative at the stationary point. In our case, the second derivative is

$$\frac{\mathrm{d}^2\ell(x)}{\mathrm{d}x^2} = 12x^2 + 42x + 10\,. \tag{7.3}$$

By substituting our visually estimated values of $x = -4.5, -1.4, 0.7$, we will observe that as expected the middle point is a maximum $\left(\frac{\mathrm{d}^2\ell(x)}{\mathrm{d}x^2} < 0\right)$ and the other two stationary points are minimums.

Note that we have avoided analytically solving for values of x in the previous discussion, although for low-order polynomials such as the preceding we could do so. In general, we are unable to find analytic solutions, and hence we need to start at some value, say $x_0 = -10$, and follow the gradient. The gradient indicates that we should go right, but not how far (this is called the step-size). Furthermore, if we had started at the right side (e.g., $x_0 = 0$) the gradient would have led us to the wrong minimum. Figure 7.2 illustrates the fact that for $x > -1$, the gradient points toward the minimum on the right of the figure, which has a larger objective value.

According to the Abel–Ruffini theorem, there is in general no algebraic solution for polynomials of degree 5 or more (Abel, 1826).

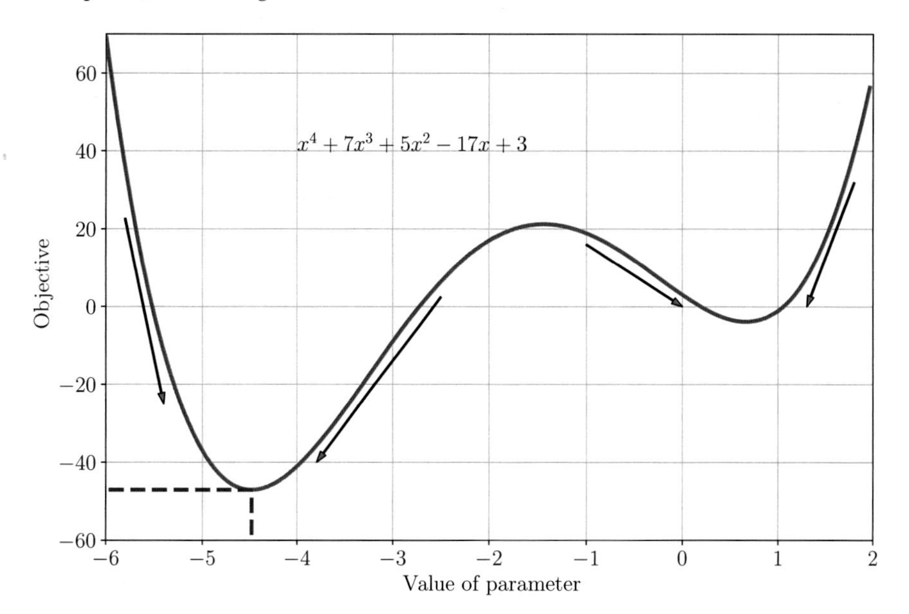

Figure 7.2 Example objective function. Gradients are indicated by arrows, and the global minimum is indicated by the dashed blue line.

In Section 7.3, we will learn about a class of functions, called convex functions, that do not exhibit this tricky dependency on the starting point of the optimization algorithm.For convex functions, all local minimums are global minimum. It turns out that many machine learning objective functions are designed such that they are convex, and we will see an example in Chapter 12.

For convex functions all local minima are global minimum.

The discussion in this chapter so far was about a one-dimensional function, where we are able to visualize the ideas of gradients, descent directions, and optimal values. In the rest of this chapter we develop the same ideas in high dimensions. Unfortunately, we can only visualize the concepts in one dimension, but some concepts do not generalize directly to higher dimensions, therefore some care needs to be taken when reading.

7.1 Optimization Using Gradient Descent

We now consider the problem of solving for the minimum of a real-valued function

$$\min_{\boldsymbol{x}} f(\boldsymbol{x}) \tag{7.4}$$

where $f : \mathbb{R}^d \to \mathbb{R}$ is an objective function that captures the machine learning problem at hand. We assume that our function f is differentiable, and we are unable to analytically find a solution in closed form.

Gradient descent is a first-order optimization algorithm. To find a local minimum of a function using gradient descent, one takes steps proportional to the negative of the gradient of the function at the current point. Recall from Section 5.1 that the gradient points in the direction of the steepest ascent. Another useful intuition is to consider the set of lines where the function is at a certain value ($f(\boldsymbol{x}) = c$ for some value $c \in \mathbb{R}$), which are known as the contour lines.

We use the convention of row vectors for gradients.

The gradient points in a direction that is orthogonal to the contour lines of the function we wish to optimize.

Let us consider multivariate functions. Imagine a surface (described by the function $f(\boldsymbol{x})$) with a ball starting at a particular location $\boldsymbol{x}_0$. When the ball is released, it will move downhill in the direction of steepest descent. Gradient descent exploits the fact that $f(\boldsymbol{x}_0)$ decreases fastest if one moves from $\boldsymbol{x}_0$ in the direction of the negative gradient $-((\nabla f)(\boldsymbol{x}_0))^\top$ of f at $\boldsymbol{x}_0$. We assume in this book that the functions are differentiable, and refer the reader to more general settings in Section 7.4. Then, if

$$\boldsymbol{x}_1 = \boldsymbol{x}_0 - \gamma((\nabla f)(\boldsymbol{x}_0))^\top \tag{7.5}$$

for a small *step-size* $\gamma \geqslant 0$, then $f(\boldsymbol{x}_1) \leqslant f(\boldsymbol{x}_0)$. Note that we use the transpose for the gradient since otherwise the dimensions will not work out.

This observation allows us to define a simple gradient descent algorithm: If we want to find a local optimum $f(\boldsymbol{x}_*)$ of a function $f : \mathbb{R}^n \to \mathbb{R}$, $\boldsymbol{x} \mapsto f(\boldsymbol{x})$, we start with an initial guess $\boldsymbol{x}_0$ of the parameters we wish to optimize and then iterate according to

$$\boldsymbol{x}_{i+1} = \boldsymbol{x}_i - \gamma_i((\nabla f)(\boldsymbol{x}_i))^\top . \tag{7.6}$$

For suitable step-size γ_i, the sequence $f(\boldsymbol{x}_0) \geqslant f(\boldsymbol{x}_1) \geqslant \ldots$ converges to a local minimum.

Example 7.1

Consider a quadratic function in two dimensions

$$f\left(\begin{bmatrix}x_1\\x_2\end{bmatrix}\right) = \frac{1}{2}\begin{bmatrix}x_1\\x_2\end{bmatrix}^\top \begin{bmatrix}2 & 1\\1 & 20\end{bmatrix}\begin{bmatrix}x_1\\x_2\end{bmatrix} - \begin{bmatrix}5\\3\end{bmatrix}^\top \begin{bmatrix}x_1\\x_2\end{bmatrix} \tag{7.7}$$

with gradient

$$\nabla f\left(\begin{bmatrix}x_1\\x_2\end{bmatrix}\right) = \begin{bmatrix}x_1\\x_2\end{bmatrix}^\top \begin{bmatrix}2 & 1\\1 & 20\end{bmatrix} - \begin{bmatrix}5\\3\end{bmatrix}^\top . \tag{7.8}$$

Starting at the initial location $\boldsymbol{x}_0 = [-3, -1]^\top$, we iteratively apply (7.6) to obtain a sequence of estimates that converge to the minimum value (illustrated in Figure 7.3). We can see (both from the figure and by plugging $\boldsymbol{x}_0$ into (7.8) with $\gamma = 0.085$) that the gradient at $\boldsymbol{x}_0$ points north and east, leading to $\boldsymbol{x}_1 = [-1.98, 1.21]^\top$. Repeating that argument gives us $\boldsymbol{x}_2 = [-1.32, -0.42]^\top$, and so on.

Remark. Gradient descent can be relatively slow close to the minimum: Its asymptotic rate of convergence is inferior to many other methods. Using the ball rolling down the hill analogy, when the surface is a long, thin valley, the problem is poorly conditioned (Trefethen and Bau III, 1997). For poorly conditioned convex problems, gradient descent increasingly “zigzags” as the gradients point nearly orthogonally to the shortest direction to a minimum point; see Figure 7.3. ◊

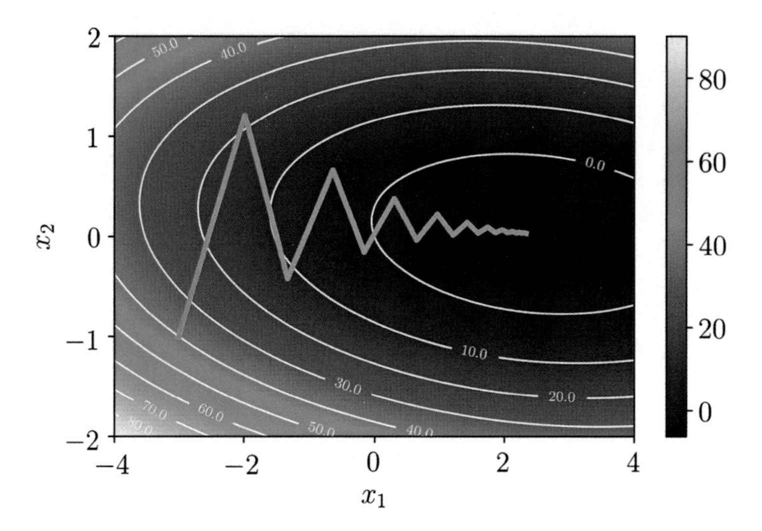

Figure 7.3 Gradient descent on a two-dimensional quadratic surface (shown as a heatmap). See Example 7.1 for a description.

7.1.1 Step-Size

As mentioned earlier, choosing a good step-size is important in gradient descent. If the step-size is too small, gradient descent can be slow. If the step-size is chosen too large, gradient descent can overshoot, fail to converge, or even diverge. We will discuss the use of momentum in the next section. It is a method that smoothes out erratic behavior of gradient updates and dampens oscillations.

The step-size is also called the learning rate.

Adaptive gradient methods rescale the step-size at each iteration, depending on local properties of the function. There are two simple heuristics (Toussaint, 2012):

- When the function value increases after a gradient step, the step-size was too large. Undo the step and decrease the step-size.
- When the function value decreases the step could have been larger. Try to increase the step-size.

Although the "undo" step seems to be a waste of resources, using this heuristic guarantees monotonic convergence.

Example 7.2 (Solving a Linear Equation System)
When we solve linear equations of the form $\boldsymbol{A}\boldsymbol{x} = \boldsymbol{b}$, in practice we solve $\boldsymbol{A}\boldsymbol{x} - \boldsymbol{b} = \boldsymbol{0}$ approximately by finding $\boldsymbol{x}_*$ that minimizes the squared error

$$\|\boldsymbol{A}\boldsymbol{x} - \boldsymbol{b}\|^2 = (\boldsymbol{A}\boldsymbol{x} - \boldsymbol{b})^\top(\boldsymbol{A}\boldsymbol{x} - \boldsymbol{b}) \tag{7.9}$$

if we use the Euclidean norm. The gradient of (7.9) with respect to $\boldsymbol{x}$ is

$$\nabla_{\boldsymbol{x}} = 2(\boldsymbol{A}\boldsymbol{x} - \boldsymbol{b})^\top \boldsymbol{A}\,. \tag{7.10}$$

We can use this gradient directly in a gradient descent algorithm. However, for this particular special case, it turns out that there is an analytic solution, which can be found by setting the gradient to zero. We will see more on solving squared error problems in Chapter 9.

Remark. When applied to the solution of linear systems of equations $\boldsymbol{A}\boldsymbol{x} = \boldsymbol{b}$, gradient descent may converge slowly. The speed of convergence of gradient descent is dependent on the *condition number* $\kappa = \frac{\sigma(\boldsymbol{A})_{\max}}{\sigma(\boldsymbol{A})_{\min}}$, which is the ratio of the maximum to the minimum singular value (Section 4.5) of $\boldsymbol{A}$. The condition number essentially measures the ratio of the most curved direction versus the least curved direction, which corresponds to our imagery that poorly conditioned problems are long, thin valleys: They are very curved in one direction, but very flat in the other. Instead of directly solving $\boldsymbol{A}\boldsymbol{x} = \boldsymbol{b}$, one could instead solve $\boldsymbol{P}^{-1}(\boldsymbol{A}\boldsymbol{x} - \boldsymbol{b}) = \boldsymbol{0}$, where $\boldsymbol{P}$ is called the *preconditioner*. The goal is to design $\boldsymbol{P}^{-1}$ such that $\boldsymbol{P}^{-1}\boldsymbol{A}$ has a better condition number, but at the same time $\boldsymbol{P}^{-1}$ is easy to compute. For further information on gradient descent, preconditioning, and convergence, we refer to Boyd and Vandenberghe (2004, chapter 9). ◇

condition number

preconditioner

7.1.2 Gradient Descent With Momentum

As illustrated in Figure 7.3, the convergence of gradient descent may be very slow if the curvature of the optimization surface is such that there are regions that are poorly scaled. The curvature is such that the gradient descent steps hops between the walls of the valley and approaches the optimum in small steps. The proposed tweak to improve convergence is to give gradient descent some memory.

Goh (2017) wrote an intuitive blog post on gradient descent with momentum.

Gradient descent with momentum (Rumelhart et al., 1986) is a method that introduces an additional term to remember what happened in the previous iteration. This memory dampens oscillations and smoothes out the gradient updates. Continuing the ball analogy, the momentum term emulates the phenomenon of a heavy ball that is reluctant to change directions. The idea is to have a gradient update with memory to implement a moving average. The momentum-based method remembers the update $\Delta\boldsymbol{x}_i$ at each iteration i and determines the next update as a linear combination of the current and previous gradients

$$\boldsymbol{x}_{i+1} = \boldsymbol{x}_i - \gamma_i((\nabla f)(\boldsymbol{x}_i))^\top + \alpha\Delta\boldsymbol{x}_i \tag{7.11}$$

$$\Delta\boldsymbol{x}_i = \boldsymbol{x}_i - \boldsymbol{x}_{i-1} = \alpha\Delta\boldsymbol{x}_{i-1} - \gamma_{i-1}((\nabla f)(\boldsymbol{x}_{i-1}))^\top, \tag{7.12}$$

where $\alpha \in [0, 1]$. Sometimes we will only know the gradient approximately. In such cases, the momentum term is useful since it averages out different noisy estimates of the gradient. One particularly useful way to obtain an approximate gradient is by using a stochastic approximation, which we discuss next.

7.1.3 Stochastic Gradient Descent

Computing the gradient can be very time consuming. However, often it is possible to find a "cheap" approximation of the gradient. Approximating the gradient is still useful as long as it points in roughly the same direction as the true gradient.

stochastic gradient descent

Stochastic gradient descent (often shortened as SGD) is a stochastic approximation of the gradient descent method for minimizing an objective function that is written as a sum of differentiable functions. The word stochastic here refers to the fact that we acknowledge that we do not know the gradient precisely, but instead only know a noisy approximation to it. By constraining the probability distribution of the approximate gradients, we can still theoretically guarantee that SGD will converge.

In machine learning, given $n = 1, \dots, N$ data points, we often consider objective functions that are the sum of the losses L_n incurred by each example n. In mathematical notation, we have the form

$$L(\boldsymbol{\theta}) = \sum_{n=1}^{N} L_n(\boldsymbol{\theta}), \tag{7.13}$$

where $\boldsymbol{\theta}$ is the vector of parameters of interest, i.e., we want to find $\boldsymbol{\theta}$ that minimizes L. An example from regression (Chapter 9) is the negative log-likelihood, which is expressed as a sum over log-likelihoods of individual examples so that

$$L(\boldsymbol{\theta}) = -\sum_{n=1}^{N} \log p(y_n | \boldsymbol{x}_n, \boldsymbol{\theta}), \tag{7.14}$$

where $\boldsymbol{x}_n \in \mathbb{R}^D$ are the training inputs, y_n are the training targets, and $\boldsymbol{\theta}$ are the parameters of the regression model.

Standard gradient descent, as introduced previously, is a "batch" optimization method, i.e., optimization is performed using the full training set by updating the vector of parameters according to

$$\boldsymbol{\theta}_{i+1} = \boldsymbol{\theta}_i - \gamma_i (\nabla L(\boldsymbol{\theta}_i))^\top = \boldsymbol{\theta}_i - \gamma_i \sum_{n=1}^{N} (\nabla L_n(\boldsymbol{\theta}_i))^\top \tag{7.15}$$

for a suitable step-size parameter γ_i. Evaluating the sum gradient may require expensive evaluations of the gradients from all individual functions L_n. When the training set is enormous and/or no simple formulas exist, evaluating the sums of gradients becomes very expensive.

Consider the term $\sum_{n=1}^{N}(\nabla L_n(\boldsymbol{\theta}_i))$ in (7.15), we can reduce the amount of computation by taking a sum over a smaller set of L_n. In contrast to batch gradient descent, which uses all L_n for $n = 1, \dots, N$, we randomly choose a subset of L_n for minibatch gradient descent. In the extreme case, we randomly select only a single L_n to estimate the gradient. The key insight about why taking a subset of data is sensible is to realize that for gradient descent to converge, we only require that the gradient is an unbiased estimate of the true gradient. In fact, the term $\sum_{n=1}^{N}(\nabla L_n(\boldsymbol{\theta}_i))$ in (7.15) is an empirical estimate of the expected value (Section 6.4.1) of the gradient. Therefore, any other unbiased empirical estimate of the expected value, for example using any subsample of the data, would suffice for convergence of gradient descent.

Remark. When the learning rate decreases at an appropriate rate, and subject to relatively mild assumptions, stochastic gradient descent converges almost surely to local minimum (Bottou, 1998). ◇

Why should one consider using an approximate gradient? A major reason is practical implementation constraints, such as the size of central processing unit (CPU)/graphics processing unit (GPU) memory or limits on computational time. We can think of the size of the subset used to estimate the gradient in the same way that we thought of the size of a sample when estimating empirical means (Section 6.4.1). Large minibatch sizes will provide accurate estimates of the gradient, reducing the variance in the parameter update. Furthermore, large minibatches take advantage of highly optimized matrix operations in vectorized implementations of the cost and gradient. The reduction in variance leads to more stable convergence, but each gradient calculation will be more expensive.

In contrast, small minibatches are quick to estimate. If we keep the minibatch size small, the noise in our gradient estimate will allow us to get out of some bad local optima, which we may otherwise get stuck in. In machine learning, optimization methods are used for training by minimizing an objective function on the training data, but the overall goal is to improve generalization performance (Chapter 8). Since the goal in machine learning does not necessarily need a precise estimate of the minimum of the objective function, approximate gradients using minibatch approaches have been widely used. Stochastic gradient descent is very effective in large-scale machine learning problems (Bottou et al., 2018), such as training deep neural networks on millions of images (Dean et al., 2012), topic models (Hoffman et al., 2013), reinforcement learning (Mnih et al., 2015), or training of large-scale Gaussian process models (Hensman et al., 2013; Gal et al., 2014).

7.2 Constrained Optimization and Lagrange Multipliers

In the previous section, we considered the problem of solving for the minimum of a function

$$\min_{\boldsymbol{x}} f(\boldsymbol{x}), \tag{7.16}$$

where $f : \mathbb{R}^D \to \mathbb{R}$.

In this section, we have additional constraints. That is, for real-valued functions $g_i : \mathbb{R}^D \to \mathbb{R}$ for $i = 1, \ldots, m$, we consider the constrained optimization problem (Figure 7.4)

$$\begin{aligned} \min_{\boldsymbol{x}} \quad & f(\boldsymbol{x}) \\ \text{subject to} \quad & g_i(\boldsymbol{x}) \leqslant 0 \quad \text{for all} \quad i = 1, \ldots, m\,. \end{aligned} \tag{7.17}$$

It is worth pointing out that the functions f and g_i could be nonconvex in general, and we will consider the convex case in the next section.

One obvious, but not very practical, way of converting the constrained problem (7.17) into an unconstrained one is to use an indicator function

$$J(\boldsymbol{x}) = f(\boldsymbol{x}) + \sum_{i=1}^{m} \mathbf{1}(g_i(\boldsymbol{x}))\,, \tag{7.18}$$

where $\mathbf{1}(z)$ is an infinite step function

$$\mathbf{1}(z) = \begin{cases} 0 & \text{if } z \leqslant 0 \\ \infty & \text{otherwise} \end{cases}. \tag{7.19}$$

This gives infinite penalty if the constraint is not satisfied, and hence would provide the same solution. However, this infinite step function is equally difficult to

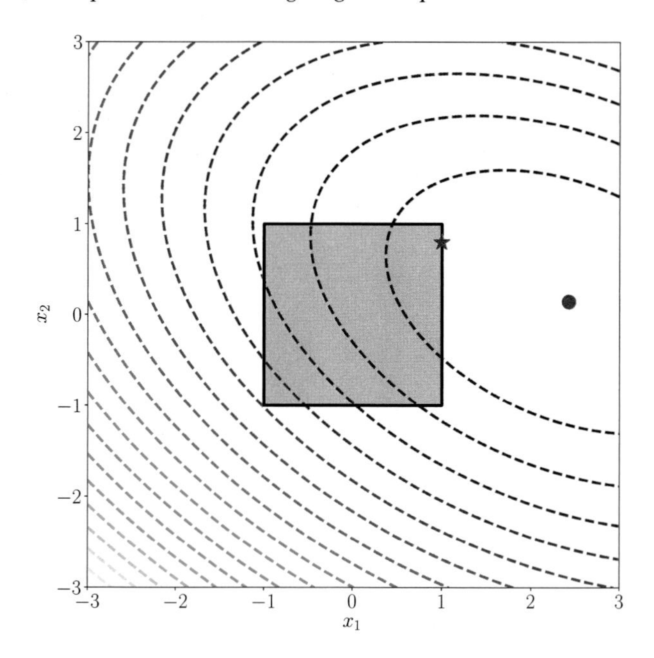

Figure 7.4 Illustration of constrained optimization. The unconstrained problem (indicated by the contour lines) has a minimum on the right side (indicated by the circle). The box constraints ($-1 \leqslant x \leqslant 1$ and $-1 \leqslant y \leqslant 1$) require that the optimal solution is within the box, resulting in an optimal value indicated by the star.

optimize. We can overcome this difficulty by introducing *Lagrange multipliers*. The idea of Lagrange multipliers is to replace the step function with a linear function.

Lagrange multiplier

We associate to problem (7.17) the *Lagrangian* by introducing the Lagrange multipliers $\lambda_i \geqslant 0$ corresponding to each inequality constraint respectively (Boyd and Vandenberghe, 2004, chapter 4) so that

Lagrangian

$$\mathfrak{L}(\boldsymbol{x}, \boldsymbol{\lambda}) = f(\boldsymbol{x}) + \sum_{i=1}^{m} \lambda_i g_i(\boldsymbol{x}) \tag{7.20a}$$

$$= f(\boldsymbol{x}) + \boldsymbol{\lambda}^\top \boldsymbol{g}(\boldsymbol{x}), \tag{7.20b}$$

where in the last line we have a concatenated all constraints $g_i(\boldsymbol{x})$ into a vector $\boldsymbol{g}(\boldsymbol{x})$, and all the Lagrange multipliers into a vector $\boldsymbol{\lambda} \in \mathbb{R}^m$.

We now introduce the idea of Lagrangian duality. In general, duality in optimization is the idea of converting an optimization problem in one set of variables $\boldsymbol{x}$ (called the primal variables) into another optimization problem in a different set of variables $\boldsymbol{\lambda}$ (called the dual variables). We introduce two different approaches to duality: In this section, we discuss Lagrangian duality; in Section 7.3.3, we discuss Legendre–Fenchel duality.

Definition 7.1. The problem in (7.17)

$$\min_{\boldsymbol{x}} \quad f(\boldsymbol{x}) \tag{7.21}$$
$$\text{subject to} \quad g_i(\boldsymbol{x}) \leqslant 0 \quad \text{for all} \quad i = 1, \dots, m$$

is known as the *primal problem*, corresponding to the primal variables x. The associated *Lagrangian dual problem* is given by

primal problem

Lagrangian dual problem

$$\max_{\boldsymbol{\lambda} \in \mathbb{R}^m} \quad \mathfrak{D}(\boldsymbol{\lambda})$$
$$\text{subject to} \quad \boldsymbol{\lambda} \geqslant \boldsymbol{0}, \tag{7.22}$$

where $\boldsymbol{\lambda}$ are the dual variables and $\mathfrak{D}(\boldsymbol{\lambda}) = \min_{\boldsymbol{x} \in \mathbb{R}^d} \mathfrak{L}(\boldsymbol{x}, \boldsymbol{\lambda})$.

Remark. In the discussion of Definition 7.1, we use two concepts that are also of independent interest (Boyd and Vandenberghe, 2004).

minimax inequality

First is the *minimax inequality*, which says that for any function with two arguments $\varphi(\boldsymbol{x}, \boldsymbol{y})$, the maximin is less than the minimax, i.e.,

$$\max_{\boldsymbol{y}} \min_{\boldsymbol{x}} \varphi(\boldsymbol{x}, \boldsymbol{y}) \leqslant \min_{\boldsymbol{x}} \max_{\boldsymbol{y}} \varphi(\boldsymbol{x}, \boldsymbol{y}) . \tag{7.23}$$

This inequality can be proved by considering the inequality

$$\text{For all } \boldsymbol{x}, \boldsymbol{y} \qquad \min_{\boldsymbol{x}} \varphi(\boldsymbol{x}, \boldsymbol{y}) \leqslant \max_{\boldsymbol{y}} \varphi(\boldsymbol{x}, \boldsymbol{y}) . \tag{7.24}$$

Note that taking the maximum over $\boldsymbol{y}$ of the left-hand side of (7.24) maintains the inequality since the inequality is true for all $\boldsymbol{y}$. Similarly, we can take the minimum over $\boldsymbol{x}$ of the right-hand side of (7.24) to obtain (7.23).

weak duality

The second concept is *weak duality*, which uses (7.23) to show that primal values are always greater than or equal to dual values. This is described in more detail in (7.27). ♢

Recall that the difference between $J(\boldsymbol{x})$ in (7.18) and the Lagrangian in (7.20b) is that we have relaxed the indicator function to a linear function. Therefore, when $\lambda \geqslant 0$, the Lagrangian $\mathfrak{L}(\boldsymbol{x}, \boldsymbol{\lambda})$ is a lower bound of $J(\boldsymbol{x})$. Hence, the maximum of $\mathfrak{L}(\boldsymbol{x}, \boldsymbol{\lambda})$ with respect to $\boldsymbol{\lambda}$ is

$$J(\boldsymbol{x}) = \max_{\boldsymbol{\lambda} \geqslant \mathbf{0}} \mathfrak{L}(\boldsymbol{x}, \boldsymbol{\lambda}). \tag{7.25}$$

Recall that the original problem was minimizing $J(\boldsymbol{x})$,

$$\min_{\boldsymbol{x} \in \mathbb{R}^d} \max_{\boldsymbol{\lambda} \geqslant \mathbf{0}} \mathfrak{L}(\boldsymbol{x}, \boldsymbol{\lambda}) . \tag{7.26}$$

By the minimax inequality (7.23), it follows that swapping the order of the minimum and maximum results in a smaller value, i.e.,

$$\min_{\boldsymbol{x} \in \mathbb{R}^d} \max_{\boldsymbol{\lambda} \geqslant \mathbf{0}} \mathfrak{L}(\boldsymbol{x}, \boldsymbol{\lambda}) \geqslant \max_{\boldsymbol{\lambda} \geqslant \mathbf{0}} \min_{\boldsymbol{x} \in \mathbb{R}^d} \mathfrak{L}(\boldsymbol{x}, \boldsymbol{\lambda}) . \tag{7.27}$$

weak duality

This is also known as *weak duality*. Note that the inner part of the right-hand side is the dual objective function $\mathfrak{D}(\boldsymbol{\lambda})$ and the definition follows.

In contrast to the original optimization problem, which has constraints, $\min_{\boldsymbol{x} \in \mathbb{R}^d} \mathfrak{L}(\boldsymbol{x}, \boldsymbol{\lambda})$ is an unconstrained optimization problem for a given value of $\boldsymbol{\lambda}$. If solving $\min_{\boldsymbol{x} \in \mathbb{R}^d} \mathfrak{L}(\boldsymbol{x}, \boldsymbol{\lambda})$ is easy, then the overall problem is easy to solve. The reason is that the outer problem (maximization over $\boldsymbol{\lambda}$) is a maximum over a set of affine functions, and hence is a concave function, even though $f(\cdot)$ and $g_i(\cdot)$ may be nonconvex. The maximum of a concave function can be efficiently computed.

Assuming $f(\cdot)$ and $g_i(\cdot)$ are differentiable, we find the Lagrange dual problem by differentiating the Lagrangian with respect to $\boldsymbol{x}$, setting the differential to zero, and solving for the optimal value. We will discuss two concrete examples in Sections 7.3.1 and 7.3.2, where $f(\cdot)$ and $g_i(\cdot)$ are convex.

Remark (Equality Constraints). Consider (7.17) with additional equality constraints

$$\begin{aligned} \min_{\boldsymbol{x}} \quad & f(\boldsymbol{x}) \\ \text{subject to} \quad & g_i(\boldsymbol{x}) \leqslant 0 \quad \text{for all} \quad i = 1, \dots, m \\ & h_j(\boldsymbol{x}) = 0 \quad \text{for all} \quad j = 1, \dots, n \,. \end{aligned} \tag{7.28}$$

We can model equality constraints by replacing them with two inequality constraints. That is, for each equality constraint $h_j(\boldsymbol{x}) = 0$ we equivalently replace it by two constraints $h_j(\boldsymbol{x}) \leqslant 0$ and $h_j(\boldsymbol{x}) \geqslant 0$. It turns out that the resulting Lagrange multipliers are then unconstrained.

Therefore, we constrain the Lagrange multipliers corresponding to the inequality constraints in (7.28) to be nonnegative, and leave the Lagrange multipliers corresponding to the equality constraints unconstrained. ◇

7.3 Convex Optimization

We focus our attention of a particularly useful class of optimization problems, where we can guarantee global optimality. When $f(\cdot)$ is a convex function, and when the constraints involving $g(\cdot)$ and $h(\cdot)$ are convex sets, this is called a *convex optimization problem*. In this setting, we have *strong duality*: The optimal solution of the dual problem is the same as the optimal solution of the primal problem. The distinction between convex functions and convex sets are often not strictly presented in machine learning literature, but one can often infer the implied meaning from context.

convex optimization problem

strong duality

Definition 7.2. A set $\mathcal{C}$ is a *convex set* if for any $x, y \in \mathcal{C}$ and for any scalar θ with $0 \leqslant \theta \leqslant 1$, we have

convex set

$$\theta x + (1 - \theta) y \in \mathcal{C} \,. \tag{7.29}$$

Convex sets are sets such that a straight line connecting any two elements of the set lie inside the set. Figures 7.5 and 7.6 illustrate convex and nonconvex sets, respectively.

Figure 7.5 Example of a convex set.

Convex functions are functions such that a straight line between any two points of the function lie above the function. Figure 7.2 shows a nonconvex function, and Figure 7.3 shows a convex function. Another convex function is shown in Figure 7.7.

Figure 7.6 Example of a nonconvex set.

Definition 7.3. Let function $f : \mathbb{R}^D \to \mathbb{R}$ be a function whose domain is a convex set. The function f is a *convex function* if for all $\boldsymbol{x}, \boldsymbol{y}$ in the domain of f, and for any scalar θ with $0 \leqslant \theta \leqslant 1$, we have

$$f(\theta \boldsymbol{x} + (1 - \theta) \boldsymbol{y}) \leqslant \theta f(\boldsymbol{x}) + (1 - \theta) f(\boldsymbol{y}) \,. \tag{7.30}$$

Remark. A *concave function* is the negative of a convex function. ◇

convex function

concave function

The constraints involving $g(\cdot)$ and $h(\cdot)$ in (7.28) truncate functions at a scalar value, resulting in sets. Another relation between convex functions and convex sets is to consider the set obtained by "filling in" a convex function. A convex function is a bowllike object, and we imagine pouring water into it to fill it

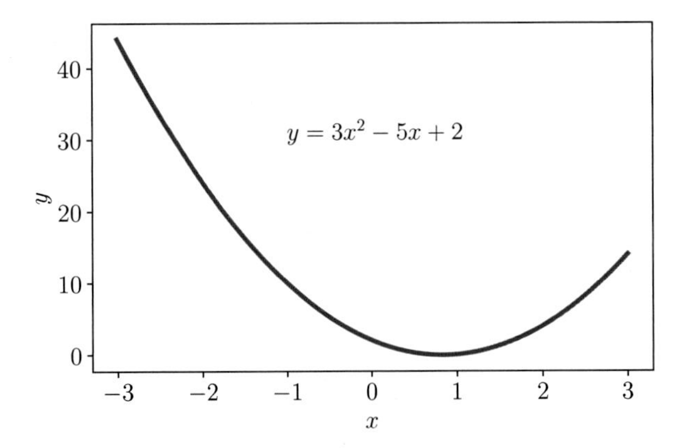

Figure 7.7 Example of a convex function.

epigraph

up. This resulting filled-in set, called the *epigraph* of the convex function, is a convex set.

If a function $f : \mathbb{R}^n \to \mathbb{R}$ is differentiable, we can specify convexity in terms of its gradient $\nabla_{\boldsymbol{x}} f(\boldsymbol{x})$ (Section 5.2). A function $f(\boldsymbol{x})$ is convex if and only if for any two points $\boldsymbol{x}, \boldsymbol{y}$ it holds that

$$f(\boldsymbol{y}) \geqslant f(\boldsymbol{x}) + \nabla_{\boldsymbol{x}} f(\boldsymbol{x})^\top (\boldsymbol{y} - \boldsymbol{x}) \,. \tag{7.31}$$

If we further know that a function $f(\boldsymbol{x})$ is twice differentiable, that is, the Hessian (5.147) exists for all values in the domain of $\boldsymbol{x}$, then the function $f(\boldsymbol{x})$ is convex if and only if $\nabla^2_{\boldsymbol{x}} f(\boldsymbol{x})$ is positive semidefinite (Boyd and Vandenberghe, 2004).

Example 7.3

The negative entropy $f(x) = x \log_2 x$ is convex for $x > 0$. A visualization of the function is shown in Figure 7.8, and we can see that the function is convex. To illustrate the previous definitions of convexity, let us check the calculations for two points $x = 2$ and $x = 4$. Note that to prove convexity of $f(x)$ we would need to check for all points $x \in \mathbb{R}$.

Recall Definition 7.3. Consider a point midway between the two points (that is, $\theta = 0.5$); then the left-hand side is $f(0.5 \cdot 2 + 0.5 \cdot 4) = 3 \log_2 3 \approx 4.75$. The right-hand side is $0.5(2 \log_2 2) + 0.5(4 \log_2 4) = 1 + 4 = 5$. And therefore the definition is satisfied.

Since $f(x)$ is differentiable, we can alternatively use (7.31). Calculating the derivative of $f(x)$, we obtain

$$\nabla_x (x \log_2 x) = 1 \cdot \log_2 x + x \cdot \frac{1}{x \log_e 2} = \log_2 x + \frac{1}{\log_e 2} \,. \tag{7.32}$$

Using the same two test points $x = 2$ and $x = 4$, the left-hand side of (7.31) is given by $f(4) = 8$. The right-hand side is

$$\begin{aligned} f(\boldsymbol{x}) + \nabla_{\boldsymbol{x}}^\top (\boldsymbol{y} - \boldsymbol{x}) &= f(2) + \nabla f(2) \cdot (4 - 2) & \text{(7.33a)} \\ &= 2 + (1 + \frac{1}{\log_e 2}) \cdot 2 \approx 6.9 \,. & \text{(7.33b)} \end{aligned}$$

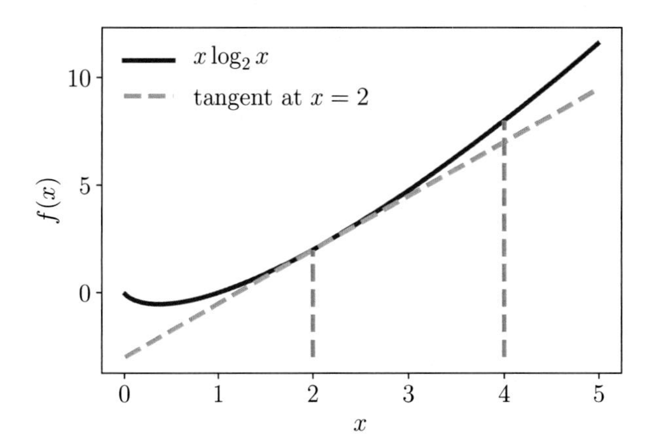

Figure 7.8 The negative entropy function (which is convex) and its tangent at $x = 2$.

We can check that a function or set is convex from first principles by recalling the definitions. In practice, we often rely on operations that preserve convexity to check that a particular function or set is convex. Although the details are vastly different, this is again the idea of closure that we introduced in Chapter 2 for vector spaces.

Example 7.4

A nonnegative weighted sum of convex functions is convex. Observe that if f is a convex function, and $\alpha \geqslant 0$ is a nonnegative scalar, then the function αf is convex. We can see this by multiplying α to both sides of the equation in Definition 7.3, and recalling that multiplying a nonnegative number does not change the inequality.

If f_1 and f_2 are convex functions, then we have by the definition

$$f_1(\theta \boldsymbol{x} + (1-\theta)\boldsymbol{y}) \leqslant \theta f_1(\boldsymbol{x}) + (1-\theta) f_1(\boldsymbol{y}) \tag{7.34}$$

$$f_2(\theta \boldsymbol{x} + (1-\theta)\boldsymbol{y}) \leqslant \theta f_2(\boldsymbol{x}) + (1-\theta) f_2(\boldsymbol{y}) . \tag{7.35}$$

Summing up both sides gives us

$$\begin{aligned} & f_1(\theta \boldsymbol{x} + (1-\theta)\boldsymbol{y}) + f_2(\theta \boldsymbol{x} + (1-\theta)\boldsymbol{y}) \\ & \leqslant \theta f_1(\boldsymbol{x}) + (1-\theta) f_1(\boldsymbol{y}) + \theta f_2(\boldsymbol{x}) + (1-\theta) f_2(\boldsymbol{y}) , \end{aligned} \tag{7.36}$$

where the right-hand side can be rearranged to

$$\theta(f_1(\boldsymbol{x}) + f_2(\boldsymbol{x})) + (1-\theta)(f_1(\boldsymbol{y}) + f_2(\boldsymbol{y})) , \tag{7.37}$$

completing the proof that the sum of convex functions is convex.

Combining the preceding two facts, we see that $\alpha f_1(\boldsymbol{x}) + \beta f_2(\boldsymbol{x})$ is convex for $\alpha, \beta \geqslant 0$. This closure property can be extended using a similar argument for nonnegative weighted sums of more than two convex functions.

Remark. The inequality in (7.30) is sometimes called *Jensen's inequality*. In fact, a whole class of inequalities for taking nonnegative weighted sums of convex functions are all called Jensen's inequality. ◇

Jensen's inequality

convex optimization problem

In summary, a constrained optimization problem is called a *convex optimization problem* if

$$\begin{aligned} &\min_{\boldsymbol{x}} f(\boldsymbol{x}) \\ &\text{subject to} g_i(\boldsymbol{x}) \leqslant 0 \quad \text{for all} \quad i = 1, \dots, m \\ &h_j(\boldsymbol{x}) = 0 \quad \text{for all} \quad j = 1, \dots, n, \end{aligned} \tag{7.38}$$

where all functions $f(\boldsymbol{x})$ and $g_i(\boldsymbol{x})$ are convex functions, and all $h_j(\boldsymbol{x}) = 0$ are convex sets. In the following, we will describe two classes of convex optimization problems that are widely used and well understood.

7.3.1 Linear Programming

Linear programs are one of the most widely used approaches in industry.

Consider the special case when all the preceding functions are linear, that is,

$$\begin{aligned} \min_{\boldsymbol{x} \in \mathbb{R}^d} \quad & \boldsymbol{c}^\top \boldsymbol{x} \\ \text{subject to} \quad & \boldsymbol{A}\boldsymbol{x} \leqslant \boldsymbol{b}, \end{aligned} \tag{7.39}$$

linear program

where $\boldsymbol{A} \in \mathbb{R}^{m \times d}$ and $\boldsymbol{b} \in \mathbb{R}^m$. This is known as a *linear program*. It has d variables and m linear constraints. The Lagrangian is given by

$$\mathfrak{L}(\boldsymbol{x}, \boldsymbol{\lambda}) = \boldsymbol{c}^\top \boldsymbol{x} + \boldsymbol{\lambda}^\top (\boldsymbol{A}\boldsymbol{x} - \boldsymbol{b}), \tag{7.40}$$

where $\boldsymbol{\lambda} \in \mathbb{R}^m$ is the vector of nonnegative Lagrange multipliers. Rearranging the terms corresponding to $\boldsymbol{x}$ yields

$$\mathfrak{L}(\boldsymbol{x}, \boldsymbol{\lambda}) = (\boldsymbol{c} + \boldsymbol{A}^\top \boldsymbol{\lambda})^\top \boldsymbol{x} - \boldsymbol{\lambda}^\top \boldsymbol{b}. \tag{7.41}$$

Taking the derivative of $\mathfrak{L}(\boldsymbol{x}, \boldsymbol{\lambda})$ with respect to $\boldsymbol{x}$ and setting it to zero gives us

$$\boldsymbol{c} + \boldsymbol{A}^\top \boldsymbol{\lambda} = \boldsymbol{0}. \tag{7.42}$$

Therefore, the dual Lagrangian is $\mathfrak{D}(\boldsymbol{\lambda}) = -\boldsymbol{\lambda}^\top \boldsymbol{b}$. Recall we would like to maximize $\mathfrak{D}(\boldsymbol{\lambda})$. In addition to the constraint due to the derivative of $\mathfrak{L}(\boldsymbol{x}, \boldsymbol{\lambda})$ being zero, we also have the fact that $\boldsymbol{\lambda} \geqslant \boldsymbol{0}$, resulting in the following dual optimization problem

It is convention to minimize the primal and maximize the dual.

$$\begin{aligned} \max_{\boldsymbol{\lambda} \in \mathbb{R}^m} \quad & -\boldsymbol{b}^\top \boldsymbol{\lambda} \\ \text{subject to} \quad & \boldsymbol{c} + \boldsymbol{A}^\top \boldsymbol{\lambda} = \boldsymbol{0} \\ & \boldsymbol{\lambda} \geqslant \boldsymbol{0}. \end{aligned} \tag{7.43}$$

This is also a linear program, but with m variables. We have the choice of solving the primal (7.39) or the dual (7.43) program depending on whether m or d is larger. Recall that d is the number of variables and m is the number of constraints in the primal linear program.

Example 7.5 (Linear Program)
Consider the linear program

$$\min_{\boldsymbol{x}\in\mathbb{R}^2} \quad -\begin{bmatrix}5\\3\end{bmatrix}^\top \begin{bmatrix}x_1\\x_2\end{bmatrix}$$

$$\text{subject to} \quad \begin{bmatrix}2 & 2\\ 2 & -4\\ -2 & 1\\ 0 & -1\\ 0 & 1\end{bmatrix}\begin{bmatrix}x_1\\x_2\end{bmatrix} \leqslant \begin{bmatrix}33\\8\\5\\-1\\8\end{bmatrix} \tag{7.44}$$

with two variables. This program is also shown in Figure 7.9. The objective function is linear, resulting in linear contour lines. The constraint set in standard form is translated into the legend. The optimal value must lie in the shaded (feasible) region and is indicated by the star.

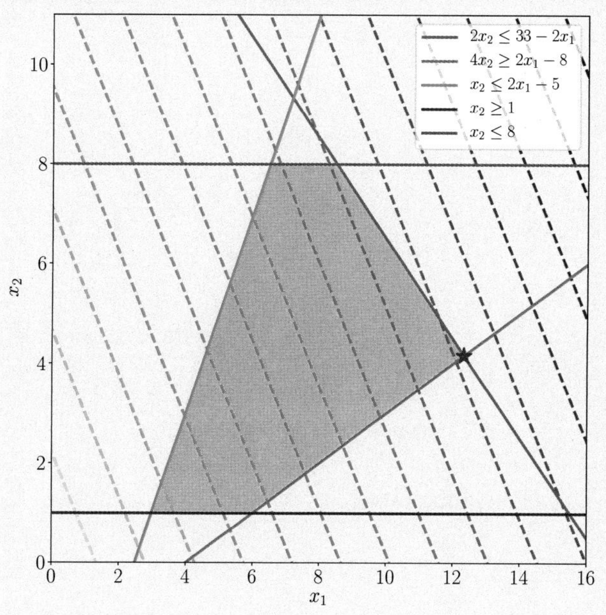

Figure 7.9 Illustration of a linear program. The unconstrained problem (indicated by the contour lines) has a minimum on the right side. The optimal value given the constraints are shown by the star.

7.3.2 Quadratic Programming

Consider the case of a convex quadratic objctive function, where the constraints are affine, i.e.,

$$\min_{\boldsymbol{x}\in\mathbb{R}^d} \quad \frac{1}{2}\boldsymbol{x}^\top \boldsymbol{Q}\boldsymbol{x} + \boldsymbol{c}^\top \boldsymbol{x} \tag{7.45}$$

$$\text{subject to} \quad \boldsymbol{A}\boldsymbol{x} \leqslant \boldsymbol{b}\,,$$

where $\boldsymbol{A} \in \mathbb{R}^{m\times d}$, $\boldsymbol{b} \in \mathbb{R}^m$, and $\boldsymbol{c} \in \mathbb{R}^d$. The square symmetric matrix $\boldsymbol{Q} \in \mathbb{R}^{d\times d}$ is positive definite, and therefore the objective function is convex. This is known as a *quadratic program*. Observe that it has d variables and m linear constraints.

Example 7.6 (Quadratic Program)
Consider the quadratic program

$$\min_{\boldsymbol{x}\in\mathbb{R}^2} \quad \frac{1}{2}\begin{bmatrix}x_1\\x_2\end{bmatrix}^\top \begin{bmatrix}2 & 1\\1 & 4\end{bmatrix}\begin{bmatrix}x_1\\x_2\end{bmatrix} + \begin{bmatrix}5\\3\end{bmatrix}^\top \begin{bmatrix}x_1\\x_2\end{bmatrix} \tag{7.46}$$

$$\text{subject to} \quad \begin{bmatrix}1 & 0\\-1 & 0\\0 & 1\\0 & -1\end{bmatrix}\begin{bmatrix}x_1\\x_2\end{bmatrix} \leqslant \begin{bmatrix}1\\1\\1\\1\end{bmatrix} \tag{7.47}$$

of two variables. The program is also illustrated in Figure 7.4. The objective function is quadratic with a positive semidefinite matrix $\boldsymbol{Q}$, resulting in elliptical contour lines. The optimal value must lie in the shaded (feasible) region and is indicated by the star.

The Lagrangian is given by

$$\begin{aligned}\mathfrak{L}(\boldsymbol{x},\boldsymbol{\lambda}) &= \frac{1}{2}\boldsymbol{x}^\top\boldsymbol{Q}\boldsymbol{x} + \boldsymbol{c}^\top\boldsymbol{x} + \boldsymbol{\lambda}^\top(\boldsymbol{A}\boldsymbol{x}-\boldsymbol{b}) & (7.48a)\\ &= \frac{1}{2}\boldsymbol{x}^\top\boldsymbol{Q}\boldsymbol{x} + (\boldsymbol{c}+\boldsymbol{A}^\top\boldsymbol{\lambda})^\top\boldsymbol{x} - \boldsymbol{\lambda}^\top\boldsymbol{b}, & (7.48b)\end{aligned}$$

where again we have rearranged the terms. Taking the derivative of $\mathfrak{L}(\boldsymbol{x},\boldsymbol{\lambda})$ with respect to $\boldsymbol{x}$ and setting it to zero gives

$$\boldsymbol{Q}\boldsymbol{x} + (\boldsymbol{c}+\boldsymbol{A}^\top\boldsymbol{\lambda}) = \boldsymbol{0}. \tag{7.49}$$

Assuming that $\boldsymbol{Q}$ is invertible, we get

$$\boldsymbol{x} = -\boldsymbol{Q}^{-1}(\boldsymbol{c}+\boldsymbol{A}^\top\boldsymbol{\lambda}). \tag{7.50}$$

Substituting (7.50) into the primal Lagrangian $\mathfrak{L}(\boldsymbol{x},\boldsymbol{\lambda})$, we get the dual Lagrangian

$$\mathfrak{D}(\boldsymbol{\lambda}) = -\frac{1}{2}(\boldsymbol{c}+\boldsymbol{A}^\top\boldsymbol{\lambda})^\top\boldsymbol{A}\boldsymbol{Q}^{-1}(\boldsymbol{c}+\boldsymbol{A}^\top\boldsymbol{\lambda}) - \boldsymbol{\lambda}^\top\boldsymbol{b}. \tag{7.51}$$

Therefore, the dual optimization problem is given by

$$\begin{aligned}\max_{\boldsymbol{\lambda}\in\mathbb{R}^m} \quad & -\frac{1}{2}(\boldsymbol{c}+\boldsymbol{A}^\top\boldsymbol{\lambda})^\top\boldsymbol{A}\boldsymbol{Q}^{-1}(\boldsymbol{c}+\boldsymbol{A}^\top\boldsymbol{\lambda}) - \boldsymbol{\lambda}^\top\boldsymbol{b}\\ \text{subject to} \quad & \boldsymbol{\lambda} \geqslant \boldsymbol{0}.\end{aligned} \tag{7.52}$$

We will see an application of quadratic programming in machine learning in Chapter 12.

7.3.3 Legendre–Fenchel Transform and Convex Conjugate

Let us revisit the idea of duality from Section 7.2, without considering constraints. One useful fact about a convex set is that it can be equivalently described by its supporting hyperplanes. A hyperplane is called a *supporting hyperplane* of a convex set if it intersects the convex set, and the convex set is contained on just one side of it. Recall that we can fill up a convex function to obtain the epigraph, which is a convex set. Therefore, we can also describe convex functions in terms of their supporting hyperplanes. Furthermore, observe that the supporting hyperplane just touches the convex function, and is in fact the tangent to the function at that point. And recall that the tangent of a function $f(\boldsymbol{x})$ at a given point $\boldsymbol{x}_0$ is the evaluation of the gradient of that function at that point $\left.\frac{\mathrm{d}f(\boldsymbol{x})}{\mathrm{d}\boldsymbol{x}}\right|_{\boldsymbol{x}=\boldsymbol{x}_0}$. In summary, because convex sets can be equivalently described by its supporting hyperplanes, convex functions can be equivalently described by a function of their gradient. The *Legendre transform* formalizes this concept.

supporting hyperplane

Legendre transform

We begin with the most general definition, which unfortunately has a counterintuitive form, and look at special cases to relate the definition to the intuition described in the preceding paragraph. The *Legendre–Fenchel transform* is a transformation (in the sense of a Fourier transform) from a convex differentiable function $f(\boldsymbol{x})$ to a function that depends on the tangents $s(\boldsymbol{x}) = \nabla_{\boldsymbol{x}} f(\boldsymbol{x})$. It is worth stressing that this is a transformation of the function $f(\cdot)$ and not the variable $\boldsymbol{x}$ or the function evaluated at $\boldsymbol{x}$. The Legendre–Fenchel transform is also known as the *convex conjugate* (for reasons we will see soon) and is closely related to duality (Hiriart-Urruty and Lemaréchal, 2001, chapter 5).

Physics students are often introduced to the Legendre transform as relating the Lagrangian and the Hamiltonian in classical mechanics.

Legendre–Fenchel transform

convex conjugate

Definition 7.4. The *convex conjugate* of a function $f : \mathbb{R}^D \to \mathbb{R}$ is a function f^* defined by

convex conjugate

$$f^*(\boldsymbol{s}) = \sup_{\boldsymbol{x} \in \mathbb{R}^D} \left(\langle \boldsymbol{s}, \boldsymbol{x} \rangle - f(\boldsymbol{x}) \right) . \tag{7.53}$$

Note that the preceding convex conjugate definition does not need the function f to be convex nor differentiable. In Definition 7.4, we have used a general inner product (Section 3.2) but in the rest of this section we will consider the standard dot product between finite-dimensional vectors ($\langle \boldsymbol{s}, \boldsymbol{x} \rangle = \boldsymbol{s}^\top \boldsymbol{x}$) to avoid too many technical details.

To understand Definition 7.4 in a geometric fashion, consider an nice simple one-dimensional convex and differentiable function, for example $f(x) = x^2$. Note that since we are looking at a one-dimensional problem, hyperplanes reduce to a line. Consider a line $y = sx + c$. Recall that we are able to describe convex functions by their supporting hyperplanes, so let us try to describe this function $f(x)$ by its supporting lines. Fix the gradient of the line $s \in \mathbb{R}$ and for each point $(x_0, f(x_0))$ on the graph of f, find the minimum value of c such that the line still intersects $(x_0, f(x_0))$. Note that the minimum value of c is the place where a line with slope s "just touches" the function $f(x) = x^2$. The line passing through $(x_0, f(x_0))$ with gradient s is given by

This derivation is easiest to understand by drawing the reasoning as it progresses.

$$y - f(x_0) = s(x - x_0) . \tag{7.54}$$

The y-intercept of this line is $-sx_0 + f(x_0)$. The minimum of c for which $y = sx + c$ intersects with the graph of f is therefore

$$\inf_{x_0} -sx_0 + f(x_0). \tag{7.55}$$

The preceding convex conjugate is by convention defined to be the negative of this. The reasoning in this paragraph did not rely on the fact that we chose a one-dimensional convex and differentiable function, and holds for $f : \mathbb{R}^D \to \mathbb{R}$, which are nonconvex and nondifferentiable.

The classical Legendre transform is defined on convex differentiable functions in $\mathbb{R}^D$.

Remark. Convex differentiable functions such as the example $f(x) = x^2$ is a nice special case, where there is no need for the supremum, and there is a one-to-one correspondence between a function and its Legendre transform. Let us derive this from first principles. For a convex differentiable function, we know that at x_0 the tangent touches $f(x_0)$ so that

$$f(x_0) = sx_0 + c. \tag{7.56}$$

Recall that we want to describe the convex function $f(x)$ in terms of its gradient $\nabla_x f(x)$, and that $s = \nabla_x f(x_0)$. We rearrange to get an expression for $-c$ to obtain

$$-c = sx_0 - f(x_0). \tag{7.57}$$

Note that $-c$ changes with x_0 and therefore with s, which is why we can think of it as a function of s, which we call

$$f^*(s) := sx_0 - f(x_0). \tag{7.58}$$

Comparing (7.58) with Definition 7.4, we see that (7.58) is a special case (without the supremum). ◇

The conjugate function has nice properties; for example, for convex functions, applying the Legendre transform again gets us back to the original function. In the same way that the slope of $f(x)$ is s, the slope of $f^*(s)$ is x. The following two examples show common uses of convex conjugates in machine learning.

Example 7.7 (Convex Conjugates)
To illustrate the application of convex conjugates, consider the quadratic function

$$f(\boldsymbol{y}) = \frac{\lambda}{2}\boldsymbol{y}^\top \boldsymbol{K}^{-1}\boldsymbol{y} \tag{7.59}$$

based on a positive definite matrix $\boldsymbol{K} \in \mathbb{R}^{n\times n}$. We denote the primal variable to be $\boldsymbol{y} \in \mathbb{R}^n$ and the dual variable to be $\boldsymbol{\alpha} \in \mathbb{R}^n$.

Applying Definition 7.4, we obtain the function

$$f^*(\boldsymbol{\alpha}) = \sup_{\boldsymbol{y}\in\mathbb{R}^n} \langle \boldsymbol{y}, \boldsymbol{\alpha}\rangle - \frac{\lambda}{2}\boldsymbol{y}^\top \boldsymbol{K}^{-1}\boldsymbol{y}. \tag{7.60}$$

Since the function is differentiable, we can find the maximum by taking the derivative and with respect to $\boldsymbol{y}$ setting it to zero.

$$\frac{\partial\left[\langle \boldsymbol{y}, \boldsymbol{\alpha}\rangle - \frac{\lambda}{2}\boldsymbol{y}^\top \boldsymbol{K}^{-1}\boldsymbol{y}\right]}{\partial \boldsymbol{y}} = (\boldsymbol{\alpha} - \lambda \boldsymbol{K}^{-1}\boldsymbol{y})^\top \tag{7.61}$$

and hence when the gradient is zero we have $\boldsymbol{y} = \frac{1}{\lambda}\boldsymbol{K}\boldsymbol{\alpha}$. Substituting into (7.60) yields

$$f^*(\boldsymbol{\alpha}) = \frac{1}{\lambda}\boldsymbol{\alpha}^\top \boldsymbol{K}\boldsymbol{\alpha} - \frac{\lambda}{2}\left(\frac{1}{\lambda}\boldsymbol{K}\boldsymbol{\alpha}\right)^\top \boldsymbol{K}^{-1}\left(\frac{1}{\lambda}\boldsymbol{K}\boldsymbol{\alpha}\right) = \frac{1}{2\lambda}\boldsymbol{\alpha}^\top \boldsymbol{K}\boldsymbol{\alpha}. \tag{7.62}$$

Example 7.8

In machine learning, we often use sums of functions; for example, the objective function of the training set includes a sum of the losses for each example in the training set. In the following, we derive the convex conjugate of a sum of losses $\ell(t)$, where $\ell : \mathbb{R} \to \mathbb{R}$. This also illustrates the application of the convex conjugate to the vector case. Let $\mathcal{L}(\boldsymbol{t}) = \sum_{i=1}^n \ell_i(t_i)$. Then,

$$\begin{aligned}
\mathcal{L}^*(\boldsymbol{z}) &= \sup_{\boldsymbol{t}\in\mathbb{R}^n} \langle \boldsymbol{z}, \boldsymbol{t}\rangle - \sum_{i=1}^n \ell_i(t_i) && (7.63a)\\
&= \sup_{\boldsymbol{t}\in\mathbb{R}^n} \sum_{i=1}^n z_i t_i - \ell_i(t_i) && \text{definition of dot product} \quad (7.63b)\\
&= \sum_{i=1}^n \sup_{\boldsymbol{t}\in\mathbb{R}^n} z_i t_i - \ell_i(t_i) && (7.63c)\\
&= \sum_{i=1}^n \ell_i^*(z_i). && \text{definition of conjugate} \quad (7.63d)
\end{aligned}$$

Recall that in Section 7.2 we derived a dual optimization problem using Lagrange multipliers. Furthermore, for convex optimization problems we have strong duality, that is, the solutions of the primal and dual problem match. The Legendre–Fenchel transform described here also can be used to derive a dual optimization problem. Furthermore, when the function is convex and differentiable, the supremum is unique. To further investigate the relation between these two approaches, let us consider a linear equality constrained convex optimization problem.

Example 7.9

Let $f(\boldsymbol{y})$ and $g(\boldsymbol{x})$ be convex functions, and $\boldsymbol{A}$ a real matrix of appropriate dimensions such that $\boldsymbol{A}\boldsymbol{x} = \boldsymbol{y}$. Then

$$\min_{\boldsymbol{x}} f(\boldsymbol{A}\boldsymbol{x}) + g(\boldsymbol{x}) = \min_{\boldsymbol{A}\boldsymbol{x}=\boldsymbol{y}} f(\boldsymbol{y}) + g(\boldsymbol{x}). \tag{7.64}$$

By introducing the Lagrange multiplier $\boldsymbol{u}$ for the constraints $\boldsymbol{A}\boldsymbol{x} = \boldsymbol{y}$,

$$\begin{aligned}
\min_{\boldsymbol{A}\boldsymbol{x}=\boldsymbol{y}} f(\boldsymbol{y}) + g(\boldsymbol{x}) &= \min_{\boldsymbol{x},\boldsymbol{y}} \max_{\boldsymbol{u}} f(\boldsymbol{y}) + g(\boldsymbol{x}) + (\boldsymbol{A}\boldsymbol{x} - \boldsymbol{y})^\top \boldsymbol{u} && (7.65a)\\
&= \max_{\boldsymbol{u}} \min_{\boldsymbol{x},\boldsymbol{y}} f(\boldsymbol{y}) + g(\boldsymbol{x}) + (\boldsymbol{A}\boldsymbol{x} - \boldsymbol{y})^\top \boldsymbol{u}, && (7.65b)
\end{aligned}$$

where the last step of swapping max and min is due to the fact that $f(\boldsymbol{y})$ and $g(\boldsymbol{x})$ are convex functions. By splitting up the dot product term and collecting $\boldsymbol{x}$ and $\boldsymbol{y}$,

$$\max_{\boldsymbol{u}} \min_{\boldsymbol{x},\boldsymbol{y}} f(\boldsymbol{y}) + g(\boldsymbol{x}) + (\boldsymbol{A}\boldsymbol{x} - \boldsymbol{y})^\top \boldsymbol{u} \tag{7.66a}$$

$$= \max_{\boldsymbol{u}} \left[\min_{\boldsymbol{y}} -\boldsymbol{y}^\top \boldsymbol{u} + f(\boldsymbol{y}) \right] + \left[\min_{\boldsymbol{x}} (\boldsymbol{A}\boldsymbol{x})^\top \boldsymbol{u} + g(\boldsymbol{x}) \right] \tag{7.66b}$$

$$= \max_{\boldsymbol{u}} \left[\min_{\boldsymbol{y}} -\boldsymbol{y}^\top \boldsymbol{u} + f(\boldsymbol{y}) \right] + \left[\min_{\boldsymbol{x}} \boldsymbol{x}^\top \boldsymbol{A}^\top \boldsymbol{u} + g(\boldsymbol{x}) \right] \tag{7.66c}$$

For general inner products, $\boldsymbol{A}^\top$ is replaced by the adjoint $\boldsymbol{A}^*$.

Recall the convex conjugate (Definition 7.4) and the fact that dot products are symmetric,

$$\max_{\boldsymbol{u}} \left[\min_{\boldsymbol{y}} -\boldsymbol{y}^\top \boldsymbol{u} + f(\boldsymbol{y}) \right] + \left[\min_{\boldsymbol{x}} \boldsymbol{x}^\top \boldsymbol{A}^\top \boldsymbol{u} + g(\boldsymbol{x}) \right] \tag{7.67a}$$

$$= \max_{\boldsymbol{u}} -f^*(\boldsymbol{u}) - g^*(-\boldsymbol{A}^\top \boldsymbol{u}). \tag{7.67b}$$

Therefore, we have shown that

$$\min_{\boldsymbol{x}} f(\boldsymbol{A}\boldsymbol{x}) + g(\boldsymbol{x}) = \max_{\boldsymbol{u}} -f^*(\boldsymbol{u}) - g^*(-\boldsymbol{A}^\top \boldsymbol{u}). \tag{7.68}$$

The Legendre–Fenchel conjugate turns out to be quite useful for machine learning problems that can be expressed as convex optimization problems. In particular, for convex loss functions that apply independently to each example, the conjugate loss is a convenient way to derive a dual problem.

7.4 Further Reading

Continuous optimization is an active area of research, and we do not try to provide a comprehensive account of recent advances.

From a gradient descent perspective, there are two major weaknesses which each have their own set of literature. The first challenge is the fact that gradient descent is a first-order algorithm, and does not use information about the curvature of the surface. When there are long valleys, the gradient points perpendicularly to the direction of interest. The idea of momentum can be generalized to a general class of acceleration methods (Nesterov, 2018). Conjugate gradient methods avoid the issues faced by gradient descent by taking previous directions into account (Shewchuk, 1994). Second-order methods such as Newton methods use the Hessian to provide information about the curvature. Many of the choices for choosing step-sizes and ideas like momentum arise by considering the curvature of the objective function (Goh, 2017; Bottou et al., 2018). Quasi-Newton methods such as L-BFGS try to use cheaper computational methods to approximate the Hessian (Nocedal and Wright, 2006). Recently there has been interest in other metrics for computing descent directions, resulting in approaches such as mirror descent (Beck and Teboulle, 2003) and natural gradient (Toussaint, 2012).

The second challenge is to handle nondifferentiable functions. Gradient methods are not well defined when there are kinks in the function. In these cases, *subgradient methods* can be used (Shor, 1985). For further information and algorithms for optimizing nondifferentiable functions, we refer to the book by Bertsekas (1999). There is a vast amount of literature on different approaches for numerically solving continuous optimization problems, including algorithms for constrained optimization problems. Good starting points to appreciate this literature are the books by Luenberger (1969) and Bonnans et al. (2006). A recent survey of continuous optimization is provided by Bubeck (2015).

Modern applications of machine learning often mean that the size of datasets prohibit the use of batch gradient descent, and hence stochastic gradient descent is the current workhorse of large-scale machine learning methods. Recent surveys of the literature include Hazan (2015) and Bottou et al. (2018).

For duality and convex optimization, the book by Boyd and Vandenberghe (2004) includes lectures and slides online. A more mathematical treatment is provided by Bertsekas (2009), and recent book by one of the key researchers in the area of optimization is Nesterov (2018). Convex optimization is based upon convex analysis, and the reader interested in more foundational results about convex functions is referred to Rockafellar (1970), Hiriart-Urruty and Lemaréchal (2001), and Borwein and Lewis (2006). Legendre–Fenchel transforms are also covered in the aforementioned books on convex analysis, but a more beginner-friendly presentation is available at Zia et al. (2009). The role of Legendre–Fenchel transforms in the analysis of convex optimization algorithms is surveyed in Polyak (2016).

Hugo Gonçalves' blog is also a good resource for an easier introduction to Legendre–Fenchel transforms: `https://tinyurl.com/ydaal7hj`

Exercises

7.1 Consider the univariate function

$$f(x) = x^3 + 6x^2 - 3x - 5.$$

Find its stationary points and indicate whether they are maximum, minimum, or saddle points.

7.2 Consider the update equation for stochastic gradient descent (7.15). Write down the update when we use a minibatch size of one.

7.3 Consider whether the following statements are true or false:

a. The intersection of any two convex sets is convex.
b. The union of any two convex sets is convex.
c. The difference of a convex set A from another convex set B is convex.

7.4 Consider whether the following statements are true or false:

a. The sum of any two convex functions is convex.
b. The difference of any two convex functions is convex.
c. The product of any two convex functions is convex.
d. The maximum of any two convex functions is convex.

7.5 Express the following optimization problem as a standard linear program in matrix notation

$$\max_{\boldsymbol{x} \in \mathbb{R}^2,\, \xi \in \mathbb{R}} \boldsymbol{p}^\top \boldsymbol{x} + \xi$$

subject to the constraints that $\xi \geqslant 0$, $x_0 \leqslant 0$ and $x_1 \leqslant 3$.

7.6 Consider the linear program illustrated in Figure 7.9,

$$\min_{\boldsymbol{x}\in\mathbb{R}^2} \; -\begin{bmatrix}5\\3\end{bmatrix}^\top \begin{bmatrix}x_1\\x_2\end{bmatrix}$$

$$\text{subject to} \quad \begin{bmatrix}2 & 2\\ 2 & -4\\ -2 & 1\\ 0 & -1\\ 0 & 1\end{bmatrix}\begin{bmatrix}x_1\\x_2\end{bmatrix} \leqslant \begin{bmatrix}33\\8\\5\\-1\\8\end{bmatrix}$$

Derive the dual linear program using Lagrange duality.

7.7 Consider the quadratic program illustrated in Figure 7.4,

$$\min_{\boldsymbol{x}\in\mathbb{R}^2} \frac{1}{2}\begin{bmatrix}x_1\\x_2\end{bmatrix}^\top \begin{bmatrix}2 & 1\\ 1 & 4\end{bmatrix}\begin{bmatrix}x_1\\x_2\end{bmatrix} + \begin{bmatrix}5\\3\end{bmatrix}^\top \begin{bmatrix}x_1\\x_2\end{bmatrix}$$

$$\text{subject to} \quad \begin{bmatrix}1 & 0\\ -1 & 0\\ 0 & 1\\ 0 & -1\end{bmatrix}\begin{bmatrix}x_1\\x_2\end{bmatrix} \leqslant \begin{bmatrix}1\\1\\1\\1\end{bmatrix}$$

Derive the dual quadratic program using Lagrange duality.

7.8 Consider the following convex optimization problem

$$\min_{\boldsymbol{w}\in\mathbb{R}^D} \quad \frac{1}{2}\boldsymbol{w}^\top\boldsymbol{w}$$

$$\text{subject to} \quad \boldsymbol{w}^\top\boldsymbol{x} \geqslant 1 \,.$$

Derive the Lagrangian dual by introducing the Lagrange multiplier λ.

7.9 Consider the negative entropy of $\boldsymbol{x} \in \mathbb{R}^D$,

$$f(\boldsymbol{x}) = \sum_{d=1}^{D} x_d \log x_d \,.$$

Derive the convex conjugate function $f^*(\boldsymbol{s})$, by assuming the standard dot product.
Hint: Take the gradient of an appropriate function and set the gradient to zero.

7.10 Consider the function

$$f(\boldsymbol{x}) = \frac{1}{2}\boldsymbol{x}^\top\boldsymbol{A}\boldsymbol{x} + \boldsymbol{b}^\top\boldsymbol{x} + c\,,$$

where $\boldsymbol{A}$ is strictly positive definite, which means that it is invertible. Derive the convex conjugate of $f(\boldsymbol{x})$.
Hint: Take the gradient of an appropriate function and set the gradient to zero.

7.11 The hinge loss (which is the loss used by the support vector machine) is given by

$$L(\alpha) = \max\{0, 1-\alpha\}\,.$$

If we are interested in applying gradient methods such as L-BFGS, and do not want to resort to subgradient methods, we need to smooth the kink in the hinge loss. Compute the convex conjugate of the hinge loss $L^*(\beta)$ where β is the dual variable. Add a ℓ_2 proximal term, and compute the conjugate of the resulting function

$$L^*(\beta) + \frac{\gamma}{2}\beta^2\,,$$

where γ is a given hyperparameter.

Part II

Central Machine Learning Problems

8

When Models Meet Data

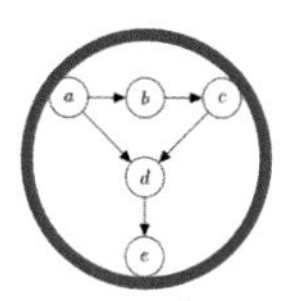

In the first part of the book, we introduced the mathematics that form the foundations of many machine learning methods. The hope is that a reader would be able to learn the rudimentary forms of the language of mathematics from the first part, which we will now use to describe and discuss machine learning. The second part of the book introduces four pillars of machine learning:

- Regression (Chapter 9)
- Dimensionality reduction (Chapter 10)
- Density estimation (Chapter 11)
- Classification (Chapter 12)

The main aim of this part of the book is to illustrate how the mathematical concepts introduced in the first part of the book can be used to design machine learning algorithms that can be used to solve tasks within the remit of the four pillars. We do not intend to introduce advanced machine learning concepts, but instead to provide a set of practical methods that allow the reader to apply the knowledge they gained from the first part of the book. It also provides a gateway to the wider machine learning literature for readers already familiar with the mathematics.

8.1 Data, Models, and Learning

It is worth at this point, to pause and consider the problem that a machine learning algorithm is designed to solve. As discussed in Chapter 1, there are three major components of a machine learning system: data, models, and learning. The main question of machine learning is "What do we mean by good models?" The word *model* has many subtleties, and we will revisit it multiple times in this chapter. It is also not entirely obvious how to objectively define the word "good." One of the guiding principles of machine learning is that good models should perform well on unseen data. This requires us to define some performance metrics, such as accuracy or distance from ground truth, as well as figuring out ways to do well under these performance metrics. This chapter covers a few necessary bits and pieces of mathematical and statistical language that are commonly used to talk about machine learning models. By doing so, we briefly outline the current best practices for training a model such that the resulting predictor does well on data that we have not yet seen.

model

As mentioned in Chapter 1, there are two different senses in which we use the phrase "machine learning algorithm": training and prediction. We will describe these ideas in this chapter, as well as the idea of selecting among different models. We will introduce the framework of empirical risk minimization in Section 8.2, the principle of maximum likelihood in Section 8.3, and the idea of probabilistic models in Section 8.4. We briefly outline a graphical language for specifying probabilistic models in Section 8.5 and finally discuss model selection in Section 8.6. The rest of this section expands upon the three main components of machine learning: data, models, and learning.

8.1.1 Data as Vectors

We assume that our data can be read by a computer, and represented adequately in a numerical format. Data is assumed to be tabular (Figure 8.1), where we think of each row of the table as representing a particular instance or example, and each column to be a particular feature. In recent years, machine learning has been applied to many types of data that do not obviously come in the tabular numerical format, for example genomic sequences, text and image contents of a webpage, and social media graphs. We do not discuss the important and challenging aspects of identifying good features. Many of these aspects depend on domain expertise and require careful engineering, and, in recent years, they have been put under the umbrella of data science (Stray, 2016; Adhikari and DeNero, 2018).

Data is assumed to be in a tidy format (Wickham, 2014; Codd, 1990).

Even when we have data in tabular format, there are still choices to be made to obtain a numerical representation. For example, in Table 8.1, the gender column (a categorical variable) may be converted into numbers 0 representing "Male" and 1 representing "Female." Alternatively, the gender could be represented by numbers $-1, +1$, respectively (as shown in Table 8.2). Furthermore, it is often important to use domain knowledge when constructing the representation, such as knowing that university degrees progress from bachelor's to master's to PhD or realizing that the postcode provided is not just a string of characters

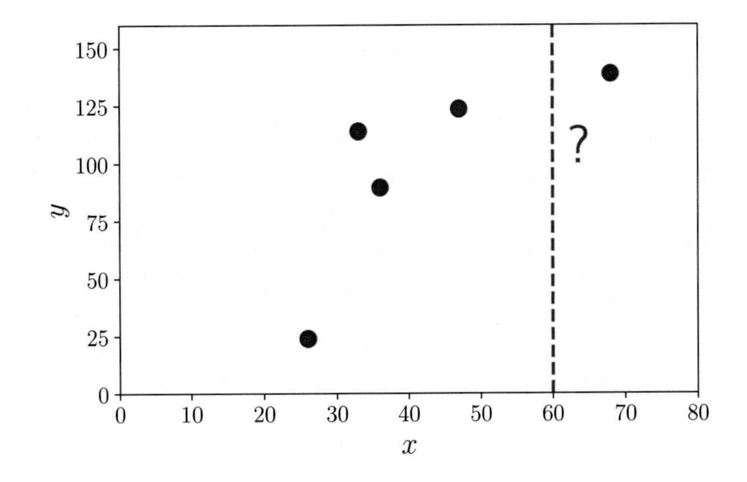

Figure 8.1 Toy data for linear regression. Training data in (x_n, y_n) pairs from the rightmost two columns of Table 8.2. We are interested in the salary of a person aged sixty ($x = 60$) illustrated as a vertical dashed red line, which is not part of the training data.

Name	Gender	Degree	Postcode	Age	Annual salary
Aditya	M	MSc	W21BG	36	89563
Bob	M	PhD	EC1A1BA	47	123543
Chloé	F	BEcon	SW1A1BH	26	23989
Daisuke	M	BSc	SE207AT	68	138769
Elisabeth	F	MBA	SE10AA	33	113888

Table 8.1 Example data from a fictitious human resource database that is not in a numerical format.

Gender ID	Degree	Latitude (in degrees)	Longitude (in degrees)	Age	Annual Salary (in thousands)
−1	2	51.5073	0.1290	36	89.563
−1	3	51.5074	0.1275	47	123.543
+1	1	51.5071	0.1278	26	23.989
−1	1	51.5075	0.1281	68	138.769
+1	2	51.5074	0.1278	33	113.888

Table 8.2 Example data from a fictitious human resource database (see Table 8.1), converted to a numerical format.

but actually encodes an area in London. In Table 8.2, we converted the data from Table 8.1 to a numerical format, and each postcode is represented as two numbers, a latitude and longitude. Even numerical data that could potentially be directly read into a machine learning algorithm should be carefully considered for units, scaling, and constraints. Without additional information, one should shift and scale all columns of the dataset such that they have an empirical mean of 0 and an empirical variance of 1. For the purposes of this book, we assume that a domain expert already converted data appropriately, i.e., each input $\boldsymbol{x}_n$ is a D-dimensional vector of real numbers, which are called *features*, *attributes*, or *covariates*. We consider a dataset to be of the form as illustrated by Table 8.2. Observe that we have dropped the Name column of Table 8.1 in the new numerical representation. There are two main reasons why this is desirable: (1) we do not expect the identifier (the Name) to be informative for a machine learning task; and (2) we may wish to anonymize the data to help protect the privacy of the employees.

feature
attribute
covariate

In this part of the book, we will use N to denote the number of examples in a dataset and index the examples with lowercase $n = 1, \dots, N$. We assume that we are given a set of numerical data, represented as an array of vectors (Table 8.2). Each row is a particular individual $\boldsymbol{x}_n$, often referred to as an *example* or *data point* in machine learning. The subscript n refers to the fact that this is the nth example out of a total of N examples in the dataset. Each column represents a particular feature of interest about the example, and we index the features as $d = 1, \dots, D$. Recall that data is represented as vectors, which means that each example (each data point) is a D-dimensional vector. The orientation of the table originates from the database community, but for some machine learning algorithms (e.g., in Chapter 10) it is more convenient to represent examples as column vectors.

example
data point

Let us consider the problem of predicting annual salary from age, based on the data in Table 8.2. This is called a supervised learning problem where we have

label

a *label* y_n (the salary) associated with each example $\boldsymbol{x}_n$ (the age). The label y_n has various other names, including target, response variable, and annotation. A dataset is written as a set of example–label pairs $\{(\boldsymbol{x}_1, y_1), \ldots, (\boldsymbol{x}_n, y_n), \ldots, (\boldsymbol{x}_N, y_N)\}$. The table of examples $\{\boldsymbol{x}_1, \ldots, \boldsymbol{x}_N\}$ is often concatenated, and written as $\boldsymbol{X} \in \mathbb{R}^{N \times D}$. Figure 8.1 illustrates the dataset consisting of the two rightmost columns of Table 8.2, where $x =$ age and $y =$ salary.

We use the concepts introduced in the first part of the book to formalize the machine learning problems such as that in the previous paragraph. Representing data as vectors $\boldsymbol{x}_n$ allows us to use concepts from linear algebra (introduced in Chapter 2). In many machine learning algorithms, we need to additionally be able to compare two vectors. As we will see in Chapters 9 and 12, computing the similarity or distance between two examples allows us to formalize the intuition that examples with similar features should have similar labels. The comparison of two vectors requires that we construct a geometry (explained in Chapter 3) and allows us to optimize the resulting learning problem using techniques from Chapter 7.

feature map

kernel

Since we have vector representations of data, we can manipulate data to find potentially better representations of it. We will discuss finding good representations in two ways: finding lower-dimensional approximations of the original feature vector, and using nonlinear higher-dimensional combinations of the original feature vector. In Chapter 10, we will see an example of finding a low-dimensional approximation of the original data space by finding the principal components. Finding principal components is closely related to concepts of eigenvalue and singular value decomposition as introduced in Chapter 4. For the high-dimensional representation, we will see an explicit *feature map* $\phi(\cdot)$ that allows us to represent inputs $\boldsymbol{x}_n$ using a higher-dimensional representation $\phi(\boldsymbol{x}_n)$. The main motivation for higher-dimensional representations is that we can construct new features as nonlinear combinations of the original features, which in turn may make the learning problem easier. We will discuss the feature map in Section 9.2 and show how this feature map leads to a *kernel* in Section 12.4. In recent years, deep learning methods (Goodfellow et al., 2016) have shown promise in using the data itself to learn new good features, and have been very successful in areas such as computer vision, speech recognition, and natural language processing. We will not cover neural networks in this part of the book, but the reader is referred to Section 5.6 for the mathematical description of backpropagation, a key concept for training neural networks.

8.1.2 Models as Functions

predictor

Once we have data in an appropriate vector representation, we can get to the business of constructing a predictive function (known as a *predictor*). In Chapter 1, we did not yet have the language to be precise about models. Using the concepts from the first part of the book, we can now introduce what "model" means. We present two major approaches in this book: a predictor as a function, and a predictor as a probabilistic model. We describe the former here and the latter in the next subsection.

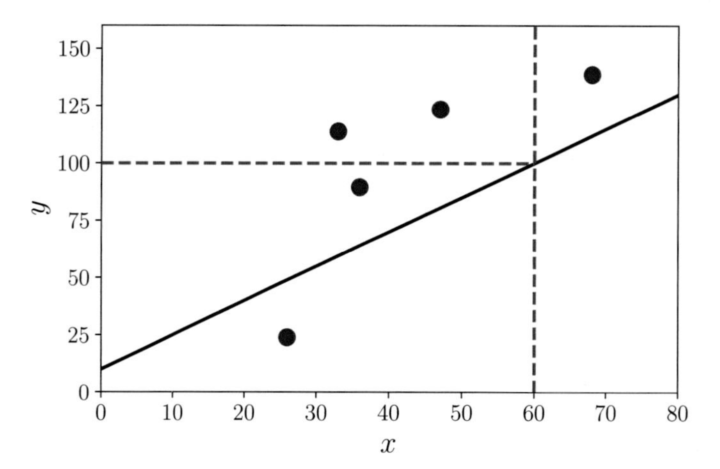

Figure 8.2 Example function (black solid diagonal line) and its prediction at $x = 60$, i.e., $f(60) = 100$.

A predictor is a function that, when given a particular input example (in our case, a vector of features), produces an output. For now, consider the output to be a single number, i.e., a real-valued scalar output. This can be written as

$$f : \mathbb{R}^D \to \mathbb{R}\,, \tag{8.1}$$

where the input vector $\boldsymbol{x}$ is D-dimensional (has D features), and the function f then applied to it (written as $f(\boldsymbol{x})$) returns a real number. Figure 8.2 illustrates a possible function that can be used to compute the value of the prediction for input values x.

In this book, we do not consider the general case of all functions, which would involve the need for functional analysis. Instead, we consider the special case of linear functions

$$f(\boldsymbol{x}) = \boldsymbol{\theta}^\top \boldsymbol{x} + \theta_0 \tag{8.2}$$

for unknown $\boldsymbol{\theta}$ and θ_0. This restriction means that the contents of Chapters 2 and 3 suffice for precisely stating the notion of a predictor for the nonprobabilistic (in contrast to the probabilistic view described next) view of machine learning. Linear functions strike a good balance between the generality of the problems that can be solved and the amount of background mathematics that is needed.

8.1.3 Models as Probability Distributions

We often consider data to be noisy observations of some true underlying effect, and hope that by applying machine learning we can identify the signal from the noise. This requires us to have a language for quantifying the effect of noise. We often would also like to have predictors that express some sort of uncertainty, e.g., to quantify the confidence we have about the value of the prediction for a particular test data point. As we have seen in Chapter 6, probability theory provides a language for quantifying uncertainty. Figure 8.3 illustrates the predictive uncertainty of the function as a Gaussian distribution.

Instead of considering a predictor as a single function, we could consider predictors to be probabilistic models, i.e., models describing the distribution of possible functions. We limit ourselves in this book to the special case of

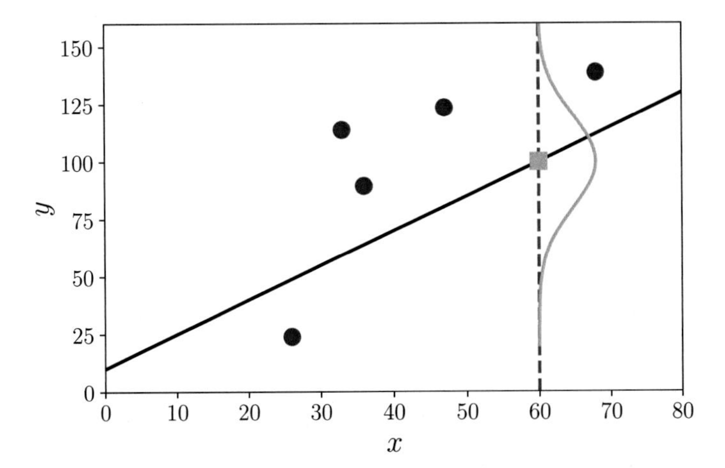

Figure 8.3 Example function (black solid diagonal line) and its predictive uncertainty at $x = 60$ (drawn as a Gaussian).

distributions with finite-dimensional parameters, which allows us to describe probabilistic models without needing stochastic processes and random measures. For this special case, we can think about probabilistic models as multivariate probability distributions, which already allow for a rich class of models.

We will introduce how to use concepts from probability (Chapter 6) to define machine learning models in Section 8.4, and introduce a graphical language for describing probabilistic models in a compact way in Section 8.5.

8.1.4 Learning Is Finding Parameters

The goal of learning is to find a model and its corresponding parameters such that the resulting predictor will perform well on unseen data. There are conceptually three distinct algorithmic phases when discussing machine learning algorithms:

1. Prediction or inference
2. Training or parameter estimation
3. Hyperparameter tuning or model selection

The prediction phase is when we use a trained predictor on previously unseen test data. In other words, the parameters and model choice are already fixed and the predictor is applied to new vectors representing new input data points. As outlined in Chapter 1 and the previous subsection, we will consider two schools of machine learning in this book, corresponding to whether the predictor is a function or a probabilistic model. When we have a probabilistic model (discussed further in Section 8.4), the prediction phase is called inference.

Remark. Unfortunately, there is no agreed upon naming for the different algorithmic phases. The word "inference" is sometimes also used to mean parameter estimation of a probabilistic model, and less often may be also used to mean prediction for nonprobabilistic models. ◊

The training or parameter estimation phase is when we adjust our predictive model based on training data. We would like to find good predictors given training data, and there are two main strategies for doing so: finding the best predictor based on some measure of quality (sometimes called finding a point estimate), or using Bayesian inference. Finding a point estimate can be applied to both types of predictors, but Bayesian inference requires probabilistic models.

For the nonprobabilistic model, we follow the principle of *empirical risk minimization*, which we describe in Section 8.2. Empirical risk minimization directly provides an optimization problem for finding good parameters. With a statistical model, the principle of *maximum likelihood* is used to find a good set of parameters (Section 8.3). We can additionally model the uncertainty of parameters using a probabilistic model, which we will look at in more detail in Section 8.4.

empirical risk minimization

maximum likelihood

We use numerical methods to find good parameters that "fit" the data, and most training methods can be thought of as hill-climbing approaches to find the maximum of an objective, for example the maximum of a likelihood. To apply hill-climbing approaches, we use the gradients described in Chapter 5 and implement numerical optimization approaches from Chapter 7.

The convention in optimization is to minimize objectives. Hence, there is often an extra minus sign in machine learning objectives.

As mentioned in Chapter 1, we are interested in learning a model based on data such that it performs well on future data. It is not enough for the model to only fit the training data well, the predictor needs to perform well on unseen data. We simulate the behavior of our predictor on future unseen data using *cross-validation* (Section 8.2.4). As we will see in this chapter, to achieve the goal of performing well on unseen data, we will need to balance between fitting well on training data and finding "simple" explanations of the phenomenon. This trade-off is achieved using regularization (Section 8.2.3) or by adding a prior (Section 8.3.2). In philosophy, this is considered to be neither induction nor deduction, but is called *abduction*. According to the *Stanford Encyclopedia of Philosophy*, abduction is the process of inference to the best explanation (Douven, 2017).

cross-validation

abduction

A good movie title is "AI abduction."

We often need to make high-level modeling decisions about the structure of the predictor, such as the number of components to use or the class of probability distributions to consider. The choice of the number of components is an example of a *hyperparameter*, and this choice can affect the performance of the model significantly. The problem of choosing among different models is called *model selection*, which we describe in Section 8.6. For nonprobabilistic models, model selection is often done using *nested cross-validation*, which is described in Section 8.6.1. We also use model selection to choose hyperparameters of our model.

hyperparameter

model selection

nested cross-validation

Remark. The distinction between parameters and hyperparameters is somewhat arbitrary, and is mostly driven by the distinction between what can be numerically optimized versus what needs to use search techniques. Another way to consider the distinction is to consider parameters as the explicit parameters of a probabilistic model, and to consider hyperparameters (higher-level parameters) as parameters that control the distribution of these explicit parameters. ◇

In the following sections, we will look at three flavors of machine learning: empirical risk minimization (Section 8.2), the principle of maximum likelihood (Section 8.3), and probabilistic modeling (Section 8.4).

8.2 Empirical Risk Minimization

After having all the mathematics under our belt, we are now in a position to introduce what it means to learn. The "learning" part of machine learning boils down to estimating parameters based on training data.

In this section, we consider the case of a predictor that is a function, and consider the case of probabilistic models in Section 8.3. We describe the idea of empirical risk minimization, which was originally popularized by the proposal of the support vector machine (described in Chapter 12). However, its general principles are widely applicable and allow us to ask the question of what is learning without explicitly constructing probabilistic models. There are four main design choices, which we will cover in detail in the following subsections:

Section 8.2.1 What is the set of functions we allow the predictor to take?

Section 8.2.2 How do we measure how well the predictor performs on the training data?

Section 8.2.3 How do we construct predictors from only training data that performs well on unseen test data?

Section 8.2.4 What is the procedure for searching over the space of models?

8.2.1 Hypothesis Class of Functions

Assume we are given N examples $\boldsymbol{x}_n \in \mathbb{R}^D$ and corresponding scalar labels $y_n \in \mathbb{R}$. We consider the supervised learning setting, where we obtain pairs $(\boldsymbol{x}_1, y_1), \ldots, (\boldsymbol{x}_N, y_N)$. Given this data, we would like to estimate a predictor $f(\cdot, \boldsymbol{\theta}) : \mathbb{R}^D \to \mathbb{R}$, parametrized by $\boldsymbol{\theta}$. We hope to be able to find a good parameter $\boldsymbol{\theta}^*$ such that we fit the data well, that is,

$$f(\boldsymbol{x}_n, \boldsymbol{\theta}^*) \approx y_n \quad \text{for all} \quad n = 1, \ldots, N . \tag{8.3}$$

In this section, we use the notation $\hat{y}_n = f(\boldsymbol{x}_n, \boldsymbol{\theta}^*)$ to represent the output of the predictor.

Remark. For ease of presentation, we will describe empirical risk minimization in terms of supervised learning (where we have labels). This simplifies the definition of the hypothesis class and the loss function. It is also common in machine learning to choose a parametrized class of functions, for example affine functions. ♢

Example 8.1

We introduce the problem of ordinary least-squares regression to illustrate empirical risk minimization. A more comprehensive account of regression is given in Chapter 9. When the label y_n is real-valued, a popular choice of function class for predictors is the set of affine functions. We choose a more compact notation for an affine function by concatenating an additional unit feature $x^{(0)} = 1$ to $\boldsymbol{x}_n$, i.e., $\boldsymbol{x}_n = [1, x_n^{(1)}, x_n^{(2)}, \ldots, x_n^{(D)}]^\top$. The parameter

Affine functions are often referred to as linear functions in machine learning.

vector is correspondingly $\boldsymbol{\theta} = [\theta_0, \theta_1, \theta_2, \ldots, \theta_D]^\top$, allowing us to write the predictor as a linear function

$$f(\boldsymbol{x}_n, \boldsymbol{\theta}) = \boldsymbol{\theta}^\top \boldsymbol{x}_n \,. \tag{8.4}$$

This linear predictor is equivalent to the affine model

$$f(\boldsymbol{x}_n, \boldsymbol{\theta}) = \theta_0 + \sum_{d=1}^{D} \theta_d x_n^{(d)} \,. \tag{8.5}$$

Observe that the predictor takes the vector of features representing a single example $\boldsymbol{x}_n$ as input and produces a real-valued output, i.e., $f : \mathbb{R}^{D+1} \to \mathbb{R}$. The previous figures in this chapter had a straight line as a predictor, which means that we have assumed an affine function.

Instead of a linear function, we may wish to consider nonlinear functions as predictors. Recent advances in neural networks allow for efficient computation of more complex nonlinear function classes.

Given the class of functions, we want to search for a good predictor. We now move on to the second ingredient of empirical risk minimization: how to measure how well the predictor fits the training data.

8.2.2 Loss Function for Training

Consider the label y_n for a particular example, and the corresponding prediction $\hat{y}_n$ that we make based on $\boldsymbol{x}_n$. To define what it means to fit the data well, we need to specify a *loss function* $\ell(y_n, \hat{y}_n)$ that takes the ground truth label and the prediction as input and produces a nonnegative number (referred to as the loss) representing how much error we have made on this particular prediction. Our goal for finding a good parameter vector $\boldsymbol{\theta}^*$ is to minimize the average loss on the set of N training examples.

loss function

The expression "error" is often used to mean loss.

One assumption that is commonly made in machine learning is that the set of examples $(\boldsymbol{x}_1, y_1), \ldots, (\boldsymbol{x}_N, y_N)$ is *independent and identically distributed*. The word independent (Section 6.4.5) means that two data points $(\boldsymbol{x}_i, y_i)$ and $(\boldsymbol{x}_j, y_j)$ do not statistically depend on each other, meaning that the empirical mean is a good estimate of the population mean (Section 6.4.1). This implies that we can use the empirical mean of the loss on the training data. For a given *training set* $\{(\boldsymbol{x}_1, y_1), \ldots, (\boldsymbol{x}_N, y_N)\}$, we introduce the notation of an example matrix $\boldsymbol{X} := [\boldsymbol{x}_1, \ldots, \boldsymbol{x}_N]^\top \in \mathbb{R}^{N \times D}$ and a label vector $\boldsymbol{y} := [y_1, \ldots, y_N]^\top \in \mathbb{R}^N$. Using this matrix notation, the average loss is given by

independent and identically distributed

training set

$$\mathbf{R}_{\text{emp}}(f, \boldsymbol{X}, \boldsymbol{y}) = \frac{1}{N} \sum_{n=1}^{N} \ell(y_n, \hat{y}_n) \,, \tag{8.6}$$

where $\hat{y}_n = f(\boldsymbol{x}_n, \boldsymbol{\theta})$. Equation (8.6) is called the *empirical risk* and depends on three arguments, the predictor f and the data $\boldsymbol{X}, \boldsymbol{y}$. This general strategy for learning is called *empirical risk minimization*.

empirical risk

empirical risk minimization

Example 8.2 (Least-Squares Loss)
Continuing the example of least-squares regression, we specify that we measure the cost of making an error during training using the squared loss $\ell(y_n, \hat{y}_n) = (y_n - \hat{y}_n)^2$. We wish to minimize the empirical risk (8.6), which is the average of the losses over the data

$$\min_{\boldsymbol{\theta} \in \mathbb{R}^D} \frac{1}{N} \sum_{n=1}^{N} (y_n - f(\boldsymbol{x}_n, \boldsymbol{\theta}))^2, \tag{8.7}$$

where we substituted the predictor $\hat{y}_n = f(\boldsymbol{x}_n, \boldsymbol{\theta})$. By using our choice of a linear predictor $f(\boldsymbol{x}_n, \boldsymbol{\theta}) = \boldsymbol{\theta}^\top \boldsymbol{x}_n$, we obtain the optimization problem

$$\min_{\boldsymbol{\theta} \in \mathbb{R}^D} \frac{1}{N} \sum_{n=1}^{N} (y_n - \boldsymbol{\theta}^\top \boldsymbol{x}_n)^2 . \tag{8.8}$$

This equation can be equivalently expressed in matrix form

$$\min_{\boldsymbol{\theta} \in \mathbb{R}^D} \frac{1}{N} \|\boldsymbol{y} - \boldsymbol{X}\boldsymbol{\theta}\|^2 . \tag{8.9}$$

least-squares problem

This is known as the *least-squares problem*. There exists a closed-form analytic solution for this by solving the normal equations, which we will discuss in Section 9.2.

We are not interested in a predictor that only performs well on the training data. Instead, we seek a predictor that performs well (has low risk) on unseen test data. More formally, we are interested in finding a predictor f (with parameters fixed) that minimizes the *expected risk*

expected risk

$$\mathbf{R}_{\text{true}}(f) = \mathbb{E}_{\boldsymbol{x},y}[\ell(y, f(\boldsymbol{x}))] , \tag{8.10}$$

where y is the label and $f(\boldsymbol{x})$ is the prediction based on the example $\boldsymbol{x}$. The notation $\mathbf{R}_{\text{true}}(f)$ indicates that this is the true risk if we had access to an infinite amount of data. The expectation is over the (infinite) set of all possible data and labels. There are two practical questions that arise from our desire to minimize expected risk, which we address in the following two subsections:

Another phrase commonly used for expected risk is "population risk."

- How should we change our training procedure to generalize well?
- How do we estimate expected risk from (finite) data?

Remark. Many machine learning tasks are specified with an associated performance measure, e.g., accuracy of prediction or root mean square error. The performance measure could be more complex, be cost sensitive, and capture details about the particular application. In principle, the design of the loss function for empirical risk minimization should correspond directly to the performance measure specified by the machine learning task. In practice, there is often a mismatch between the design of the loss function and the performance measure. This could be due to issues such as ease of implementation or efficiency of optimization. ♢

8.2.3 Regularization to Reduce Overfitting

This section describes an addition to empirical risk minimization that allows it to generalize well (approximately minimizing expected risk). Recall that the aim of training a machine learning predictor is so that we can perform well on unseen data, i.e., the predictor generalizes well. We simulate this unseen data by holding out a proportion of the whole dataset. This holdout set is referred to as the *test set*. Given a sufficiently rich class of functions for the predictor f, we can essentially memorize the training data to obtain zero empirical risk. While this is great to minimize the loss (and therefore the risk) on the training data, we would not expect the predictor to generalize well to unseen data. In practice, we have only a finite set of data, and hence we split our data into a training and a test set. The training set is used to fit the model, and the test set (not seen by the machine learning algorithm during training) is used to evaluate generalization performance. It is important for the user to not cycle back to a new round of training after having observed the test set. We use the subscripts $_{\text{train}}$ and $_{\text{test}}$ to denote the training and test sets, respectively. We will revisit this idea of using a finite dataset to evaluate expected risk in Section 8.2.4.

test set

Even knowing only the performance of the predictor on the test set leaks information (Blum and Hardt, 2015).

It turns out that empirical risk minimization can lead to *overfitting*, i.e., the predictor fits too closely to the training data and does not generalize well to new data (Mitchell, 1997). This general phenomenon of having very small average loss on the training set but large average loss on the test set tends to occur when we have little data and a complex hypothesis class. For a particular predictor f (with parameters fixed), the phenomenon of overfitting occurs when the risk estimate from the training data $\mathbf{R}_{\text{emp}}(f, \boldsymbol{X}_{\text{train}}, \boldsymbol{y}_{\text{train}})$ underestimates the expected risk $\mathbf{R}_{\text{true}}(f)$. Since we estimate the expected risk $\mathbf{R}_{\text{true}}(f)$ by using the empirical risk on the test set $\mathbf{R}_{\text{emp}}(f, \boldsymbol{X}_{\text{test}}, \boldsymbol{y}_{\text{test}})$ if the test risk is much larger than the training risk, this is an indication of overfitting. We revisit the idea of overfitting in Section 8.3.3.

overfitting

Therefore, we need to somehow bias the search for the minimizer of empirical risk by introducing a penalty term, which makes it harder for the optimizer to return an overly flexible predictor. In machine learning, the penalty term is referred to as *regularization*. Regularization is a way to compromise between accurate solution of empirical risk minimization and the size or complexity of the solution.

regularization

Example 8.3 (Regularized Least Squares)

Regularization is an approach that discourages complex or extreme solutions to an optimization problem. The simplest regularization strategy is to replace the least-squares problem

$$\min_{\boldsymbol{\theta}} \frac{1}{N} \|\boldsymbol{y} - \boldsymbol{X}\boldsymbol{\theta}\|^2 . \tag{8.11}$$

in the previous example with the "regularized" problem by adding a penalty term involving only $\boldsymbol{\theta}$:

$$\min_{\boldsymbol{\theta}} \frac{1}{N} \|\boldsymbol{y} - \boldsymbol{X}\boldsymbol{\theta}\|^2 + \lambda \|\boldsymbol{\theta}\|^2 . \tag{8.12}$$

regularizer

regularization parameter

The additional term $\|\boldsymbol{\theta}\|^2$ is called the *regularizer*, and the parameter λ is the *regularization parameter*. The regularization parameter trades off minimizing the loss on the training set and the magnitude of the parameters $\boldsymbol{\theta}$. It often happens that the magnitude of the parameter values becomes relatively large if we run into overfitting (Bishop, 2006).

penalty term

The regularization term is sometimes called the *penalty term*, which biases the vector $\boldsymbol{\theta}$ to be closer to the origin. The idea of regularization also appears in probabilistic models as the prior probability of the parameters. Recall from Section 6.6 that for the posterior distribution to be of the same form as the prior distribution, the prior and the likelihood need to be conjugate. We will revisit this idea in Section 8.3.2. We will see in Chapter 12 that the idea of the regularizer is equivalent to the idea of a large margin.

8.2.4 Cross-Validation to Assess the Generalization Performance

validation set

cross-validation

We mentioned in the previous section that we measure the generalization error by estimating it by applying the predictor on test data. This data is also sometimes referred to as the *validation set*. The validation set is a subset of the available training data that we keep aside. A practical issue with this approach is that the amount of data is limited, and ideally we would use as much of the data available to train the model. This would require us to keep our validation set $\mathcal{V}$ small, which then would lead to a noisy estimate (with high variance) of the predictive performance. One solution to these contradictory objectives (large training set, large validation set) is to use *cross-validation*. Kfold cross-validation effectively partitions the data into K chunks, $K-1$ of which form the training set $\mathcal{R}$, and the last chunk serves as the validation set $\mathcal{V}$ (similar to the idea outlined previously). Cross-validation iterates through (ideally) all combinations of assignments of chunks to $\mathcal{R}$ and $\mathcal{V}$; see Figure 8.4. This procedure is repeated for all K choices for the validation set, and the performance of the model from the K runs is averaged.

We partition our dataset into two sets $\mathcal{D} = \mathcal{R} \cup \mathcal{V}$, such that they do not overlap ($\mathcal{R} \cap \mathcal{V} = \emptyset$), where $\mathcal{V}$ is the validation set, and train our model on $\mathcal{R}$. After training, we assess the performance of the predictor f on the validation set $\mathcal{V}$ (e.g., by computing root mean square error (RMSE) of the trained model

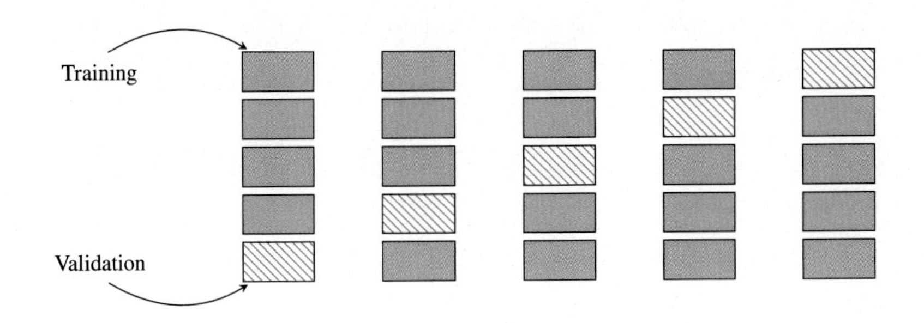

Figure 8.4 K-fold cross-validation. The dataset is divided into $K = 5$ chunks, $K - 1$ of which serve as the training set (blue) and one as the validation set (orange hatch).

on the validation set). More precisely, for each partition k the training data $\mathcal{R}^{(k)}$ produces a predictor $f^{(k)}$, which is then applied to validation set $\mathcal{V}^{(k)}$ to compute the empirical risk $R(f^{(k)}, \mathcal{V}^{(k)})$. We cycle through all possible partitionings of validation and training sets and compute the average generalization error of the predictor. Cross-validation approximates the expected generalization error

$$\mathbb{E}_{\mathcal{V}}[R(f, \mathcal{V})] \approx \frac{1}{K} \sum_{k=1}^{K} R(f^{(k)}, \mathcal{V}^{(k)}), \tag{8.13}$$

where $R(f^{(k)}, \mathcal{V}^{(k)})$ is the risk (e.g., RMSE) on the validation set $\mathcal{V}^{(k)}$ for predictor $f^{(k)}$. The approximation has two sources: first, due to the finite training set, which results in not the best possible $f^{(k)}$; and second, due to the finite validation set, which results in an inaccurate estimation of the risk $R(f^{(k)}, \mathcal{V}^{(k)})$. A potential disadvantage of Kfold cross-validation is the computational cost of training the model K times, which can be burdensome if the training cost is computationally expensive. In practice, it is often not sufficient to look at the direct parameters alone. For example, we need to explore multiple complexity parameters (e.g., multiple regularization parameters), which may not be direct parameters of the model. Evaluating the quality of the model, depending on these hyperparameters, may result in a number of training runs that is exponential in the number of model parameters. One can use nested cross-validation (Section 8.6.1) to search for good hyperparameters.

However, cross-validation is an *embarrassingly parallel* problem, i.e., little effort is needed to separate the problem into a number of parallel tasks. Given sufficient computing resources (e.g., cloud computing, server farms), cross-validation does not require longer than a single performance assessment.

embarrassingly parallel

In this section, we saw that empirical risk minimization is based on the following concepts: the hypothesis class of functions, the loss function and regularization. In Section 8.3, we will see the effect of using a probability distribution to replace the idea of loss functions and regularization.

8.2.5 Further Reading

Due to the fact that the original development of empirical risk minimization (Vapnik, 1998) was couched in heavily theoretical language, many of the subsequent developments have been theoretical. The area of study is called *statistical learning theory* (Vapnik, 1999; Evgeniou et al., 2000; Hastie et al., 2001; von Luxburg and Schölkopf, 2011). A recent machine learning textbook that builds on the theoretical foundations and develops efficient learning algorithms is Shalev-Shwartz and Ben-David (2014).

statistical learning theory

The concept of regularization has its roots in the solution of ill-posed inverse problems (Neumaier, 1998). The approach presented here is called *Tikhonov regularization*, and there is a closely related constrained version called Ivanov regularization. Tikhonov regularization has deep relationships to the bias-variance trade-off and feature selection (Bühlmann and Van De Geer, 2011). An alternative to cross-validation is bootstrap and jackknife (Hall, 1992; Efron and Tibshirani, 1993; Davidson and Hinkley, 1997).

Tikhonov regularization

Thinking about empirical risk minimization (Section 8.2) as "probability free" is incorrect. There is an underlying unknown probability distribution $p(\boldsymbol{x}, y)$ that governs the data generation. However, the approach of empirical risk minimization is agnostic to that choice of distribution. This is in contrast to standard statistical approaches that explicitly require the knowledge of $p(\boldsymbol{x}, y)$. Furthermore, since the distribution is a joint distribution on both examples $\boldsymbol{x}$ and labels y, the labels can be nondeterministic. In contrast to standard statistics we do not need to specify the noise distribution for the labels y.

8.3 Parameter Estimation

In Section 8.2, we did not explicitly model our problem using probability distributions. In this section, we will see how to use probability distributions to model our uncertainty due to the observation process and our uncertainty in the parameters of our predictors. In Section 8.3.1, we introduce the likelihood, which is analogous to the concept of loss functions (Section 8.2.2) in empirical risk minimization. The concept of priors (Section 8.3.2) is analogous to the concept of regularization (Section 8.2.3).

8.3.1 Maximum Likelihood Estimation

maximum likelihood estimation

likelihood

negative log-likelihood

The idea behind *maximum likelihood estimation* (MLE) is to define a function of the parameters that enables us to find a model that fits the data well. The estimation problem is focused on the *likelihood* function, or more precisely its negative logarithm. For data represented by a random variable $\boldsymbol{x}$ and for a family of probability densities $p(\boldsymbol{x} \mid \boldsymbol{\theta})$ parametrized by $\boldsymbol{\theta}$, the *negative log-likelihood* is given by

$$\mathcal{L}_{\boldsymbol{x}}(\boldsymbol{\theta}) = -\log p(\boldsymbol{x} \mid \boldsymbol{\theta}). \tag{8.14}$$

The notation $\mathcal{L}_{\boldsymbol{x}}(\boldsymbol{\theta})$ emphasizes the fact that the parameter $\boldsymbol{\theta}$ is varying and the data $\boldsymbol{x}$ is fixed. We very often drop the reference to $\boldsymbol{x}$ when writing the negative log-likelihood, as it is really a function of $\boldsymbol{\theta}$, and write it as $\mathcal{L}(\boldsymbol{\theta})$ when the random variable representing the uncertainty in the data is clear from the context.

Let us interpret what the probability density $p(\boldsymbol{x} \mid \boldsymbol{\theta})$ is modeling for a fixed value of $\boldsymbol{\theta}$. It is a distribution that models the uncertainty of the data. In other words, once we have chosen the type of function we want as a predictor, the likelihood provides the probability of observing data $\boldsymbol{x}$.

In a complementary view, if we consider the data to be fixed (because it has been observed), and we vary the parameters $\boldsymbol{\theta}$, what does $\mathcal{L}(\boldsymbol{\theta})$ tell us? It tells us how likely a particular setting of $\boldsymbol{\theta}$ is for the observations $\boldsymbol{x}$. Based on this second view, the maximum likelihood estimator gives us the most likely parameter $\boldsymbol{\theta}$ for the set of data.

We consider the supervised learning setting, where we obtain pairs $(\boldsymbol{x}_1, y_1), \ldots, (\boldsymbol{x}_N, y_N)$ with $\boldsymbol{x}_n \in \mathbb{R}^D$ and labels $y_n \in \mathbb{R}$. We are interested in constructing a predictor that takes a feature vector $\boldsymbol{x}_n$ as input and produces a prediction y_n (or something close to it), i.e., given a vector $\boldsymbol{x}_n$ we want the probability distribution of the label y_n. In other words, we specify the conditional probability distribution of the labels given the examples for the particular parameter setting $\boldsymbol{\theta}$.

Example 8.4

The first example that is often used is to specify that the conditional probability of the labels given the examples is a Gaussian distribution. In other words, we assume that we can explain our observation uncertainty by independent Gaussian noise (refer to Section 6.5) with zero mean, $\varepsilon_n \sim \mathcal{N}(0, \sigma^2)$. We further assume that the linear model $\boldsymbol{x}_n^\top \boldsymbol{\theta}$ is used for prediction. This means we specify a Gaussian likelihood for each example label pair $\boldsymbol{x}_n, y_n$,

$$p(y_n \mid \boldsymbol{x}_n, \boldsymbol{\theta}) = \mathcal{N}(y_n \mid \boldsymbol{x}_n^\top \boldsymbol{\theta}, \sigma^2) . \tag{8.15}$$

An illustration of a Gaussian likelihood for a given parameter $\boldsymbol{\theta}$ is shown in Figure 8.3. We will see in Section 9.2 how to explicitly expand the preceding expression out in terms of the Gaussian distribution.

We assume that the set of examples $(x_1, y_1), \ldots, (x_N, y_N)$ are *independent and identically distributed* (i.i.d.). The word "independent" (Section 6.4.5) implies that the likelihood of the whole dataset ($\mathcal{Y} = \{y_1, \ldots, y_N\}$ and $\mathcal{X} = \{\boldsymbol{x}_1, \ldots, \boldsymbol{x}_N\}$ factorizes into a product of the likelihoods of each individual example

independent and identically distributed

$$p(\mathcal{Y} \mid \mathcal{X}, \boldsymbol{\theta}) = \prod_{n=1}^{N} p(y_n \mid \boldsymbol{x}_n, \boldsymbol{\theta}) , \tag{8.16}$$

where $p(y_n \mid \boldsymbol{x}_n, \boldsymbol{\theta})$ is a particular distribution (which was Gaussian in Example 8.4). The expression "identically distributed" means that each term in the product (8.16) is of the same distribution, and all of them share the same parameters. It is often easier from an optimization viewpoint to compute functions that can be decomposed into sums of simpler functions. Hence, in machine learning we often consider the negative log-likelihood

Recall $\log(ab) = \log(a) + \log(b)$

$$\mathcal{L}(\boldsymbol{\theta}) = -\log p(\mathcal{Y} \mid \mathcal{X}, \boldsymbol{\theta}) = -\sum_{n=1}^{N} \log p(y_n \mid \boldsymbol{x}_n, \boldsymbol{\theta}) . \tag{8.17}$$

While it is temping to interpret the fact that $\boldsymbol{\theta}$ is on the right of the conditioning in $p(y_n|\boldsymbol{x}_n, \boldsymbol{\theta})$ (8.15), and hence should be interpreted as observed and fixed, this interpretation is incorrect. The negative log-likelihood $\mathcal{L}(\boldsymbol{\theta})$ is a function of $\boldsymbol{\theta}$. Therefore, to find a good parameter vector $\boldsymbol{\theta}$ that explains the data $(\boldsymbol{x}_1, y_1), \ldots, (\boldsymbol{x}_N, y_N)$ well, minimize the negative log-likelihood $\mathcal{L}(\boldsymbol{\theta})$ with respect to $\boldsymbol{\theta}$.

Remark. The negative sign in (8.17) is a historical artifact that is due to the convention that we want to maximize likelihood, but numerical optimization literature tends to study minimization of functions. ◇

Example 8.5

Continuing on our example of Gaussian likelihoods (8.15), the negative log-likelihood can be rewritten as

$$\mathcal{L}(\boldsymbol{\theta}) = -\sum_{n=1}^{N} \log p(y_n \mid \boldsymbol{x}_n, \boldsymbol{\theta}) = -\sum_{n=1}^{N} \log \mathcal{N}\big(y_n \mid \boldsymbol{x}_n^\top \boldsymbol{\theta}, \sigma^2\big) \tag{8.18a}$$

$$= -\sum_{n=1}^{N} \log \frac{1}{\sqrt{2\pi\sigma^2}} \exp\left(-\frac{(y_n - \boldsymbol{x}_n^\top \boldsymbol{\theta})^2}{2\sigma^2}\right) \tag{8.18b}$$

$$= -\sum_{n=1}^{N} \log \exp\left(-\frac{(y_n - \boldsymbol{x}_n^\top \boldsymbol{\theta})^2}{2\sigma^2}\right) - \sum_{n=1}^{N} \log \frac{1}{\sqrt{2\pi\sigma^2}} \tag{8.18c}$$

$$= \frac{1}{2\sigma^2} \sum_{n=1}^{N} (y_n - \boldsymbol{x}_n^\top \boldsymbol{\theta})^2 - \sum_{n=1}^{N} \log \frac{1}{\sqrt{2\pi\sigma^2}} . \tag{8.18d}$$

As σ is given, the second term in (8.18d) is constant, and minimizing $\mathcal{L}(\boldsymbol{\theta})$ corresponds to solving the least-squares problem (compare with (8.8)) expressed in the first term.

It turns out that for Gaussian likelihoods the resulting optimization problem corresponding to maximum likelihood estimation has a closed-form solution. We will see more details on this in Chapter 9. Figure 8.5 shows a regression dataset and the function that is induced by the maximum-likelihood parameters. Maximum likelihood estimation may suffer from overfitting (Section 8.3.3), analogous to unregularized empirical risk minimization (Section 9.2.3). For other likelihood functions, i.e., if we model our noise with non-Gaussian distributions, maximum likelihood estimation may not have a closed-form analytic solution. In this case, we resort to numerical optimization methods discussed in Chapter 7.

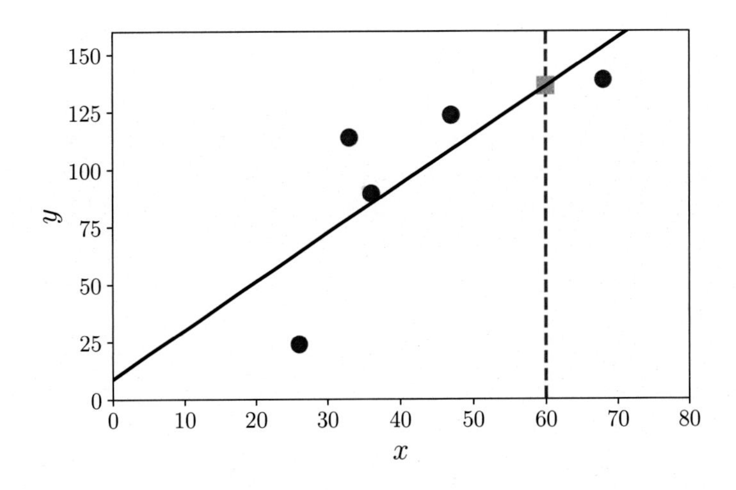

Figure 8.5 For the given data, the maximum likelihood estimate of the parameters results in the black diagonal line. The orange square shows the value of the maximum likelihood prediction at $x = 60$.

8.3.2 Maximum A Posteriori Estimation

If we have prior knowledge about the distribution of the parameters $\boldsymbol{\theta}$, we can multiply an additional term to the likelihood. This additional term is a prior probability distribution on parameters $p(\boldsymbol{\theta})$. For a given prior, after observing some data $\boldsymbol{x}$, how should we update the distribution of $\boldsymbol{\theta}$? In other words, how should we represent the fact that we have more specific knowledge of $\boldsymbol{\theta}$ after observing data $\boldsymbol{x}$? Bayes' theorem, as discussed in Section 6.3, gives us a principled tool to update our probability distributions of random variables. It allows us to compute a *posterior* distribution $p(\boldsymbol{\theta} \mid \boldsymbol{x})$ (the more specific knowledge) on the parameters $\boldsymbol{\theta}$ from general *prior* statements (prior distribution) $p(\boldsymbol{\theta})$ and the function $p(\boldsymbol{x} \mid \boldsymbol{\theta})$ that links the parameters $\boldsymbol{\theta}$ and the observed data $\boldsymbol{x}$ (called the *likelihood*):

posterior

prior

likelihood

$$p(\boldsymbol{\theta} \mid \boldsymbol{x}) = \frac{p(\boldsymbol{x} \mid \boldsymbol{\theta})p(\boldsymbol{\theta})}{p(\boldsymbol{x})} . \tag{8.19}$$

Recall that we are interested in finding the parameter $\boldsymbol{\theta}$ that maximizes the posterior. Since the distribution $p(\boldsymbol{x})$ does not depend on $\boldsymbol{\theta}$, we can ignore the value of the denominator for the optimization and obtain

$$p(\boldsymbol{\theta} \mid \boldsymbol{x}) \propto p(\boldsymbol{x} \mid \boldsymbol{\theta})p(\boldsymbol{\theta}) . \tag{8.20}$$

The preceding proportion relation hides the density of the data $p(\boldsymbol{x})$, which may be difficult to estimate. Instead of estimating the minimum of the negative log-likelihood, we now estimate the minimum of the negative log-posterior, which is referred to as *maximum a posteriori estimation* (*MAP estimation*). An illustration of the effect of adding a zero-mean Gaussian prior is shown in Figure 8.6.

maximum a posteriori estimation
MAP estimation

Example 8.6
In addition to the assumption of Gaussian likelihood in the previous example, we assume that the parameter vector is distributed as a multivariate Gaussian with zero mean, i.e., $p(\boldsymbol{\theta}) = \mathcal{N}(\mathbf{0}, \boldsymbol{\Sigma})$, where $\boldsymbol{\Sigma}$ is the covariance matrix (Section 6.5). Note that the conjugate prior of a Gaussian is also a Gaussian (Section 6.6.1), and therefore we expect the posterior distribution to also be a Gaussian. We will see the details of maximum a posteriori estimation in Chapter 9.

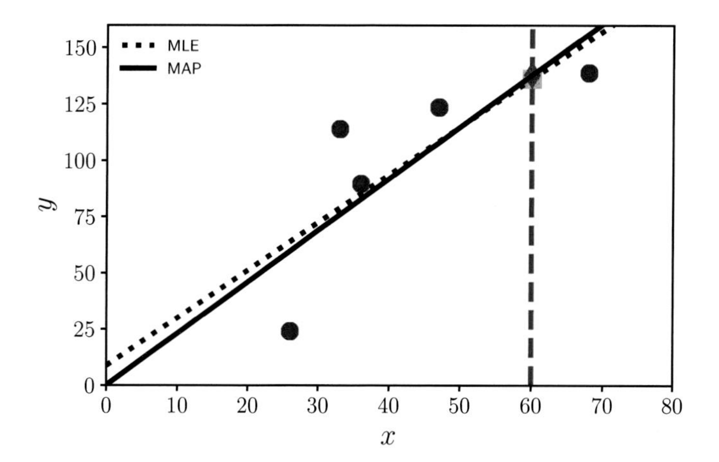

Figure 8.6 Comparing the predictions with the maximum likelihood estimate and the MAP estimate at $x = 60$. The prior biases the slope to be less steep and the intercept to be closer to zero. In this example, the bias that moves the intercept closer to zero actually increases the slope.

The idea of including prior knowledge about where good parameters lie is widespread in machine learning. An alternative view, which we saw in Section 8.2.3, is the idea of regularization, which introduces an additional term that biases the resulting parameters to be close to the origin. Maximum a posteriori estimation can be considered to bridge the non-probabilistic and probabilistic worlds as it explicitly acknowledges the need for a prior distribution but it still only produces a point estimate of the parameters.

Remark. The maximum likelihood estimate $\boldsymbol{\theta}_{\text{ML}}$ possesses the following properties (Lehmann and Casella, 1998; Efron and Hastie, 2016):

- Asymptotic consistency: The MLE converges to the true value in the limit of infinitely many observations, plus a random error that is approximately normal.
- The size of the samples necessary to achieve these properties can be quite large.
- The error's variance decays in $1/N$, where N is the number of data points.
- Especially, in the "small" data regime, maximum likelihood estimation can lead to *overfitting*.

◇

The principle of maximum likelihood estimation (and maximum a posteriori estimation) uses probabilistic modeling to reason about the uncertainty in the data and model parameters. However, we have not yet taken probabilistic modeling to its full extent. In this section, the resulting training procedure still produces a point estimate of the predictor, i.e., training returns one single set of parameter values that represent the best predictor. In Section 8.4, we will take the view that the parameter values should also be treated as random variables, and instead of estimating "best" values of that distribution, we will use the full parameter distribution when making predictions.

8.3.3 Model Fitting

Consider the setting where we are given a dataset, and we are interested in fitting a parametrized model to the data. When we talk about "fitting," we typically mean optimizing/learning model parameters so that they minimize some loss function, e.g., the negative log-likelihood. With maximum likelihood (Section 8.3.1) and maximum a posteriori estimation (Section 8.3.2), we already discussed two commonly used algorithms for model fitting.

The parametrization of the model defines a model class $M_{\boldsymbol{\theta}}$ with which we can operate. For example, in a linear regression setting, we may define the relationship between inputs x and (noise-free) observations y to be $y = ax + b$, where $\boldsymbol{\theta} := \{a, b\}$ are the model parameters. In this case, the model parameters $\boldsymbol{\theta}$ describe the family of affine functions, i.e., straight lines with slope a, which are offset from 0 by b. Assume the data comes from a model M^*, which is unknown to us. For a given training dataset, we optimize $\boldsymbol{\theta}$ so that $M_{\boldsymbol{\theta}}$ is as close as possible to M^*, where the "closeness" is defined by the objective function we

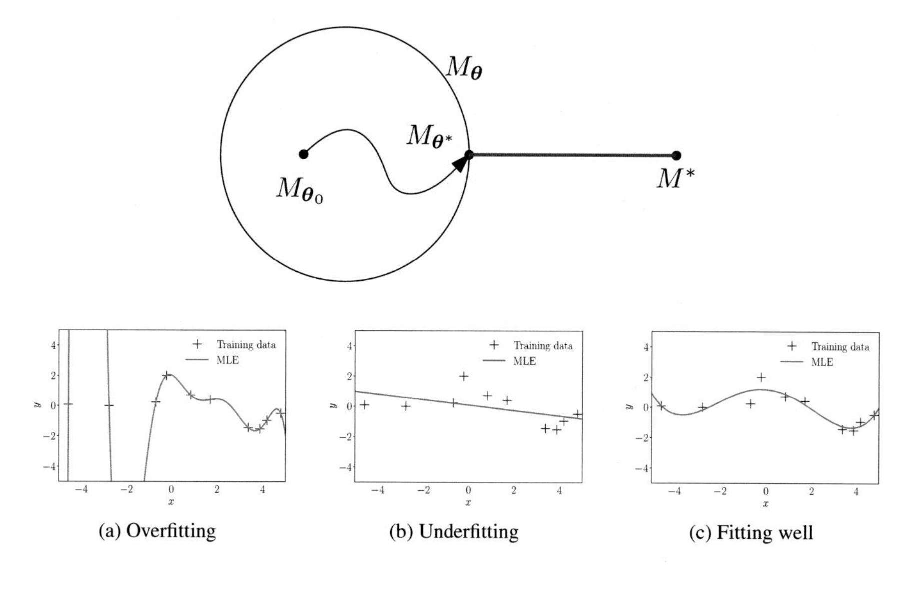

Figure 8.7 Model fitting. In a parametrized class $M_{\boldsymbol{\theta}}$ of models, we optimize the model parameters $\boldsymbol{\theta}$ to minimize the distance to the true (unknown) model M^*.

Figure 8.8 Fitting (by maximum likelihood) of different model classes to a regression dataset.

optimize (e.g., squared loss on the training data). Figure 8.7 illustrates a setting where we have a small model class (indicated by the circle $M_{\boldsymbol{\theta}}$), and the data generation model M^* lies outside the set of considered models. We begin our parameter search at $M_{\boldsymbol{\theta}_0}$. After the optimization, i.e., when we obtain the best possible parameters $\boldsymbol{\theta}^*$, we distinguish three different cases: (i) overfitting, (ii) underfitting, and (iii) fitting well. We will give a high-level intuition of what these three concepts mean.

overfitting

Roughly speaking, *overfitting* refers to the situation where the parametrized model class is too rich to model the dataset generated by M^*, i.e., $M_{\boldsymbol{\theta}}$ could model much more complicated datasets. For instance, if the dataset was generated by a linear function, and we define $M_{\boldsymbol{\theta}}$ to be the class of seventh-order polynomials, we could model not only linear functions, but also polynomials of degree two, three, etc. Models that overfit typically have a large number of parameters. An observation we often make is that the overly flexible model class $M_{\boldsymbol{\theta}}$ uses all its modeling power to reduce the training error. If the training data is noisy, it will therefore find some useful signal in the noise itself. This will cause enormous problems when we predict away from the training data. Figure 8.8(a) gives an example of overfitting in the context of regression where the model parameters are learned by means of maximum likelihood (see Section 8.3.1). We will discuss overfitting in regression more in Section 9.2.2.

One way to detect overfitting in practice is to observe that the model has low training risk but high test risk during cross-validation (Section 8.2.4).

underfitting

When we run into *underfitting*, we encounter the opposite problem where the model class $M_{\boldsymbol{\theta}}$ is not rich enough. For example, if our dataset was generated by a sinusoidal function, but $\boldsymbol{\theta}$ only parametrizes straight lines, the best optimization procedure will not get us close to the true model. However, we still optimize the parameters and find the best straight line that models the dataset. Figure 8.8(b) shows an example of a model that underfits because it is insufficiently flexible. Models that underfit typically have few parameters.

The third case is when the parametrized model class is about right. Then, our model fits well, i.e., it neither overfits nor underfits. This means our model class is just rich enough to describe the dataset we are given. Figure 8.8(c) shows a model that fits the given dataset fairly well. Ideally, this is the model class we would want to work with since it has good generalization properties.

In practice, we often define very rich model classes $M_{\boldsymbol{\theta}}$ with many parameters, such as deep neural networks. To mitigate the problem of overfitting, we can use regularization (Section 8.2.3) or priors (Section 8.3.2). We will discuss how to choose the model class in Section 8.6.

8.3.4 Further Reading

When considering probabilistic models, the principle of maximum likelihood estimation generalizes the idea of least-squares regression for linear models, which we will discuss in detail in Chapter 9. When restricting the predictor to have linear form with an additional nonlinear function φ applied to the output, i.e.,

$$p(y_n|\boldsymbol{x}_n, \boldsymbol{\theta}) = \varphi(\boldsymbol{\theta}^\top \boldsymbol{x}_n), \tag{8.21}$$

we can consider other models for other prediction tasks, such as binary classification or modeling count data (McCullagh and Nelder, 1989). An alternative view of this is to consider likelihoods that are from the exponential family (Section 6.6). The class of models, which have linear dependence between parameters and data, and have potentially nonlinear transformation φ (called a *link function*), is referred to as *generalized linear models* (Agresti, 2002, chapter 4).

link function

generalized linear model

Maximum likelihood estimation has a rich history, and was originally proposed by Sir Ronald Fisher in the 1930s. We will expand upon the idea of a probabilistic model in Section 8.4. One debate among researchers who use probabilistic models is the discussion between Bayesian and frequentist statistics. As mentioned in Section 6.1.1, it boils down to the definition of probability. Recall from Section 6.1 that one can consider probability to be a generalization (by allowing uncertainty) of logical reasoning (Cheeseman, 1985; Jaynes, 2003). The method of maximum likelihood estimation is frequentist in nature, and the interested reader is pointed to Efron and Hastie (2016) for a balanced view of both Bayesian and frequentist statistics.

There are some probabilistic models where maximum likelihood estimation may not be possible. The reader is referred to more advanced statistical textbooks, e.g., Casella and Berger (2002), for approaches, such as method of moments, M-estimation, and estimating equations.

8.4 Probabilistic Modeling and Inference

In machine learning, we are frequently concerned with the interpretation and analysis of data, e.g., for prediction of future events and decision making. To make this task more tractable, we often build models that describe the *generative process* that generates the observed data.

generative process

For example, we can describe the outcome of a coin-flip experiment ("heads" or "tails") in two steps. First, we define a parameter μ, which describes the probability of "heads" as the parameter of a Bernoulli distribution (Chapter 6);

second, we can sample an outcome $x \in \{\text{head}, \text{tail}\}$ from the Bernoulli distribution $p(x \mid \mu) = \text{Ber}(\mu)$. The parameter μ gives rise to a specific dataset $\mathcal{X}$ and depends on the coin used. Since μ is unknown in advance and can never be observed directly, we need mechanisms to learn something about μ given observed outcomes of coin-flip experiments. In the following, we will discuss how probabilistic modeling can be used for this purpose.

8.4.1 Probabilistic Models

Probabilistic models represent the uncertain aspects of an experiment as probability distributions. The benefit of using probabilistic models is that they offer a unified and consistent set of tools from probability theory (Chapter 6) for modeling, inference, prediction, and model selection.

A probabilistic model is specified by the joint distribution of all random variables.

In probabilistic modeling, the joint distribution $p(\boldsymbol{x}, \boldsymbol{\theta})$ of the observed variables $\boldsymbol{x}$ and the hidden parameters $\boldsymbol{\theta}$ is of central importance: It encapsulates information from the following:

- The prior and the likelihood (product rule, Section 6.3).
- The marginal likelihood $p(\boldsymbol{x})$, which will play an important role in model selection (Section 8.6), can be computed by taking the joint distribution and integrating out the parameters (sum rule, Section 6.3).
- The posterior, which can be obtained by dividing the joint by the marginal likelihood.

Only the joint distribution has this property. Therefore, a probabilistic model is specified by the joint distribution of all its random variables.

8.4.2 Bayesian Inference

A key task in machine learning is to take a model and the data to uncover the values of the model's hidden variables $\boldsymbol{\theta}$ given the observed variables $\boldsymbol{x}$. In Section 8.3.1, we already discussed two ways for estimating model parameters $\boldsymbol{\theta}$ using maximum likelihood or maximum a posteriori estimation. In both cases, we obtain a single-best value for $\boldsymbol{\theta}$ so that the key algorithmic problem of parameter estimation is solving an optimization problem. Once these point estimates $\boldsymbol{\theta}^*$ are known, we use them to make predictions. More specifically, the predictive distribution will be $p(\boldsymbol{x} \mid \boldsymbol{\theta}^*)$, where we use $\boldsymbol{\theta}^*$ in the likelihood function.

Parameter estimation can be phrased as an optimization problem.

As discussed in Section 6.3, focusing solely on some statistic of the posterior distribution (such as the parameter $\boldsymbol{\theta}^*$ that maximizes the posterior) leads to loss of information, which can be critical in a system that uses the prediction $p(\boldsymbol{x} \mid \boldsymbol{\theta}^*)$ to make decisions. These decision-making systems typically have different objective functions than the likelihood, a squared-error loss or a misclassification error. Therefore, having the full posterior distribution around can be extremely useful and leads to more robust decisions. *Bayesian inference* is about finding this posterior distribution (Gelman et al., 2004). For a dataset $\mathcal{X}$, a parameter prior $p(\boldsymbol{\theta})$, and a likelihood function, the posterior

Bayesian inference is about learning the distribution of random variables.
Bayesian inference

$$p(\boldsymbol{\theta} \mid \mathcal{X}) = \frac{p(\mathcal{X} \mid \boldsymbol{\theta})p(\boldsymbol{\theta})}{p(\mathcal{X})}, \qquad p(\mathcal{X}) = \int p(\mathcal{X} \mid \boldsymbol{\theta})p(\boldsymbol{\theta})\mathrm{d}\boldsymbol{\theta}, \tag{8.22}$$

Bayesian inference inverts the relationship between parameters and the data.

is obtained by applying Bayes' theorem. The key idea is to exploit Bayes' theorem to invert the relationship between the parameters $\boldsymbol{\theta}$ and the data $\mathcal{X}$ (given by the likelihood) to obtain the posterior distribution $p(\boldsymbol{\theta} \mid \mathcal{X})$.

The implication of having a posterior distribution on the parameters is that it can be used to propagate uncertainty from the parameters to the data. More specifically, with a distribution $p(\boldsymbol{\theta})$ on the parameters our predictions will be

$$p(\boldsymbol{x}) = \int p(\boldsymbol{x} \mid \boldsymbol{\theta})p(\boldsymbol{\theta})\mathrm{d}\boldsymbol{\theta} = \mathbb{E}_{\boldsymbol{\theta}}[p(\boldsymbol{x} \mid \boldsymbol{\theta})], \tag{8.23}$$

and they no longer depend on the model parameters $\boldsymbol{\theta}$, which have been marginalized/integrated out. Equation (8.23) reveals that the prediction is an average over all plausible parameter values $\boldsymbol{\theta}$, where the plausibility is encapsulated by the parameter distribution $p(\boldsymbol{\theta})$.

Having discussed parameter estimation in Section 8.3 and Bayesian inference here, let us compare these two approaches to learning. Parameter estimation via maximum likelihood or MAP estimation yields a consistent point estimate $\boldsymbol{\theta}^*$ of the parameters, and the key computational problem to be solved is optimization. In contrast, Bayesian inference yields a (posterior) distribution, and the key computational problem to be solved is integration. Predictions with point estimates are straightforward, whereas predictions in the Bayesian framework require solving another integration problem; see (8.23). However, Bayesian inference gives us a principled way to incorporate prior knowledge, account for side information, and incorporate structural knowledge, all of which is not easily done in the context of parameter estimation. Moreover, the propagation of parameter uncertainty to the prediction can be valuable in decision-making systems for risk assessment and exploration in the context of data-efficient learning (Deisenroth et al., 2015; Kamthe and Deisenroth, 2018).

While Bayesian inference is a mathematically principled framework for learning about parameters and making predictions, there are some practical challenges that come with it because of the integration problems we need to solve; see (8.22) and (8.23). More specifically, if we do not choose a conjugate prior on the parameters (Section 6.6.1), the integrals in (8.22) and (8.23) are not analytically tractable, and we cannot compute the posterior, the predictions, or the marginal likelihood in closed form. In these cases, we need to resort to approximations. Here, we can use stochastic approximations, such as Markov chain Monte Carlo (MCMC) (Gilks et al., 1996), or deterministic approximations, such as the Laplace approximation (Bishop, 2006; Barber, 2012; Murphy, 2012), variational inference (Jordan et al., 1999; Blei et al., 2017), or expectation propagation (Minka, 2001a).

Despite these challenges, Bayesian inference has been successfully applied to a variety of problems, including large-scale topic modeling (Hoffman et al., 2013), click-through-rate prediction (Graepel et al., 2010), data-efficient reinforcement learning in control systems (Deisenroth et al., 2015), online ranking systems (Herbrich et al., 2007), and large-scale recommender systems. There are generic tools, such as Bayesian optimization (Brochu et al., 2009; Snoek et al., 2012; Shahriari et al., 2016), that are very useful ingredients for an efficient search of metaparameters of models or algorithms.

Remark. In the machine learning literature, there can be a somewhat arbitrary separation between (random) "variables" and "parameters." While parameters are estimated (e.g., via maximum likelihood), variables are usually marginalized out. In this book, we are not so strict with this separation because, in principle, we can place a prior on any parameter and integrate it out, which would then turn the parameter into a random variable according to the aforementioned separation. 

8.4.3 Latent-Variable Models

In practice, it is sometimes useful to have additional *latent variables* $\boldsymbol{z}$ (besides the model parameters $\boldsymbol{\theta}$) as part of the model (Moustaki et al., 2015). These latent variables are different from the model parameters $\boldsymbol{\theta}$ as they do not parametrize the model explicitly. Latent variables may describe the data-generating process, thereby contributing to the interpretability of the model. They also often simplify the structure of the model and allow us to define simpler and richer model structures. Simplification of the model structure often goes hand in hand with a smaller number of model parameters (Paquet, 2008; Murphy, 2012). Learning in latent-variable models (at least via maximum likelihood) can be done in a principled way using the expectation maximization (EM) algorithm (Dempster et al., 1977; Bishop, 2006). Examples, where such latent variables are helpful, are principal component analysis for dimensionality reduction (Chapter 10), Gaussian mixture models for density estimation (Chapter 11), hidden Markov models (Maybeck, 1979), or dynamical systems (Ghahramani and Roweis, 1999; Ljung, 1999) for time-series modeling, and meta learning and task generalization (Hausman et al., 2018; Sæmundsson et al., 2018). Although the introduction of these latent variables may make the model structure and the generative process easier, learning in latent-variable models is generally hard, as we will see in Chapter 11.

latent variable

Since latent-variable models also allow us to define the process that generates data from parameters, let us have a look at this generative process. Denoting data by $\boldsymbol{x}$, the model parameters by $\boldsymbol{\theta}$ and the latent variables by $\boldsymbol{z}$, we obtain the conditional distribution

$$p(\boldsymbol{x} \mid \boldsymbol{\theta}, \boldsymbol{z}) \tag{8.24}$$

that allows us to generate data for any model parameters and latent variables. Given that $\boldsymbol{z}$ are latent variables, we place a prior $p(\boldsymbol{z})$ on them.

As with the models we discussed previously, models with latent variables can be used for parameter learning and inference within the frameworks we discussed in Sections 8.3 and 8.4.2. To facilitate learning (e.g., by means of maximum likelihood estimation or Bayesian inference), we follow a two-step procedure. First, we compute the likelihood $p(\boldsymbol{x} \mid \boldsymbol{\theta})$ of the model, which does not depend on the latent variables. Second, we use this likelihood for parameter estimation or

Bayesian inference, where we use exactly the same expressions as in Sections 8.3 and 8.4.2, respectively.

Since the likelihood function $p(\boldsymbol{x} \mid \boldsymbol{\theta})$ is the predictive distribution of the data given the model parameters, we need to marginalize out the latent variables so that

$$p(\boldsymbol{x} \mid \boldsymbol{\theta}) = \int p(\boldsymbol{x} \mid \boldsymbol{\theta}, \boldsymbol{z}) p(\boldsymbol{z}) \mathrm{d}\boldsymbol{z}\,, \tag{8.25}$$

where $p(\boldsymbol{x} \mid \boldsymbol{z}, \boldsymbol{\theta})$ is given in (8.24) and $p(\boldsymbol{z})$ is the prior on the latent variables. Note that the likelihood must not depend on the latent variables $\boldsymbol{z}$, but it is only a function of the data $\boldsymbol{x}$ and the model parameters $\boldsymbol{\theta}$.

The likelihood is a function of the data and the model parameters, but is independent of the latent variables.

The likelihood in (8.25) directly allows for parameter estimation via maximum likelihood. MAP estimation is also straightforward with an additional prior on the model parameters $\boldsymbol{\theta}$ as discussed in Section 8.3.2. Moreover, with the likelihood (8.25) Bayesian inference (Section 8.4.2) in a latent-variable model works in the usual way: We place a prior $p(\boldsymbol{\theta})$ on the model parameters and use Bayes' theorem to obtain a posterior distribution

$$p(\boldsymbol{\theta} \mid \mathcal{X}) = \frac{p(\mathcal{X} \mid \boldsymbol{\theta}) p(\boldsymbol{\theta})}{p(\mathcal{X})} \tag{8.26}$$

over the model parameters given a dataset $\mathcal{X}$. The posterior in (8.26) can be used for predictions within a Bayesian inference framework; see (8.23).

One challenge we have in this latent-variable model is that the likelihood $p(\mathcal{X} \mid \boldsymbol{\theta})$ requires the marginalization of the latent variables according to (8.25). Except when we choose a conjugate prior $p(\boldsymbol{z})$ for $p(\boldsymbol{x} \mid \boldsymbol{z}, \boldsymbol{\theta})$, the marginalization in (8.25) is not analytically tractable, and we need to resort to approximations (Bishop, 2006; Paquet, 2008; Murphy, 2012; Moustaki et al., 2015).

Similar to the parameter posterior (8.26), we can compute a posterior on the latent variables according to

$$p(\boldsymbol{z} \mid \mathcal{X}) = \frac{p(\mathcal{X} \mid \boldsymbol{z}) p(\boldsymbol{z})}{p(\mathcal{X})}\,, \qquad p(\mathcal{X} \mid \boldsymbol{z}) = \int p(\mathcal{X} \mid \boldsymbol{z}, \boldsymbol{\theta}) p(\boldsymbol{\theta}) \mathrm{d}\boldsymbol{\theta}\,, \tag{8.27}$$

where $p(\boldsymbol{z})$ is the prior on the latent variables and $p(\mathcal{X} \mid \boldsymbol{z})$ requires us to integrate out the model parameters $\boldsymbol{\theta}$.

Given the difficulty of solving integrals analytically, it is clear that marginalizing out both the latent variables and the model parameters at the same time is not possible in general (Bishop, 2006; Murphy, 2012). A quantity that is easier to compute is the posterior distribution on the latent variables, but conditioned on the model parameters, i.e.,

$$p(\boldsymbol{z} \mid \mathcal{X}, \boldsymbol{\theta}) = \frac{p(\mathcal{X} \mid \boldsymbol{z}, \boldsymbol{\theta}) p(\boldsymbol{z})}{p(\mathcal{X} \mid \boldsymbol{\theta})}\,, \tag{8.28}$$

where $p(\boldsymbol{z})$ is the prior on the latent variables and $p(\mathcal{X} \mid \boldsymbol{z}, \boldsymbol{\theta})$ is given in (8.24).

In Chapters 10 and 11, we derive the likelihood functions for PCA and Gaussian mixture models, respectively. Moreover, we compute the posterior distributions (8.28) on the latent variables for both PCA and Gaussian mixture models.

Remark. In the following chapters, we may not be drawing such a clear distinction between latent variables z and uncertain model parameters θ and call the model parameters "latent" or "hidden" as well because they are unobserved. In Chapters 10 and 11, where we use the latent variables z, we will pay attention to the difference as we will have two different types of hidden variables: model parameters θ and latent variables z. ♢

We can exploit the fact that all the elements of a probabilistic model are random variables to define a unified language for representing them. In Section 8.5, we will see a concise graphical language for representing the structure of probabilistic models. We will use this graphical language to describe the probabilistic models in the subsequent chapters.

8.4.4 Further Reading

Probabilistic models in machine learning (Bishop, 2006; Barber, 2012; Murphy, 2012) provide a way for users to capture uncertainty about data and predictive models in a principled fashion. Ghahramani (2015) presents a short review of probabilistic models in machine learning. Given a probabilistic model, we may be lucky enough to be able to compute parameters of interest analytically. However, in general, analytic solutions are rare, and computational methods such as sampling (Gilks et al., 1996; Brooks et al., 2011) and variational inference (Jordan et al., 1999; Blei et al., 2017) are used. Moustaki et al. (2015) and Paquet (2008) provide a good overview of Bayesian inference in latent-variable models.

In recent years, several programming languages have been proposed that aim to treat the variables defined in software as random variables corresponding to probability distributions. The objective is to be able to write complex functions of probability distributions, while under the hood the compiler automatically takes care of the rules of Bayesian inference. This rapidly changing field is called *probabilistic programming*.

probabilistic programming

8.5 Directed Graphical Models

In this section, we introduce a graphical language for specifying a probabilistic model, called the *directed graphical model*. It provides a compact and succinct way to specify probabilistic models, and allows the reader to visually parse dependencies between random variables. A graphical model visually captures the way in which the joint distribution over all random variables can be decomposed into a product of factors depending only on a subset of these variables. In Section 8.4, we identified the joint distribution of a probabilistic model as the key quantity of interest because it comprises information about the prior, the likelihood, and the posterior. However, the joint distribution by itself can be quite complicated, and it does not tell us anything about structural properties of the probabilistic model. For example, the joint distribution $p(a, b, c)$ does not tell us anything about independence relations. This is the point where graphical

directed graphical model

Directed graphical models are also known as Bayesian networks.

Figure 8.9 Examples of directed graphical models.

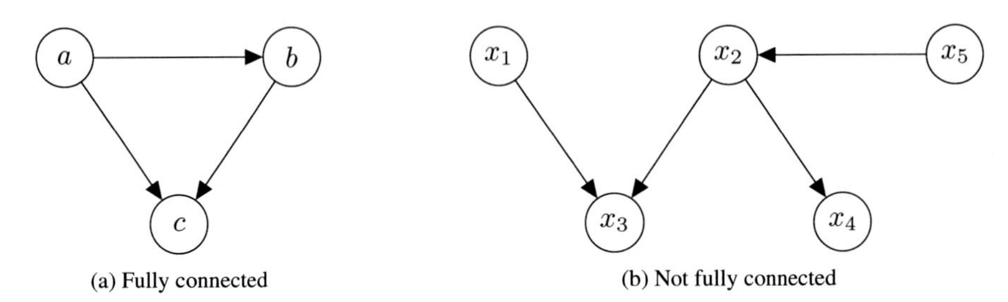

(a) Fully connected

(b) Not fully connected

models come into play. This section relies on the concepts of independence and conditional independence, as described in Section 6.4.5.

graphical model

In a *graphical model*, nodes are random variables. In Figure 8.9(a), the nodes represent the random variables a, b, c. Edges represent probabilistic relations between variables, e.g., conditional probabilities.

Remark. Not every distribution can be represented in a particular choice of graphical model. A discussion of this can be found in Bishop (2006). $\diamondsuit$

Probabilistic graphical models have some convenient properties:

- They are a simple way to visualize the structure of a probabilistic model.
- They can be used to design or motivate new kinds of statistical models.
- Inspection of the graph alone gives us insight into properties, e.g., conditional independence.
- Complex computations for inference and learning in statistical models can be expressed in terms of graphical manipulations.

8.5.1 Graph Semantics

directed graphical model/Bayesian network

With additional assumptions, the arrows can be used to indicate causal relationships (Pearl, 2009).

Directed graphical models/Bayesian networks are a method for representing conditional dependencies in a probabilistic model. They provide a visual description of the conditional probabilities, hence providing a simple language for describing complex interdependence. The modular description also entails computational simplification. Directed links (arrows) between two nodes (random variables) indicate conditional probabilities. For example, the arrow between a and b in Figure 8.9(a) gives the conditional probability $p(b \mid a)$ of b given a.

Directed graphical models can be derived from joint distributions if we know something about their factorization.

Example 8.7

Consider the joint distribution

$$p(a, b, c) = p(c \mid a, b)p(b \mid a)p(a) \tag{8.29}$$

of three random variables a, b, c. The factorization of the joint distribution in (8.29) tells us something about the relationship between the random variables:

- c depends directly on a and b.
- b depends directly on a.
- a depends neither on b nor on c.

For the factorization in (8.29), we obtain the directed graphical model in Figure 8.9(a).

In general, we can construct the corresponding directed graphical model from a factorized joint distribution as follows:

1. Create a node for all random variables.
2. For each conditional distribution, we add a directed link (arrow) to the graph from the nodes corresponding to the variables on which the distribution is conditioned.

The graph layout depends on the choice of factorization of the joint distribution.

The graph layout depends on the factorization of the joint distribution.

We discussed how to get from a known factorization of the joint distribution to the corresponding directed graphical model. Now, we will do exactly the opposite and describe how to extract the joint distribution of a set of random variables from a given graphical model.

Example 8.8

Looking at the graphical model in Figure 8.9(b), we exploit two properties:

- The joint distribution $p(x_1, \dots, x_5)$ we seek is the product of a set of conditionals, one for each node in the graph. In this particular example, we will need five conditionals.
- Each conditional depends only on the parents of the corresponding node in the graph. For example, x_4 will be conditioned on x_2.

These two properties yield the desired factorization of the joint distribution

$$p(x_1, x_2, x_3, x_4, x_5) = p(x_1)p(x_5)p(x_2 \mid x_5)p(x_3 \mid x_1, x_2)p(x_4 \mid x_2) . \tag{8.30}$$

In general, the joint distribution $p(\boldsymbol{x}) = p(x_1, \dots, x_K)$ is given as

$$p(\boldsymbol{x}) = \prod_{k=1}^{K} p(x_k \mid \mathrm{Pa}_k) , \tag{8.31}$$

where Pa_k means "the parent nodes of x_k." Parent nodes of x_k are nodes that have arrows pointing to x_k.

We conclude this subsection with a concrete example of the coin-flip experiment. Consider a Bernoulli experiment (Example 6.8) where the probability that the outcome x of this experiment is "heads" is

$$p(x \mid \mu) = \mathrm{Ber}(\mu) . \tag{8.32}$$

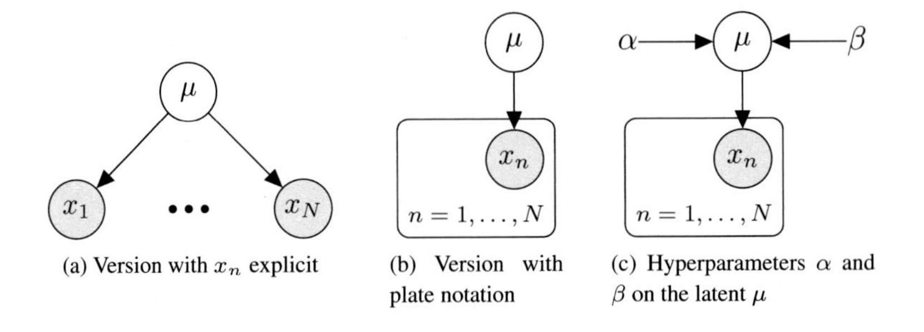

Figure 8.10 Graphical models for a repeated Bernoulli experiment.

(a) Version with x_n explicit

(b) Version with plate notation

(c) Hyperparameters α and β on the latent μ

We now repeat this experiment N times and observe outcomes $x_1, \ldots, x_N$ so that we obtain the joint distribution

$$p(x_1, \ldots, x_N \mid \mu) = \prod_{n=1}^{N} p(x_n \mid \mu) . \tag{8.33}$$

The expression on the right-hand side is a product of Bernoulli distributions on each individual outcome because the experiments are independent. Recall from Section 6.4.5 that statistical independence means that the distribution factorizes. To write the graphical model down for this setting, we make the distinction between unobserved/latent variables and observed variables. Graphically, observed variables are denoted by shaded nodes so that we obtain the graphical model in Figure 8.10(a). We see that the single parameter μ is the same for all x_n, $n = 1, \ldots, N$ as the outcomes x_n are identically distributed. A more compact, but equivalent, graphical model for this setting is given in Figure 8.10(b), where we use the *plate* notation. The plate (box) repeats everything inside (in this case, the observations x_n) N times. Therefore, both graphical models are equivalent, but the plate notation is more compact. Graphical models immediately allow us to place a hyperprior on μ. A *hyperprior* is a second layer of prior distributions on the parameters of the first layer of priors. Figure 8.10(c) places a Beta(α, β) prior on the latent variable μ. If we treat α and β as deterministic parameters, i.e., not random variables, we omit the circle around it.

plate

hyperprior

8.5.2 Conditional Independence and d-Separation

Directed graphical models allow us to find conditional independence (Section 6.4.5) relationship properties of the joint distribution only by looking at the graph. A concept called *d-separation* (Pearl, 1988) is key to this.

d-separation

Consider a general directed graph in which $\mathcal{A}, \mathcal{B}, \mathcal{C}$ are arbitrary nonintersecting sets of nodes (whose union may be smaller than the complete set of nodes in the graph). We wish to ascertain whether a particular conditional independence statement, "$\mathcal{A}$ is conditionally independent of $\mathcal{B}$ given $\mathcal{C}$," denoted by

$$\mathcal{A} \perp\!\!\!\perp \mathcal{B} \mid \mathcal{C} , \tag{8.34}$$

is implied by a given directed acyclic graph. To do so, we consider all possible trails (paths that ignore the direction of the arrows) from any node in $\mathcal{A}$ to any nodes in $\mathcal{B}$. Any such path is said to be blocked if it includes any node such that either of the following are true:

- The arrows on the path meet either head to tail or tail to tail at the node, and the node is in the set $\mathcal{C}$.
- The arrows meet head to head at the node, and neither the node nor any of its descendants is in the set $\mathcal{C}$.

If all paths are blocked, then $\mathcal{A}$ is said to be *d-separated* from $\mathcal{B}$ by $\mathcal{C}$, and the joint distribution over all of the variables in the graph will satisfy $\mathcal{A} \perp\!\!\!\perp \mathcal{B} \,|\, \mathcal{C}$.

Example 8.9 (Conditional Independence)
Consider the graphical model in Figure 8.11. By visual inspection, we see that

$$b \perp\!\!\!\perp d \,|\, a, c \tag{8.35}$$
$$a \perp\!\!\!\perp c \,|\, b \tag{8.36}$$
$$b \not\perp\!\!\!\perp d \,|\, c \tag{8.37}$$
$$a \not\perp\!\!\!\perp c \,|\, b, e\,. \tag{8.38}$$

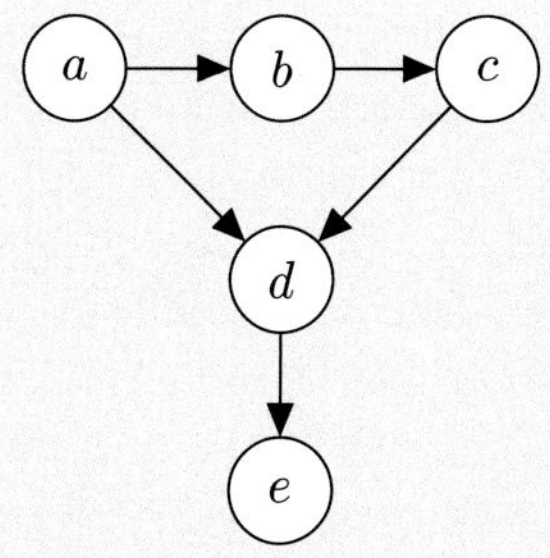

Figure 8.11 D-separation example.

Directed graphical models allow a compact representation of probabilistic models, and we will see examples of directed graphical models in Chapters 9 through 11. The representation, along with the concept of conditional independence, allows us to factorize the respective probabilistic models into expressions that are easier to optimize.

The graphical representation of the probabilistic model allows us to visually see the impact of design choices we have made on the structure of the model. We often need to make high-level assumptions about the structure of the model. These modeling assumptions (hyperparameters) affect the prediction performance, but cannot be selected directly using the approaches we have seen so far. We will discuss different ways to choose the structure in Section 8.6.

8.5.3 Further Reading

An introduction to probabilistic graphical models can be found in Bishop (2006, chapter 8), and an extensive description of the different applications and corresponding algorithmic implications can be found in the book by Koller and Friedman (2009).

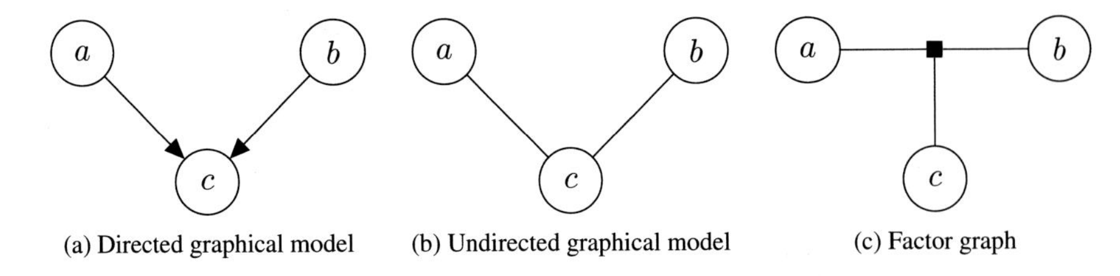

Figure 8.12 Three types of graphical models: (a) directed graphical models (Bayesian networks); (b) undirected graphical models (Markov random fields); (c) factor graphs.

There are three main types of probabilistic graphical models:

directed graphical model
Bayesian network
undirected graphical model
Markov random field
factor graph

- *Directed graphical models* (*Bayesian networks*); see Figure 8.12(a)
- *Undirected graphical models* (*Markov random fields*); see Figure 8.12(b)
- *Factor graphs*; see Figure 8.12(c)

Graphical models allow for graph-based algorithms for inference and learning, e.g., via local message passing. Applications range from ranking in online games (Herbrich et al., 2007) and computer vision (e.g., image segmentation, semantic labeling, image denoising, image restoration (Kittler and Föglein, 1984; Sucar and Gillies, 1994; Shotton et al., 2006; Szeliski et al., 2008)) to coding theory (McEliece et al., 1998), solving linear equation systems (Shental et al., 2008), and iterative Bayesian state estimation in signal processing (Bickson et al., 2007; Deisenroth and Mohamed, 2012).

One topic that is particularly important in real applications that we do not discuss in this book is the idea of structured prediction (Bakir et al., 2007; Nowozin et al., 2014), which allows machine learning models to tackle predictions that are structured, for example sequences, trees, and graphs. The popularity of neural network models has allowed more flexible probabilistic models to be used, resulting in many useful applications of structured models (Goodfellow et al., 2016, chapter 16). In recent years, there has been a renewed interest in graphical models due to their applications to causal inference (Pearl, 2009; Imbens and Rubin, 2015; Peters et al., 2017; Rosenbaum, 2017).

8.6 Model Selection

In machine learning, we often need to make high-level modeling decisions that critically influence the performance of the model. The choices we make (e.g., the functional form of the likelihood) influence the number and type of free parameters in the model and thereby also the flexibility and expressivity of the model. More complex models are more flexible in the sense that they can be used to describe more datasets. For instance, a polynomial of degree 1 (a line $y = a_0 + a_1x$) can only be used to describe linear relations between inputs x and observations y. A polynomial of degree 2 can additionally describe quadratic relationships between inputs and observations.

A polynomial $y = a_0 + a_1x + a_2x^2$ can also describe linear functions by setting $a_2 = 0$, i.e., it is strictly more expressive than a first-order polynomial.

One would now think that very flexible models are generally preferable to simple models because they are more expressive. A general problem is that at training time we can only use the training set to evaluate the performance of the model and learn its parameters. However, the performance on the training set is

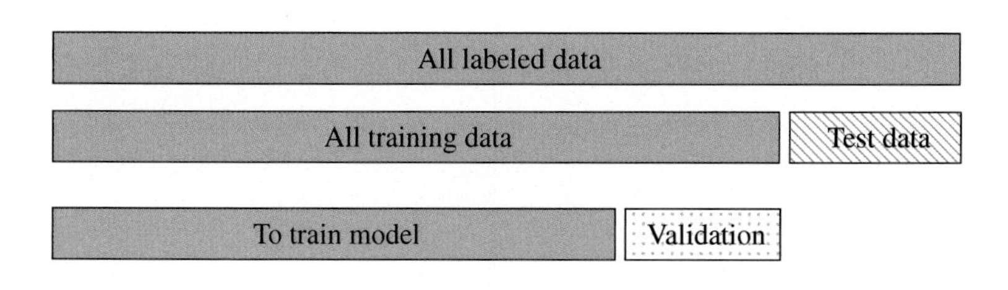

Figure 8.13 Nested cross-validation. We perform two levels of K-fold cross-validation.

not really what we are interested in. In Section 8.3, we have seen that maximum likelihood estimation can lead to overfitting, especially when the training dataset is small. Ideally, our model (also) works well on the test set (which is not available at training time). Therefore, we need some mechanisms for assessing how a model *generalizes* to unseen test data. *Model selection* is concerned with exactly this problem.

8.6.1 Nested Cross-Validation

We have already seen an approach (cross-validation in Section 8.2.4) that can be used for model selection. Recall that cross-validation provides an estimate of the generalization error by repeatedly splitting the dataset into training and validation sets. We can apply this idea one more time, i.e., for each split, we can perform another round of cross-validation. This is sometimes referred to as *nested cross-validation* (see Figure 8.13). The inner level is used to estimate the performance of a particular choice of model or hyperparameter on a internal validation set. The outer level is used to estimate generalization performance for the best choice of model chosen by the inner loop. We can test different model and hyperparameter choices in the inner loop. To distinguish the two levels, the set used to estimate the generalization performance is often called the *test set* and the set used for choosing the best model is called the *validation set*. The inner loop estimates the expected value of the generalization error for a given model (8.39), by approximating it using the empirical error on the validation set, i.e.,

nested cross-validation

test set

validation set

The standard error is defined as $\frac{\sigma}{\sqrt{K}}$, where K is the number of experiments and σ is the standard deviation of the risk of each experiment.

$$\mathbb{E}_{\mathcal{V}}[\mathbf{R}(\mathcal{V} \mid M)] \approx \frac{1}{K}\sum_{k=1}^{K}\mathbf{R}(\mathcal{V}^{(k)} \mid M)\,, \tag{8.39}$$

where $\mathbf{R}(\mathcal{V} \mid M)$ is the empirical risk (e.g., root mean square error) on the validation set $\mathcal{V}$ for model M. We repeat this procedure for all models and choose the model that performs best. Note that cross-validation not only gives us the expected generalization error, but we can also obtain high-order statistics, e.g., the standard error, an estimate of how uncertain the mean estimate is. Once the model is chosen, we can evaluate the final performance on the test set.

8.6.2 Bayesian Model Selection

There are many approaches to model selection, some of which are covered in this section. Generally, they all attempt to trade off model complexity and data fit. We assume that simpler models are less prone to overfitting than complex

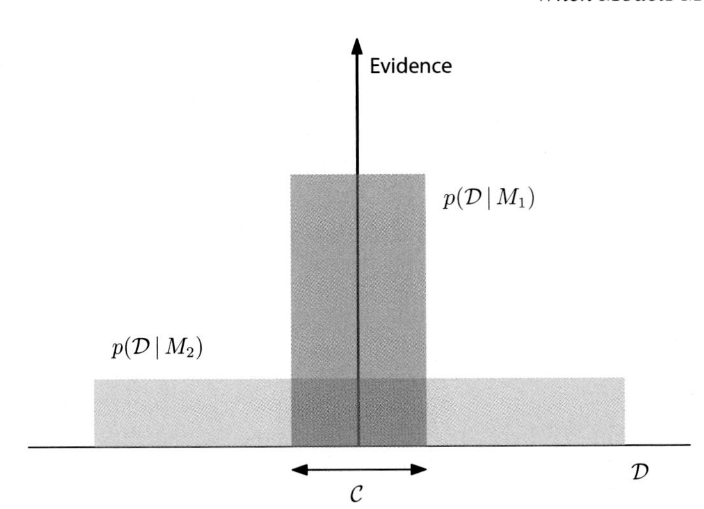

Figure 8.14 Bayesian inference embodies Occam's razor. The horizontal axis describes the space of all possible datasets $\mathcal{D}$. The evidence (vertical axis) evaluates how well a model predicts available data. Since $p(\mathcal{D} \mid M_i)$ needs to integrate to 1, we should choose the model with the greatest evidence. Adapted from MacKay (2003).

models, and hence the objective of model selection is to find the simplest model that explains the data reasonably well. This concept is also known as *Occam's razor*.

Occam's razor

Remark. If we treat model selection as a hypothesis testing problem, we are looking for the simplest hypothesis that is consistent with the data (Murphy, 2012). ◇

One may consider placing a prior on models that favors simpler models. However, it is not necessary to do this: An "automatic Occam's Razor" is quantitatively embodied in the application of Bayesian probability (Smith and Spiegelhalter, 1980; Jefferys and Berger, 1992; MacKay, 1992). Figure 8.14, adapted from MacKay (2003), gives us the basic intuition why complex and very expressive models may turn out to be a less probable choice for modeling a given dataset $\mathcal{D}$. Let us think of the horizontal axis representing the space of all possible datasets $\mathcal{D}$. If we are interested in the posterior probability $p(M_i \mid \mathcal{D})$ of model M_i given the data $\mathcal{D}$, we can employ Bayes' theorem. Assuming a uniform prior $p(M)$ over all models, Bayes' theorem rewards models in proportion to how much they predicted the data that occurred. This prediction of the data given model M_i, $p(\mathcal{D} \mid M_i)$, is called the *evidence* for M_i. A simple model M_1 can only predict a small number of datasets, which is shown by $p(\mathcal{D} \mid M_1)$; a more powerful model M_2 that has, e.g., more free parameters than M_1, is able to predict a greater variety of datasets. This means, however, that M_2 does not predict the datasets in region C as well as M_1. Suppose that equal prior probabilities have been assigned to the two models. Then, if the dataset falls into region C, the less powerful model M_1 is the more probable model.

These predictions are quantified by a normalized probability distribution on $\mathcal{D}$, i.e., it needs to integrate/sum to 1.

evidence

Earlier in this chapter, we argued that models need to be able to explain the data, i.e., there should be a way to generate data from a given model. Furthermore, if the model has been appropriately learned from the data, then we expect that the generated data should be similar to the empirical data. For this, it is helpful to phrase model selection as a hierarchical inference problem, which allows us to compute the posterior distribution over models.

Let us consider a finite number of models $M = \{M_1, \dots, M_K\}$, where each model M_k possesses parameters $\boldsymbol{\theta}_k$. In *Bayesian model selection*, we place a prior $p(M)$ on the set of models. The corresponding *generative process* that allows us to generate data from this model is

Bayesian model selection

generative process

$$M_k \sim p(M) \tag{8.40}$$

$$\boldsymbol{\theta}_k \sim p(\boldsymbol{\theta} \mid M_k) \tag{8.41}$$

$$\mathcal{D} \sim p(\mathcal{D} \mid \boldsymbol{\theta}_k) \tag{8.42}$$

and illustrated in Figure 8.15. Given a training set $\mathcal{D}$, we apply Bayes' theorem and compute the posterior distribution over models as

$$p(M_k \mid \mathcal{D}) \propto p(M_k)p(\mathcal{D} \mid M_k) . \tag{8.43}$$

Note that this posterior no longer depends on the model parameters $\boldsymbol{\theta}_k$ because they have been integrated out in the Bayesian setting since

$$p(\mathcal{D} \mid M_k) = \int p(\mathcal{D} \mid \boldsymbol{\theta}_k)p(\boldsymbol{\theta}_k \mid M_k)d\boldsymbol{\theta}_k , \tag{8.44}$$

where $p(\boldsymbol{\theta}_k \mid M_k)$ is the prior distribution of the model parameters $\boldsymbol{\theta}_k$ of model M_k. The term (8.44) is referred to as the *model evidence* or *marginal likelihood*. From the posterior in (8.43), we determine the MAP estimate

Figure 8.15 Illustration of the hierarchical generative process in Bayesian model selection. We place a prior $p(M)$ on the set of models. For each model, there is a distribution $p(\boldsymbol{\theta} \mid M)$ on the corresponding model parameters, which is used to generate the data $\mathcal{D}$.

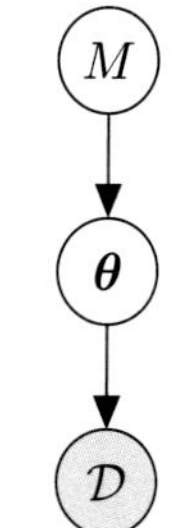

model evidence

marginal likelihood

$$M^* = \arg\max_{M_k} p(M_k \mid \mathcal{D}) . \tag{8.45}$$

With a uniform prior $p(M_k) = \frac{1}{K}$, which gives every model equal (prior) probability, determining the MAP estimate over models amounts to picking the model that maximizes the model evidence (8.44).

Remark (Likelihood and Marginal Likelihood). There are some important differences between a likelihood and a marginal likelihood (evidence): While the likelihood is prone to overfitting, the marginal likelihood is typically not as the model parameters have been marginalized out (i.e., we no longer have to fit the parameters). Furthermore, the marginal likelihood automatically embodies a trade-off between model complexity and data fit (Occam's razor). $\diamondsuit$

8.6.3 Bayes Factors for Model Comparison

Consider the problem of comparing two probabilistic models M_1, M_2, given a dataset $\mathcal{D}$. If we compute the posteriors $p(M_1 \mid \mathcal{D})$ and $p(M_2 \mid \mathcal{D})$, we can compute the ratio of the posteriors

$$\underbrace{\frac{p(M_1 \mid \mathcal{D})}{p(M_2 \mid \mathcal{D})}}_{\text{Posterior odds}} = \frac{\frac{p(\mathcal{D} \mid M_1)p(M_1)}{p(\mathcal{D})}}{\frac{p(\mathcal{D} \mid M_2)p(M_2)}{p(\mathcal{D})}} = \underbrace{\frac{p(M_1)}{p(M_2)}}_{\text{Prior odds}} \underbrace{\frac{p(\mathcal{D} \mid M_1)}{p(\mathcal{D} \mid M_2)}}_{\text{Bayes factor}} . \tag{8.46}$$

The ratio of the posteriors is also called the *posterior odds*. The first fraction on the right-hand side of (8.46), the *prior odds*, measures how much our prior (initial) beliefs favor M_1 over M_2. The ratio of the marginal likelihoods (second

posterior odds

prior odds

Bayes factor

fraction on the right-hand side) is called the *Bayes factor* and measures how well the data $\mathcal{D}$ is predicted by M_1 compared to M_2.

Jeffreys–Lindley paradox

Remark. The *Jeffreys–Lindley paradox* states that the "Bayes factor always favors the simpler model since the probability of the data under a complex model with a diffuse prior will be very small" (Murphy, 2012). Here, a diffuse prior refers to a prior that does not favor specific models, i.e., many models are a priori plausible under this prior. ◇

If we choose a uniform prior over models, the prior odds term in (8.46) is 1, i.e., the posterior odds is the ratio of the marginal likelihoods (Bayes factor)

$$\frac{p(\mathcal{D} \mid M_1)}{p(\mathcal{D} \mid M_2)}. \tag{8.47}$$

If the Bayes factor is greater than 1, we choose model M_1, otherwise model M_2. In a similar way to frequentist statistics, there are guidelines on the size of the ratio that one should consider before "significance" of the result (Jeffreys, 1961).

Remark (Computing the Marginal Likelihood). The marginal likelihood plays an important role in model selection: We need to compute Bayes factors (8.46) and posterior distributions over models (8.43).

Unfortunately, computing the marginal likelihood requires us to solve an integral (8.44). This integration is generally analytically intractable, and we will have to resort to approximation techniques, e.g., numerical integration (Stoer and Burlirsch, 2002), stochastic approximations using Monte Carlo (Murphy, 2012), or Bayesian Monte Carlo techniques (O'Hagan, 1991; Rasmussen and Ghahramani, 2003).

However, there are special cases in which we can solve it. In Section 6.6.1, we discussed conjugate models. If we choose a conjugate parameter prior $p(\boldsymbol{\theta})$, we can compute the marginal likelihood in closed form. In Chapter 9, we will do exactly this in the context of linear regression. ◇

We have seen a brief introduction to the basic concepts of machine learning in this chapter. For the rest of this part of the book, we will see how the three different flavors of learning in Sections 8.2 through 8.4 are applied to the four pillars of machine learning (regression, dimensionality reduction, density estimation, and classification).

8.6.4 Further Reading

We mentioned at the start of the section that there are high-level modeling choices that influence the performance of the model. Examples include the following:

- The degree of a polynomial in a regression setting
- The number of components in a mixture model
- The network architecture of a (deep) neural network
- The type of kernel in a support vector machine
- The dimensionality of the latent space in PCA
- The learning rate (schedule) in an optimization algorithm

Rasmussen and Ghahramani (2001) showed necessarily penalize the number of parameters in a model, but it is active in terms of thc complexity of functions. They also showed that the automatic Occam's razor also holds for Bayesian nonparametric models with many parameters, e.g., Gaussian processes.

In parametric models, the number of parameters is often related to the complexity of the model class.

If we focus on the maximum likelihood estimate, there exist a number of heuristics for model selection that discourage overfitting. They are called information criteria, and we choose the model with the largest value. The *Akaike information criterion* (AIC) (Akaike, 1974)

Akaike information criterion

$$\log p(\boldsymbol{x} \mid \boldsymbol{\theta}) - M \tag{8.48}$$

corrects for the bias of the maximum likelihood estimator by addition of a penalty term to compensate for the overfitting of more complex models with lots of parameters. Here, M is the number of model parameters. The AIC estimates the relative information lost by a given model.

The *Bayesian information criterion* (BIC) (Schwarz, 1978)

Bayesian information criterion

$$\log p(\boldsymbol{x}) = \log \int p(\boldsymbol{x} \mid \boldsymbol{\theta}) p(\boldsymbol{\theta}) \mathrm{d}\boldsymbol{\theta} \approx \log p(\boldsymbol{x} \mid \boldsymbol{\theta}) - \frac{1}{2} M \log N \tag{8.49}$$

can be used for exponential family distributions. Here, N is the number of data points and M is the number of parameters. BIC penalizes model complexity more heavily than AIC.

9

Linear Regression

In the following, we will apply the mathematical concepts from Chapters 2 and 5 through 7 to solve linear regression (curve fitting) problems. In *regression*, we aim to find a function f that maps inputs $\boldsymbol{x} \in \mathbb{R}^D$ to corresponding function values $f(\boldsymbol{x}) \in \mathbb{R}$. We assume we are given a set of training inputs $\boldsymbol{x}_n$ and corresponding noisy observations $y_n = f(\boldsymbol{x}_n) + \epsilon$, where ϵ is an i.i.d. random variable that describes measurement/observation noise and potentially unmodeled processes (which we will not consider further in this chapter). Throughout this chapter, we assume zero-mean Gaussian noise. Our task is to find a function that not only models the training data, but generalizes well to predicting function values at input locations that are not part of the training data (see Chapter 8). An illustration of such a regression problem is given in Figure 9.1. A typical regression setting is given in Figure 9.2(a): For some input values x_n, we observe (noisy) function values $y_n = f(x_n) + \epsilon$. The task is to infer the function f that generated the data and generalizes well to function values at new input locations. A possible solution is given in Figure 9.2(b), where we also show three distributions centered at the function values $f(x)$ that represent the noise in the data.

regression

Regression is a fundamental problem in machine learning, and regression problems appear in a diverse range of research areas and applications, including time-series analysis (e.g., system identification), control and robotics (e.g., reinforcement learning, forward/inverse model learning), optimization (e.g., line searches, global optimization), and deep-learning applications (e.g.,

Figure 9.1 (a) Dataset; (b) possible solution to the regression problem.

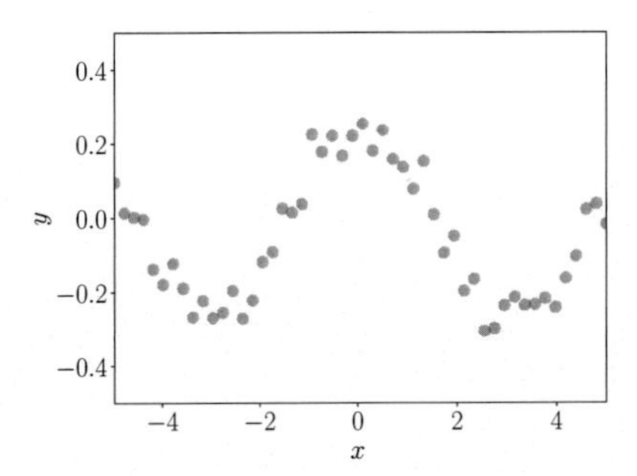

(a) Regression problem: observed noisy function values from which we wish to infer the underlying function that generated the data.

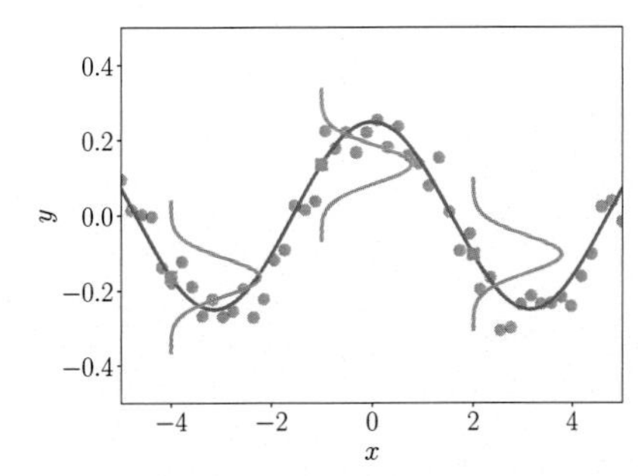

(b) Regression solution: possible function that could have generated the data (blue) with indication of the measurement noise of the function value at the corresponding inputs (orange distributions).

computer games, speech-to-text translation, image recognition, automatic video annotation). Regression is also a key ingredient of classification algorithms. Finding a regression function requires solving a variety of problems, including the following:

- **Choice of the model (type) and the parametrization** of the regression function. Given a dataset, what function classes (e.g., polynomials) are good candidates for modeling the data, and what particular parametrization (e.g., degree of the polynomial) should we choose? Model selection, as discussed in Section 8.6, allows us to compare various models to find the simplest model that explains the training data reasonably well.
- **Finding good parameters.** Having chosen a model of the regression function, how do we find good model parameters? Here, we will need to look at different loss/objective functions (they determine what a "good" fit is) and optimization algorithms that allow us to minimize this loss.
- **Overfitting and model selection.** Overfitting is a problem when the regression function fits the training data "too well" but does not generalize to unseen test data. Overfitting typically occurs if the underlying model (or its parametrization) is overly flexible and expressive; see Section 8.6. We will look at the underlying reasons and discuss ways to mitigate the effect of overfitting in the context of linear regression.
- **Relationship between loss functions and parameter priors.** Loss functions (optimization objectives) are often motivated and induced by probabilistic models. We will look at the connection between loss functions and the underlying prior assumptions that induce these losses.
- **Uncertainty modeling.** In any practical setting, we have access to only a finite, potentially large, amount of (training) data for selecting the model class and the corresponding parameters. Given that this finite amount of training data does not cover all possible scenarios, we may want to describe the remaining parameter uncertainty to obtain a measure of confidence of the model's prediction at test time; the smaller the training set, the more important uncertainty modeling. Consistent modeling of uncertainty equips model predictions with confidence bounds.

Normally, the type of noise could also be a "model choice," but we fix the noise to be Gaussian in this chapter.

In the following, we will be using the mathematical tools from Chapters 3 and 5 through 7 to solve linear regression problems. We will discuss maximum likelihood and maximum a posteriori (MAP) estimation to find optimal model parameters. Using these parameter estimates, we will have a brief look at generalization errors and overfitting. Toward the end of this chapter, we will discuss Bayesian linear regression, which allows us to reason about model parameters at a higher level, thereby removing some of the problems encountered in maximum likelihood and MAP estimation.

9.1 Problem Formulation

Because of the presence of observation noise, we will adopt a probabilistic approach and explicitly model the noise using a likelihood function. More specifically, throughout this chapter, we consider a regression problem with the likelihood function

$$p(y \mid \boldsymbol{x}) = \mathcal{N}\big(y \mid f(\boldsymbol{x}),\, \sigma^2\big)\,. \tag{9.1}$$

Here, $\boldsymbol{x} \in \mathbb{R}^D$ are inputs and $y \in \mathbb{R}$ are noisy function values (targets). With (9.1), the functional relationship between $\boldsymbol{x}$ and y is given as

$$y = f(\boldsymbol{x}) + \epsilon\,, \tag{9.2}$$

where $\epsilon \sim \mathcal{N}\big(0,\, \sigma^2\big)$ is independent, identically distributed (i.i.d.) Gaussian measurement noise with mean 0 and variance σ^2. Our objective is to find a function that is close (similar) to the unknown function f that generated the data and that generalizes well.

In this chapter, we focus on parametric models, i.e., we choose a parametrized function and find parameters $\boldsymbol{\theta}$ that "work well" for modeling the data. For the time being, we assume that the noise variance σ^2 is known and focus on learning the model parameters $\boldsymbol{\theta}$. In linear regression, we consider the special case that the parameters $\boldsymbol{\theta}$ appear linearly in our model. An example of linear regression is given by

$$p(y \mid \boldsymbol{x}, \boldsymbol{\theta}) = \mathcal{N}\big(y \mid \boldsymbol{x}^\top \boldsymbol{\theta},\, \sigma^2\big) \tag{9.3}$$

$$\iff y = \boldsymbol{x}^\top \boldsymbol{\theta} + \epsilon\,, \quad \epsilon \sim \mathcal{N}\big(0,\, \sigma^2\big)\,, \tag{9.4}$$

where $\boldsymbol{\theta} \in \mathbb{R}^D$ are the parameters we seek. The class of functions described by (9.4) are straight lines that pass through the origin. In (9.4), we chose a parametrization $f(\boldsymbol{x}) = \boldsymbol{x}^\top \boldsymbol{\theta}$.

A Dirac delta (delta function) is zero everywhere except at a single point, and its integral is 1. It can be considered a Gaussian in the limit of $\sigma^2 \to 0$.

likelihood

The *likelihood* in (9.3) is the probability density function of y evaluated at $\boldsymbol{x}^\top \boldsymbol{\theta}$. Note that the only source of uncertainty originates from the observation noise (as $\boldsymbol{x}$ and $\boldsymbol{\theta}$ are assumed known in (9.3)). Without observation noise, the relationship between $\boldsymbol{x}$ and y would be deterministic and (9.3) would be a Dirac delta.

Example 9.1

For $x, \theta \in \mathbb{R}$, the linear regression model in (9.4) describes straight lines (linear functions), and the parameter θ is the slope of the line. Figure 9.2(a) shows some example functions for different values of θ.

Linear regression refers to models that are linear in the parameters.

The linear regression model in (9.3) and (9.4) is not only linear in the parameters, but also linear in the inputs x. Figure 9.2(a) shows examples of such functions. We will see later that $y = \boldsymbol{\phi}^\top(\boldsymbol{x})\boldsymbol{\theta}$ for nonlinear transformations $\boldsymbol{\phi}$ is also a linear regression model because "linear regression" refers to models that

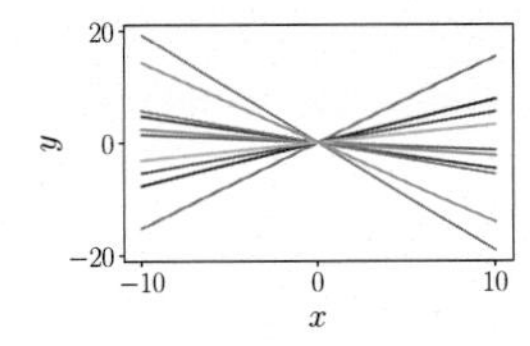

(a) Example functions (straight lines) that can be described using the linear model in (9.4)

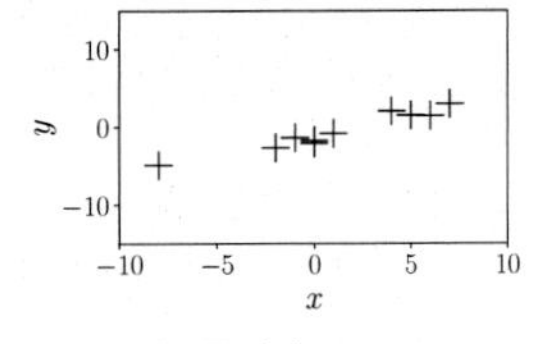

(b) Training set

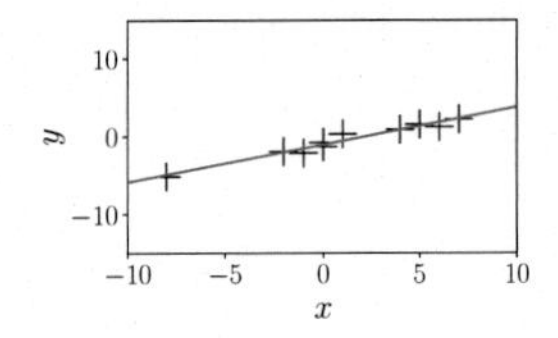

(c) Maximum likelihood estimate

Figure 9.2 Linear regression example. (a) Example functions that fall into this category, (b) training set, (c) maximum likelihood estimate.

are "linear in the parameters," i.e., models that describe a function by a linear combination of input features. Here, a "feature" is a representation $\phi(\boldsymbol{x})$ of the inputs $\boldsymbol{x}$.

In the following, we will discuss in more detail how to find good parameters $\boldsymbol{\theta}$ and how to evaluate whether a parameter set "works well." For the time being, we assume that the noise variance σ^2 is known.

9.2 Parameter Estimation

Consider the linear regression setting (9.4) and assume we are given a *training set* $\mathcal{D} := \{(\boldsymbol{x}_1, y_1), \ldots, (\boldsymbol{x}_N, y_N)\}$ consisting of N inputs $\boldsymbol{x}_n \in \mathbb{R}^D$ and corresponding observations/targets $y_n \in \mathbb{R}$, $n = 1, \ldots, N$. The corresponding graphical model is given in Figure 9.3. Note that y_i and y_j are conditionally independent given their respective inputs $\boldsymbol{x}_i, \boldsymbol{x}_j$ so that the likelihood factorizes according to

training set

$$p(\mathcal{Y} \mid \mathcal{X}, \boldsymbol{\theta}) = p(y_1, \ldots, y_N \mid \boldsymbol{x}_1, \ldots, \boldsymbol{x}_N, \boldsymbol{\theta}) \tag{9.5a}$$

$$= \prod_{n=1}^{N} p(y_n \mid \boldsymbol{x}_n, \boldsymbol{\theta}) = \prod_{n=1}^{N} \mathcal{N}\big(y_n \mid \boldsymbol{x}_n^\top \boldsymbol{\theta},\, \sigma^2\big), \tag{9.5b}$$

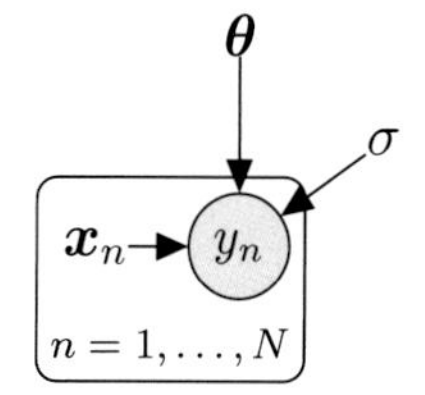

Figure 9.3 Probabilistic graphical model for linear regression. Observed random variables are shaded, deterministic/known values are without circles.

where we defined $\mathcal{X} := \{\boldsymbol{x}_1, \ldots, \boldsymbol{x}_N\}$ and $\mathcal{Y} := \{y_1, \ldots, y_N\}$ as the sets of training inputs and corresponding targets, respectively. The likelihood and the factors $p(y_n \mid \boldsymbol{x}_n, \boldsymbol{\theta})$ are Gaussian due to the noise distribution; see (9.3).

In the following, we will discuss how to find optimal parameters $\boldsymbol{\theta}^* \in \mathbb{R}^D$ for the linear regression model (9.4). Once the parameters $\boldsymbol{\theta}^*$ are found, we can predict function values by using this parameter estimate in (9.4) so that at an arbitrary test input $\boldsymbol{x}_*$ the distribution of the corresponding target y_* is

$$p(y_* \mid \boldsymbol{x}_*, \boldsymbol{\theta}^*) = \mathcal{N}\big(y_* \mid \boldsymbol{x}_*^\top \boldsymbol{\theta}^*,\, \sigma^2\big). \tag{9.6}$$

In the following, we will have a look at parameter estimation by maximizing the likelihood, a topic that we already covered to some degree in Section 8.3.

9.2.1 Maximum Likelihood Estimation

A widely used approach to finding the desired parameters $\boldsymbol{\theta}_{\mathrm{ML}}$ is *maximum likelihood estimation*, where we find parameters $\boldsymbol{\theta}_{\mathrm{ML}}$ that maximize the likelihood (9.5b). Intuitively, maximizing the likelihood means maximizing the predictive distribution of the training data given the model parameters. We obtain the maximum likelihood parameters as

maximum likelihood estimation

Maximizing the likelihood means maximizing the predictive distribution of the (training) data given the parameters.

$$\boldsymbol{\theta}_{\mathrm{ML}} = \arg\max_{\boldsymbol{\theta}} p(\mathcal{Y} \mid \mathcal{X}, \boldsymbol{\theta}). \tag{9.7}$$

The likelihood is not a probability distribution in the parameters.

Remark. The likelihood $p(\boldsymbol{y} \mid \boldsymbol{x}, \boldsymbol{\theta})$ is not a probability distribution in $\boldsymbol{\theta}$: It is simply a function of the parameters $\boldsymbol{\theta}$ but does not integrate to 1 (i.e., it is unnormalized), and may not even be integrable with respect to $\boldsymbol{\theta}$. However, the likelihood in (9.7) is a normalized probability distribution in $\boldsymbol{y}$. ◇

To find the desired parameters $\boldsymbol{\theta}_{\text{ML}}$ that maximize the likelihood, we typically perform gradient ascent (or gradient descent on the negative likelihood). In the case of linear regression we consider here, however, a closed-form solution exists, which makes iterative gradient descent unnecessary. In practice, instead of maximizing the likelihood directly, we apply the log-transformation to the likelihood function and minimize the negative log-likelihood.

Since the logarithm is a (strictly) monotonically increasing function, the optimum of a function f is identical to the optimum of $\log f$.

Remark (Log-Transformation). Since the likelihood (9.5b) is a product of N Gaussian distributions, the log-transformation is useful since (a) it does not suffer from numerical underflow, and (b) the differentiation rules will turn out simpler. More specifically, numerical underflow will be a problem when we multiply N probabilities, where N is the number of data points, since we cannot represent very small numbers, such as 10^{-256}. Furthermore, the log-transform will turn the product into a sum of log-probabilities such that the corresponding gradient is a sum of individual gradients, instead of a repeated application of the product rule (5.46) to compute the gradient of a product of N terms. ◇

To find the optimal parameters $\boldsymbol{\theta}_{\text{ML}}$ of our linear regression problem, we minimize the negative log-likelihood

$$-\log p(\mathcal{Y} \mid \mathcal{X}, \boldsymbol{\theta}) = -\log \prod_{n=1}^{N} p(y_n \mid \boldsymbol{x}_n, \boldsymbol{\theta}) = -\sum_{n=1}^{N} \log p(y_n \mid \boldsymbol{x}_n, \boldsymbol{\theta}), \quad (9.8)$$

where we exploited that the likelihood (9.5b) factorizes over the number of data points due to our independence assumption on the training set.

In the linear regression model (9.4), the likelihood is Gaussian (due to the Gaussian additive noise term), such that we arrive at

$$\log p(y_n \mid \boldsymbol{x}_n, \boldsymbol{\theta}) = -\frac{1}{2\sigma^2}(y_n - \boldsymbol{x}_n^\top \boldsymbol{\theta})^2 + \text{const}, \quad (9.9)$$

where the constant includes all terms independent of $\boldsymbol{\theta}$. Using (9.9) in the negative log-likelihood (9.8), we obtain (ignoring the constant terms)

$$\mathcal{L}(\boldsymbol{\theta}) := \frac{1}{2\sigma^2}\sum_{n=1}^{N}(y_n - \boldsymbol{x}_n^\top \boldsymbol{\theta})^2 \quad (9.10a)$$

$$= \frac{1}{2\sigma^2}(\boldsymbol{y} - \boldsymbol{X}\boldsymbol{\theta})^\top(\boldsymbol{y} - \boldsymbol{X}\boldsymbol{\theta}) = \frac{1}{2\sigma^2}\|\boldsymbol{y} - \boldsymbol{X}\boldsymbol{\theta}\|^2, \quad (9.10b)$$

The negative log-likelihood function is also called the *error function*.

design matrix

The squared error is often used as a measure of distance. Recall from Section 3.1 that $\|\boldsymbol{x}\|^2 = \boldsymbol{x}^\top \boldsymbol{x}$ if we choose the dot product as the inner product.

where we define the *design matrix* $\boldsymbol{X} := [\boldsymbol{x}_1, \ldots, \boldsymbol{x}_N]^\top \in \mathbb{R}^{N \times D}$ as the collection of training inputs and $\boldsymbol{y} := [y_1, \ldots, y_N]^\top \in \mathbb{R}^N$ as a vector that collects all training targets. Note that the nth row in the design matrix $\boldsymbol{X}$ corresponds to the training input $\boldsymbol{x}_n$. In (9.10b), we used the fact that the sum of squared errors between the observations y_n and the corresponding model prediction $\boldsymbol{x}_n^\top \boldsymbol{\theta}$ equals the squared distance between $\boldsymbol{y}$ and $\boldsymbol{X}\boldsymbol{\theta}$.

With (9.10b), we have now a concrete form of the negative log-likelihood function we need to optimize. We immediately see that (9.10b) is quadratic in $\boldsymbol{\theta}$. This means that we can find a unique global solution $\boldsymbol{\theta}_{\text{ML}}$ for minimizing the negative log-likelihood $\mathcal{L}$. We can find the global optimum by computing the gradient of $\mathcal{L}$, setting it to $\boldsymbol{0}$ and solving for $\boldsymbol{\theta}$.

Using the results from Chapter 5, we compute the gradient of $\mathcal{L}$ with respect to the parameters as

$$\frac{\mathrm{d}\mathcal{L}}{\mathrm{d}\boldsymbol{\theta}} = \frac{\mathrm{d}}{\mathrm{d}\boldsymbol{\theta}}\left(\frac{1}{2\sigma^2}(\boldsymbol{y} - \boldsymbol{X}\boldsymbol{\theta})^\top(\boldsymbol{y} - \boldsymbol{X}\boldsymbol{\theta})\right) \tag{9.11a}$$

$$= \frac{1}{2\sigma^2}\frac{\mathrm{d}}{\mathrm{d}\boldsymbol{\theta}}\left(\boldsymbol{y}^\top\boldsymbol{y} - 2\boldsymbol{y}^\top\boldsymbol{X}\boldsymbol{\theta} + \boldsymbol{\theta}^\top\boldsymbol{X}^\top\boldsymbol{X}\boldsymbol{\theta}\right) \tag{9.11b}$$

$$= \frac{1}{\sigma^2}(-\boldsymbol{y}^\top\boldsymbol{X} + \boldsymbol{\theta}^\top\boldsymbol{X}^\top\boldsymbol{X}) \in \mathbb{R}^{1\times D}\,. \tag{9.11c}$$

The maximum likelihood estimator $\boldsymbol{\theta}_{\mathrm{ML}}$ solves $\frac{\mathrm{d}\mathcal{L}}{\mathrm{d}\boldsymbol{\theta}} = \boldsymbol{0}^\top$ (necessary optimality condition) and we obtain

Ignoring the possibility of duplicate data points, $\mathrm{rk}(\boldsymbol{X}) = D$ if $N \geqslant D$, i.e., we do not have more parameters than data points.

$$\frac{\mathrm{d}\mathcal{L}}{\mathrm{d}\boldsymbol{\theta}} = \boldsymbol{0}^\top \overset{(9.11c)}{\iff} \boldsymbol{\theta}_{\mathrm{ML}}^\top\boldsymbol{X}^\top\boldsymbol{X} = \boldsymbol{y}^\top\boldsymbol{X} \tag{9.12a}$$

$$\iff \boldsymbol{\theta}_{\mathrm{ML}}^\top = \boldsymbol{y}^\top\boldsymbol{X}(\boldsymbol{X}^\top\boldsymbol{X})^{-1} \tag{9.12b}$$

$$\iff \boldsymbol{\theta}_{\mathrm{ML}} = (\boldsymbol{X}^\top\boldsymbol{X})^{-1}\boldsymbol{X}^\top\boldsymbol{y}\,. \tag{9.12c}$$

We could right-multiply the first equation by $(\boldsymbol{X}^\top\boldsymbol{X})^{-1}$ because $\boldsymbol{X}^\top\boldsymbol{X}$ is positive definite if $\mathrm{rk}(\boldsymbol{X}) = D$, where $\mathrm{rk}(\boldsymbol{X})$ denotes the rank of $\boldsymbol{X}$.

Remark. Setting the gradient to $\boldsymbol{0}^\top$ is a necessary and sufficient condition, and we obtain a global minimum since the Hessian $\nabla^2_{\boldsymbol{\theta}}\mathcal{L}(\boldsymbol{\theta}) = \boldsymbol{X}^\top\boldsymbol{X} \in \mathbb{R}^{D\times D}$ is positive definite. ♢

Remark. The maximum likelihood solution in (9.12c) requires us to solve a system of linear equations of the form $\boldsymbol{A}\boldsymbol{\theta} = \boldsymbol{b}$ with $\boldsymbol{A} = (\boldsymbol{X}^\top\boldsymbol{X})$ and $\boldsymbol{b} = \boldsymbol{X}^\top\boldsymbol{y}$. ♢

Example 9.2 (Fitting Lines)
Let us have a look at Figure 9.2, where we aim to fit a straight line $f(x) = \theta x$, where θ is an unknown slope, to a dataset using maximum likelihood estimation. Examples of functions in this model class (straight lines) are shown in Figure 9.2(a). For the dataset shown in Figure 9.2(b), we find the maximum likelihood estimate of the slope parameter θ using (9.12c) and obtain the maximum likelihood linear function in Figure 9.2(c).

Maximum Likelihood Estimation with Features

So far, we considered the linear regression setting described in (9.4), which allowed us to fit straight lines to data using maximum likelihood estimation. However, straight lines are not sufficiently expressive when it comes to fitting more interesting data. Fortunately, linear regression offers us a way to fit nonlinear functions within the linear regression framework: Since "linear regression" only refers to "linear in the parameters," we can perform an arbitrary nonlinear transformation $\boldsymbol{\phi}(\boldsymbol{x})$ of the inputs $\boldsymbol{x}$ and then linearly combine the components of this transformation. The corresponding linear regression model is

Linear regression refers to "linear-in-the-parameters" regression models, but the inputs can undergo any nonlinear transformation.

$$\begin{aligned} p(y\,|\,\boldsymbol{x},\boldsymbol{\theta}) &= \mathcal{N}\big(y\,|\,\boldsymbol{\phi}^\top(\boldsymbol{x})\boldsymbol{\theta},\,\sigma^2\big) \\ \iff y &= \boldsymbol{\phi}^\top(\boldsymbol{x})\boldsymbol{\theta} + \epsilon = \sum_{k=0}^{K-1}\theta_k\phi_k(\boldsymbol{x}) + \epsilon\,, \end{aligned} \tag{9.13}$$

where $\boldsymbol{\phi} : \mathbb{R}^D \to \mathbb{R}^K$ is a (nonlinear) transformation of the inputs $\boldsymbol{x}$ and $\phi_k : \mathbb{R}^D \to \mathbb{R}$ is the kth component of the *feature vector* $\boldsymbol{\phi}$. Note that the model parameters $\boldsymbol{\theta}$ still appear only linearly.

feature vector

Example 9.3 (Polynomial Regression)
We are concerned with a regression problem $y = \boldsymbol{\phi}^\top(x)\boldsymbol{\theta} + \epsilon$, where $x \in \mathbb{R}$ and $\boldsymbol{\theta} \in \mathbb{R}^K$. A transformation that is often used in this context is

$$\boldsymbol{\phi}(x) = \begin{bmatrix} \phi_0(x) \\ \phi_1(x) \\ \vdots \\ \phi_{K-1}(x) \end{bmatrix} = \begin{bmatrix} 1 \\ x \\ x^2 \\ x^3 \\ \vdots \\ x^{K-1} \end{bmatrix} \in \mathbb{R}^K . \tag{9.14}$$

This means that we "lift" the original one-dimensional input space into a K-dimensional feature space consisting of all monomials x^k for $k = 0, \ldots, K-1$. With these features, we can model polynomials of degree $\leqslant K-1$ within the framework of linear regression: A polynomial of degree $K-1$ is

$$f(x) = \sum_{k=0}^{K-1} \theta_k x^k = \boldsymbol{\phi}^\top(x)\boldsymbol{\theta} , \tag{9.15}$$

where $\boldsymbol{\phi}$ is defined in (9.14) and $\boldsymbol{\theta} = [\theta_0, \ldots, \theta_{K-1}]^\top \in \mathbb{R}^K$ contains the (linear) parameters θ_k.

Let us now have a look at maximum likelihood estimation of the parameters $\boldsymbol{\theta}$ in the linear regression model (9.13). We consider training inputs $\boldsymbol{x}_n \in \mathbb{R}^D$ and targets $y_n \in \mathbb{R}$, $n = 1, \ldots, N$, and define the *feature matrix* (*design matrix*) as

feature matrix
design matrix

$$\boldsymbol{\Phi} := \begin{bmatrix} \boldsymbol{\phi}^\top(\boldsymbol{x}_1) \\ \vdots \\ \boldsymbol{\phi}^\top(\boldsymbol{x}_N) \end{bmatrix} = \begin{bmatrix} \phi_0(\boldsymbol{x}_1) & \cdots & \phi_{K-1}(\boldsymbol{x}_1) \\ \phi_0(\boldsymbol{x}_2) & \cdots & \phi_{K-1}(\boldsymbol{x}_2) \\ \vdots & & \vdots \\ \phi_0(\boldsymbol{x}_N) & \cdots & \phi_{K-1}(\boldsymbol{x}_N) \end{bmatrix} \in \mathbb{R}^{N \times K} , \tag{9.16}$$

where $\Phi_{ij} = \phi_j(\boldsymbol{x}_i)$ and $\phi_j : \mathbb{R}^D \to \mathbb{R}$.

Example 9.4 (Feature Matrix for Second-Order Polynomials)
For a second-order polynomial and N training points $x_n \in \mathbb{R}, n = 1, \ldots, N$, the feature matrix is

$$\boldsymbol{\Phi} = \begin{bmatrix} 1 & x_1 & x_1^2 \\ 1 & x_2 & x_2^2 \\ \vdots & \vdots & \vdots \\ 1 & x_N & x_N^2 \end{bmatrix} . \tag{9.17}$$

With the feature matrix $\boldsymbol{\Phi}$ defined in (9.16), the negative log-likelihood for the linear regression model (9.13) can be written as

$$-\log p(\mathcal{Y} \mid \mathcal{X}, \boldsymbol{\theta}) = \frac{1}{2\sigma^2}(\boldsymbol{y} - \boldsymbol{\Phi}\boldsymbol{\theta})^\top(\boldsymbol{y} - \boldsymbol{\Phi}\boldsymbol{\theta}) + \text{const}\,. \tag{9.18}$$

Comparing (9.18) with the negative log-likelihood in (9.10b) for the "feature-free" model, we immediately see we just need to replace $\boldsymbol{X}$ with $\boldsymbol{\Phi}$. Since both $\boldsymbol{X}$ and $\boldsymbol{\Phi}$ are independent of the parameters $\boldsymbol{\theta}$ that we wish to optimize, we arrive immediately at the *maximum likelihood estimate*

maximum likelihood estimate

$$\boldsymbol{\theta}_{\text{ML}} = (\boldsymbol{\Phi}^\top\boldsymbol{\Phi})^{-1}\boldsymbol{\Phi}^\top\boldsymbol{y} \tag{9.19}$$

for the linear regression problem with nonlinear features defined in (9.13).

Remark. When we were working without features, we required $\boldsymbol{X}^\top\boldsymbol{X}$ to be invertible, which is the case when the rows of $\boldsymbol{X}$ are linearly independent. In (9.19), we therefore require $\boldsymbol{\Phi}^\top\boldsymbol{\Phi} \in \mathbb{R}^{K\times K}$ to be invertible. This is the case if and only if $\text{rk}(\boldsymbol{\Phi}) = K$. ◊

Example 9.5 (Maximum Likelihood Polynomial Fit)
Consider the dataset in Figure 9.4(a). The dataset consists of $N = 10$ pairs (x_n, y_n), where $x_n \sim \mathcal{U}[-5, 5]$ and $y_n = -\sin(x_n/5) + \cos(x_n) + \epsilon$, where $\epsilon \sim \mathcal{N}(0,\, 0.2^2)$.

We fit a polynomial of degree 4 using maximum likelihood estimation, i.e., parameters $\boldsymbol{\theta}_{\text{ML}}$ are given in (9.19). The maximum likelihood estimate yields function values $\boldsymbol{\phi}^\top(x_*)\boldsymbol{\theta}_{\text{ML}}$ at any test location x_*. The result is shown in Figure 9.4(b).

Estimating the Noise Variance

Thus far, we assumed that the noise variance σ^2 is known. However, we can also use the principle of maximum likelihood estimation to obtain the maximum likelihood estimator σ^2_{ML} for the noise variance. To do this, we follow the standard procedure: We write down the log-likelihood, compute its derivative with respect to $\sigma^2 > 0$, set it to 0, and solve. The log-likelihood is given by

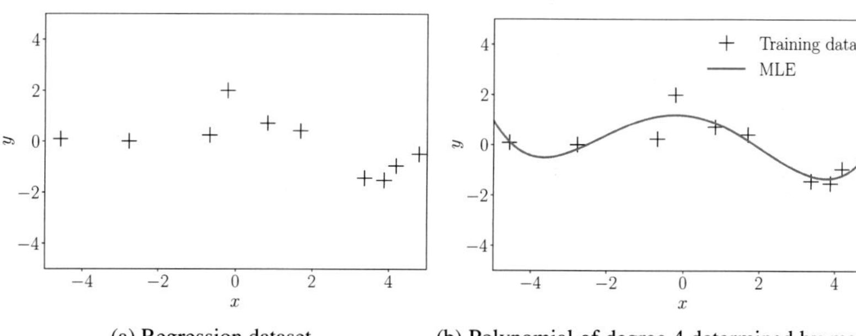

(a) Regression dataset

(b) Polynomial of degree 4 determined by maximum likelihood estimation

Figure 9.4 Polynomial regression: (a) dataset consisting of (x_n, y_n) pairs, $n = 1, \ldots, 10$; (b) maximum likelihood polynomial of degree 4.

$$\log p(\mathcal{Y} \mid \mathcal{X}, \boldsymbol{\theta}, \sigma^2) = \sum_{n=1}^{N} \log \mathcal{N}\big(y_n \mid \boldsymbol{\phi}^\top(\boldsymbol{x}_n)\boldsymbol{\theta},\, \sigma^2\big) \tag{9.20a}$$

$$= \sum_{n=1}^{N} \left(-\frac{1}{2}\log(2\pi) - \frac{1}{2}\log\sigma^2 - \frac{1}{2\sigma^2}(y_n - \boldsymbol{\phi}^\top(\boldsymbol{x}_n)\boldsymbol{\theta})^2\right) \tag{9.20b}$$

$$= -\frac{N}{2}\log\sigma^2 - \frac{1}{2\sigma^2}\underbrace{\sum_{n=1}^{N}(y_n - \boldsymbol{\phi}^\top(\boldsymbol{x}_n)\boldsymbol{\theta})^2}_{=:s} + \text{const}\,. \tag{9.20c}$$

The partial derivative of the log-likelihood with respect to σ^2 is then

$$\frac{\partial \log p(\mathcal{Y} \mid \mathcal{X}, \boldsymbol{\theta}, \sigma^2)}{\partial \sigma^2} = -\frac{N}{2\sigma^2} + \frac{1}{2\sigma^4}s = 0 \tag{9.21a}$$

$$\iff \frac{N}{2\sigma^2} = \frac{s}{2\sigma^4} \tag{9.21b}$$

so that we identify

$$\sigma^2_{\text{ML}} = \frac{s}{N} = \frac{1}{N}\sum_{n=1}^{N}(y_n - \boldsymbol{\phi}^\top(\boldsymbol{x}_n)\boldsymbol{\theta})^2\,. \tag{9.22}$$

Therefore, the maximum likelihood estimate of the noise variance is the empirical mean of the squared distances between the noise-free function values $\boldsymbol{\phi}^\top(\boldsymbol{x}_n)\boldsymbol{\theta}$ and the corresponding noisy observations y_n at input locations $\boldsymbol{x}_n$.

9.2.2 Overfitting in Linear Regression

We just discussed how to use maximum likelihood estimation to fit linear models (e.g., polynomials) to data. We can evaluate the quality of the model by computing the error/loss incurred. One way of doing this is to compute the negative log-likelihood (9.10b), which we minimized to determine the maximum likelihood estimator. Alternatively, given that the noise parameter σ^2 is not a free model parameter, we can ignore the scaling by $1/\sigma^2$, so that we end up with a squared-error-loss function $\|\boldsymbol{y} - \boldsymbol{\Phi}\|^2$. Instead of using this squared loss, we often use the *root mean square error* (*RMSE*)

root mean square error
RMSE

$$\sqrt{\|\boldsymbol{y} - \boldsymbol{\Phi}\boldsymbol{\theta}\|^2 / N} = \sqrt{\frac{1}{N}\sum_{n=1}^{N}(y_n - \boldsymbol{\phi}^\top(\boldsymbol{x}_n)\boldsymbol{\theta})^2}\,, \tag{9.23}$$

The RMSE is normalized.

which (a) allows us to compare errors of datasets with different sizes and (b) has the same scale and the same units as the observed function values y_n. For example, if we fit a model that maps postcodes ($\boldsymbol{x}$ is given in latitude, longitude) to house prices (y-values are EUR) then the RMSE is also measured in EUR, whereas the squared error is given in EUR^2. If we choose to include the factor σ^2 from the original negative log-likelihood (9.10b), then we end up with a unitless objective, i.e., in the preceding example, our objective would no longer be in EUR or EUR^2.

The negative log-likelihood is unitless.

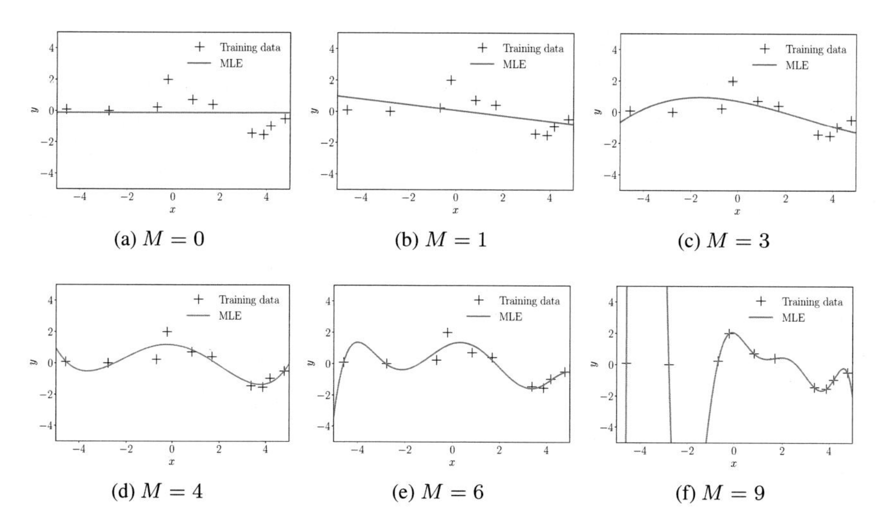

(a) $M = 0$ (b) $M = 1$ (c) $M = 3$

(d) $M = 4$ (e) $M = 6$ (f) $M = 9$

Figure 9.5 Maximum likelihood fits for different polynomial degrees M.

For model selection (see Section 8.6), we can use the RMSE (or the negative log-likelihood) to determine the best degree of the polynomial by finding the polynomial degree M that minimizes the objective. Given that the polynomial degree is a natural number, we can perform a brute-force search and enumerate all (reasonable) values of M. For a training set of size N, it is sufficient to test $0 \leqslant M \leqslant N - 1$. For $M < N$, the maximum likelihood estimator is unique. For $M \geqslant N$, we have more parameters than data points, and would need to solve an underdetermined system of linear equations ($\boldsymbol{\Phi}^\top \boldsymbol{\Phi}$ in (9.19) would also no longer be invertible) so that there are infinitely many possible maximum likelihood estimators.

Figure 9.5 shows a number of polynomial fits determined by maximum likelihood for the dataset from Figure 9.4(a) with $N = 10$ observations. We notice that polynomials of low degree (e.g., constants ($M = 0$) or linear ($M = 1$) fit the data poorly and hence are poor representations of the true underlying function. For degrees $M = 3, \ldots, 5$, the fits look plausible and smoothly interpolate the data. When we go to higher-degree polynomials, we notice that they fit the data better and better. In the extreme case of $M = N - 1 = 9$, the function will pass through every single data point. However, these high-degree polynomials oscillate wildly and are a poor representation of the underlying function that generated the data, such that we suffer from *overfitting*.

The case of $M = N - 1$ is extreme in the sense that otherwise the null space of the corresponding system of linear equations would be nontrivial, and we would have infinitely many optimal solutions to the linear regression problem.

overfitting

Note that the noise variance $\sigma^2 > 0$.

Remember that the goal is to achieve good generalization by making accurate predictions for new (unseen) data. We obtain some quantitative insight into the dependence of the generalization performance on the polynomial of degree M by considering a separate test set comprising 200 data points generated using exactly the same procedure used to generate the training set. As test inputs, we chose a linear grid of 200 points in the interval of $[-5, 5]$. For each choice of M, we evaluate the RMSE (9.23) for both the training data and the test data.

Looking now at the test error, which is a qualitive measure of the generalization properties of the corresponding polynomial, we notice that initially the test error decreases; see Figure 9.6 (orange). For fourth-order polynomials, the test

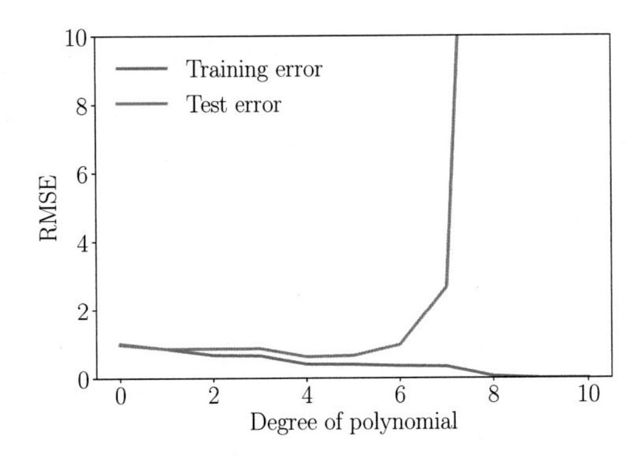

Figure 9.6 Training and test error.

error is relatively low and stays relatively constant up to degree 5. However, from degree 6 onward the test error increases significantly, and high-order polynomials have very bad generalization properties. In this particular example, this also is evident from the corresponding maximum likelihood fits in Figure 9.5. Note that the *training error* (blue curve in Figure 9.6) never increases when the degree of the polynomial increases. In our example, the best generalization (the point of the smallest *test error*) is obtained for a polynomial of degree $M = 4$.

training error

test error

9.2.3 Maximum A Posteriori Estimation

We just saw that maximum likelihood estimation is prone to overfitting. We often observe that the magnitude of the parameter values becomes relatively large if we run into overfitting (Bishop, 2006).

To mitigate the effect of huge parameter values, we can place a prior distribution $p(\boldsymbol{\theta})$ on the parameters. The prior distribution explicitly encodes what parameter values are plausible (before having seen any data). For example, a Gaussian prior $p(\theta) = \mathcal{N}(0, 1)$ on a single parameter θ encodes that parameter values are expected lie in the interval $[-2, 2]$ (two standard deviations around the mean value). Once a dataset $\mathcal{X}$, $\mathcal{Y}$ is available, instead of maximizing the likelihood we seek parameters that maximize the posterior distribution $p(\boldsymbol{\theta} \mid \mathcal{X}, \mathcal{Y})$. This procedure is called *maximum a posteriori* (*MAP*) estimation.

maximum a posteriori

MAP

The posterior over the parameters $\boldsymbol{\theta}$, given the training data $\mathcal{X}, \mathcal{Y}$, is obtained by applying Bayes' theorem (Section 6.3) as

$$p(\boldsymbol{\theta} \mid \mathcal{X}, \mathcal{Y}) = \frac{p(\mathcal{Y} \mid \mathcal{X}, \boldsymbol{\theta}) p(\boldsymbol{\theta})}{p(\mathcal{Y} \mid \mathcal{X})} . \tag{9.24}$$

Since the posterior explicitly depends on the parameter prior $p(\boldsymbol{\theta})$, the prior will have an effect on the parameter vector we find as the maximizer of the posterior. We will see this more explicitly in the following. The parameter vector $\boldsymbol{\theta}_{\text{MAP}}$ that maximizes the posterior (9.24) is the MAP estimate.

To find the MAP estimate, we follow steps that are similar in flavor to maximum likelihood estimation. We start with the log-transform and compute the log-posterior as

$$\log p(\boldsymbol{\theta} \mid \mathcal{X}, \mathcal{Y}) = \log p(\mathcal{Y} \mid \mathcal{X}, \boldsymbol{\theta}) + \log p(\boldsymbol{\theta}) + \text{const} , \tag{9.25}$$

where the constant comprises the terms that are independent of $\boldsymbol{\theta}$. We see that the log-posterior in (9.25) is the sum of the log-likelihood $p(\mathcal{Y} \mid \mathcal{X}, \boldsymbol{\theta})$ and the log-prior $\log p(\boldsymbol{\theta})$ so that the MAP estimate will be a "compromise" between the prior (our suggestion for plausible parameter values before observing data) and the data-dependent likelihood.

To find the MAP estimate $\boldsymbol{\theta}_{\text{MAP}}$, we minimize the negative log-posterior distribution with respect to $\boldsymbol{\theta}$, i.e., we solve

$$\boldsymbol{\theta}_{\text{MAP}} \in \arg\min_{\boldsymbol{\theta}} \{-\log p(\mathcal{Y} \mid \mathcal{X}, \boldsymbol{\theta}) - \log p(\boldsymbol{\theta})\} . \tag{9.26}$$

The gradient of the negative log-posterior with respect to $\boldsymbol{\theta}$ is

$$-\frac{\mathrm{d} \log p(\boldsymbol{\theta} \mid \mathcal{X}, \mathcal{Y})}{\mathrm{d}\boldsymbol{\theta}} = -\frac{\mathrm{d} \log p(\mathcal{Y} \mid \mathcal{X}, \boldsymbol{\theta})}{\mathrm{d}\boldsymbol{\theta}} - \frac{\mathrm{d} \log p(\boldsymbol{\theta})}{\mathrm{d}\boldsymbol{\theta}} , \tag{9.27}$$

where we identify the first term on the right-hand side as the gradient of the negative log-likelihood from (9.11c).

With a (conjugate) Gaussian prior $p(\boldsymbol{\theta}) = \mathcal{N}(\mathbf{0},\, b^2\boldsymbol{I})$ on the parameters $\boldsymbol{\theta}$, the negative log-posterior for the linear regression setting (9.13), we obtain the negative log posterior

$$-\log p(\boldsymbol{\theta} \mid \mathcal{X}, \mathcal{Y}) = \frac{1}{2\sigma^2}(\boldsymbol{y} - \boldsymbol{\Phi}\boldsymbol{\theta})^\top(\boldsymbol{y} - \boldsymbol{\Phi}\boldsymbol{\theta}) + \frac{1}{2b^2}\boldsymbol{\theta}^\top\boldsymbol{\theta} + \text{const} . \tag{9.28}$$

Here, the first term corresponds to the contribution from the log-likelihood, and the second term originates from the log-prior. The gradient of the log-posterior with respect to the parameters $\boldsymbol{\theta}$ is then

$$-\frac{\mathrm{d} \log p(\boldsymbol{\theta} \mid \mathcal{X}, \mathcal{Y})}{\mathrm{d}\boldsymbol{\theta}} = \frac{1}{\sigma^2}(\boldsymbol{\theta}^\top\boldsymbol{\Phi}^\top\boldsymbol{\Phi} - \boldsymbol{y}^\top\boldsymbol{\Phi}) + \frac{1}{b^2}\boldsymbol{\theta}^\top . \tag{9.29}$$

We will find the MAP estimate $\boldsymbol{\theta}_{\text{MAP}}$ by setting this gradient to $\mathbf{0}^\top$ and solving for $\boldsymbol{\theta}_{\text{MAP}}$. We obtain

$$\frac{1}{\sigma^2}(\boldsymbol{\theta}^\top\boldsymbol{\Phi}^\top\boldsymbol{\Phi} - \boldsymbol{y}^\top\boldsymbol{\Phi}) + \frac{1}{b^2}\boldsymbol{\theta}^\top = \mathbf{0}^\top \tag{9.30a}$$

$$\iff \boldsymbol{\theta}^\top\left(\frac{1}{\sigma^2}\boldsymbol{\Phi}^\top\boldsymbol{\Phi} + \frac{1}{b^2}\boldsymbol{I}\right) - \frac{1}{\sigma^2}\boldsymbol{y}^\top\boldsymbol{\Phi} = \mathbf{0}^\top \tag{9.30b}$$

$$\iff \boldsymbol{\theta}^\top\left(\boldsymbol{\Phi}^\top\boldsymbol{\Phi} + \frac{\sigma^2}{b^2}\boldsymbol{I}\right) = \boldsymbol{y}^\top\boldsymbol{\Phi} \tag{9.30c}$$

$$\iff \boldsymbol{\theta}^\top = \boldsymbol{y}^\top\boldsymbol{\Phi}\left(\boldsymbol{\Phi}^\top\boldsymbol{\Phi} + \frac{\sigma^2}{b^2}\boldsymbol{I}\right)^{-1} \tag{9.30d}$$

so that the MAP estimate is (by transposing both sides of the last equality)

$$\boldsymbol{\theta}_{\text{MAP}} = \left(\boldsymbol{\Phi}^\top\boldsymbol{\Phi} + \frac{\sigma^2}{b^2}\boldsymbol{I}\right)^{-1}\boldsymbol{\Phi}^\top\boldsymbol{y} . \tag{9.31}$$

$\boldsymbol{\Phi}^\top\boldsymbol{\Phi}$ is symmetric, positive semidefinite. The additional term in (9.31) is strictly positive definite so that the inverse exists.

Comparing the MAP estimate in (9.31) with the maximum likelihood estimate in (9.19), we see that the only difference between both solutions is the additional term $\frac{\sigma^2}{b^2}\boldsymbol{I}$ in the inverse matrix. This term ensures that $\boldsymbol{\Phi}^\top\boldsymbol{\Phi} + \frac{\sigma^2}{b^2}\boldsymbol{I}$ is symmetric and strictly positive definite (i.e., its inverse exists and the MAP estimate is the unique solution of a system of linear equations). Moreover, it reflects the impact of the regularizer.

Example 9.6 (MAP Estimation for Polynomial Regression)
In the polynomial regression example from Section 9.2.1, we place a Gaussian prior $p(\boldsymbol{\theta}) = \mathcal{N}(\mathbf{0}, \boldsymbol{I})$ on the parameters $\boldsymbol{\theta}$ and determine the MAP estimates according to (9.31). In Figure 9.7, we show both the maximum likelihood and the MAP estimates for polynomials of degree 6 (left) and degree 8 (right). The prior (regularizer) does not play a significant role for the low-degree polynomial, but keeps the function relatively smooth for higher-degree polynomials. Although the MAP estimate can push the boundaries of overfitting, it is not a general solution to this problem, so we need a more principled approach to tackle overfitting.

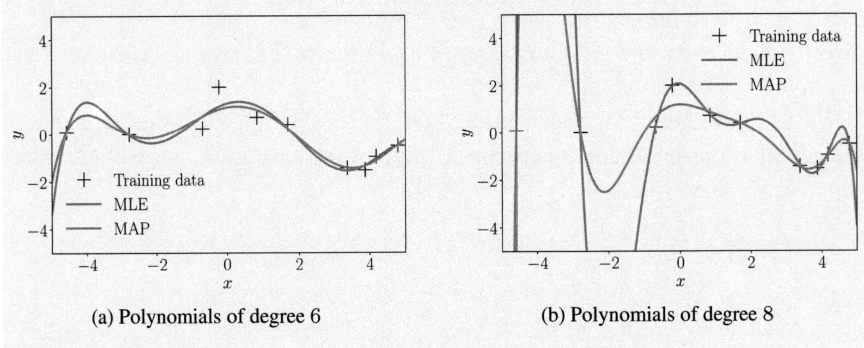

(a) Polynomials of degree 6 (b) Polynomials of degree 8

Figure 9.7 Polynomial regression: maximum likelihood and MAP estimates. (a) Polynomials of degree 6; (b) polynomials of degree 8.

9.2.4 MAP Estimation as Regularization

Instead of placing a prior distribution on the parameters $\boldsymbol{\theta}$, it is also possible to mitigate the effect of overfitting by penalizing the amplitude of the parameter by means of *regularization*. In *regularized least squares*, we consider the loss function

regularization
regularized least squares

$$\|\boldsymbol{y} - \boldsymbol{\Phi}\boldsymbol{\theta}\|^2 + \lambda \|\boldsymbol{\theta}\|_2^2 \,, \tag{9.32}$$

which we minimize with respect to $\boldsymbol{\theta}$ (see Section 8.2.3). Here, the first term is a *data-fit term* (also called *misfit term*), which is proportional to the negative log-likelihood; see (9.10b). The second term is called the *regularizer*, and the *regularization parameter* $\lambda \geqslant 0$ controls the "strictness" of the regularization.

data-fit term
misfit term
regularizer
regularization parameter

Remark. Instead of the Euclidean norm $\|\cdot\|_2$, we can choose any p-norm $\|\cdot\|_p$ in (9.32). In practice, smaller values for p lead to sparser solutions. Here, "sparse" means that many parameter values $\theta_d = 0$, which is also useful for variable selection. For $p = 1$, the regularizer is called *LASSO* (least absolute shrinkage and selection operator) and was proposed by Tibshirani (1996). ◇

LASSO

The regularizer $\lambda \|\boldsymbol{\theta}\|_2^2$ in (9.32) can be interpreted as a negative log-Gaussian prior, which we use in MAP estimation; see (9.26). More specifically, with a Gaussian prior $p(\boldsymbol{\theta}) = \mathcal{N}(\mathbf{0}, b^2\boldsymbol{I})$, we obtain the negative log-Gaussian prior

$$-\log p(\boldsymbol{\theta}) = \frac{1}{2b^2} \|\boldsymbol{\theta}\|_2^2 + \text{const} \tag{9.33}$$

so that for $\lambda = \frac{1}{2b^2}$ the regularization term and the negative log-Gaussian prior are identical.

Given that the regularized least-squares loss function in (9.32) consists of terms that are closely related to the negative log-likelihood plus a negative log-prior, it is not surprising that, when we minimize this loss, we obtain a solution that closely resembles the MAP estimate in (9.31). More specifically, minimizing the regularized least-squares loss function yields

$$\boldsymbol{\theta}_{\text{RLS}} = (\boldsymbol{\Phi}^\top \boldsymbol{\Phi} + \lambda \boldsymbol{I})^{-1} \boldsymbol{\Phi}^\top \boldsymbol{y}, \tag{9.34}$$

which is identical to the MAP estimate in (9.31) for $\lambda = \frac{\sigma^2}{b^2}$, where σ^2 is the noise variance and b^2 the variance of the (isotropic) Gaussian prior $p(\boldsymbol{\theta}) = \mathcal{N}(\boldsymbol{0},\, b^2\boldsymbol{I})$.

A point estimate is a single specific parameter value, unlike a distribution over plausible parameter settings.

So far, we have covered parameter estimation using maximum likelihood and MAP estimation where we found point estimates $\boldsymbol{\theta}^*$ that optimize an objective function (likelihood or posterior). We saw that both maximum likelihood and MAP estimation can lead to overfitting. In the next section, we will discuss Bayesian linear regression, where we use Bayesian inference (Section 8.4) to find a posterior distribution over the unknown parameters, which we subsequently use to make predictions. More specifically, for predictions we will average over all plausible sets of parameters instead of focusing on a point estimate.

9.3 Bayesian Linear Regression

Previously, we looked at linear regression models where we estimated the model parameters $\boldsymbol{\theta}$, e.g., by means of maximum likelihood or MAP estimation. We discovered that MLE can lead to severe overfitting, in particular, in the small-data regime. MAP addresses this issue by placing a prior on the parameters that plays the role of a regularizer.

Bayesian linear regression

Bayesian linear regression pushes the idea of the parameter prior a step further and does not even attempt to compute a point estimate of the parameters, but instead the full posterior distribution over the parameters is taken into account when making predictions. This means we do not fit any parameters, but we compute a mean over all plausible parameters settings (according to the posterior).

9.3.1 Model

In Bayesian linear regression, we consider the model

$$\begin{aligned} &\text{prior} && p(\boldsymbol{\theta}) = \mathcal{N}(\boldsymbol{m}_0,\, \boldsymbol{S}_0)\,, \\ &\text{likelihood} && p(y \,|\, \boldsymbol{x}, \boldsymbol{\theta}) = \mathcal{N}(y \,|\, \boldsymbol{\phi}^\top(\boldsymbol{x})\boldsymbol{\theta},\, \sigma^2)\,, \end{aligned} \tag{9.35}$$

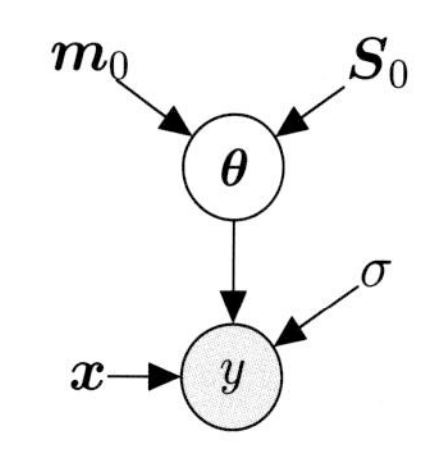

Figure 9.8 Graphical model for Bayesian linear regression.

where we now explicitly place a Gaussian prior $p(\boldsymbol{\theta}) = \mathcal{N}(\boldsymbol{m}_0,\, \boldsymbol{S}_0)$ on $\boldsymbol{\theta}$, which turns the parameter vector into a random variable. This allows us to write down the corresponding graphical model in Figure 9.8, where we made the parameters of the Gaussian prior on $\boldsymbol{\theta}$ explicit. The full probabilistic model, i.e., the joint distribution of observed and unobserved random variables, y and $\boldsymbol{\theta}$, respectively, is

$$p(y, \boldsymbol{\theta} \,|\, \boldsymbol{x}) = p(y \,|\, \boldsymbol{x}, \boldsymbol{\theta})p(\boldsymbol{\theta})\,. \tag{9.36}$$

9.3.2 Prior Predictions

In practice, we are usually not so much interested in the parameter values $\boldsymbol{\theta}$ themselves. Instead, our focus often lies in the predictions we make with those parameter values. In a Bayesian setting, we take the parameter distribution and average over all plausible parameter settings when we make predictions. More specifically, to make predictions at an input $\boldsymbol{x}_*$, we integrate out $\boldsymbol{\theta}$ and obtain

$$p(y_* \,|\, \boldsymbol{x}_*) = \int p(y_* \,|\, \boldsymbol{x}_*, \boldsymbol{\theta}) p(\boldsymbol{\theta}) \mathrm{d}\boldsymbol{\theta} = \mathbb{E}_{\boldsymbol{\theta}}[p(y_* \,|\, \boldsymbol{x}_*, \boldsymbol{\theta})] , \tag{9.37}$$

which we can interpret as the average prediction of $y_* \,|\, \boldsymbol{x}_*, \boldsymbol{\theta}$ for all plausible parameters $\boldsymbol{\theta}$ according to the prior distribution $p(\boldsymbol{\theta})$. Note that predictions using the prior distribution only require us to specify the input $\boldsymbol{x}_*$, but no training data.

In our model (9.35), we chose a conjugate (Gaussian) prior on $\boldsymbol{\theta}$ so that the predictive distribution is Gaussian as well (and can be computed in closed form): With the prior distribution $p(\boldsymbol{\theta}) = \mathcal{N}\big(\boldsymbol{m}_0, \boldsymbol{S}_0\big)$, we obtain the predictive distribution as

$$p(y_* \,|\, \boldsymbol{x}_*) = \mathcal{N}\big(\boldsymbol{\phi}^\top(\boldsymbol{x}_*)\boldsymbol{m}_0, \, \boldsymbol{\phi}^\top(\boldsymbol{x}_*)\boldsymbol{S}_0\boldsymbol{\phi}(\boldsymbol{x}_*) + \sigma^2\big) , \tag{9.38}$$

where we exploited that (i) the prediction is Gaussian due to conjugacy (see Section 6.6) and the marginalization property of Gaussians (see Section 6.5), (ii) the Gaussian noise is independent so that

$$\mathbb{V}[y_*] = \mathbb{V}_{\boldsymbol{\theta}}[\boldsymbol{\phi}^\top(\boldsymbol{x}_*)\boldsymbol{\theta}] + \mathbb{V}_\epsilon[\epsilon] , \tag{9.39}$$

and (iii) y_* is a linear transformation of $\boldsymbol{\theta}$ so that we can apply the rules for computing the mean and covariance of the prediction analytically by using (6.50) and (6.51), respectively. In (9.38), the term $\boldsymbol{\phi}^\top(\boldsymbol{x}_*)\boldsymbol{S}_0\boldsymbol{\phi}(\boldsymbol{x}_*)$ in the predictive variance explicitly accounts for the uncertainty associated with the parameters $\boldsymbol{\theta}$, whereas σ^2 is the uncertainty contribution due to the measurement noise.

If we are interested in predicting noise-free function values $f(\boldsymbol{x}_*) = \boldsymbol{\phi}^\top(\boldsymbol{x}_*)\boldsymbol{\theta}$ instead of the noise-corrupted targets y_* we obtain

$$p(f(\boldsymbol{x}_*)) = \mathcal{N}\big(\boldsymbol{\phi}^\top(\boldsymbol{x}_*)\boldsymbol{m}_0, \, \boldsymbol{\phi}^\top(\boldsymbol{x}_*)\boldsymbol{S}_0\boldsymbol{\phi}(\boldsymbol{x}_*)\big) , \tag{9.40}$$

which only differs from (9.38) in the omission of the noise variance σ^2 in the predictive variance.

The parameter distribution $p(\boldsymbol{\theta})$ induces a distribution over functions.

Remark (Distribution over Functions). Since we can represent the distribution $p(\boldsymbol{\theta})$ using a set of samples $\boldsymbol{\theta}_i$ and every sample $\boldsymbol{\theta}_i$ gives rise to a function $f_i(\cdot) = \boldsymbol{\theta}_i^\top\boldsymbol{\phi}(\cdot)$, it follows that the parameter distribution $p(\boldsymbol{\theta})$ induces a distribution $p(f(\cdot))$ over functions. Here we use the notation $(\cdot)$ to explicitly denote a functional relationship. ◇

Example 9.7 (Prior over Functions)
Let us consider a Bayesian linear regression problem with polynomials of degree 5. We choose a parameter prior $p(\boldsymbol{\theta}) = \mathcal{N}\big(\boldsymbol{0}, \frac{1}{4}\boldsymbol{I}\big)$. Figure 9.9 visualizes the induced prior distribution over functions (shaded area: dark gray: 67% confidence bound; light gray: 95% confidence bound) induced by this parameter prior, including some function samples from this prior.

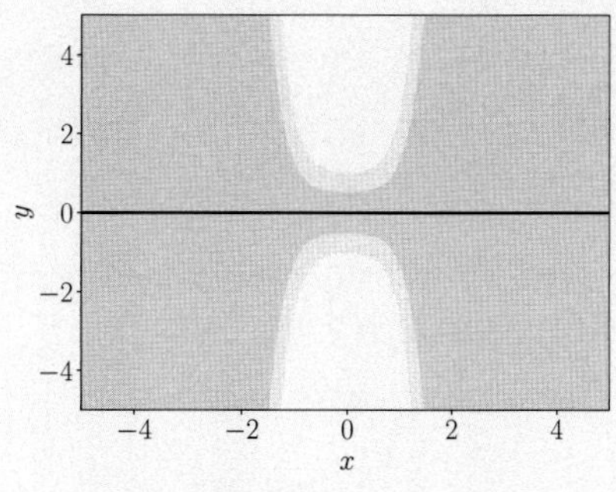

(a) Prior distribution over functions

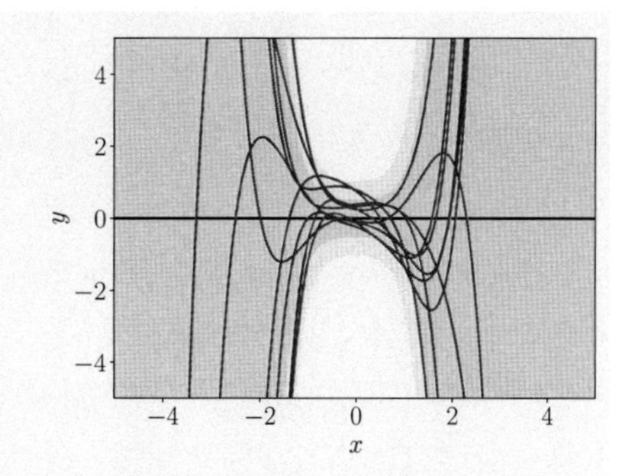

(b) Samples from the prior distribution over functions.

Figure 9.9 Prior over functions. (a) Distribution over functions represented by the mean function (black line) and the marginal uncertainties (shaded), representing the 67% and 95% confidence bounds, respectively; (b) samples from the prior over functions, which are induced by the samples from the parameter prior.

A function sample is obtained by first sampling a parameter vector $\boldsymbol{\theta}_i \sim p(\boldsymbol{\theta})$ and then computing $f_i(\cdot) = \boldsymbol{\theta}_i^\top \boldsymbol{\phi}(\cdot)$. We used 200 input locations $x_* \in [-5, 5]$ to which we apply the feature function $\boldsymbol{\phi}(\cdot)$. The uncertainty (represented by the shaded area) in Figure 9.9 is solely due to the parameter uncertainty because we considered the noise-free predictive distribution (9.40).

So far, we looked at computing predictions using the parameter prior $p(\boldsymbol{\theta})$. However, when we have a parameter posterior (given some training data $\mathcal{X}$, $\mathcal{Y}$), the same principles for prediction and inference hold as in (9.37) – we just need to replace the prior $p(\boldsymbol{\theta})$ with the posterior $p(\boldsymbol{\theta} \mid \mathcal{X}, \mathcal{Y})$. In the following, we will derive the posterior distribution in detail before using it to make predictions.

9.3.3 Posterior Distribution

Given a training set of inputs $\boldsymbol{x}_n \in \mathbb{R}^D$ and corresponding observations $y_n \in \mathbb{R}$, $n = 1, \dots, N$, we compute the posterior over the parameters using Bayes' theorem as

$$p(\boldsymbol{\theta} \mid \mathcal{X}, \mathcal{Y}) = \frac{p(\mathcal{Y} \mid \mathcal{X}, \boldsymbol{\theta}) p(\boldsymbol{\theta})}{p(\mathcal{Y} \mid \mathcal{X})}, \tag{9.41}$$

where $\mathcal{X}$ is the set of training inputs and $\mathcal{Y}$ the collection of corresponding training targets. Furthermore, $p(\mathcal{Y} \mid \mathcal{X}, \boldsymbol{\theta})$ is the likelihood, $p(\boldsymbol{\theta})$ the parameter prior, and

$$p(\mathcal{Y} \mid \mathcal{X}) = \int p(\mathcal{Y} \mid \mathcal{X}, \boldsymbol{\theta}) p(\boldsymbol{\theta}) \mathrm{d}\boldsymbol{\theta} = \mathbb{E}_{\boldsymbol{\theta}}[p(\mathcal{Y} \mid \mathcal{X}, \boldsymbol{\theta})] \tag{9.42}$$

the *marginal likelihood/evidence*, which is independent of the parameters $\boldsymbol{\theta}$ and ensures that the posterior is normalized, i.e., it integrates to 1. We can think of the marginal likelihood as the likelihood averaged over all possible parameter settings (with respect to the prior distribution $p(\boldsymbol{\theta})$).

marginal likelihood
evidence
The marginal likelihood is the expected likelihood under the parameter prior.

Theorem 9.1 (Parameter Posterior). *In our model* (9.35)*, the parameter posterior* (9.41) *can be computed in closed form as*

$$p(\boldsymbol{\theta} \mid \mathcal{X}, \mathcal{Y}) = \mathcal{N}\big(\boldsymbol{\theta} \mid \boldsymbol{m}_N,\, \boldsymbol{S}_N\big)\,, \tag{9.43a}$$

$$\boldsymbol{S}_N = (\boldsymbol{S}_0^{-1} + \sigma^{-2}\boldsymbol{\Phi}^\top\boldsymbol{\Phi})^{-1}\,, \tag{9.43b}$$

$$\boldsymbol{m}_N = \boldsymbol{S}_N(\boldsymbol{S}_0^{-1}\boldsymbol{m}_0 + \sigma^{-2}\boldsymbol{\Phi}^\top\boldsymbol{y})\,, \tag{9.43c}$$

where the subscript N indicates the size of the training set.

Proof Bayes' theorem tells us that the posterior $p(\boldsymbol{\theta} \mid \mathcal{X}, \mathcal{Y})$ is proportional to the product of the likelihood $p(\mathcal{Y} \mid \mathcal{X}, \boldsymbol{\theta})$ and the prior $p(\boldsymbol{\theta})$:

$$\text{Posterior} \quad p(\boldsymbol{\theta} \mid \mathcal{X}, \mathcal{Y}) = \frac{p(\mathcal{Y} \mid \mathcal{X}, \boldsymbol{\theta})p(\boldsymbol{\theta})}{p(\mathcal{Y} \mid \mathcal{X})} \tag{9.44a}$$

$$\text{Likelihood} \quad p(\mathcal{Y} \mid \mathcal{X}, \boldsymbol{\theta}) = \mathcal{N}\big(\boldsymbol{y} \mid \boldsymbol{\Phi\theta},\, \sigma^2\boldsymbol{I}\big) \tag{9.44b}$$

$$\text{Prior} \quad p(\boldsymbol{\theta}) = \mathcal{N}\big(\boldsymbol{\theta} \mid \boldsymbol{m}_0,\, \boldsymbol{S}_0\big)\,. \tag{9.44c}$$

Instead of looking at the product of the prior and the likelihood, we can transform the problem into log-space and solve for the mean and covariance of the posterior by completing the squares.

The sum of the log-prior and the log-likelihood is

$$\log \mathcal{N}\big(\boldsymbol{y} \mid \boldsymbol{\Phi\theta},\, \sigma^2\boldsymbol{I}\big) + \log \mathcal{N}\big(\boldsymbol{\theta} \mid \boldsymbol{m}_0,\, \boldsymbol{S}_0\big) \tag{9.45a}$$

$$= -\frac{1}{2}\big(\sigma^{-2}(\boldsymbol{y} - \boldsymbol{\Phi\theta})^\top(\boldsymbol{y} - \boldsymbol{\Phi\theta}) + (\boldsymbol{\theta} - \boldsymbol{m}_0)^\top\boldsymbol{S}_0^{-1}(\boldsymbol{\theta} - \boldsymbol{m}_0)\big) + \text{const} \tag{9.45b}$$

where the constant contains terms independent of $\boldsymbol{\theta}$. We will ignore the constant in the following. We now factorize (9.45b), which yields

$$-\frac{1}{2}\big(\sigma^{-2}\boldsymbol{y}^\top\boldsymbol{y} - 2\sigma^{-2}\boldsymbol{y}^\top\boldsymbol{\Phi\theta} + \boldsymbol{\theta}^\top\sigma^{-2}\boldsymbol{\Phi}^\top\boldsymbol{\Phi\theta} + \boldsymbol{\theta}^\top\boldsymbol{S}_0^{-1}\boldsymbol{\theta} - 2\boldsymbol{m}_0^\top\boldsymbol{S}_0^{-1}\boldsymbol{\theta} + \boldsymbol{m}_0^\top\boldsymbol{S}_0^{-1}\boldsymbol{m}_0\big) \tag{9.46a}$$

$$= -\frac{1}{2}\big(\boldsymbol{\theta}^\top(\sigma^{-2}\boldsymbol{\Phi}^\top\boldsymbol{\Phi} + \boldsymbol{S}_0^{-1})\boldsymbol{\theta} - 2(\sigma^{-2}\boldsymbol{\Phi}^\top\boldsymbol{y} + \boldsymbol{S}_0^{-1}\boldsymbol{m}_0)^\top\boldsymbol{\theta}\big) + \text{const}\,, \tag{9.46b}$$

where the constant contains the black terms in (9.46a), which are independent of $\boldsymbol{\theta}$. The orange terms are terms that are linear in $\boldsymbol{\theta}$, and the blue terms are the ones that are quadratic in $\boldsymbol{\theta}$. Inspecting (9.46b), we find that this equation is quadratic in $\boldsymbol{\theta}$. The fact that the unnormalized log-posterior distribution is a (negative) quadratic form implies that the posterior is Gaussian, i.e.,

$$p(\boldsymbol{\theta} \mid \mathcal{X}, \mathcal{Y}) = \exp(\log p(\boldsymbol{\theta} \mid \mathcal{X}, \mathcal{Y})) \propto \exp(\log p(\mathcal{Y} \mid \mathcal{X}, \boldsymbol{\theta}) + \log p(\boldsymbol{\theta})) \tag{9.47a}$$

$$\propto \exp\Big(-\frac{1}{2}\big(\boldsymbol{\theta}^\top(\sigma^{-2}\boldsymbol{\Phi}^\top\boldsymbol{\Phi} + \boldsymbol{S}_0^{-1})\boldsymbol{\theta} - 2(\sigma^{-2}\boldsymbol{\Phi}^\top\boldsymbol{y} + \boldsymbol{S}_0^{-1}\boldsymbol{m}_0)^\top\boldsymbol{\theta}\big)\Big)\,, \tag{9.47b}$$

where we used (9.46b) in the last expression.

The remaining task is it to bring this (unnormalized) Gaussian into the form that is proportional to $\mathcal{N}\big(\boldsymbol{\theta} \mid \boldsymbol{m}_N,\, \boldsymbol{S}_N\big)$, i.e., we need to identify the mean $\boldsymbol{m}_N$ and the covariance matrix $\boldsymbol{S}_N$. To do this, we use the concept of *completing the squares*. The desired log-posterior is

completing the squares

$$\log\mathcal{N}(\boldsymbol{\theta}\,|\,\boldsymbol{m}_N,\,\boldsymbol{S}_N) = -\frac{1}{2}(\boldsymbol{\theta}-\boldsymbol{m}_N)^\top\boldsymbol{S}_N^{-1}(\boldsymbol{\theta}-\boldsymbol{m}_N) + \text{const} \tag{9.48a}$$

$$= -\frac{1}{2}\big(\boldsymbol{\theta}^\top\boldsymbol{S}_N^{-1}\boldsymbol{\theta} - 2\boldsymbol{m}_N^\top\boldsymbol{S}_N^{-1}\boldsymbol{\theta} + \boldsymbol{m}_N^\top\boldsymbol{S}_N^{-1}\boldsymbol{m}_N\big)\,. \tag{9.48b}$$

Here, we factorized the quadratic form $(\boldsymbol{\theta}-\boldsymbol{m}_N)^\top\boldsymbol{S}_N^{-1}(\boldsymbol{\theta}-\boldsymbol{m}_N)$ into a term that is quadratic in $\boldsymbol{\theta}$ alone (blue), a term that is linear in $\boldsymbol{\theta}$ (orange), and a constant term (black). This allows us now to find $\boldsymbol{S}_N$ and $\boldsymbol{m}_N$ by matching the colored expressions in (9.46b) and (9.48b), which yields

$$\boldsymbol{S}_N^{-1} = \boldsymbol{\Phi}^\top\sigma^{-2}\boldsymbol{I}\boldsymbol{\Phi} + \boldsymbol{S}_0^{-1} \tag{9.49a}$$

$$\iff \boldsymbol{S}_N = (\sigma^{-2}\boldsymbol{\Phi}^\top\boldsymbol{\Phi} + \boldsymbol{S}_0^{-1})^{-1} \tag{9.49b}$$

and

$$\boldsymbol{m}_N^\top\boldsymbol{S}_N^{-1} = (\sigma^{-2}\boldsymbol{\Phi}^\top\boldsymbol{y} + \boldsymbol{S}_0^{-1}\boldsymbol{m}_0)^\top \tag{9.50a}$$

$$\iff \boldsymbol{m}_N = \boldsymbol{S}_N(\sigma^{-2}\boldsymbol{\Phi}^\top\boldsymbol{y} + \boldsymbol{S}_0^{-1}\boldsymbol{m}_0)\,. \tag{9.50b}$$

□

Remark (General Approach to Completing the Squares). If we are given an equation

$$\boldsymbol{x}^\top\boldsymbol{A}\boldsymbol{x} - 2\boldsymbol{a}^\top\boldsymbol{x} + \text{const}_1\,, \tag{9.51}$$

where $\boldsymbol{A}$ is symmetric and positive definite, which we wish to bring into the form

$$(\boldsymbol{x}-\boldsymbol{\mu})^\top\boldsymbol{\Sigma}(\boldsymbol{x}-\boldsymbol{\mu}) + \text{const}_2\,, \tag{9.52}$$

we can do this by setting

$$\boldsymbol{\Sigma} := \boldsymbol{A}\,, \tag{9.53}$$

$$\boldsymbol{\mu} := \boldsymbol{\Sigma}^{-1}\boldsymbol{a} \tag{9.54}$$

and $\text{const}_2 = \text{const}_1 - \boldsymbol{\mu}^\top\boldsymbol{\Sigma}\boldsymbol{\mu}$. ◊

We can see that the terms inside the exponential in (9.47b) are of the form (9.51) with

$$\boldsymbol{A} := \sigma^{-2}\boldsymbol{\Phi}^\top\boldsymbol{\Phi} + \boldsymbol{S}_0^{-1}\,, \tag{9.55}$$

$$\boldsymbol{a} := \sigma^{-2}\boldsymbol{\Phi}^\top\boldsymbol{y} + \boldsymbol{S}_0^{-1}\boldsymbol{m}_0\,. \tag{9.56}$$

Since $\boldsymbol{A}, \boldsymbol{a}$ can be difficult to identify in equations like (9.46a), it is often helpful to bring these equations into the form (9.51) that decouples quadratic term, linear terms, and constants, which simplifies finding the desired solution.

9.3.4 Posterior Predictions

In (9.37), we computed the predictive distribution of y_* at a test input $\boldsymbol{x}_*$ using the parameter prior $p(\boldsymbol{\theta})$. In principle, predicting with the parameter posterior $p(\boldsymbol{\theta}\,|\,\mathcal{X},\mathcal{Y})$ is not fundamentally different given that in our conjugate model the prior and posterior are both Gaussian (with different parameters). Therefore, by following the same reasoning as in Section 9.3.2, we obtain the (posterior) predictive distribution

$$\begin{aligned} p(y_* \mid \mathcal{X}, \mathcal{Y}, \boldsymbol{x}_*) &= \int p(y_* \mid \boldsymbol{x}_*, \boldsymbol{\theta}) p(\boldsymbol{\theta} \mid \mathcal{X}, \mathcal{Y}) \mathrm{d}\boldsymbol{\theta} && (9.57a) \\ &= \int \mathcal{N}\big(y_* \mid \boldsymbol{\phi}^\top(\boldsymbol{x}_*)\boldsymbol{\theta}, \sigma^2\big) \mathcal{N}\big(\boldsymbol{\theta} \mid \boldsymbol{m}_N, \boldsymbol{S}_N\big) \mathrm{d}\boldsymbol{\theta} && (9.57b) \\ &= \mathcal{N}\big(y_* \mid \boldsymbol{\phi}^\top(\boldsymbol{x}_*)\boldsymbol{m}_N, \boldsymbol{\phi}^\top(\boldsymbol{x}_*)\boldsymbol{S}_N\boldsymbol{\phi}(\boldsymbol{x}_*) + \sigma^2\big). && (9.57c) \end{aligned}$$

The term $\boldsymbol{\phi}^\top(\boldsymbol{x}_*)\boldsymbol{S}_N\boldsymbol{\phi}(\boldsymbol{x}_*)$ reflects the posterior uncertainty associated with the parameters $\boldsymbol{\theta}$. Note that $\boldsymbol{S}_N$ depends on the training inputs through $\boldsymbol{\Phi}$; see (9.43b). The predictive mean $\boldsymbol{\phi}^\top(\boldsymbol{x}_*)\boldsymbol{m}_N$ coincides with the MAP estimate.

Remark (Marginal Likelihood and Posterior Predictive Distribution). By replacing the integral in (9.57a), the predictive distribution can be equivalently written as the expectation $\mathbb{E}_{\boldsymbol{\theta} \mid \mathcal{X}, \mathcal{Y}}[p(y_* \mid \boldsymbol{x}_*, \boldsymbol{\theta})]$, where the expectation is taken with respect to the parameter posterior $p(\boldsymbol{\theta} \mid \mathcal{X}, \mathcal{Y})$.

Writing the posterior predictive distribution in this way highlights a close resemblance to the marginal likelihood (9.42). The key difference between the marginal likelihood and the posterior predictive distribution are (i) the marginal likelihood can be thought of predicting the training targets $\boldsymbol{y}$ and not the test targets y_*, and (ii) the marginal likelihood averages with respect to the parameter prior and not the parameter posterior. ◇

Remark (Mean and Variance of Noise-Free Function Values). In many cases, we are not interested in the predictive distribution $p(y_* \mid \mathcal{X}, \mathcal{Y}, \boldsymbol{x}_*)$ of a (noisy) observation y_*. Instead, we would like to obtain the distribution of the (noise-free) function values $f(\boldsymbol{x}_*) = \boldsymbol{\phi}^\top(\boldsymbol{x}_*)\boldsymbol{\theta}$. We determine the corresponding moments by exploiting the properties of means and variances, which yields

$$\begin{aligned} \mathbb{E}[f(\boldsymbol{x}_*) \mid \mathcal{X}, \mathcal{Y}] &= \mathbb{E}_{\boldsymbol{\theta}}[\boldsymbol{\phi}^\top(\boldsymbol{x}_*)\boldsymbol{\theta} \mid \mathcal{X}, \mathcal{Y}] = \boldsymbol{\phi}^\top(\boldsymbol{x}_*)\mathbb{E}_{\boldsymbol{\theta}}[\boldsymbol{\theta} \mid \mathcal{X}, \mathcal{Y}] \\ &= \boldsymbol{\phi}^\top(\boldsymbol{x}_*)\boldsymbol{m}_N = \boldsymbol{m}_N^\top\boldsymbol{\phi}(\boldsymbol{x}_*), \end{aligned} \tag{9.58}$$

$$\begin{aligned} \mathbb{V}_{\boldsymbol{\theta}}[f(\boldsymbol{x}_*) \mid \mathcal{X}, \mathcal{Y}] &= \mathbb{V}_{\boldsymbol{\theta}}[\boldsymbol{\phi}^\top(\boldsymbol{x}_*)\boldsymbol{\theta} \mid \mathcal{X}, \mathcal{Y}] \\ &= \boldsymbol{\phi}^\top(\boldsymbol{x}_*)\mathbb{V}_{\boldsymbol{\theta}}[\boldsymbol{\theta} \mid \mathcal{X}, \mathcal{Y}]\boldsymbol{\phi}(\boldsymbol{x}_*) \\ &= \boldsymbol{\phi}^\top(\boldsymbol{x}_*)\boldsymbol{S}_N\boldsymbol{\phi}(\boldsymbol{x}_*). \end{aligned} \tag{9.59}$$

We see that the predictive mean is the same as the predictive mean for noisy observations as the noise has mean 0, and the predictive variance only differs by σ^2, which is the variance of the measurement noise: When we predict noisy function values, we need to include σ^2 as a source of uncertainty, but this term is not needed for noise-free predictions. Here, the only remaining uncertainty stems from the parameter posterior. ◇

Integrating out parameters induces a distribution over functions.

mean function

Remark (Distribution over Functions). The fact that we integrate out the parameters $\boldsymbol{\theta}$ induces a distribution over functions: If we sample $\boldsymbol{\theta}_i \sim p(\boldsymbol{\theta} \mid \mathcal{X}, \mathcal{Y})$ from the parameter posterior, we obtain a single function realization $\boldsymbol{\theta}_i^\top\boldsymbol{\phi}(\cdot)$. The *mean function*, i.e., the set of all expected function values $\mathbb{E}_{\boldsymbol{\theta}}[f(\cdot) \mid \boldsymbol{\theta}, \mathcal{X}, \mathcal{Y}]$, of this distribution over functions is $\boldsymbol{m}_N^\top\boldsymbol{\phi}(\cdot)$. The (marginal) variance, i.e., the variance of the function $f(\cdot)$, is given by $\boldsymbol{\phi}^\top(\cdot)\boldsymbol{S}_N\boldsymbol{\phi}(\cdot)$. ◇

Example 9.8 (Posterior over Functions)
Let us revisit the Bayesian linear regression problem with polynomials of degree 5. We choose a parameter prior $p(\boldsymbol{\theta}) = \mathcal{N}(\mathbf{0}, \frac{1}{4}\boldsymbol{I})$. Figure 9.9 visualizes the prior over functions induced by the parameter prior and sample functions from this prior.

Figure 9.10 shows the posterior over functions that we obtain via Bayesian linear regression. The training dataset is shown in panel (a); panel (b) shows the posterior distribution over functions, including the functions we would obtain via maximum likelihood and MAP estimation. The function we obtain using the MAP estimate also corresponds to the posterior mean function in the Bayesian linear regression setting. Panel (c) shows some plausible realizations (samples) of functions under that posterior over functions.

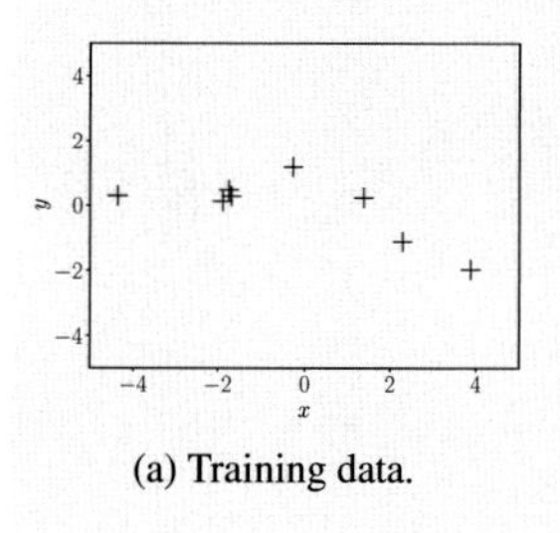

(a) Training data.

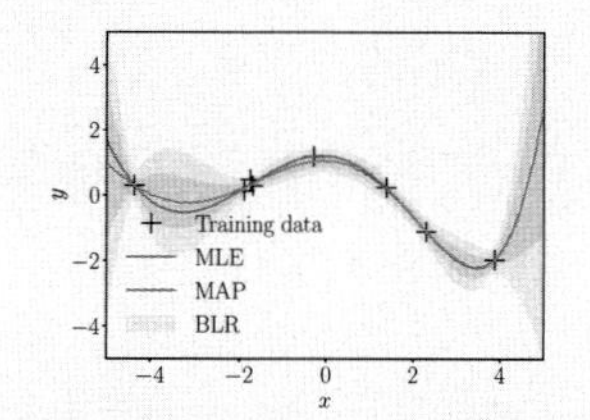

(b) Posterior over functions represented by the marginal uncertainties (shaded) showing the 67% and 95% predictive confidence bounds, the maximum likelihood estimate (MLE) and the MAP estimate (MAP), the latter of which is identical to the posterior mean function.

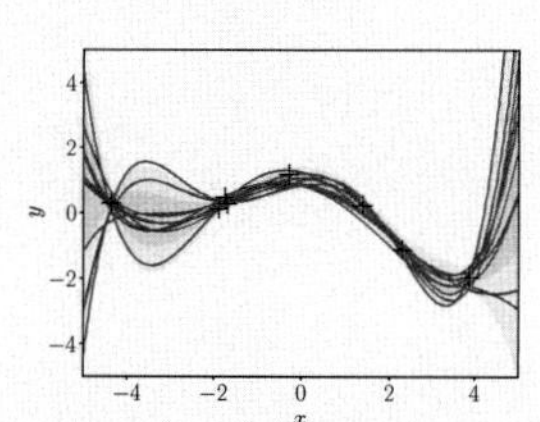

(c) Samples from the posterior over functions, which are induced by the samples from the parameter posterior.

Figure 9.10 Bayesian linear regression and posterior over functions. (a) training data; (b) posterior distribution over functions; (c) samples from the posterior over functions.

Figure 9.11 shows some posterior distributions over functions induced by the parameter posterior. For different polynomial degrees M, the left panels show the maximum likelihood function $\boldsymbol{\theta}_{\text{ML}}^\top \boldsymbol{\phi}(\cdot)$, the MAP function $\boldsymbol{\theta}_{\text{MAP}}^\top \boldsymbol{\phi}(\cdot)$ (which is identical to the posterior mean function), and the 67% and 95% predictive confidence bounds obtained by Bayesian linear regression, represented by the shaded areas.

The right panels show samples from the posterior over functions: Here, we sampled parameters $\boldsymbol{\theta}_i$ from the parameter posterior and computed the function $\boldsymbol{\phi}^\top(\boldsymbol{x}_*)\boldsymbol{\theta}_i$, which is a single realization of a function under the posterior distribution over functions. For low-order polynomials, the parameter posterior does not allow the parameters to vary much: The sampled functions are nearly identical. When we make the model more flexible by adding more parameters (i.e., we end up with a higher-order polynomial), these parameters are not sufficiently constrained by the posterior, and the sampled functions can be easily visually separated. We also see in the corresponding panels on the left how the uncertainty increases, especially at the boundaries.

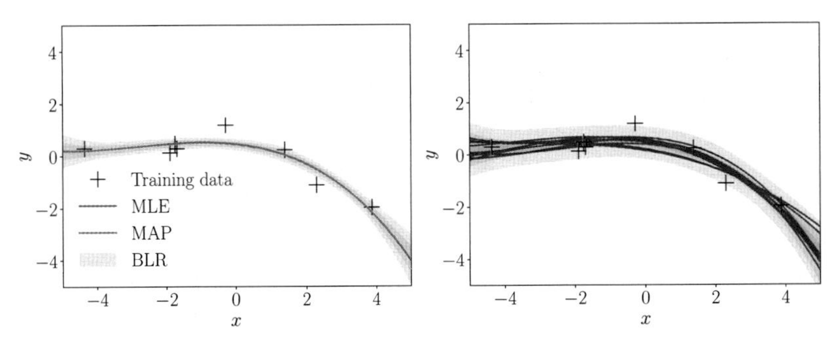

(a) Posterior distribution for polynomials of degree $M = 3$ (left) and samples from the posterior over functions (right).

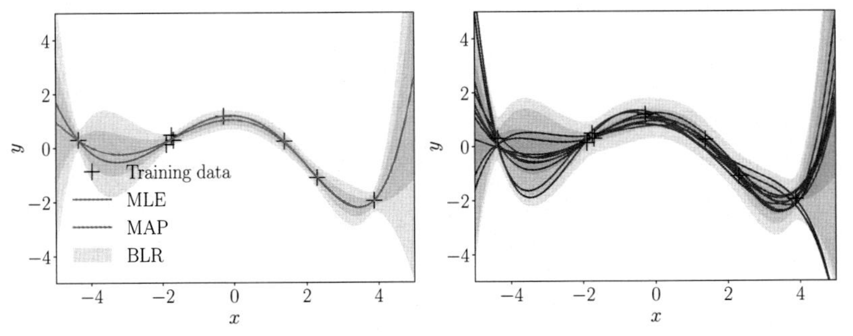

(b) Posterior distribution for polynomials of degree $M = 5$ (left) and samples from the posterior over functions (right).

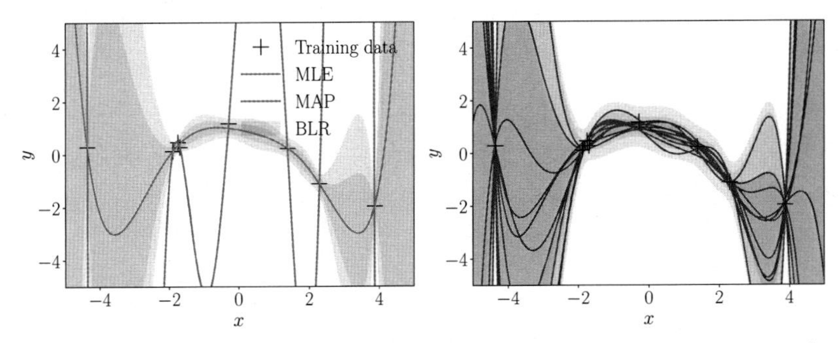

(c) Posterior distribution for polynomials of degree $M = 7$ (left) and samples from the posterior over functions (right).

Figure 9.11 Bayesian linear regression. Left panels: Shaded areas indicate the 67% (dark gray) and 95% (light gray) predictive confidence bounds. The mean of the Bayesian linear regression model coincides with the MAP estimate. The predictive uncertainty is the sum of the noise term and the posterior parameter uncertainty, which depends on the location of the test input. Right panels: sampled functions from the posterior distribution.

Although for a seventh-order polynomial the MAP estimate yields a reasonable fit, the Bayesian linear regression model additionally tells us that the posterior uncertainty is huge. This information can be critical when we use these predictions in a decision-making system, where bad decisions can have significant consequences (e.g., in reinforcement learning or robotics).

9.3.5 Computing the Marginal Likelihood

In Section 8.6.2, we highlighted the importance of the marginal likelihood for Bayesian model selection. In the following, we compute the marginal likelihood for Bayesian linear regression with a conjugate Gaussian prior on the parameters, i.e., exactly the setting we have been discussing in this chapter.

Just to recap, we consider the following generative process:

$$\boldsymbol{\theta} \sim \mathcal{N}\big(\boldsymbol{m}_0,\, \boldsymbol{S}_0\big) \tag{9.60a}$$

$$y_n \,|\, \boldsymbol{x}_n, \boldsymbol{\theta} \sim \mathcal{N}\big(\boldsymbol{x}_n^\top \boldsymbol{\theta},\, \sigma^2\big)\,, \tag{9.60b}$$

$n = 1, \ldots, N$. The marginal likelihood is given by

The marginal likelihood can be interpreted as the expected likelihood under the prior, i.e., $\mathbb{E}_{\boldsymbol{\theta}}[p(\mathcal{Y} \,|\, \mathcal{X}, \boldsymbol{\theta})]$.

$$p(\mathcal{Y} \,|\, \mathcal{X}) = \int p(\mathcal{Y} \,|\, \mathcal{X}, \boldsymbol{\theta}) p(\boldsymbol{\theta}) \mathrm{d}\boldsymbol{\theta} \tag{9.61a}$$

$$= \int \mathcal{N}\big(\boldsymbol{y} \,|\, \boldsymbol{X}\boldsymbol{\theta},\, \sigma^2 \boldsymbol{I}\big) \mathcal{N}\big(\boldsymbol{\theta} \,|\, \boldsymbol{m}_0,\, \boldsymbol{S}_0\big) \mathrm{d}\boldsymbol{\theta}\,, \tag{9.61b}$$

where we integrate out the model parameters $\boldsymbol{\theta}$. We compute the marginal likelihood in two steps: First, we show that the marginal likelihood is Gaussian (as a distribution in $\boldsymbol{y}$); second, we compute the mean and covariance of this Gaussian.

1. The marginal likelihood is Gaussian: From Section 6.5.2, we know that (i) the product of two Gaussian random variables is an (unnormalized) Gaussian distribution, and (ii) a linear transformation of a Gaussian random variable is Gaussian distributed. In (9.61b), we require a linear transformation to bring $\mathcal{N}\big(\boldsymbol{y} \,|\, \boldsymbol{X}\boldsymbol{\theta},\, \sigma^2 \boldsymbol{I}\big)$ into the form $\mathcal{N}\big(\boldsymbol{\theta} \,|\, \boldsymbol{\mu},\, \boldsymbol{\Sigma}\big)$ for some $\boldsymbol{\mu}, \boldsymbol{\Sigma}$. Once this is done, the integral can be solved in closed form. The result is the normalizing constant of the product of the two Gaussians. The normalizing constant itself has Gaussian shape; see (6.76).
2. Mean and covariance. We compute the mean and covariance matrix of the marginal likelihood by exploiting the standard results for means and covariances of affine transformations of random variables; see Section 6.4.4. The mean of the marginal likelihood is computed as

$$\mathbb{E}_{\boldsymbol{\theta}}[\mathcal{Y} \,|\, \mathcal{X}] = \mathbb{E}_{\boldsymbol{\theta}}[\boldsymbol{X}\boldsymbol{\theta} + \boldsymbol{\epsilon}] = \boldsymbol{X}\mathbb{E}_{\boldsymbol{\theta}}[\boldsymbol{\theta}] = \boldsymbol{X}\boldsymbol{m}_0\,. \tag{9.62}$$

Note that $\boldsymbol{\epsilon} \sim \mathcal{N}\big(\boldsymbol{0},\, \sigma^2 \boldsymbol{I}\big)$ is a vector of i.i.d. random variables. The covariance matrix is given as

$$\mathrm{Cov}_{\boldsymbol{\theta}}[\mathcal{Y}|\mathcal{X}] = \mathrm{Cov}[\boldsymbol{X}\boldsymbol{\theta}] + \sigma^2 \boldsymbol{I} = \boldsymbol{X}\,\mathrm{Cov}_{\boldsymbol{\theta}}[\boldsymbol{\theta}]\boldsymbol{X}^\top + \sigma^2 \boldsymbol{I} \tag{9.63a}$$

$$= \boldsymbol{X}\boldsymbol{S}_0\boldsymbol{X}^\top + \sigma^2 \boldsymbol{I}\,. \tag{9.63b}$$

Hence, the marginal likelihood is

$$p(\mathcal{Y} \,|\, \mathcal{X}) = (2\pi)^{-\frac{N}{2}} \det(\boldsymbol{X}\boldsymbol{S}_0\boldsymbol{X}^\top + \sigma^2 \boldsymbol{I})^{-\frac{1}{2}} \tag{9.64a}$$

$$\cdot \exp\big(-\tfrac{1}{2}(\boldsymbol{y} - \boldsymbol{X}\boldsymbol{m}_0)^\top (\boldsymbol{X}\boldsymbol{S}_0\boldsymbol{X}^\top + \sigma^2 \boldsymbol{I})^{-1} (\boldsymbol{y} - \boldsymbol{X}\boldsymbol{m}_0)\big)$$

$$= \mathcal{N}\big(\boldsymbol{y} \,|\, \boldsymbol{X}\boldsymbol{m}_0,\, \boldsymbol{X}\boldsymbol{S}_0\boldsymbol{X}^\top + \sigma^2 \boldsymbol{I}\big)\,. \tag{9.64b}$$

Given the close connection with the posterior predictive distribution (see Remark on "Marginal Likelihood and Posterior Predictive Distribution" earlier in this section), the functional form of the marginal likelihood should not be too surprising.

Figure 9.12 Geometric interpretation of least squares. (a) Dataset; (b) maximum likelihood solution interpreted as a projection.

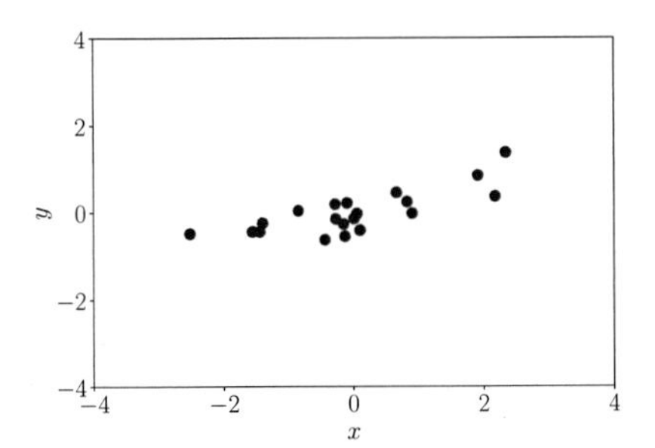

(a) Regression dataset consisting of noisy observations y_n (blue) of function values $f(x_n)$ at input locations x_n.

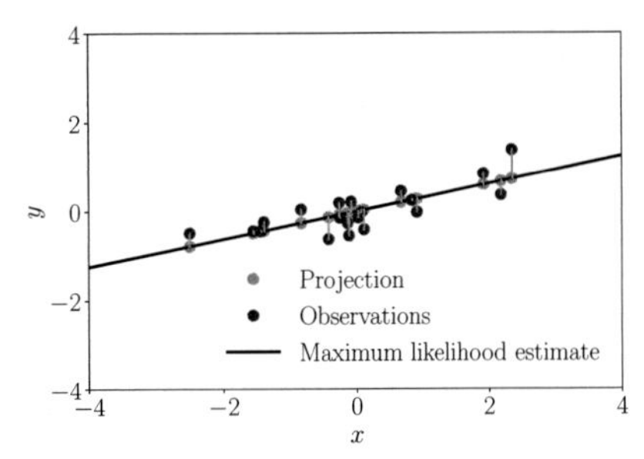

(b) The orange dots are the projections of the noisy observations (blue dots) onto the line $\theta_{\mathrm{ML}}x$. The maximum likelihood solution to a linear regression problem finds a subspace (line) onto which the overall projection error (orange lines) of the observations is minimized.

9.4 Maximum Likelihood as Orthogonal Projection

Having crunched through much algebra to derive maximum likelihood and MAP estimates, we will now provide a geometric interpretation of maximum likelihood estimation. Let us consider a simple linear regression setting

$$y = x\theta + \epsilon, \quad \epsilon \sim \mathcal{N}(0,\, \sigma^2)\,, \tag{9.65}$$

in which we consider linear functions $f : \mathbb{R} \to \mathbb{R}$ that go through the origin (we omit features here for clarity). The parameter θ determines the slope of the line. Figure 9.12(a) shows a one-dimensional dataset.

With a training data set $\{(x_1, y_1), \ldots, (x_N, y_N)\}$ we recall the results from Section 9.2.1 and obtain the maximum likelihood estimator for the slope parameter as

$$\theta_{\mathrm{ML}} = (\boldsymbol{X}^\top\boldsymbol{X})^{-1}\boldsymbol{X}^\top\boldsymbol{y} = \frac{\boldsymbol{X}^\top\boldsymbol{y}}{\boldsymbol{X}^\top\boldsymbol{X}} \in \mathbb{R}\,, \tag{9.66}$$

where $\boldsymbol{X} = [x_1, \ldots, x_N]^\top \in \mathbb{R}^N$, $\boldsymbol{y} = [y_1, \ldots, y_N]^\top \in \mathbb{R}^N$.

This means for the training inputs $\boldsymbol{X}$ we obtain the optimal (maximum likelihood) reconstruction of the training targets as

$$\boldsymbol{X}\theta_{\mathrm{ML}} = \boldsymbol{X}\frac{\boldsymbol{X}^\top\boldsymbol{y}}{\boldsymbol{X}^\top\boldsymbol{X}} = \frac{\boldsymbol{X}\boldsymbol{X}^\top}{\boldsymbol{X}^\top\boldsymbol{X}}\boldsymbol{y}\,, \tag{9.67}$$

i.e., we obtain the approximation with the minimum least-squares error between $\boldsymbol{y}$ and $\boldsymbol{X}\theta$.

Linear regression can be thought of as a method for solving systems of linear equations.

Maximum likelihood linear regression performs an orthogonal projection.

As we are looking for a solution of $\boldsymbol{y} = \boldsymbol{X}\theta$, we can think of linear regression as a problem for solving systems of linear equations. Therefore, we can relate to concepts from linear algebra and analytic geometry that we discussed in Chapters 2 and 3. In particular, looking carefully at (9.67) we see that the maximum likelihood estimator θ_{ML} in our example from (9.65) effectively does an orthogonal projection of $\boldsymbol{y}$ onto the one-dimensional subspace spanned by $\boldsymbol{X}$. Recalling the results on orthogonal projections from Section 3.8, we identify $\frac{\boldsymbol{X}\boldsymbol{X}^\top}{\boldsymbol{X}^\top\boldsymbol{X}}$ as the projection matrix, θ_{ML} as the coordinates of the projection onto the one-dimensional subspace of $\mathbb{R}^N$ spanned by $\boldsymbol{X}$ and $\boldsymbol{X}\theta_{\text{ML}}$ as the orthogonal projection of $\boldsymbol{y}$ onto this subspace.

Therefore, the maximum likelihood solution provides also a geometrically optimal solution by finding the vectors in the subspace spanned by $\boldsymbol{X}$ that are "closest" to the corresponding observations $\boldsymbol{y}$, where "closest" means the smallest (squared) distance of the function values y_n to $x_n\theta$. This is achieved by orthogonal projections. Figure 9.12(b) shows the projection of the noisy observations onto the subspace that minimizes the squared distance between the original dataset and its projection (note that the x-coordinate is fixed), which corresponds to the maximum likelihood solution.

In the general linear regression case where

$$y = \boldsymbol{\phi}^\top(\boldsymbol{x})\boldsymbol{\theta} + \epsilon, \quad \epsilon \sim \mathcal{N}(0, \sigma^2) \tag{9.68}$$

with vector-valued features $\boldsymbol{\phi}(\boldsymbol{x}) \in \mathbb{R}^K$, we again can interpret the maximum likelihood result

$$\boldsymbol{y} \approx \boldsymbol{\Phi}\boldsymbol{\theta}_{\text{ML}}\,, \tag{9.69}$$

$$\boldsymbol{\theta}_{\text{ML}} = (\boldsymbol{\Phi}^\top\boldsymbol{\Phi})^{-1}\boldsymbol{\Phi}^\top\boldsymbol{y} \tag{9.70}$$

as a projection onto a K-dimensional subspace of $\mathbb{R}^N$, which is spanned by the columns of the feature matrix $\boldsymbol{\Phi}$; see Section 3.8.2.

If the feature functions ϕ_k that we use to construct the feature matrix $\boldsymbol{\Phi}$ are orthonormal (see Section 3.7), we obtain a special case where the columns of $\boldsymbol{\Phi}$ form an orthonormal basis (see Section 3.5), such that $\boldsymbol{\Phi}^\top\boldsymbol{\Phi} = \boldsymbol{I}$. This will then lead to the projection

$$\boldsymbol{\Phi}(\boldsymbol{\Phi}^\top\boldsymbol{\Phi})^{-1}\boldsymbol{\Phi}\boldsymbol{y} = \boldsymbol{\Phi}\boldsymbol{\Phi}^\top\boldsymbol{y} = \left(\sum_{k=1}^{K}\boldsymbol{\phi}_k\boldsymbol{\phi}_k^\top\right)\boldsymbol{y} \tag{9.71}$$

so that the coupling between different features has disappeared and the maximum likelihood projection is simply the sum of projections of $\boldsymbol{y}$ onto the individual basis vectors $\boldsymbol{\phi}_k$, i.e., the columns of $\boldsymbol{\Phi}$. Many popular basis functions in signal processing, such as wavelets and Fourier bases, are orthogonal basis functions. When the basis is not orthogonal, one can convert a set of linearly independent basis functions to an orthogonal basis by using the Gram–Schmidt process (Strang, 2003).

9.5 Further Reading

In this chapter, we discussed linear regression for Gaussian likelihoods and conjugate Gaussian priors on the parameters of the model. This allowed for closed-form Bayesian inference. However, in some applications we may want to choose a different likelihood function. For example, in a binary *classification* setting, we observe only two possible (categorical) outcomes, and a Gaussian likelihood is inappropriate in this setting. Instead, we can choose a Bernoulli likelihood that will return a probability of the predicted label to be 1 (or 0). We refer to the books by Bishop (2006), Murphy (2012), and Barber (2012) for an in-depth

classification

introduction to classification problems. A different example where non-Gaussian likelihoods are important is count data. Counts are nonnegative integers, and in this case a Binomial or Poisson likelihood would be a better choice than a Gaussian. All these examples fall into the category of *generalized linear models*, a flexible generalization of linear regression that allows for response variables that have error distributions other than a Gaussian distribution. The GLM generalizes linear regression by allowing the linear model to be related to the observed values via a smooth and invertible function $\sigma(\cdot)$ that may be nonlinear so that $y = \sigma\backslash(f(\boldsymbol{x}))$, where $f(\boldsymbol{x}) = \boldsymbol{\theta}^\top \boldsymbol{\phi}(\boldsymbol{x})$ is the linear regression model from (9.13). We can therefore think of a generalized linear model in terms of function composition $y = \sigma \circ f$, where f is a linear regression model and σ the activation function. Note that although we are talking about "generalized linear models," the outputs y are no longer linear in the parameters $\boldsymbol{\theta}$. In *logistic regression*, we choose the *logistic sigmoid* $\sigma(f) = \frac{1}{1+\exp(-f)} \in [0, 1]$, which can be interpreted as the probability of observing $y = 1$ of a Bernoulli random variable $y \in \{0, 1\}$. The function $\sigma(\cdot)$ is called *transfer function* or *activation function*, and its inverse is called the *canonical link function*. From this perspective, it is also clear that generalized linear models are the building blocks of (deep) feedforward neural networks: If we consider a generalized linear model $\boldsymbol{y} = \sigma(\boldsymbol{A}\boldsymbol{x} + \boldsymbol{b})$, where $\boldsymbol{A}$ is a weight matrix and $\boldsymbol{b}$ a bias vector, we identify this generalized linear model as a single-layer neural network with activation function $\sigma(\cdot)$. We can now recursively compose these functions via

generalized linear model

Generalized linear models are the building blocks of deep neural networks.

logistic regression

logistic sigmoid

transfer function

activation function

canonical link function

For ordinary linear regression, the activation function would simply be the identity.

$$\begin{aligned} \boldsymbol{x}_{k+1} &= \boldsymbol{f}_k(\boldsymbol{x}_k) \\ \boldsymbol{f}_k(\boldsymbol{x}_k) &= \sigma_k(\boldsymbol{A}_k \boldsymbol{x}_k + \boldsymbol{b}_k) \end{aligned} \tag{9.72}$$

for $k = 0, \ldots, K-1$, where $\boldsymbol{x}_0$ are the input features and $\boldsymbol{x}_K = \boldsymbol{y}$ are the observed outputs, such that $\boldsymbol{f}_{K-1} \circ \cdots \circ \boldsymbol{f}_0$ is a K-layer deep neural network. Therefore, the building blocks of this deep neural network are the generalized linear models defined in (9.72). Neural networks (Bishop, 1995; Goodfellow et al., 2016) are significantly more expressive and flexible than linear regression models. However, maximum likelihood parameter estimation is a nonconvex optimization problem, and marginalization of the parameters in a fully Bayesian setting is analytically intractable.

A great post on the relation between GLMs and deep networks is available at `https://tinyurl.com/glm-dnn`.

We briefly hinted at the fact that a distribution over parameters induces a distribution over regression functions. *Gaussian processes* (Rasmussen and Williams, 2006) are regression models where the concept of a distribution over function is central. Instead of placing a distribution over parameters, a Gaussian process places a distribution directly on the space of functions without the "detour" via the parameters. To do so, the Gaussian process exploits the *kernel trick* (Schölkopf and Smola, 2002), which allows us to compute inner products between two function values $f(\boldsymbol{x}_i), f(\boldsymbol{x}_j)$ only by looking at the corresponding input $\boldsymbol{x}_i, \boldsymbol{x}_j$. A Gaussian process is closely related to both Bayesian linear regression and support vector regression but can also be interpreted as a Bayesian neural network with a single hidden layer where the number of units tends to infinity (Neal, 1996; Williams, 1997). Excellent introductions to Gaussian processes can be found in MacKay (1998) and Rasmussen and Williams (2006).

Gaussian process

kernel trick

We focused on Gaussian parameter priors in the discussions in this chapter because they allow for closed-form inference in linear regression models. However, even in a regression setting with Gaussian likelihoods, we may choose a non-Gaussian prior. Consider a setting, where the inputs are $\boldsymbol{x} \in \mathbb{R}^D$ and our training set is small and of size $N \ll D$. This means that the regression problem is under determined. In this case, we can choose a parameter prior that enforces sparsity, i.e., a prior that tries to set as many parameters to 0 as possible (*variable selection*). This prior provides a stronger regularizer than the Gaussian prior, which often leads to an increased prediction accuracy and interpretability of the model. The Laplace prior is one example that is frequently used for this purpose. A linear regression model with the Laplace prior on the parameters is equivalent to linear regression with L1 regularization (*LASSO*) (Tibshirani, 1996). The Laplace distribution is sharply peaked at zero (its first derivative is discontinuous) and it concentrates its probability mass closer to zero than the Gaussian distribution, which encourages parameters to be 0. Therefore, the nonzero parameters are relevant for the regression problem, which is the reason why we also speak of "variable selection."

variable selection

LASSO

10

Dimensionality Reduction with Principal Component Analysis

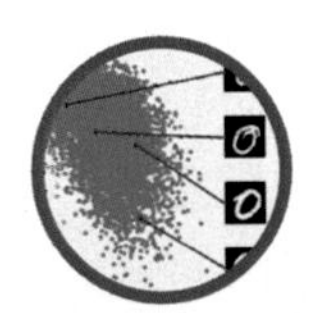

A 640 × 480 pixel color image is a data point in a million-dimensional space, where every pixel responds to three dimensions, one for each color channel (red, green, blue).

Working directly with high-dimensional data, such as images, comes with some difficulties: It is hard to analyze, interpretation is difficult, visualization is nearly impossible, and (from a practical point of view) storage of the data vectors can be expensive. However, high-dimensional data often has properties that we can exploit. For example, high-dimensional data is often overcomplete, i.e., many dimensions are redundant and can be explained by a combination of other dimensions. Furthermore, dimensions in high-dimensional data are often correlated so that the data possesses an intrinsic lower-dimensional structure. Dimensionality reduction exploits structure and correlation and allows us to work with a more compact representation of the data, ideally without losing information. We can think of dimensionality reduction as a compression technique, similar to jpeg or mp3, which are compression algorithms for images and music.

principal component analysis
PCA
dimensionality reduction

In this chapter, we will discuss *principal component analysis (PCA)*, an algorithm for linear *dimensionality reduction*. PCA, proposed by Pearson (1901) and Hotelling (1933), has been around for more than 100 years and is still one of the most commonly used techniques for data compression and data visualization. It is also used for the identification of simple patterns, latent factors, and structures of high-dimensional data. In the signal processing community, PCA is also known as the *Karhunen–Loève transform*. In this chapter, we derive PCA from first principles, drawing on our understanding of basis and basis change (Sections 2.6.1 and 2.7.2), projections (Section 3.8), eigenvalues (Section 4.2), Gaussian distributions (Section 6.5), and constrained optimization (Section 7.2).

Karhunen–Loève transform

Dimensionality reduction generally exploits a property of high-dimensional data (e.g., images) that it often lies on a low-dimensional subspace. Figure 10.1 gives an illustrative example in two dimensions. Although the data in Figure 10.1(a) does not quite lie on a line, the data does not vary much in the x_2-direction, so that we can express it as if it were on a line – with nearly no loss; see Figure 10.1(b). To describe the data in Figure 10.1(b), only the x_1-coordinate is required, and the data lies in a one-dimensional subspace of $\mathbb{R}^2$.

10.1 Problem Setting

In PCA, we are interested in finding projections $\tilde{\boldsymbol{x}}_n$ of data points $\boldsymbol{x}_n$ that are as similar to the original data points as possible, but which have a significantly

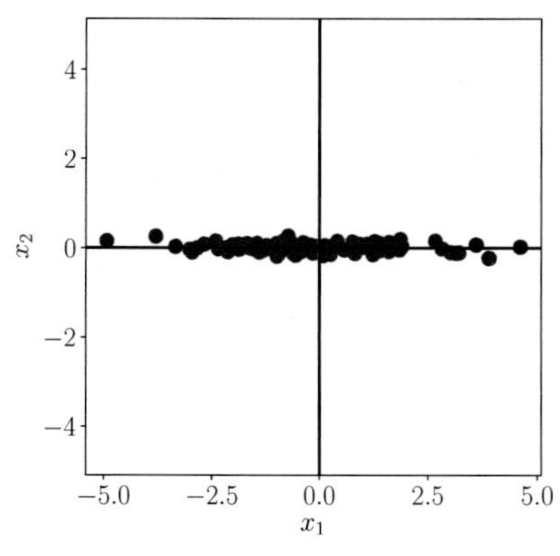

(a) Dataset with x_1 and x_2 coordinates

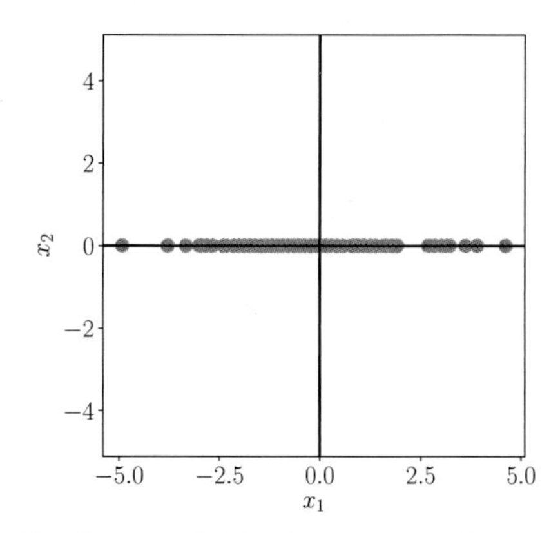

(b) Compressed dataset where only the x_1 coordinate is relevant

Figure 10.1 Illustration: dimensionality reduction. (a) The original dataset does not vary much along the x_2 direction. (b) The data from (a) can be represented using the x_1-coordinate alone with nearly no loss.

lower intrinsic dimensionality. Figure 10.1 gives an illustration of what this could look like.

More concretely, we consider an i.i.d. dataset $\mathcal{X} = \{\boldsymbol{x}_1, \ldots, \boldsymbol{x}_N\}$, $\boldsymbol{x}_n \in \mathbb{R}^D$, with mean $\mathbf{0}$ that possesses the *data covariance matrix* (6.42)

data covariance matrix

$$\boldsymbol{S} = \frac{1}{N} \sum_{n=1}^{N} \boldsymbol{x}_n \boldsymbol{x}_n^\top . \tag{10.1}$$

Furthermore, we assume there exists a low-dimensional compressed representation (code)

$$\boldsymbol{z}_n = \boldsymbol{B}^\top \boldsymbol{x}_n \in \mathbb{R}^M \tag{10.2}$$

of $\boldsymbol{x}_n$, where we define the projection matrix

$$\boldsymbol{B} := [\boldsymbol{b}_1, \ldots, \boldsymbol{b}_M] \in \mathbb{R}^{D \times M} . \tag{10.3}$$

We assume that the columns of $\boldsymbol{B}$ are orthonormal (Definition 3.7) so that $\boldsymbol{b}_i^\top \boldsymbol{b}_j = 0$ if and only if $i \neq j$ and $\boldsymbol{b}_i^\top \boldsymbol{b}_i = 1$. We seek an M-dimensional subspace $U \subseteq \mathbb{R}^D$, $\dim(U) = M < D$ onto which we project the data. We denote the projected data by $\tilde{\boldsymbol{x}}_n \in U$, and their coordinates (with respect to the basis vectors $\boldsymbol{b}_1, \ldots, \boldsymbol{b}_M$ of U) by $\boldsymbol{z}_n$. Our aim is to find projections $\tilde{\boldsymbol{x}}_n \in \mathbb{R}^D$ (or equivalently the codes $\boldsymbol{z}_n$ and the basis vectors $\boldsymbol{b}_1, \ldots, \boldsymbol{b}_M$) so that they are as similar to the original data $\boldsymbol{x}_n$ and minimize the loss due to compression.

The columns $\boldsymbol{b}_1, \ldots, \boldsymbol{b}_M$ of $\boldsymbol{B}$ form a basis of the M-dimensional subspace in which the projected data $\tilde{\boldsymbol{x}} = \boldsymbol{B}\boldsymbol{B}^\top \boldsymbol{x} \in \mathbb{R}^D$ live.

Example 10.1 (Coordinate Representation/Code)
Consider $\mathbb{R}^2$ with the canonical basis $\boldsymbol{e}_1 = [1, 0]^\top$, $\boldsymbol{e}_2 = [0, 1]^\top$. From Chapter 2, we know that $\boldsymbol{x} \in \mathbb{R}^2$ can be represented as a linear combination of these basis vectors, e.g.,

$$\begin{bmatrix} 5 \\ 3 \end{bmatrix} = 5\boldsymbol{e}_1 + 3\boldsymbol{e}_2 . \tag{10.4}$$

However, when we consider vectors of the form

$$\tilde{\boldsymbol{x}} = \begin{bmatrix} 0 \\ z \end{bmatrix} \in \mathbb{R}^2 , \quad z \in \mathbb{R} , \tag{10.5}$$

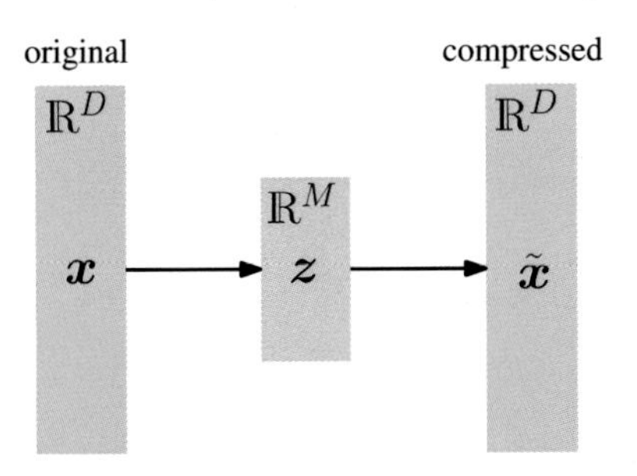

Figure 10.2 Graphical illustration of PCA. In PCA, we find a compressed version $\boldsymbol{z}$ of original data $\boldsymbol{x}$. The compressed data can be reconstructed into $\tilde{\boldsymbol{x}}$, which lives in the original data space, but has an intrinsic lower-dimensional representation than $\boldsymbol{x}$.

Figure 10.3 Examples of handwritten digits from the MNIST dataset. `http://yann.lecun.com/exdb/mnist/`.

they can always be written as $0\boldsymbol{e}_1 + z\boldsymbol{e}_2$. To represent these vectors, it is sufficient to remember/store the *coordinate/code* z of $\tilde{\boldsymbol{x}}$ with respect to the $\boldsymbol{e}_2$ vector.

More precisely, the set of $\tilde{\boldsymbol{x}}$ vectors (with the standard vector addition and scalar multiplication) forms a vector subspace U (see Section 2.4) with $\dim(U) = 1$ because $U = \operatorname{span}[\boldsymbol{e}_2]$.

The dimension of a vector space corresponds to the number of its basis vectors (see Section 2.6.1).

In Section 10.2, we will find low-dimensional representations that retain as much information as possible and minimize the compression loss. An alternative derivation of PCA is given in Section 10.3, where we will be looking at minimizing the squared reconstruction error $\|\boldsymbol{x}_n - \tilde{\boldsymbol{x}}_n\|^2$ between the original data $\boldsymbol{x}_n$ and its projection $\tilde{\boldsymbol{x}}_n$.

Figure 10.2 illustrates the setting we consider in PCA, where $\boldsymbol{z}$ represents the lower-dimensional representation of the compressed data $\tilde{\boldsymbol{x}}$ and plays the role of a bottleneck, which controls how much information can flow between $\boldsymbol{x}$ and $\tilde{\boldsymbol{x}}$. In PCA, we consider a linear relationship between the original data $\boldsymbol{x}$ and its low-dimensional code $\boldsymbol{z}$ so that $\boldsymbol{z} = \boldsymbol{B}^\top \boldsymbol{x}$ and $\tilde{\boldsymbol{x}} = \boldsymbol{B}\boldsymbol{z}$ for a suitable matrix $\boldsymbol{B}$. Based the motivation of thinking of PCA as a data compression technique, we can interpret the arrows in Figure 10.2 as a pair of operations representing encoders and decoders. The linear mapping represented by $\boldsymbol{B}$ can be thought of a decoder, which maps the low-dimensional code $\boldsymbol{z} \in \mathbb{R}^M$ back into the original data space $\mathbb{R}^D$. Similarly, $\boldsymbol{B}^\top$ can be thought of an encoder, which encodes the original data $\boldsymbol{x}$ as a low-dimensional (compressed) code $\boldsymbol{z}$.

Throughout this chapter, we will use the MNIST digits dataset as a reoccurring example, which contains 60,000 examples of handwritten digits 0 through 9. Each digit is a grayscale image of size 28×28, i.e., it contains 784 pixels so that we can interpret every image in this dataset as a vector $\boldsymbol{x} \in \mathbb{R}^{784}$. Examples of these digits are shown in Figure 10.3.

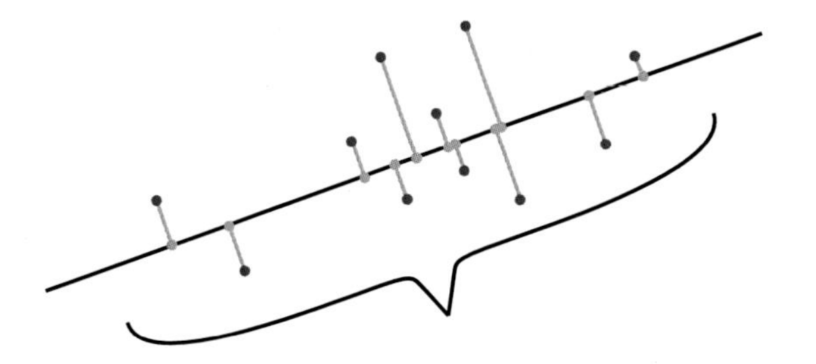

Figure 10.4 PCA finds a lower-dimensional subspace (line) that maintains as much variance (spread of the data) as possible when the data (blue) is projected onto this subspace (orange).

10.2 Maximum Variance Perspective

Figure 10.1 gave an example of how a two-dimensional dataset can be represented using a single coordinate. In Figure 10.1(b), we chose to ignore the x_2-coordinate of the data because it did not add too much information so that the compressed data is similar to the original data in Figure 10.1(a). We could have chosen to ignore the x_1-coordinate, but then the compressed data had been very dissimilar from the original data, and much information in the data would have been lost.

If we interpret information content in the data as how "space filling" the dataset is, then we can describe the information contained in the data by looking at the spread of the data. From Section 6.4.1, we know that the variance is an indicator of the spread of the data, and we can derive PCA as a dimensionality reduction algorithm that maximizes the variance in the low-dimensional representation of the data to retain as much information as possible. Figure 10.4 illustrates this.

Considering the setting discussed in Section 10.1, our aim is to find a matrix $\boldsymbol{B}$ (see (10.3)) that retains as much information as possible when compressing data by projecting it onto the subspace spanned by the columns $\boldsymbol{b}_1, \ldots, \boldsymbol{b}_M$ of $\boldsymbol{B}$. Retaining most information after data compression is equivalent to capturing the largest amount of variance in the low-dimensional code (Hotelling, 1933).

Remark. (Centered Data) For the data covariance matrix in (10.1), we assumed centered data. We can make this assumption without loss of generality: Let us assume that $\boldsymbol{\mu}$ is the mean of the data. Using the properties of the variance, which we discussed in Section 6.4.4, we obtain

$$\mathbb{V}_{\boldsymbol{z}}[\boldsymbol{z}] = \mathbb{V}_{\boldsymbol{x}}[\boldsymbol{B}^\top(\boldsymbol{x} - \boldsymbol{\mu})] = \mathbb{V}_{\boldsymbol{x}}[\boldsymbol{B}^\top\boldsymbol{x} - \boldsymbol{B}^\top\boldsymbol{\mu}] = \mathbb{V}_{\boldsymbol{x}}[\boldsymbol{B}^\top\boldsymbol{x}], \tag{10.6}$$

i.e., the variance of the low-dimensional code does not depend on the mean of the data. Therefore, we assume without loss of generality that the data has mean $\boldsymbol{0}$ for the remainder of this section. With this assumption, the mean of the low-dimensional code is also $\boldsymbol{0}$ since $\mathbb{E}_{\boldsymbol{z}}[\boldsymbol{z}] = \mathbb{E}_{\boldsymbol{x}}[\boldsymbol{B}^\top\boldsymbol{x}] = \boldsymbol{B}^\top\mathbb{E}_{\boldsymbol{x}}[\boldsymbol{x}] = \boldsymbol{0}$. ♢

10.2.1 Direction with Maximal Variance

We maximize the variance of the low-dimensional code using a sequential approach. We start by seeking a single vector $\boldsymbol{b}_1 \in \mathbb{R}^D$ that maximizes the variance of the projected data, i.e., we aim to maximize the variance of the first coordinate z_1 of $\boldsymbol{z} \in \mathbb{R}^M$ so that

The vector $\boldsymbol{b}_1$ will be the first column of the matrix $\boldsymbol{B}$ and therefore the first of M orthonormal basis vectors that span the lower-dimensional subspace.

$$V_1 := \mathbb{V}[z_1] = \frac{1}{N}\sum_{n=1}^{N} z_{1n}^2 \tag{10.7}$$

is maximized, where we exploited the i.i.d. assumption of the data and defined z_{1n} as the first coordinate of the low-dimensional representation $\boldsymbol{z}_n \in \mathbb{R}^M$ of $\boldsymbol{x}_n \in \mathbb{R}^D$. Note that first component of $\boldsymbol{z}_n$ is given by

$$z_{1n} = \boldsymbol{b}_1^\top \boldsymbol{x}_n\,, \tag{10.8}$$

i.e., it is the coordinate of the orthogonal projection of $\boldsymbol{x}_n$ onto the one-dimensional subspace spanned by $\boldsymbol{b}_1$ (Section 3.8). We substitute (10.8) into (10.7), which yields

$$V_1 = \frac{1}{N}\sum_{n=1}^{N}(\boldsymbol{b}_1^\top \boldsymbol{x}_n)^2 = \frac{1}{N}\sum_{n=1}^{N} \boldsymbol{b}_1^\top \boldsymbol{x}_n \boldsymbol{x}_n^\top \boldsymbol{b}_1 \tag{10.9a}$$

$$= \boldsymbol{b}_1^\top \left(\frac{1}{N}\sum_{n=1}^{N} \boldsymbol{x}_n \boldsymbol{x}_n^\top\right) \boldsymbol{b}_1 = \boldsymbol{b}_1^\top \boldsymbol{S} \boldsymbol{b}_1\,, \tag{10.9b}$$

where $\boldsymbol{S}$ is the data covariance matrix defined in (10.1). In (10.9a), we have used the fact that the dot product of two vectors is symmetric with respect to its arguments, that is, $\boldsymbol{b}_1^\top \boldsymbol{x}_n = \boldsymbol{x}_n^\top \boldsymbol{b}_1$.

Notice that arbitrarily increasing the magnitude of the vector $\boldsymbol{b}_1$ increases V_1, that is, a vector $\boldsymbol{b}_1$ that is two times longer can result in V_1 that is potentially four times larger. Therefore, we restrict all solutions to $\|\boldsymbol{b}_1\|^2 = 1$, which results in a constrained optimization problem in which we seek the direction along which the data varies most.

$\|\boldsymbol{b}_1\|^2 = 1$ $\iff \|\boldsymbol{b}_1\| = 1$.

With the restriction of the solution space to unit vectors the vector $\boldsymbol{b}_1$ that points in the direction of maximum variance can be found by the constrained optimization problem

$$\begin{aligned} &\max_{\boldsymbol{b}_1} \boldsymbol{b}_1^\top \boldsymbol{S} \boldsymbol{b}_1 \\ &\text{subject to } \|\boldsymbol{b}_1\|^2 = 1\,. \end{aligned} \tag{10.10}$$

Following Section 7.2, we obtain the Lagrangian

$$\mathfrak{L}(\boldsymbol{b}_1, \lambda) = \boldsymbol{b}_1^\top \boldsymbol{S} \boldsymbol{b}_1 + \lambda_1(1 - \boldsymbol{b}_1^\top \boldsymbol{b}_1) \tag{10.11}$$

to solve this constrained optimization problem. The partial derivatives of $\mathfrak{L}$ with respect to $\boldsymbol{b}_1$ and λ_1 are

$$\frac{\partial \mathfrak{L}}{\partial \boldsymbol{b}_1} = 2\boldsymbol{b}_1^\top \boldsymbol{S} - 2\lambda_1 \boldsymbol{b}_1^\top\,, \qquad \frac{\partial \mathfrak{L}}{\partial \lambda_1} = 1 - \boldsymbol{b}_1^\top \boldsymbol{b}_1\,, \tag{10.12}$$

respectively. Setting these partial derivatives to $\boldsymbol{0}$ gives us the relations

$$\boldsymbol{S}\boldsymbol{b}_1 = \lambda_1 \boldsymbol{b}_1\,, \tag{10.13}$$

$$\boldsymbol{b}_1^\top \boldsymbol{b}_1 = 1\,. \tag{10.14}$$

By comparing this with the definition of an eigenvalue decomposition (Section 4.4), we see that $\boldsymbol{b}_1$ is an eigenvector of the data covariance matrix $\boldsymbol{S}$, and the Lagrange multiplier λ_1 plays the role of the corresponding eigenvalue. This eigenvector property (10.13) allows us to rewrite our variance objective (10.10) as

$$V_1 = \boldsymbol{b}_1^\top \boldsymbol{S} \boldsymbol{b}_1 = \lambda_1 \boldsymbol{b}_1^\top \boldsymbol{b}_1 = \lambda_1 , \tag{10.15}$$

i.e., the variance of the data projected onto a one-dimensional subspace equals the eigenvalue that is associated with the basis vector $\boldsymbol{b}_1$ that spans this subspace. Therefore, to maximize the variance of the low-dimensional code, we choose the basis vector associated with the largest eigenvalue of the data covariance matrix. This eigenvector is called the first *principal component*. We can determine the effect/contribution of the principal component $\boldsymbol{b}_1$ in the original data space by mapping the coordinate z_{1n} back into data space, which gives us the projected data point

The quantity $\sqrt{\lambda_1}$ is also called the *loading* of the unit vector $\boldsymbol{b}_1$ and represents the standard deviation of the data accounted for by the principal subspace $\text{span}[\boldsymbol{b}_1]$.

principal component

$$\tilde{\boldsymbol{x}}_n = \boldsymbol{b}_1 z_{1n} = \boldsymbol{b}_1 \boldsymbol{b}_1^\top \boldsymbol{x}_n \in \mathbb{R}^D \tag{10.16}$$

in the original data space.

Remark. Although $\tilde{\boldsymbol{x}}_n$ is a D-dimensional vector, it only requires a single coordinate z_{1n} to represent it with respect to the basis vector $\boldsymbol{b}_1 \in \mathbb{R}^D$. ◊

10.2.2 M-dimensional Subspace with Maximal Variance

Assume we have found the first $m-1$ principal components as the $m-1$ eigenvectors of $\boldsymbol{S}$ that are associated with the largest $m-1$ eigenvalues. Since $\boldsymbol{S}$ is symmetric, the spectral theorem (Theorem 4.15) states that we can use these eigenvectors to construct an orthonormal eigenbasis of an $(m-1)$-dimensional subspace of $\mathbb{R}^D$. Generally, the mth principal component can be found by subtracting the effect of the first $m-1$ principal components $\boldsymbol{b}_1, \dots, \boldsymbol{b}_{m-1}$ from the data, thereby trying to find principal components that compress the remaining information. We then arrive at the new data matrix

$$\hat{\boldsymbol{X}} := \boldsymbol{X} - \sum_{i=1}^{m-1} \boldsymbol{b}_i \boldsymbol{b}_i^\top \boldsymbol{X} = \boldsymbol{X} - \boldsymbol{B}_{m-1} \boldsymbol{X} , \tag{10.17}$$

where $\boldsymbol{X} = [\boldsymbol{x}_1, \dots, \boldsymbol{x}_N] \in \mathbb{R}^{D \times N}$ contains the data points as column vectors and $\boldsymbol{B}_{m-1} := \sum_{i=1}^{m-1} \boldsymbol{b}_i \boldsymbol{b}_i^\top$ is a projection matrix that projects onto the subspace spanned by $\boldsymbol{b}_1, \dots, \boldsymbol{b}_{m-1}$.

The matrix $\hat{\boldsymbol{X}} := [\hat{\boldsymbol{x}}_1, \dots, \hat{\boldsymbol{x}}_N] \in \mathbb{R}^{D \times N}$ in (10.17) contains the information in the data that has not yet been compressed.

Remark (Notation). Throughout this chapter, we do not follow the convention of collecting data $\boldsymbol{x}_1, \dots, \boldsymbol{x}_N$ as the rows of the data matrix, but we define them to be the columns of $\boldsymbol{X}$. This means that our data matrix $\boldsymbol{X}$ is a $D \times N$ matrix instead of the conventional $N \times D$ matrix. The reason for our choice is that the algebra operations work out smoothly without the need to either transpose the matrix or to redefine vectors as row vectors that are left-multiplied onto matrices. ◊

To find the mth principal component, we maximize the variance

$$V_m = \mathbb{V}[z_m] = \frac{1}{N} \sum_{n=1}^{N} z_{mn}^2 = \frac{1}{N} \sum_{n=1}^{N} (\boldsymbol{b}_m^\top \hat{\boldsymbol{x}}_n)^2 = \boldsymbol{b}_m^\top \hat{\boldsymbol{S}} \boldsymbol{b}_m , \tag{10.18}$$

subject to $\|\boldsymbol{b}_m\|^2 = 1$, where we followed the same steps as in (10.9b) and defined $\hat{\boldsymbol{S}}$ as the data covariance matrix of the transformed dataset $\hat{\mathcal{X}} := \{\hat{\boldsymbol{x}}_1, \ldots, \hat{\boldsymbol{x}}_N\}$. As previously, when we looked at the first principal component alone, we solve a constrained optimization problem and discover that the optimal solution $\boldsymbol{b}_m$ is the eigenvector of $\hat{\boldsymbol{S}}$ that is associated with the largest eigenvalue of $\hat{\boldsymbol{S}}$.

It turns out that $\boldsymbol{b}_m$ is also an eigenvector of $\boldsymbol{S}$. More generally, the sets of eigenvectors of $\boldsymbol{S}$ and $\hat{\boldsymbol{S}}$ are identical. Since both $\boldsymbol{S}$ and $\hat{\boldsymbol{S}}$ are symmetric, we can find an ONB of eigenvectors (spectral theorem 4.15), i.e., there exist D distinct eigenvectors for both $\boldsymbol{S}$ and $\hat{\boldsymbol{S}}$. Next, we show that every eigenvector of $\boldsymbol{S}$ is an eigenvector of $\hat{\boldsymbol{S}}$. Assume we have already found eigenvectors $\boldsymbol{b}_1, \ldots, \boldsymbol{b}_{m-1}$ of $\hat{\boldsymbol{S}}$. Consider an eigenvector $\boldsymbol{b}_i$ of $\boldsymbol{S}$, i.e., $\boldsymbol{S}\boldsymbol{b}_i = \lambda_i \boldsymbol{b}_i$. In general,

$$\hat{\boldsymbol{S}}\boldsymbol{b}_i = \frac{1}{N}\hat{\boldsymbol{X}}\hat{\boldsymbol{X}}^\top \boldsymbol{b}_i = \frac{1}{N}(\boldsymbol{X} - \boldsymbol{B}_{m-1}\boldsymbol{X})(\boldsymbol{X} - \boldsymbol{B}_{m-1}\boldsymbol{X})^\top \boldsymbol{b}_i \tag{10.19a}$$

$$= (\boldsymbol{S} - \boldsymbol{S}\boldsymbol{B}_{m-1} - \boldsymbol{B}_{m-1}\boldsymbol{S} + \boldsymbol{B}_{m-1}\boldsymbol{S}\boldsymbol{B}_{m-1})\boldsymbol{b}_i\,. \tag{10.19b}$$

We distinguish between two cases. If $i \geqslant m$, i.e., $\boldsymbol{b}_i$ is an eigenvector that is not among the first $m-1$ principal components, then $\boldsymbol{b}_i$ is orthogonal to the first $m-1$ principal components and $\boldsymbol{B}_{m-1}\boldsymbol{b}_i = \boldsymbol{0}$. If $i < m$, i.e., $\boldsymbol{b}_i$ is among the first $m-1$ principal components, then $\boldsymbol{b}_i$ is a basis vector of the principal subspace onto which $\boldsymbol{B}_{m-1}$ projects. Since $\boldsymbol{b}_1, \ldots, \boldsymbol{b}_{m-1}$ are an ONB of this principal subspace, we obtain $\boldsymbol{B}_{m-1}\boldsymbol{b}_i = \boldsymbol{b}_i$. The two cases can be summarized as follows:

$$\boldsymbol{B}_{m-1}\boldsymbol{b}_i = \boldsymbol{b}_i \quad \text{if } i < m\,, \qquad \boldsymbol{B}_{m-1}\boldsymbol{b}_i = \boldsymbol{0} \quad \text{if } i \geqslant m\,. \tag{10.20}$$

In the case $i \geqslant m$, by using (10.20) in (10.19b), we obtain $\hat{\boldsymbol{S}}\boldsymbol{b}_i = (\boldsymbol{S} - \boldsymbol{B}_{m-1}\boldsymbol{S})\boldsymbol{b}_i = \boldsymbol{S}\boldsymbol{b}_i = \lambda_i \boldsymbol{b}_i$, i.e., $\boldsymbol{b}_i$ is also an eigenvector of $\hat{\boldsymbol{S}}$ with eigenvalue λ_i. Specifically,

$$\hat{\boldsymbol{S}}\boldsymbol{b}_m = \boldsymbol{S}\boldsymbol{b}_m = \lambda_m \boldsymbol{b}_m\,. \tag{10.21}$$

Equation (10.21) reveals that $\boldsymbol{b}_m$ is not only an eigenvector of $\boldsymbol{S}$ but also of $\hat{\boldsymbol{S}}$. Specifically, λ_m is the largest eigenvalue of $\hat{\boldsymbol{S}}$ and λ_m is the mth largest eigenvalue of $\boldsymbol{S}$, and both have the associated eigenvector $\boldsymbol{b}_m$.

In the case $i < m$, by using (10.20) in (10.19b), we obtain

$$\hat{\boldsymbol{S}}\boldsymbol{b}_i = (\boldsymbol{S} - \boldsymbol{S}\boldsymbol{B}_{m-1} - \boldsymbol{B}_{m-1}\boldsymbol{S} + \boldsymbol{B}_{m-1}\boldsymbol{S}\boldsymbol{B}_{m-1})\boldsymbol{b}_i = \boldsymbol{0} = 0\boldsymbol{b}_i \tag{10.22}$$

This means that $\boldsymbol{b}_1, \ldots, \boldsymbol{b}_{m-1}$ are also eigenvectors of $\hat{\boldsymbol{S}}$, but they are associated with eigenvalue 0 so that $\boldsymbol{b}_1, \ldots, \boldsymbol{b}_{m-1}$ span the null space of $\hat{\boldsymbol{S}}$.

Overall, every eigenvector of $\boldsymbol{S}$ is also an eigenvector of $\hat{\boldsymbol{S}}$. However, if the eigenvectors of $\boldsymbol{S}$ are part of the $(m-1)$ dimensional principal subspace, then the associated eigenvalue of $\hat{\boldsymbol{S}}$ is 0.

This derivation shows that there is an intimate connection between the M-dimensional subspace with maximal variance and the eigenvalue decomposition. We will revisit this connection in Section 10.4.

With the relation (10.21) and $\boldsymbol{b}_m^\top \boldsymbol{b}_m = 1$, the variance of the data projected onto the mth principal component is

$$V_m = \boldsymbol{b}_m^\top \boldsymbol{S}\boldsymbol{b}_m \overset{(10.21)}{=} \lambda_m \boldsymbol{b}_m^\top \boldsymbol{b}_m = \lambda_m\,. \tag{10.23}$$

This means that the variance of the data, when projected onto an M-dimensional subspace, equals the sum of the eigenvalues that are associated with the corresponding eigenvectors of the data covariance matrix.

Example 10.2 (Eigenvalues of MNIST "8")
Taking all digits "8" in the MNIST training data, we compute the eigenvalues of the data covariance matrix. Figure 10.5(a) shows the 200 largest eigenvalues of the data covariance matrix. We see that only a few of them have a value that differs significantly from 0. Therefore, most of the variance, when projecting data onto the subspace spanned by the corresponding eigenvectors, is captured by only a few principal components, as shown in Figure 10.5(b).

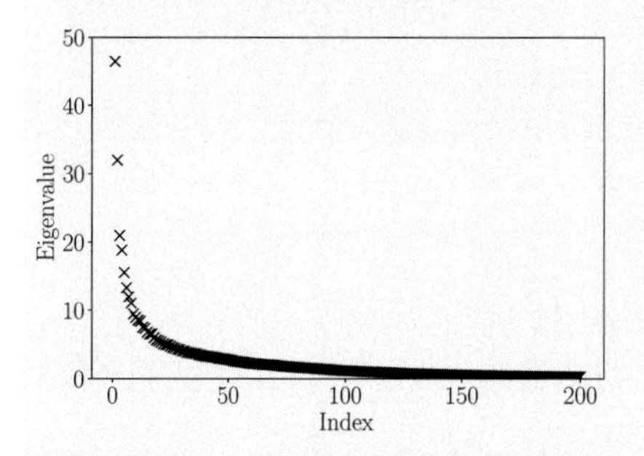

(a) Eigenvalues (sorted in descending order) of the data covariance matrix of all digits "8" in the MNIST training set

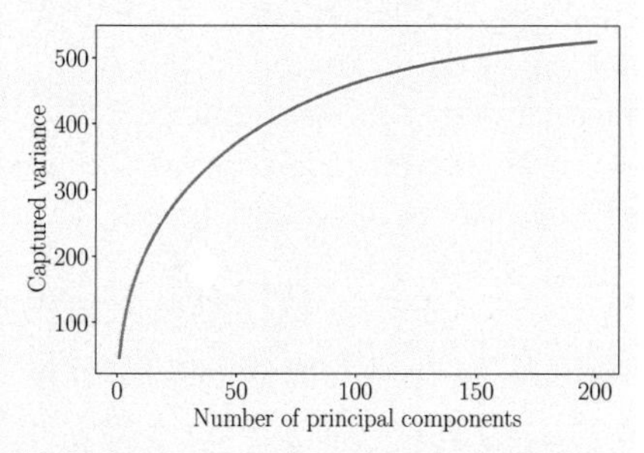

(b) Variance captured by the principal components

Figure 10.5 Properties of the training data of MNIST "8". (a) Eigenvalues sorted in descending order; (b) variance captured by the principal components associated with the largest eigenvalues.

Overall, to find an M-dimensional subspace of $\mathbb{R}^D$ that retains as much information as possible, PCA tells us to choose the columns of the matrix $\boldsymbol{B}$ in (10.3) as the M eigenvectors of the data covariance matrix $\boldsymbol{S}$ that are associated with the M largest eigenvalues. The maximum amount of variance PCA can capture with the first M principal components is

$$V_M = \sum_{m=1}^{M} \lambda_m \,, \tag{10.24}$$

where the λ_m are the M largest eigenvalues of the data covariance matrix $\boldsymbol{S}$. Consequently, the variance lost by data compression via PCA is

$$J_M := \sum_{j=M+1}^{D} \lambda_j = V_D - V_M \,. \tag{10.25}$$

Instead of these absolute quantities, we can define the relative variance captured as $\frac{V_M}{V_D}$, and the relative variance lost by compression as $1 - \frac{V_M}{V_D}$.

10.3 Projection Perspective

In the following, we will derive PCA as an algorithm that directly minimizes the average reconstruction error. This perspective allows us to interpret PCA as implementing an optimal linear auto-encoder. We will draw heavily from Chapters 2 and 3.

In the previous section, we derived PCA by maximizing the variance in the projected space to retain as much information as possible. In the following,

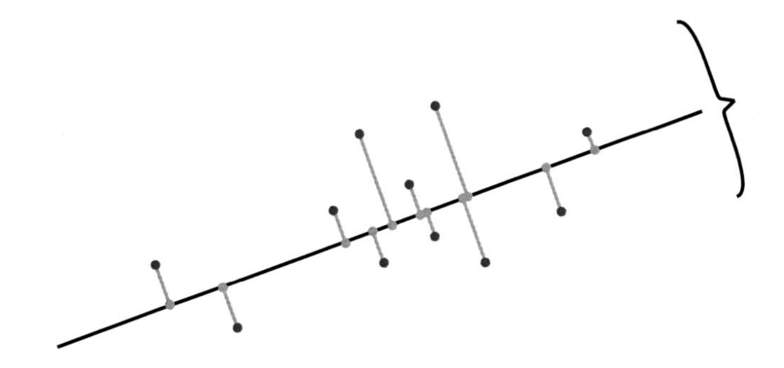

Figure 10.6 Illustration of the projection approach: Find a subspace (line) that minimizes the length of the difference vector between projected (orange) and original (blue) data.

we will look at the difference vectors between the original data $\boldsymbol{x}_n$ and their reconstruction $\tilde{\boldsymbol{x}}_n$ and minimize this distance so that $\boldsymbol{x}_n$ and $\tilde{\boldsymbol{x}}_n$ are as close as possible. Figure 10.6 illustrates this setting.

10.3.1 Setting and Objective

Assume an (ordered) orthonormal basis (ONB) $B = (\boldsymbol{b}_1, \dots, \boldsymbol{b}_D)$ of $\mathbb{R}^D$, i.e., $\boldsymbol{b}_i^\top \boldsymbol{b}_j = 1$ if and only if $i = j$ and 0 otherwise.

From Section 2.5 we know that for a basis $(\boldsymbol{b}_1, \dots, \boldsymbol{b}_D)$ of $\mathbb{R}^D$ any $\boldsymbol{x} \in \mathbb{R}^D$ can be written as a linear combination of the basis vectors of $\mathbb{R}^D$, i.e.,

Vectors $\tilde{\boldsymbol{x}} \in U$ could be vectors on a plane in $\mathbb{R}^3$. The dimensionality of the plane is 2, but the vectors still have three coordinates with respect to the standard basis of $\mathbb{R}^3$.

$$\boldsymbol{x} = \sum_{d=1}^{D} \zeta_d \boldsymbol{b}_d = \sum_{m=1}^{M} \zeta_m \boldsymbol{b}_m + \sum_{j=M+1}^{D} \zeta_j \boldsymbol{b}_j \tag{10.26}$$

for suitable coordinates $\zeta_d \in \mathbb{R}$.

We are interested in finding vectors $\tilde{\boldsymbol{x}} \in \mathbb{R}^D$, which live in lower-dimensional subspace $U \subseteq \mathbb{R}^D$, $\dim(U) = M$, so that

$$\tilde{\boldsymbol{x}} = \sum_{m=1}^{M} z_m \boldsymbol{b}_m \in U \subseteq \mathbb{R}^D \tag{10.27}$$

is as similar to $\boldsymbol{x}$ as possible. Note that at this point we need to assume that the coordinates z_m of $\tilde{\boldsymbol{x}}$ and ζ_m of $\boldsymbol{x}$ are not identical.

In the following, we use exactly this kind of representation of $\tilde{\boldsymbol{x}}$ to find optimal coordinates $\boldsymbol{z}$ and basis vectors $\boldsymbol{b}_1, \dots, \boldsymbol{b}_M$ such that $\tilde{\boldsymbol{x}}$ is as similar to the original data point $\boldsymbol{x}$, i.e., we aim to minimize the (Euclidean) distance $\|\boldsymbol{x} - \tilde{\boldsymbol{x}}\|$. Figure 10.7 illustrates this setting.

Without loss of generality, we assume that the dataset $\mathcal{X} = \{\boldsymbol{x}_1, \dots, \boldsymbol{x}_N\}$, $\boldsymbol{x}_n \in \mathbb{R}^D$, is centered at $\boldsymbol{0}$, i.e., $\mathbb{E}[\mathcal{X}] = \boldsymbol{0}$. Without the zero-mean assumption, we would arrive at exactly the same solution, but the notation would be substantially more cluttered.

We are interested in finding the best linear projection of $\mathcal{X}$ onto a lower-dimensional subspace U of $\mathbb{R}^D$ with $\dim(U) = M$ and orthonormal basis vectors $\boldsymbol{b}_1, \dots, \boldsymbol{b}_M$. We will call this subspace U the *principal subspace*. The projections of the data points are denoted by

principal subspace

$$\tilde{\boldsymbol{x}}_n := \sum_{m=1}^{M} z_{mn} \boldsymbol{b}_m = \boldsymbol{B}\boldsymbol{z}_n \in \mathbb{R}^D , \tag{10.28}$$

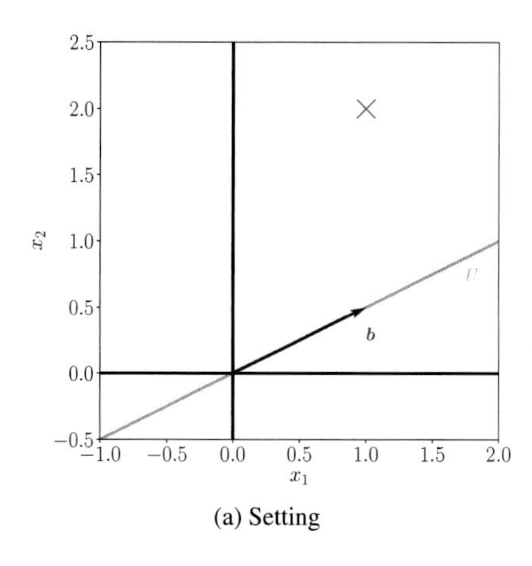

(a) Setting

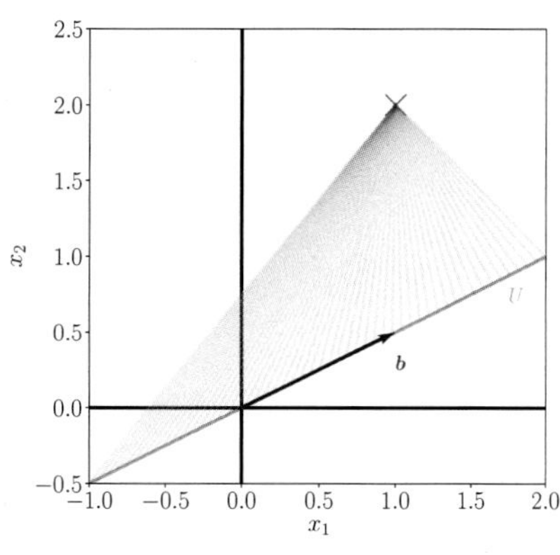

(b) Differences $\boldsymbol{x} - \tilde{\boldsymbol{x}}_i$ for 50 different $\tilde{\boldsymbol{x}}_i$ are shown by the red lines

Figure 10.7 Simplified projection setting. (a) A vector $\boldsymbol{x} \in \mathbb{R}^2$ (red cross) shall be projected onto a one-dimensional subspace $U \subseteq \mathbb{R}^2$ spanned by $\boldsymbol{b}$. (b) shows the difference vectors between $\boldsymbol{x}$ and some candidates $\tilde{\boldsymbol{x}}$.

where $\boldsymbol{z}_n := [z_{1n}, \ldots, z_{Mn}]^\top \in \mathbb{R}^M$ is the coordinate vector of $\tilde{\boldsymbol{x}}_n$ with respect to the basis $(\boldsymbol{b}_1, \ldots, \boldsymbol{b}_M)$. More specifically, we are interested in having the $\tilde{\boldsymbol{x}}_n$ as similar to $\boldsymbol{x}_n$ as possible.

The similarity measure we use in the following is the squared Euclidean norm $\|\boldsymbol{x} - \tilde{\boldsymbol{x}}\|^2$ between $\boldsymbol{x}$ and $\tilde{\boldsymbol{x}}$. We therefore define our objective as the minimizing the average squared Euclidean distance (*reconstruction error*) (Pearson, 1901)

reconstruction error

$$J_M := \frac{1}{N} \sum_{n=1}^{N} \|\boldsymbol{x}_n - \tilde{\boldsymbol{x}}_n\|^2, \tag{10.29}$$

where we make it explicit that the dimension of the subspace onto which we project the data is M. In order to find this optimal linear projection, we need to find the orthonormal basis of the principal subspace and the coordinates $\boldsymbol{z}_n \in \mathbb{R}^M$ of the projections with respect to this basis.

To find the coordinates $\boldsymbol{z}_n$ and the ONB of the principal subspace, we follow a two-step approach. First, we optimize the coordinates $\boldsymbol{z}_n$ for a given ONB $(\boldsymbol{b}_1, \ldots, \boldsymbol{b}_M)$; second, we find the optimal ONB.

10.3.2 Finding Optimal Coordinates

Let us start by finding the optimal coordinates $z_{1n}, \ldots, z_{Mn}$ of the projections $\tilde{\boldsymbol{x}}_n$ for $n = 1, \ldots, N$. Consider Figure 10.8(b), where the principal subspace is spanned by a single vector $\boldsymbol{b}$. Geometrically speaking, finding the optimal coordinates z corresponds to finding the representation of the linear projection $\tilde{\boldsymbol{x}}$ with respect to $\boldsymbol{b}$ that minimizes the distance between $\tilde{\boldsymbol{x}} - \boldsymbol{x}$. From Figure 10.8(b), it is clear that this will be the orthogonal projection, and in the following we will show exactly this.

We assume an ONB $(\boldsymbol{b}_1, \ldots, \boldsymbol{b}_M)$ of $U \subseteq \mathbb{R}^D$. To find the optimal coordinates $\boldsymbol{z}_m$ with respect to this basis, we require the partial derivatives

$$\frac{\partial J_M}{\partial z_{in}} = \frac{\partial J_M}{\partial \tilde{\boldsymbol{x}}_n} \frac{\partial \tilde{\boldsymbol{x}}_n}{\partial z_{in}}, \tag{10.30a}$$

$$\frac{\partial J_M}{\partial \tilde{\boldsymbol{x}}_n} = -\frac{2}{N} (\boldsymbol{x}_n - \tilde{\boldsymbol{x}}_n)^\top \in \mathbb{R}^{1 \times D}, \tag{10.30b}$$

Figure 10.8 Optimal projection of a vector $\boldsymbol{x} \in \mathbb{R}^2$ onto a one-dimensional subspace (continuation from Figure 10.7). (a) Distances $\|\boldsymbol{x} - \tilde{\boldsymbol{x}}\|$ for some $\tilde{\boldsymbol{x}} \in U$. (b) Orthogonal projection and optimal coordinates.

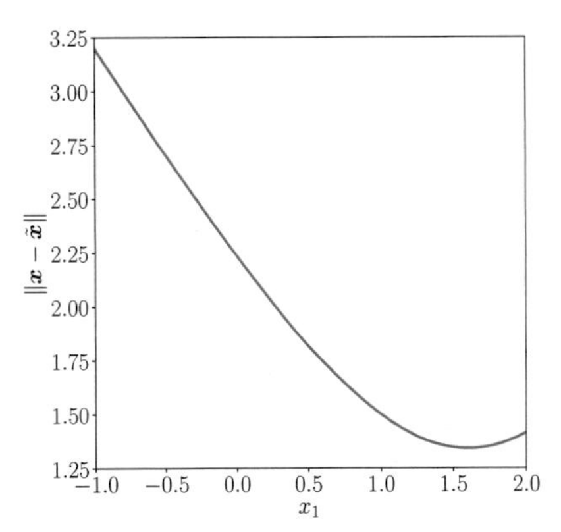

(a) Distances $\|\boldsymbol{x} - \tilde{\boldsymbol{x}}\|$ for some $\tilde{\boldsymbol{x}} = z_1\boldsymbol{b} \in U = \text{span}[\boldsymbol{b}]$; see panel (b) for the setting.

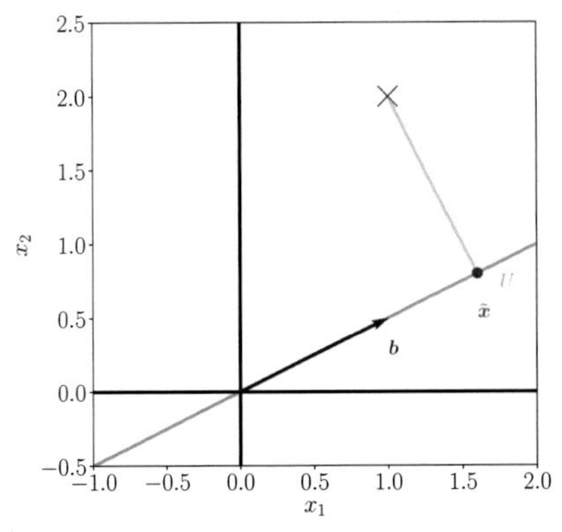

(b) The vector $\tilde{\boldsymbol{x}}$ that minimizes the distance in panel (a) is its orthogonal projection onto U. The coordinate of the projection $\tilde{\boldsymbol{x}}$ with respect to the basis vector $\boldsymbol{b}$ that spans U is the factor we need to scale $\boldsymbol{b}$ in order to "reach" $\tilde{\boldsymbol{x}}$.

$$\frac{\partial \tilde{\boldsymbol{x}}_n}{\partial z_{in}} \overset{(10.28)}{=} \frac{\partial}{\partial z_{in}}\left(\sum_{m=1}^{M} z_{mn}\boldsymbol{b}_m\right) = \boldsymbol{b}_i \tag{10.30c}$$

for $i = 1, \ldots, M$, such that we obtain

$$\frac{\partial J_M}{\partial z_{in}} \overset{\substack{(10.30b)\\(10.30c)}}{=} -\frac{2}{N}(\boldsymbol{x}_n - \tilde{\boldsymbol{x}}_n)^\top \boldsymbol{b}_i \overset{(10.28)}{=} -\frac{2}{N}\left(\boldsymbol{x}_n - \sum_{m=1}^{M} z_{mn}\boldsymbol{b}_m\right)^\top \boldsymbol{b}_i \tag{10.31a}$$

$$\overset{\text{ONB}}{=} -\frac{2}{N}(\boldsymbol{x}_n^\top \boldsymbol{b}_i - z_{in}\boldsymbol{b}_i^\top \boldsymbol{b}_i) = -\frac{2}{N}(\boldsymbol{x}_n^\top \boldsymbol{b}_i - z_{in}). \tag{10.31b}$$

since $\boldsymbol{b}_i^\top \boldsymbol{b}_i = 1$. Setting this partial derivative to 0 yields immediately the optimal coordinates

The coordinates of the optimal projection of $\boldsymbol{x}_n$ with respect to the basis vectors $\boldsymbol{b}_1, \ldots, \boldsymbol{b}_M$ are the coordinates of the orthogonal projection of $\boldsymbol{x}_n$ onto the principal subspace.

$$z_{in} = \boldsymbol{x}_n^\top \boldsymbol{b}_i = \boldsymbol{b}_i^\top \boldsymbol{x}_n \tag{10.32}$$

for $i = 1, \ldots, M$ and $n = 1, \ldots, N$. This means that the optimal coordinates z_{in} of the projection $\tilde{\boldsymbol{x}}_n$ are the coordinates of the orthogonal projection (see Section 3.8) of the original data point $\boldsymbol{x}_n$ onto the one-dimensional subspace that is spanned by $\boldsymbol{b}_i$. Consequently:

- The optimal linear projection $\tilde{\boldsymbol{x}}_n$ of $\boldsymbol{x}_n$ is an orthogonal projection.
- The coordinates of $\tilde{\boldsymbol{x}}_n$ with respect to the basis $(\boldsymbol{b}_1, \ldots, \boldsymbol{b}_M)$ are the coordinates of the orthogonal projection of $\boldsymbol{x}_n$ onto the principal subspace.
- An orthogonal projection is the best linear mapping given the objective (10.29).
- The coordinates ζ_m of $\boldsymbol{x}$ in (10.26) and the coordinates z_m of $\tilde{\boldsymbol{x}}$ in (10.27) must be identical for $m = 1, \ldots, M$ since $U^\perp = \text{span}[\boldsymbol{b}_{M+1}, \ldots, \boldsymbol{b}_D]$ is the orthogonal complement (see Section 3.6) of $U = \text{span}[\boldsymbol{b}_1, \ldots, \boldsymbol{b}_M]$.

$\boldsymbol{b}_j^\top \boldsymbol{x}$ is the coordinate of the orthogonal projection of $\boldsymbol{x}$ onto the subspace spanned by $\boldsymbol{b}_j$.

Remark (Orthogonal Projections with Orthonormal Basis Vectors). Let us briefly recap orthogonal projections from Section 3.8. If $(\boldsymbol{b}_1, \ldots, \boldsymbol{b}_D)$ is an orthonormal basis of $\mathbb{R}^D$, then

$$\tilde{\boldsymbol{x}} = \boldsymbol{b}_j(\boldsymbol{b}_j^\top \boldsymbol{b}_j)^{-1}\boldsymbol{b}_j^\top \boldsymbol{x} = \boldsymbol{b}_j\boldsymbol{b}_j^\top \boldsymbol{x} \in \mathbb{R}^D \tag{10.33}$$

is the orthogonal projection of $\boldsymbol{x}$ onto the subspace spanned by the jth basis vector, and $z_j = \boldsymbol{b}_j^\top \boldsymbol{x}$ is the coordinate of this projection with respect to the basis vector $\boldsymbol{b}_j$ that spans that subspace since $z_j \boldsymbol{b}_j = \tilde{\boldsymbol{x}}$. Figure 10.8b illustrates this setting.

More generally, if we aim to project onto an M-dimensional subspace of $\mathbb{R}^D$, we obtain the orthogonal projection of $\boldsymbol{x}$ onto the M-dimensional subspace with orthonormal basis vectors $\boldsymbol{b}_1, \ldots, \boldsymbol{b}_M$ as

$$\tilde{\boldsymbol{x}} = \boldsymbol{B}(\underbrace{\boldsymbol{B}^\top \boldsymbol{B}}_{=\boldsymbol{I}})^{-1}\boldsymbol{B}^\top \boldsymbol{x} = \boldsymbol{B}\boldsymbol{B}^\top \boldsymbol{x}\,, \tag{10.34}$$

where we defined $\boldsymbol{B} := [\boldsymbol{b}_1, \ldots, \boldsymbol{b}_M] \in \mathbb{R}^{D\times M}$. The coordinates of this projection with respect to the ordered basis $(\boldsymbol{b}_1, \ldots, \boldsymbol{b}_M)$ are $\boldsymbol{z} := \boldsymbol{B}^\top \boldsymbol{x}$ as discussed in Section 3.8.

We can think of the coordinates as a representation of the projected vector in a new coordinate system defined by $(\boldsymbol{b}_1, \ldots, \boldsymbol{b}_M)$. Note that although $\tilde{\boldsymbol{x}} \in \mathbb{R}^D$, we only need M coordinates $z_1, \ldots, z_M$ to represent this vector; the other $D - M$ coordinates with respect to the basis vectors $(\boldsymbol{b}_{M+1}, \ldots, \boldsymbol{b}_D)$ are always 0. $\diamondsuit$

So far we have shown that for a given ONB we can find the optimal coordinates of $\tilde{\boldsymbol{x}}$ by an orthogonal projection onto the principal subspace. In the following, we will determine what the best basis is.

10.3.3 Finding the Basis of the Principal Subspace

To determine the basis vectors $\boldsymbol{b}_1, \ldots, \boldsymbol{b}_M$ of the principal subspace, we rephrase the loss function (10.29) using the results we have so far. This will make it easier to find the basis vectors. To reformulate the loss function, we exploit our results from before and obtain

$$\tilde{\boldsymbol{x}}_n = \sum_{m=1}^{M} z_{mn}\boldsymbol{b}_m \overset{(10.32)}{=} \sum_{m=1}^{M} (\boldsymbol{x}_n^\top \boldsymbol{b}_m)\boldsymbol{b}_m\,. \tag{10.35}$$

We now exploit the symmetry of the dot product, which yields

$$\tilde{\boldsymbol{x}}_n = \left(\sum_{m=1}^{M} \boldsymbol{b}_m \boldsymbol{b}_m^\top\right) \boldsymbol{x}_n\,. \tag{10.36}$$

Since we can generally write the original data point $\boldsymbol{x}_n$ as a linear combination of all basis vectors, it holds that

$$\boldsymbol{x}_n = \sum_{d=1}^{D} z_{dn}\boldsymbol{b}_d \overset{(10.32)}{=} \sum_{d=1}^{D} (\boldsymbol{x}_n^\top \boldsymbol{b}_d)\boldsymbol{b}_d = \left(\sum_{d=1}^{D} \boldsymbol{b}_d \boldsymbol{b}_d^\top\right) \boldsymbol{x}_n \tag{10.37a}$$

$$= \left(\sum_{m=1}^{M} \boldsymbol{b}_m \boldsymbol{b}_m^\top\right) \boldsymbol{x}_n + \left(\sum_{j=M+1}^{D} \boldsymbol{b}_j \boldsymbol{b}_j^\top\right) \boldsymbol{x}_n\,, \tag{10.37b}$$

where we split the sum with D terms into a sum over M and a sum over $D - M$ terms. With this result, we find that the displacement vector $\boldsymbol{x}_n - \tilde{\boldsymbol{x}}_n$, i.e., the difference vector between the original data point and its projection, is

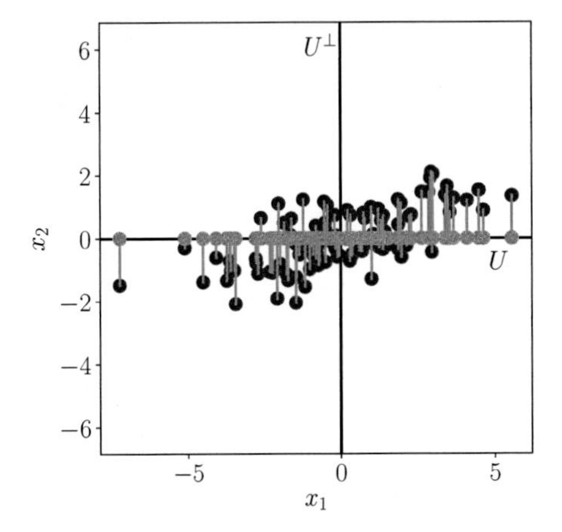

Figure 10.9 Orthogonal projection and displacement vectors. When projecting data points $\boldsymbol{x}_n$ (blue) onto subspace U_1, we obtain $\tilde{\boldsymbol{x}}_n$ (orange). The displacement vector $\tilde{\boldsymbol{x}}_n - \boldsymbol{x}_n$ lies completely in the orthogonal complement U_2 of U_1.

$$\boldsymbol{x}_n - \tilde{\boldsymbol{x}}_n = \left(\sum_{j=M+1}^{D} \boldsymbol{b}_j \boldsymbol{b}_j^\top \right) \boldsymbol{x}_n \tag{10.38a}$$

$$= \sum_{j=M+1}^{D} (\boldsymbol{x}_n^\top \boldsymbol{b}_j) \boldsymbol{b}_j \,. \tag{10.38b}$$

This means the difference is exactly the projection of the data point onto the orthogonal complement of the principal subspace: We identify the matrix $\sum_{j=M+1}^{D} \boldsymbol{b}_j \boldsymbol{b}_j^\top$ in (10.38a) as the projection matrix that performs this projection. Hence the displacement vector $\boldsymbol{x}_n - \tilde{\boldsymbol{x}}_n$ lies in the subspace that is orthogonal to the principal subspace as illustrated in Figure 10.9.

Remark (Low-Rank Approximation). In (10.38a), we saw that the projection matrix, which projects $\boldsymbol{x}$ onto $\tilde{\boldsymbol{x}}$, is given by

$$\sum_{m=1}^{M} \boldsymbol{b}_m \boldsymbol{b}_m^\top = \boldsymbol{B}\boldsymbol{B}^\top \,. \tag{10.39}$$

By construction as a sum of rank-one matrices $\boldsymbol{b}_m \boldsymbol{b}_m^\top$, we see that $\boldsymbol{B}\boldsymbol{B}^\top$ is symmetric and has rank M. Therefore, the average squared reconstruction error can also be written as

$$\frac{1}{N} \sum_{n=1}^{N} \|\boldsymbol{x}_n - \tilde{\boldsymbol{x}}_n\|^2 = \frac{1}{N} \sum_{n=1}^{N} \left\| \boldsymbol{x}_n - \boldsymbol{B}\boldsymbol{B}^\top \boldsymbol{x}_n \right\|^2 \tag{10.40a}$$

$$= \frac{1}{N} \sum_{n=1}^{N} \left\| (\boldsymbol{I} - \boldsymbol{B}\boldsymbol{B}^\top) \boldsymbol{x}_n \right\|^2 . \tag{10.40b}$$

PCA finds the best rank-M approximation of the identity matrix.

Finding orthonormal basis vectors $\boldsymbol{b}_1, \ldots, \boldsymbol{b}_M$, which minimize the difference between the original data $\boldsymbol{x}_n$ and their projections $\tilde{\boldsymbol{x}}_n$, is equivalent to finding the best rank-M approximation $\boldsymbol{B}\boldsymbol{B}^\top$ of the identity matrix $\boldsymbol{I}$ (see Section 4.6). ◇

Now we have all the tools to reformulate the loss function (10.29):

$$J_M = \frac{1}{N} \sum_{n=1}^{N} \|\boldsymbol{x}_n - \tilde{\boldsymbol{x}}_n\|^2 \overset{(10.38b)}{=} \frac{1}{N} \sum_{n=1}^{N} \left\| \sum_{j=M+1}^{D} (\boldsymbol{b}_j^\top \boldsymbol{x}_n) \boldsymbol{b}_j \right\|^2 . \tag{10.41}$$

We now explicitly compute the squared norm and exploit the fact that the $\boldsymbol{b}_j$ form an ONB, which yields

$$J_M = \frac{1}{N}\sum_{n=1}^{N}\sum_{j=M+1}^{D}(\boldsymbol{b}_j^\top \boldsymbol{x}_n)^2 = \frac{1}{N}\sum_{n=1}^{N}\sum_{j=M+1}^{D}\boldsymbol{b}_j^\top \boldsymbol{x}_n \boldsymbol{b}_j^\top \boldsymbol{x}_n \tag{10.42a}$$

$$= \frac{1}{N}\sum_{n=1}^{N}\sum_{j=M+1}^{D}\boldsymbol{b}_j^\top \boldsymbol{x}_n \boldsymbol{x}_n^\top \boldsymbol{b}_j\,, \tag{10.42b}$$

where we exploited the symmetry of the dot product in the last step to write $\boldsymbol{b}_j^\top \boldsymbol{x}_n = \boldsymbol{x}_n^\top \boldsymbol{b}_j$. We now swap the sums and obtain

$$J_M = \sum_{j=M+1}^{D}\boldsymbol{b}_j^\top \underbrace{\left(\frac{1}{N}\sum_{n=1}^{N}\boldsymbol{x}_n\boldsymbol{x}_n^\top\right)}_{=:\boldsymbol{S}}\boldsymbol{b}_j = \sum_{j=M+1}^{D}\boldsymbol{b}_j^\top \boldsymbol{S}\boldsymbol{b}_j \tag{10.43a}$$

$$= \sum_{j=M+1}^{D}\operatorname{tr}(\boldsymbol{b}_j^\top \boldsymbol{S}\boldsymbol{b}_j) = \sum_{j=M+1}^{D}\operatorname{tr}(\boldsymbol{S}\boldsymbol{b}_j\boldsymbol{b}_j^\top) = \operatorname{tr}\Big(\underbrace{\Big(\sum_{j=M+1}^{D}\boldsymbol{b}_j\boldsymbol{b}_j^\top\Big)}_{\text{projection matrix}}\boldsymbol{S}\Big)\,, \tag{10.43b}$$

where we exploited the property that the trace operator $\operatorname{tr}(\cdot)$ (see (4.18)) is linear and invariant to cyclic permutations of its arguments. Since we assumed that our dataset is centered, i.e., $\mathbb{E}[\mathcal{X}] = \boldsymbol{0}$, we identify $\boldsymbol{S}$ as the data covariance matrix. Since the projection matrix in (10.43b) is constructed as a sum of rank-one matrices $\boldsymbol{b}_j\boldsymbol{b}_j^\top$ it itself is of rank $D - M$.

Equation (10.43a) implies that we can formulate the average squared reconstruction error equivalently as the covariance matrix of the data, projected onto the orthogonal complement of the principal subspace. Minimizing the average squared reconstruction error is therefore equivalent to minimizing the variance of the data when projected onto the subspace we ignore, i.e., the orthogonal complement of the principal subspace. Equivalently, we maximize the variance of the projection that we retain in the principal subspace, which links the projection loss immediately to the maximum-variance formulation of PCA discussed in Section 10.2. But this then also means that we will obtain the same solution that we obtained for the maximum-variance perspective. Therefore, we omit a derivation that is identical to the one presented in Section 10.2 and summarize the results from earlier in the light of the projection perspective.

Minimizing the average squared reconstruction error is equivalent to minimizing the projection of the data covariance matrix onto the orthogonal complement of the principal subspace.

Minimizing the average squared reconstruction error is equivalent to maximizing the variance of the projected data.

The average squared reconstruction error, when projecting onto the M-dimensional principal subspace, is

$$J_M = \sum_{j=M+1}^{D}\lambda_j\,, \tag{10.44}$$

where λ_j are the eigenvalues of the data covariance matrix. Therefore, to minimize (10.44) we need to select the smallest $D - M$ eigenvalues, which then implies that their corresponding eigenvectors are the basis of the orthogonal complement of the principal subspace. Consequently, this means that the basis of the principal subspace comprises the eigenvectors $\boldsymbol{b}_1, \ldots, \boldsymbol{b}_M$ that are associated with the largest M eigenvalues of the data covariance matrix.

Example 10.3 (MNIST Digits Embedding)
Figure 10.10 visualizes the training data of the MMIST digits "0" and "1" embedded in the vector subspace spanned by the first two principal components. We observe a relatively clear separation between "0"s (blue dots) and "1"s (orange dots), and we see the variation within each individual cluster. Four embeddings of the digits "0" and "1" in the principal subspace are highlighted in red with their corresponding original digit. The figure reveals that the variation within the set of "0" is significantly greater than the variation within the set of "1."

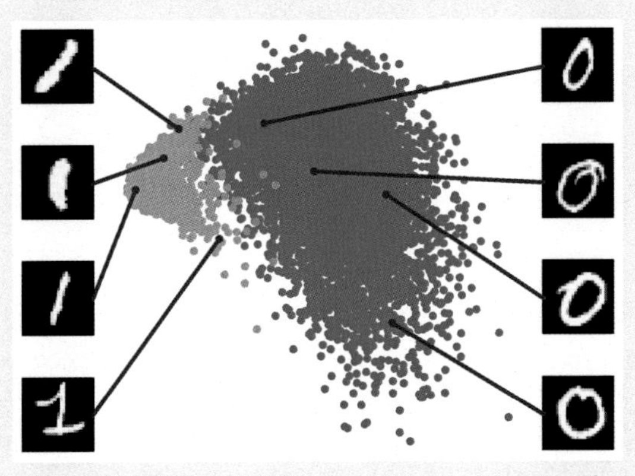

Figure 10.10 Embedding of MNIST digits 0 (blue) and 1 (orange) in a two-dimensional principal subspace using PCA. Four embeddings of the digits "0" and "1" in the principal subspace are highlighted in red with their corresponding original digit.

10.4 Eigenvector Computation and Low-Rank Approximations

In the previous sections, we obtained the basis of the principal subspace as the eigenvectors that are associated with the largest eigenvalues of the data covariance matrix

$$\boldsymbol{S} = \frac{1}{N}\sum_{n=1}^{N} \boldsymbol{x}_n \boldsymbol{x}_n^\top = \frac{1}{N}\boldsymbol{X}\boldsymbol{X}^\top, \tag{10.45}$$

$$\boldsymbol{X} = [\boldsymbol{x}_1, \ldots, \boldsymbol{x}_N] \in \mathbb{R}^{D\times N}. \tag{10.46}$$

Note that $\boldsymbol{X}$ is a $D \times N$ matrix, i.e., it is the transpose of the "typical" data matrix (Bishop, 2006; Murphy, 2012). To get the eigenvalues (and the corresponding eigenvectors) of $\boldsymbol{S}$, we can follow two approaches:

Use eigendecomposition or SVD to compute eigenvectors.

- We perform an eigendecomposition (see Section 4.2) and compute the eigenvalues and eigenvectors of $\boldsymbol{S}$ directly.
- We use a singular value decomposition (see Section 4.5). Since $\boldsymbol{S}$ is symmetric and factorizes into $\boldsymbol{X}\boldsymbol{X}^\top$ (ignoring the factor $\frac{1}{N}$), the eigenvalues of $\boldsymbol{S}$ are the squared singular values of $\boldsymbol{X}$.

More specifically, the SVD of $\boldsymbol{X}$ is given by

$$\underbrace{\boldsymbol{X}}_{D\times N} = \underbrace{\boldsymbol{U}}_{D\times D}\underbrace{\boldsymbol{\Sigma}}_{D\times N}\underbrace{\boldsymbol{V}^\top}_{N\times N}, \tag{10.47}$$

where $\boldsymbol{U} \in \mathbb{R}^{D\times D}$ and $\boldsymbol{V}^\top \in \mathbb{R}^{N\times N}$ are orthogonal matrices and $\boldsymbol{\Sigma} \in \mathbb{R}^{D\times N}$ is a matrix whose only nonzero entries are the singular values $\sigma_{ii} \geqslant 0$. It then follows that

$$\boldsymbol{S} = \frac{1}{N}\boldsymbol{X}\boldsymbol{X}^\top = \frac{1}{N}\boldsymbol{U}\boldsymbol{\Sigma}\underbrace{\boldsymbol{V}^\top\boldsymbol{V}}_{=\boldsymbol{I}_N}\boldsymbol{\Sigma}^\top\boldsymbol{U}^\top = \frac{1}{N}\boldsymbol{U}\boldsymbol{\Sigma}\boldsymbol{\Sigma}^\top\boldsymbol{U}^\top. \tag{10.48}$$

With the results from Section 4.5, we get that the columns of $\boldsymbol{U}$ are the eigenvectors of $\boldsymbol{X}\boldsymbol{X}^\top$ (and therefore $\boldsymbol{S}$). Furthermore, the eigenvalues λ_d of $\boldsymbol{S}$ are related to the singular values of $\boldsymbol{X}$ via

The columns of $\boldsymbol{U}$ are the eigenvectors of $\boldsymbol{S}$.

$$\lambda_d = \frac{\sigma_d^2}{N}. \tag{10.49}$$

This relationship between the eigenvalues of $\boldsymbol{S}$ and the singular values of $\boldsymbol{X}$ provides the connection between the maximum variance view (Section 10.2) and the singular value decomposition.

10.4.1 PCA Using Low-Rank Matrix Approximations

To maximize the variance of the projected data (or minimize the average squared reconstruction error), PCA chooses the columns of $\boldsymbol{U}$ in (10.48) to be the eigenvectors that are associated with the M largest eigenvalues of the data covariance matrix $\boldsymbol{S}$ so that we identify $\boldsymbol{U}$ as the projection matrix $\boldsymbol{B}$ in (10.3), which projects the original data onto a lower-dimensional subspace of dimension M. The *Eckart–Young theorem* (Theorem 4.25 in Section 4.6) offers a direct way to estimate the low-dimensional representation. Consider the best rank-M approximation

Eckart–Young theorem

$$\tilde{\boldsymbol{X}}_M := \operatorname{argmin}_{\operatorname{rk}(\boldsymbol{A})\leqslant M} \|\boldsymbol{X} - \boldsymbol{A}\|_2 \in \mathbb{R}^{D\times N} \tag{10.50}$$

of $\boldsymbol{X}$, where $\|\cdot\|_2$ is the spectral norm defined in (4.93). The Eckart–Young theorem states that $\tilde{\boldsymbol{X}}_M$ is given by truncating the SVD at the top-M singular value. In other words, we obtain

$$\tilde{\boldsymbol{X}}_M = \underbrace{\boldsymbol{U}_M}_{D\times M}\underbrace{\boldsymbol{\Sigma}_M}_{M\times M}\underbrace{\boldsymbol{V}_M^\top}_{M\times N} \in \mathbb{R}^{D\times N} \tag{10.51}$$

with orthogonal matrices $\boldsymbol{U}_M := [\boldsymbol{u}_1, \ldots, \boldsymbol{u}_M] \in \mathbb{R}^{D\times M}$ and $\boldsymbol{V}_M := [\boldsymbol{v}_1, \ldots, \boldsymbol{v}_M] \in \mathbb{R}^{N\times M}$ and a diagonal matrix $\boldsymbol{\Sigma}_M \in \mathbb{R}^{M\times M}$ whose diagonal entries are the M largest singular values of $\boldsymbol{X}$.

10.4.2 Practical Aspects

Finding eigenvalues and eigenvectors is also important in other fundamental machine learning methods that require matrix decompositions. In theory, as we discussed in Section 4.2, we can solve for the eigenvalues as roots of the characteristic polynomial. However, for matrices larger than 4×4 this is not possible because we would need to find the roots of a polynomial of degree 5 or higher. However, the *Abel–Ruffini theorem* (Ruffini, 1799; Abel, 1826) states that there exists no algebraic solution to this problem for polynomials of degree 5 or more. Therefore, in practice, we solve for eigenvalues or singular values using iterative methods, which are implemented in all modern packages for linear algebra.

Abel–Ruffini theorem

`np.linalg.eigh` or `np.linalg.svd`

In many applications (such as PCA presented in this chapter), we only require a few eigenvectors. It would be wasteful to compute the full decomposition, and then discard all eigenvectors with eigenvalues that are beyond the first few. It turns out that if we are interested in only the first few eigenvectors (with the largest eigenvalues), then iterative processes, which directly optimize these eigenvectors, are computationally more efficient than a full eigendecomposition (or SVD). In the extreme case of only needing the first eigenvector, a simple method called the *power iteration* is very efficient. Power iteration chooses a random vector $\boldsymbol{x}_0$ that is not in the null space of $\boldsymbol{S}$ and follows the iteration

power iteration

$$\boldsymbol{x}_{k+1} = \frac{\boldsymbol{S}\boldsymbol{x}_k}{\|\boldsymbol{S}\boldsymbol{x}_k\|}, \quad k = 0, 1, \dots . \tag{10.52}$$

This means the vector $\boldsymbol{x}_k$ is multiplied by $\boldsymbol{S}$ in every iteration and then normalized, i.e., we always have $\|\boldsymbol{x}_k\| = 1$. This sequence of vectors converges to the eigenvector associated with the largest eigenvalue of $\boldsymbol{S}$. The original Google PageRank algorithm (Page et al., 1999) uses such an algorithm for ranking web pages based on their hyperlinks.

If $\boldsymbol{S}$ is invertible, it is sufficient to ensure that $\boldsymbol{x}_0 \neq \boldsymbol{0}$.

10.5 PCA in High Dimensions

In order to do PCA, we need to compute the data covariance matrix. In D dimensions, the data covariance matrix is a $D \times D$ matrix. Computing the eigenvalues and eigenvectors of this matrix is computationally expensive as it scales cubically in D. Therefore, PCA, as we discussed earlier, will be infeasible in very high dimensions. For example, if our $\boldsymbol{x}_n$ are images with 10,000 pixels (e.g., 100×100 pixel images), we would need to compute the eigendecomposition of a $10{,}000 \times 10{,}000$ covariance matrix. In the following, we provide a solution to this problem for the case that we have substantially fewer data points than dimensions, i.e., $N \ll D$.

Assume we have a centered dataset $\boldsymbol{x}_1, \dots, \boldsymbol{x}_N$, $\boldsymbol{x}_n \in \mathbb{R}^D$. Then the data covariance matrix is given as

$$\boldsymbol{S} = \frac{1}{N}\boldsymbol{X}\boldsymbol{X}^\top \in \mathbb{R}^{D \times D}, \tag{10.53}$$

where $\boldsymbol{X} = [\boldsymbol{x}_1, \dots, \boldsymbol{x}_N]$ is a $D \times N$ matrix whose columns are the data points.

We now assume that $N \ll D$, i.e., the number of data points is smaller than the dimensionality of the data. If there are no duplicate data points, the rank of the covariance matrix $\boldsymbol{S}$ is N, so it has $D - N + 1$ many eigenvalues that are 0. Intuitively, this means that there are some redundancies. In the following, we will exploit this and turn the $D \times D$ covariance matrix into an $N \times N$ covariance matrix whose eigenvalues are all positive.

In PCA, we ended up with the eigenvector equation

$$\boldsymbol{S}\boldsymbol{b}_m = \lambda_m \boldsymbol{b}_m\,, \quad m = 1, \ldots, M\,, \tag{10.54}$$

where $\boldsymbol{b}_m$ is a basis vector of the principal subspace. Let us rewrite this equation a bit: With $\boldsymbol{S}$ defined in (10.53), we obtain

$$\boldsymbol{S}\boldsymbol{b}_m = \frac{1}{N}\boldsymbol{X}\boldsymbol{X}^\top \boldsymbol{b}_m = \lambda_m \boldsymbol{b}_m\,. \tag{10.55}$$

We now multiply $\boldsymbol{X}^\top \in \mathbb{R}^{N\times D}$ from the left-hand side, which yields

$$\frac{1}{N}\underbrace{\boldsymbol{X}^\top\boldsymbol{X}}_{N\times N}\underbrace{\boldsymbol{X}^\top\boldsymbol{b}_m}_{=:\boldsymbol{c}_m} = \lambda_m \boldsymbol{X}^\top\boldsymbol{b}_m \iff \frac{1}{N}\boldsymbol{X}^\top\boldsymbol{X}\boldsymbol{c}_m = \lambda_m \boldsymbol{c}_m\,, \tag{10.56}$$

and we get a new eigenvector/eigenvalue equation: λ_m remains eigenvalue, which confirms our results from Section 4.5.3 that the nonzero eigenvalues of $\boldsymbol{X}\boldsymbol{X}^\top$ equal the nonzero eigenvalues of $\boldsymbol{X}^\top\boldsymbol{X}$. We obtain the eigenvector of the matrix $\frac{1}{N}\boldsymbol{X}^\top\boldsymbol{X} \in \mathbb{R}^{N\times N}$ associated with λ_m as $\boldsymbol{c}_m := \boldsymbol{X}^\top\boldsymbol{b}_m$. Assuming we have no duplicate data points, this matrix has rank N and is invertible. This also implies that $\frac{1}{N}\boldsymbol{X}^\top\boldsymbol{X}$ has the same (nonzero) eigenvalues as the data covariance matrix $\boldsymbol{S}$. But this is now an $N \times N$ matrix, so that we can compute the eigenvalues and eigenvectors much more efficiently than for the original $D \times D$ data covariance matrix.

Now that we have the eigenvectors of $\frac{1}{N}\boldsymbol{X}^\top\boldsymbol{X}$, we are going to recover the original eigenvectors, which we still need for PCA. Currently, we know the eigenvectors of $\frac{1}{N}\boldsymbol{X}^\top\boldsymbol{X}$. If we left-multiply our eigenvalue/eigenvector equation with $\boldsymbol{X}$, we get

$$\underbrace{\frac{1}{N}\boldsymbol{X}\boldsymbol{X}^\top}_{\boldsymbol{S}}\boldsymbol{X}\boldsymbol{c}_m = \lambda_m \boldsymbol{X}\boldsymbol{c}_m \tag{10.57}$$

and we recover the data covariance matrix again. This now also means that we recover $\boldsymbol{X}\boldsymbol{c}_m$ as an eigenvector of $\boldsymbol{S}$.

Remark. If we want to apply the PCA algorithm that we discussed in Section 10.6, we need to normalize the eigenvectors $\boldsymbol{X}\boldsymbol{c}_m$ of $\boldsymbol{S}$ so that they have norm 1. ♢

10.6 Key Steps of PCA in Practice

In the following, we will go through the individual steps of PCA using a running example, which is summarized in Figure 10.11. We are given a two-dimensional dataset (Figure 10.11(a)), and we want to use PCA to project it onto a one-dimensional subspace.

1. **Mean subtraction** We start by centering the data by computing the mean $\boldsymbol{\mu}$ of the dataset and subtracting it from every single data point. This ensures that the dataset has mean $\boldsymbol{0}$ (Figure 10.11(b)). Mean subtraction is not strictly necessary but reduces the risk of numerical problems.

Figure 10.11 Steps of PCA. (a) Original dataset; (b) centering; (c) divide by standard deviation; (d) eigencomposition; (e) projection; (f) mapping back to the original data space.

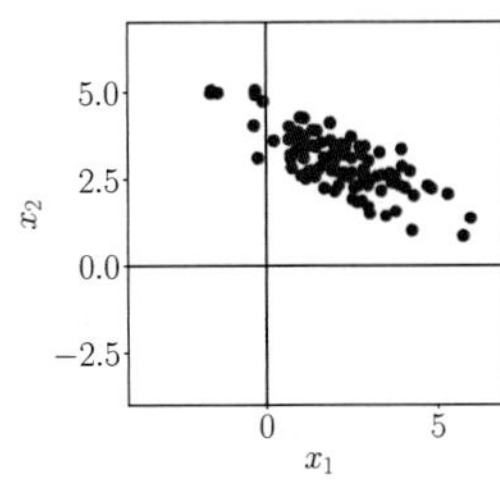

(a) Original dataset.

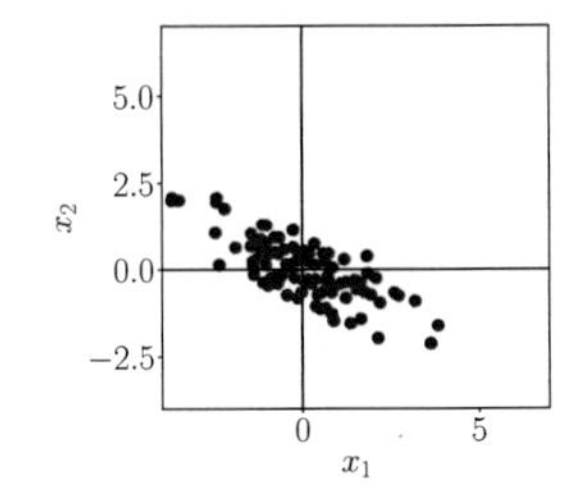

(b) Step 1: Centering by subtracting the mean from each data point.

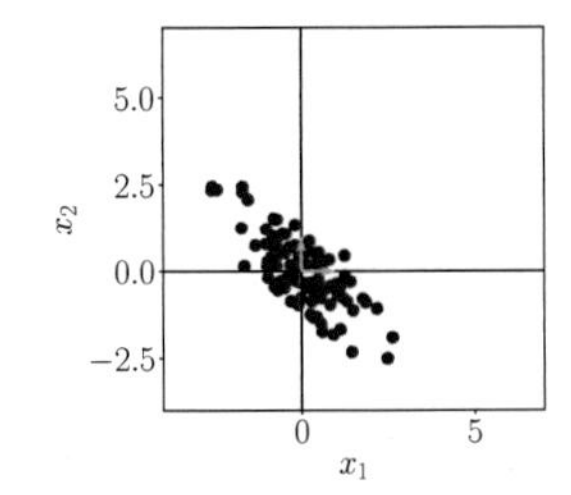

(c) Step 2: Dividing by the standard deviation to make the data unit free. Data has variance 1 along each axis.

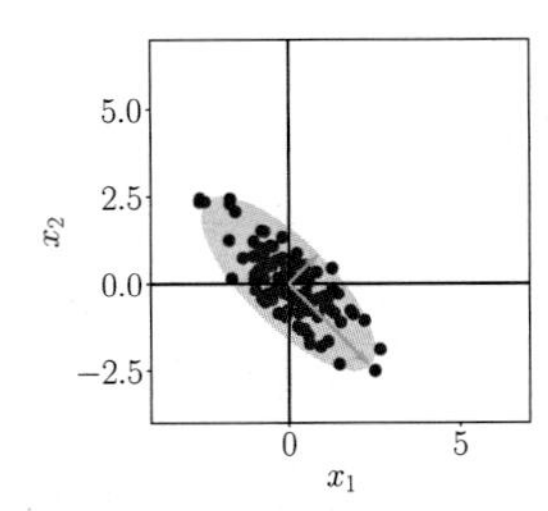

(d) Step 3: Compute eigenvalues and eigenvectors (arrows) of the data covariance matrix (ellipse).

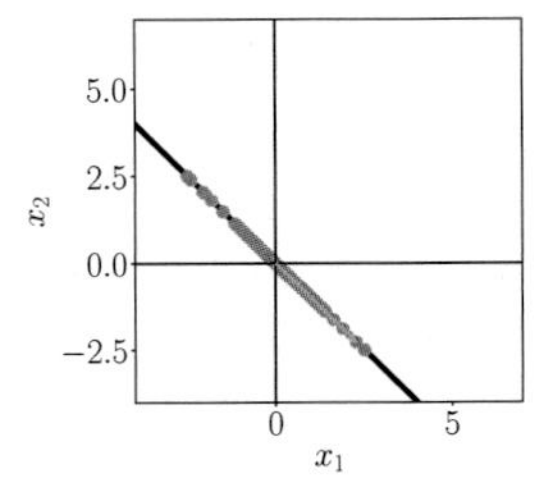

(e) Step 4: Project data onto the principal subspace.

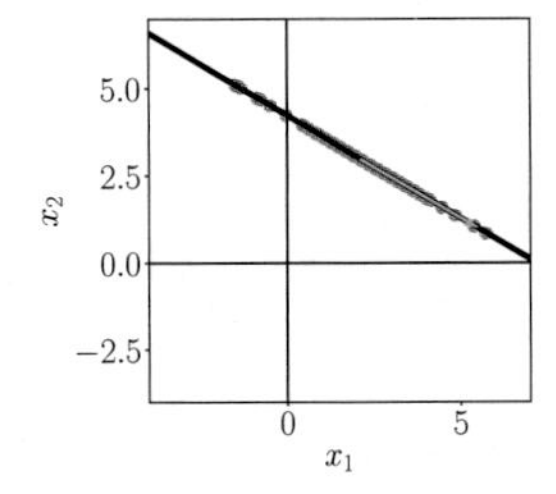

(f) Undo the standardization and move projected data back into the original data space from (a).

standardization

2. **Standardization** Divide the data points by the standard deviation σ_d of the dataset for every dimension $d = 1, \ldots, D$. Now the data is unit free, and it has variance 1 along each axis, which is indicated by the two arrows in Figure 10.11(c). This step completes the *standardization* of the data.
3. **Eigendecomposition of the covariance matrix** Compute the data covariance matrix and its eigenvalues and corresponding eigenvectors. Since the covariance matrix is symmetric, the spectral theorem (Theorem 4.15) states that we can find an ONB of eigenvectors. In Figure 10.11(d), the eigenvectors are scaled by the magnitude of the corresponding eigenvalue. The longer vector spans the principal subspace, which we denote by U. The data covariance matrix is represented by the ellipse.
4. **Projection** We can project any data point $\boldsymbol{x}_* \in \mathbb{R}^D$ onto the principal subspace: To get this right, we need to standardize $\boldsymbol{x}_*$ using the mean μ_d and standard deviation σ_d of the training data in the dth dimension, respectively, so that

$$x_*^{(d)} \leftarrow \frac{x_*^{(d)} - \mu_d}{\sigma_d}, \quad d = 1, \ldots, D, \tag{10.58}$$

where $x_*^{(d)}$ is the dth component of $\boldsymbol{x}_*$. We obtain the projection as

$$\tilde{\boldsymbol{x}}_* = \boldsymbol{B}\boldsymbol{B}^\top \boldsymbol{x}_* \tag{10.59}$$

with coordinates

$$\boldsymbol{z}_* = \boldsymbol{B}^\top \boldsymbol{x}_* \tag{10.60}$$

with respect to the basis of the principal subspace. Here, $\boldsymbol{B}$ is the matrix that contains the eigenvectors that are associated with the largest eigenvalues of the data covariance matrix as columns. PCA returns the coordinates (10.60), not the projections $\boldsymbol{x}_*$.

Having standardized our dataset, (10.59) only yields the projections in the context of the standardized dataset. To obtain our projection in the original data space (i.e., before standardization), we need to undo the standardization (10.58) and multiply by the standard deviation before adding the mean so that we obtain

$$\tilde{x}_*^{(d)} \leftarrow \tilde{x}_*^{(d)} \sigma_d + \mu_d \,, \quad d = 1, \dots, D \,. \tag{10.61}$$

Figure 10.11(f) illustrates the projection in the original data space.

Example 10.4 (MNIST Digits: Reconstruction)
In the following, we will apply PCA to the MNIST digits dataset, which contains 60,000 examples of handwritten digits 0 through 9. Each digit is an image of size 28×28, i.e., it contains 784 pixels so that we can interpret every image in this dataset as a vector $\boldsymbol{x} \in \mathbb{R}^{784}$. Examples of these digits are shown in Figure 10.3.

For illustration purposes, we apply PCA to a subset of the MNIST digits, and we focus on the digit "8." We used 5,389 training images of the digit "8" and determined the principal subspace as detailed in this chapter. We then used the learned projection matrix to reconstruct a set of test images, which is illustrated in Figure 10.12. The first row of Figure 10.12 shows a set of four original digits from the test set. The following rows show reconstructions of exactly these digits when using a principal subspace of dimensions 1, 10, 100, and 500, respectively. We see that even with a single-dimensional principal subspace we get a halfway decent reconstruction of the original digits, which, however, is blurry and generic. With an increasing number of principal components (PCs), the reconstructions become sharper and more details are accounted for. With 500 principal components, we effectively obtain a near-perfect reconstruction. If we were to choose 784 PCs, we would recover the exact digit without any compression loss.

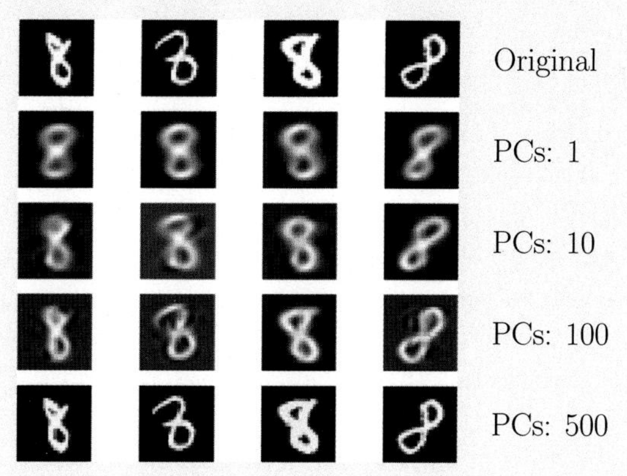

Figure 10.12 Effect of increasing the number of principal components on reconstruction.

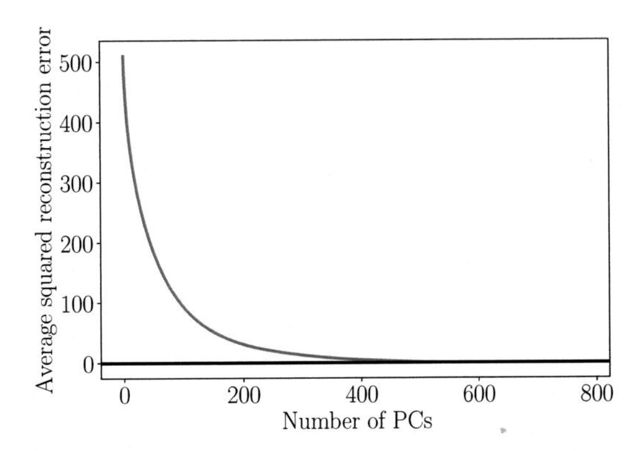

Figure 10.13 Average squared reconstruction error as a function of the number of principal components. The average squared reconstruction error is the sum of the eigenvalues in the orthogonal complement of the principal subspace.

Figure 10.13 shows the average squared reconstruction error, which is

$$\frac{1}{N}\sum_{n=1}^{N}\|\boldsymbol{x}_n - \tilde{\boldsymbol{x}}_n\|^2 = \sum_{i=M+1}^{D}\lambda_i\,, \tag{10.62}$$

as a function of the number M of principal components. We can see that the importance of the principal components drops off rapidly, and only marginal gains can be achieved by adding more PCs. This matches exactly our observation in Figure 10.5, where we discovered that most of the variance of the projected data is captured by only a few principal components. With about 550 PCs, we can essentially fully reconstruct the training data that contains the digit "8" (some pixels around the boundaries show no variation across the dataset as they are always black).

10.7 Latent Variable Perspective

In the previous sections, we derived PCA without any notion of a probabilistic model using the maximum-variance and the projection perspectives. On the one hand, this approach may be appealing as it allows us to sidestep all the mathematical difficulties that come with probability theory, but on the other hand, a probabilistic model would offer us more flexibility and useful insights. More specifically, a probabilistic model would

- Come with a likelihood function, and we can explicitly deal with noisy observations (which we did not even discuss earlier)
- Allow us to do Bayesian model comparison via the marginal likelihood as discussed in Section 8.6
- View PCA as a generative model, which allows us to simulate new data
- Allow us to make straightforward connections to related algorithms
- Deal with data dimensions that are missing at random by applying Bayes' theorem
- Give us a notion of the novelty of a new data point
- Give us a principled way to extend the model, e.g., to a mixture of PCA models
- Have the PCA we derived in earlier sections as a special case

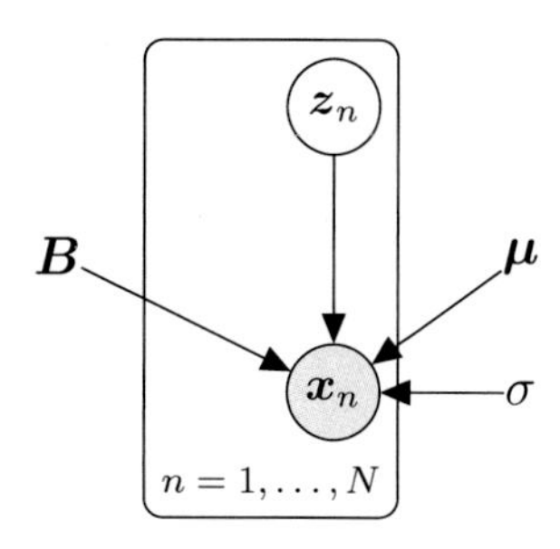

Figure 10.14 Graphical model for probabilistic PCA. The observations x_n explicitly depend on corresponding latent variables $z_n \sim \mathcal{N}(\mathbf{0}, \boldsymbol{I})$. The model parameters $\boldsymbol{B}, \boldsymbol{\mu}$ and the likelihood parameter σ are shared across the dataset.

- Allow for a fully Bayesian treatment by marginalizing out the model parameters

By introducing a continuous-valued latent variable $\boldsymbol{z} \in \mathbb{R}^M$, it is possible to phrase PCA as a probabilistic latent-variable model. Tipping and Bishop (1999) proposed this latent-variable model as *probabilistic PCA (PPCA)*. PPCA addresses most of the aforementioned issues, and the PCA solution that we obtained by maximizing the variance in the projected space or by minimizing the reconstruction error is obtained as the special case of maximum likelihood estimation in a noise-free setting.

probabilistic PCA
PPCA

10.7.1 Generative Process and Probabilistic Model

In PPCA, we explicitly write down the probabilistic model for linear dimensionality reduction. For this we assume a continuous latent variable $\boldsymbol{z} \in \mathbb{R}^M$ with a standard-normal prior $p(\boldsymbol{z}) = \mathcal{N}(\mathbf{0}, \boldsymbol{I})$ and a linear relationship between the latent variables and the observed $\boldsymbol{x}$ data where

$$\boldsymbol{x} = \boldsymbol{B}\boldsymbol{z} + \boldsymbol{\mu} + \boldsymbol{\epsilon} \in \mathbb{R}^D , \tag{10.63}$$

where $\boldsymbol{\epsilon} \sim \mathcal{N}(\mathbf{0}, \sigma^2 \boldsymbol{I})$ is Gaussian observation noise and $\boldsymbol{B} \in \mathbb{R}^{D \times M}$ and $\boldsymbol{\mu} \in \mathbb{R}^D$ describe the linear/affine mapping from latent to observed variables. Therefore, PPCA links latent and observed variables via

$$p(\boldsymbol{x}|\boldsymbol{z}, \boldsymbol{B}, \boldsymbol{\mu}, \sigma^2) = \mathcal{N}(\boldsymbol{x} \,|\, \boldsymbol{B}\boldsymbol{z} + \boldsymbol{\mu}, \sigma^2 \boldsymbol{I}) . \tag{10.64}$$

Overall, PPCA induces the following generative process:

$$\boldsymbol{z}_n \sim \mathcal{N}(\boldsymbol{z} \,|\, \mathbf{0}, \boldsymbol{I}) \tag{10.65}$$

$$\boldsymbol{x}_n \,|\, \boldsymbol{z}_n \sim \mathcal{N}(\boldsymbol{x} \,|\, \boldsymbol{B}\boldsymbol{z}_n + \boldsymbol{\mu}, \sigma^2 \boldsymbol{I}). \tag{10.66}$$

To generate a data point that is typical given the model parameters, we follow an *ancestral sampling* scheme: We first sample a latent variable $\boldsymbol{z}_n$ from $p(\boldsymbol{z})$. Then we use $\boldsymbol{z}_n$ in (10.64) to sample a data point conditioned on the sampled $\boldsymbol{z}_n$, i.e., $\boldsymbol{x}_n \sim p(\boldsymbol{x} \,|\, \boldsymbol{z}_n, \boldsymbol{B}, \boldsymbol{\mu}, \sigma^2)$.

ancestral sampling

This generative process allows us to write down the probabilistic model (i.e., the joint distribution of all random variables; see Section 8.4) as

$$p(\boldsymbol{x}, \boldsymbol{z}|\boldsymbol{B}, \boldsymbol{\mu}, \sigma^2) = p(\boldsymbol{x}|\boldsymbol{z}, \boldsymbol{B}, \boldsymbol{\mu}, \sigma^2)p(\boldsymbol{z}) , \tag{10.67}$$

which immediately gives rise to the graphical model in Figure 10.14 using the results from Section 8.5.

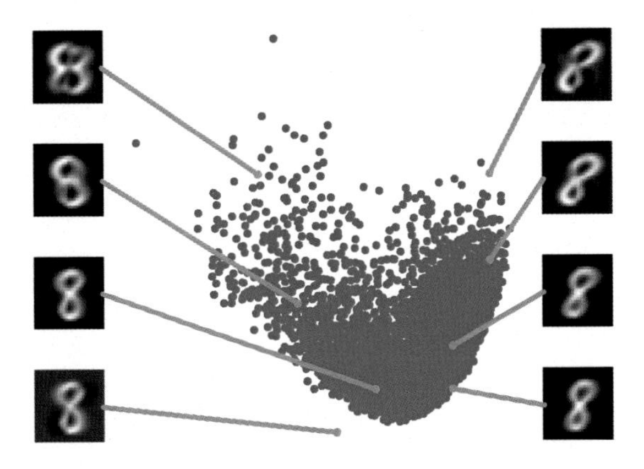

Figure 10.15 Generating new MNIST digits. The latent variables $\boldsymbol{z}$ can be used to generate new data $\tilde{\boldsymbol{x}} = \boldsymbol{B}\boldsymbol{z}$. The closer we stay to the training data, the more realistic the generated data.

Remark. Note the direction of the arrow that connects the latent variables $\boldsymbol{z}$ and the observed data $\boldsymbol{x}$: The arrow points from $\boldsymbol{z}$ to $\boldsymbol{x}$, which means that the PPCA model assumes a lower-dimensional latent cause $\boldsymbol{z}$ for high-dimensional observations $\boldsymbol{x}$. In the end, we are obviously interested in finding something out about $\boldsymbol{z}$ given some observations. To get there, we will apply Bayesian inference to "invert" the arrow implicitly and go from observations to latent variables. $\diamondsuit$

Example 10.5 (Generating New Data Using Latent Variables)
Figure 10.15 shows the latent coordinates of the MNIST digits "8" found by PCA when using a two-dimensional principal subspace (blue dots). We can query any vector $\boldsymbol{z}_*$ in this latent space and generate an image $\tilde{\boldsymbol{x}}_* = \boldsymbol{B}\boldsymbol{z}_*$ that resembles the digit "8." We show eight of such generated images with their corresponding latent space representation. Depending on where we query the latent space, the generated images look different (shape, rotation, size, etc.). If we query away from the training data, we see more and more artifacts, e.g., the top-left and top-right digits. Note that the intrinsic dimensionality of these generated images is only two.

10.7.2 Likelihood and Joint Distribution

The likelihood does not depend on the latent variables $\boldsymbol{z}$.

Using the results from Chapter 6, we obtain the likelihood of this probabilistic model by integrating out the latent variable $\boldsymbol{z}$ (see Section 8.4.3) so that

$$p(\boldsymbol{x}\,|\,\boldsymbol{B},\boldsymbol{\mu},\sigma^2) = \int p(\boldsymbol{x}\,|\,\boldsymbol{z},\boldsymbol{\mu},\sigma^2)p(\boldsymbol{z})\mathrm{d}\boldsymbol{z} \tag{10.68a}$$

$$= \int \mathcal{N}\big(\boldsymbol{x}\,|\,\boldsymbol{B}\boldsymbol{z}+\boldsymbol{\mu},\,\sigma^2\boldsymbol{I}\big)\mathcal{N}\big(\boldsymbol{z}\,|\,\boldsymbol{0},\,\boldsymbol{I}\big)\mathrm{d}\boldsymbol{z}\,. \tag{10.68b}$$

From Section 6.5, we know that the solution to this integral is a Gaussian distribution with mean

$$\mathbb{E}_{\boldsymbol{x}}[\boldsymbol{x}] = \mathbb{E}_{\boldsymbol{z}}[\boldsymbol{B}\boldsymbol{z}+\boldsymbol{\mu}] + \mathbb{E}_{\boldsymbol{\epsilon}}[\boldsymbol{\epsilon}] = \boldsymbol{\mu} \tag{10.69}$$

and with covariance matrix

$$\mathbb{V}[\boldsymbol{x}] = \mathbb{V}_{\boldsymbol{z}}[\boldsymbol{B}\boldsymbol{z} + \boldsymbol{\mu}] + \mathbb{V}_{\boldsymbol{\epsilon}}[\boldsymbol{\epsilon}] = \mathbb{V}_{\boldsymbol{z}}[\boldsymbol{B}\boldsymbol{z}] + \sigma^2\boldsymbol{I} \tag{10.70a}$$

$$= \boldsymbol{B}\mathbb{V}_{\boldsymbol{z}}[\boldsymbol{z}]\boldsymbol{B}^\top + \sigma^2\boldsymbol{I} = \boldsymbol{B}\boldsymbol{B}^\top + \sigma^2\boldsymbol{I}\,. \tag{10.70b}$$

The likelihood in (10.68b) can be used for maximum likelihood or MAP estimation of the model parameters.

Remark. We cannot use the conditional distribution in (10.64) for maximum likelihood estimation as it still depends on the latent variables. The likelihood function we require for maximum likelihood (or MAP) estimation should only be a function of the data $\boldsymbol{x}$ and the model parameters, but must not depend on the latent variables. ◇

From Section 6.5, we know that a Gaussian random variable $\boldsymbol{z}$ and a linear/affine transformation $\boldsymbol{x} = \boldsymbol{B}\boldsymbol{z}$ of it are jointly Gaussian distributed. We already know the marginals $p(\boldsymbol{z}) = \mathcal{N}\big(\boldsymbol{z}\,|\,\boldsymbol{0},\, \boldsymbol{I}\big)$ and $p(\boldsymbol{x}) = \mathcal{N}\big(\boldsymbol{x}\,|\,\boldsymbol{\mu},\, \boldsymbol{B}\boldsymbol{B}^\top + \sigma^2\boldsymbol{I}\big)$. The missing cross-covariance is given as

$$\mathrm{Cov}[\boldsymbol{x}, \boldsymbol{z}] = \mathrm{Cov}_{\boldsymbol{z}}[\boldsymbol{B}\boldsymbol{z} + \boldsymbol{\mu}] = \boldsymbol{B}\,\mathrm{Cov}_{\boldsymbol{z}}[\boldsymbol{z}, \boldsymbol{z}] = \boldsymbol{B}\,. \tag{10.71}$$

Therefore, the probabilistic model of PPCA, i.e., the joint distribution of latent and observed random variables is explicitly given by

$$p(\boldsymbol{x}, \boldsymbol{z}\,|\,\boldsymbol{B}, \boldsymbol{\mu}, \sigma^2) = \mathcal{N}\left(\begin{bmatrix}\boldsymbol{x}\\ \boldsymbol{z}\end{bmatrix} \,\middle|\, \begin{bmatrix}\boldsymbol{\mu}\\ \boldsymbol{0}\end{bmatrix},\, \begin{bmatrix}\boldsymbol{B}\boldsymbol{B}^\top + \sigma^2\boldsymbol{I} & \boldsymbol{B}\\ \boldsymbol{B}^\top & \boldsymbol{I}\end{bmatrix}\right), \tag{10.72}$$

with a mean vector of length $D + M$ and a covariance matrix of size $(D + M) \times (D + M)$.

10.7.3 Posterior Distribution

The joint Gaussian distribution $p(\boldsymbol{x}, \boldsymbol{z}\,|\,\boldsymbol{B}, \boldsymbol{\mu}, \sigma^2)$ in (10.72) allows us to determine the posterior distribution $p(\boldsymbol{z}\,|\,\boldsymbol{x})$ immediately by applying the rules of Gaussian conditioning from Section 6.5.1. The posterior distribution of the latent variable given an observation $\boldsymbol{x}$ is then

$$p(\boldsymbol{z}\,|\,\boldsymbol{x}) = \mathcal{N}\big(\boldsymbol{z}\,|\,\boldsymbol{m},\, \boldsymbol{C}\big)\,, \tag{10.73}$$

$$\boldsymbol{m} = \boldsymbol{B}^\top(\boldsymbol{B}\boldsymbol{B}^\top + \sigma^2\boldsymbol{I})^{-1}(\boldsymbol{x} - \boldsymbol{\mu})\,, \tag{10.74}$$

$$\boldsymbol{C} = \boldsymbol{I} - \boldsymbol{B}^\top(\boldsymbol{B}\boldsymbol{B}^\top + \sigma^2\boldsymbol{I})^{-1}\boldsymbol{B}\,. \tag{10.75}$$

Note that the posterior covariance does not depend on the observed data $\boldsymbol{x}$. For a new observation $\boldsymbol{x}_*$ in data space, we use (10.73) to determine the posterior distribution of the corresponding latent variable $\boldsymbol{z}_*$. The covariance matrix $\boldsymbol{C}$ allows us to assess how confident the embedding is. A covariance matrix $\boldsymbol{C}$ with a small determinant (which measures volumes) tells us that the latent embedding $\boldsymbol{z}_*$ is fairly certain. If we obtain a posterior distribution $p(\boldsymbol{z}_*\,|\,\boldsymbol{x}_*)$ with much variance, we may be faced with an outlier. However, we can explore this posterior distribution to understand what other data points $\boldsymbol{x}$ are plausible under this posterior. To do this, we exploit the generative process underlying PPCA, which allows us to explore the posterior distribution on the latent variables by generating new data that is plausible under this posterior:

1. Sample a latent variable $z_* \sim p(z \mid x_*)$ from the posterior distribution over the latent variables (10.73).
2. Sample a reconstructed vector $\tilde{x}_* \sim p(x \mid z_*, B, \mu, \sigma^2)$ from (10.64).

If we repeat this process many times, we can explore the posterior distribution (10.73) on the latent variables z_* and its implications on the observed data. The sampling process effectively hypothesizes data, which is plausible under the posterior distribution.

10.8 Further Reading

We derived PCA from two perspectives: (a) maximizing the variance in the projected space; and (b) minimizing the average reconstruction error. However, PCA can also be interpreted from different perspectives. Let us recap what we have done: We took high-dimensional data $x \in \mathbb{R}^D$ and used a matrix $B^\top$ to find a lower-dimensional representation $z \in \mathbb{R}^M$. The columns of B are the eigenvectors of the data covariance matrix S that are associated with the largest eigenvalues. Once we have a low-dimensional representation z, we can get a high-dimensional version of it (in the original data space) as $x \approx \tilde{x} = Bz = BB^\top x \in \mathbb{R}^D$, where $BB^\top$ is a projection matrix.

auto-encoder

code

encoder

decoder

We can also think of PCA as a linear *auto-encoder* as illustrated in Figure 10.16. An auto-encoder encodes the data $x_n \in \mathbb{R}^D$ to a *code* $z_n \in \mathbb{R}^M$ and decodes it to a $\tilde{x}_n$ similar to x_n. The mapping from the data to the code is called the *encoder*, and the mapping from the code back to the original data space is called the *decoder*. If we consider linear mappings where the code is given by $z_n = B^\top x_n \in \mathbb{R}^M$ and we are interested in minimizing the average squared error between the data x_n and its reconstruction $\tilde{x}_n = Bz_n$, $n = 1, \ldots, N$, we obtain

$$\frac{1}{N}\sum_{n=1}^{N} \|x_n - \tilde{x}_n\|^2 = \frac{1}{N}\sum_{n=1}^{N} \left\| x_n - B^\top B x_n \right\|^2 . \tag{10.76}$$

This means we end up with the same objective function as in (10.29) that we discussed in Section 10.3 so that we obtain the PCA solution when we minimize the squared auto-encoding loss. If we replace the linear mapping of PCA with

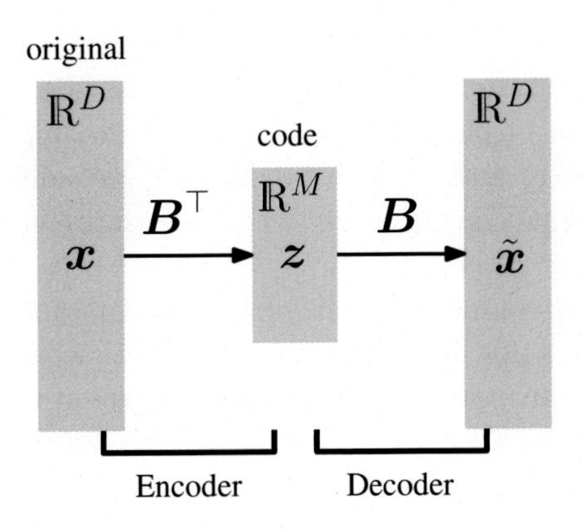

Figure 10.16 PCA can be viewed as a linear auto-encoder. It encodes the high-dimensional data x into a lower-dimensional representation (code) $z \in \mathbb{R}^M$ and decodes z using a decoder. The decoded vector $\tilde{x}$ is the orthogonal projection of the original data x onto the M-dimensional principal subspace.

a nonlinear mapping, we get a nonlinear auto-encoder. A prominent example of this is a deep auto-encoder where the linear functions are replaced with deep neural networks. In this context, the encoder is also known as a *recognition network* or *inference network*, whereas the decoder is also called a *generator*.

recognition network
inference network
generator

Another interpretation of PCA is related to information theory. We can think of the code as a smaller or compressed version of the original data point. When we reconstruct our original data using the code, we do not get the exact data point back, but a slightly distorted or noisy version of it. This means that our compression is "lossy." Intuitively, we want to maximize the correlation between the original data and the lower-dimensional code. More formally, this is related to the mutual information. We would then get the same solution to PCA we discussed in Section 10.3 by maximizing the mutual information, a core concept in information theory (MacKay, 2003).

The code is a compressed version of the original data.

In our discussion on PPCA, we assumed that the parameters of the model, i.e., $\boldsymbol{B}, \boldsymbol{\mu}$, and the likelihood parameter σ^2, are known. Tipping and Bishop (1999) describe how to derive maximum likelihood estimates for these parameters in the PPCA setting (note that we use a different notation in this chapter). The maximum likelihood parameters, when projecting D-dimensional data onto an M-dimensional subspace, are

$$\boldsymbol{\mu}_{\text{ML}} = \frac{1}{N}\sum_{n=1}^{N} \boldsymbol{x}_n \,, \tag{10.77}$$

$$\boldsymbol{B}_{\text{ML}} = \boldsymbol{T}(\boldsymbol{\Lambda} - \sigma^2\boldsymbol{I})^{\frac{1}{2}}\boldsymbol{R} \,, \tag{10.78}$$

$$\sigma^2_{\text{ML}} = \frac{1}{D-M}\sum_{j=M+1}^{D} \lambda_j \,, \tag{10.79}$$

where $\boldsymbol{T} \in \mathbb{R}^{D\times M}$ contains M eigenvectors of the data covariance matrix, $\boldsymbol{\Lambda} = \text{diag}(\lambda_1, \dots, \lambda_M) \in \mathbb{R}^{M\times M}$ is a diagonal matrix with the eigenvalues associated with the principal axes on its diagonal, and $\boldsymbol{R} \in \mathbb{R}^{M\times M}$ is an arbitrary orthogonal matrix. The maximum likelihood solution $\boldsymbol{B}_{\text{ML}}$ is unique up to an arbitrary orthogonal transformation, e.g., we can right-multiply $\boldsymbol{B}_{\text{ML}}$ with any rotation matrix $\boldsymbol{R}$ so that (10.78) essentially is a singular value decomposition (see Section 4.5). An outline of the proof is given by Tipping and Bishop (1999).

The matrix $\boldsymbol{\Lambda} - \sigma^2\boldsymbol{I}$ in (10.78) is guaranteed to be positive semidefinite as the smallest eigenvalue of the data covariance matrix is bounded from below by the noise variance σ^2.

The maximum likelihood estimate for $\boldsymbol{\mu}$ given in (10.77) is the sample mean of the data. The maximum likelihood estimator for the observation noise variance σ^2 given in (10.79) is the average variance in the orthogonal complement of the principal subspace, i.e., the average leftover variance that we cannot capture with the first M principal components is treated as observation noise.

In the noise-free limit where $\sigma \to 0$, PPCA and PCA provide identical solutions: Since the data covariance matrix $\boldsymbol{S}$ is symmetric, it can be diagonalized (see Section 4.4), i.e., there exists a matrix $\boldsymbol{T}$ of eigenvectors of $\boldsymbol{S}$ so that

$$\boldsymbol{S} = \boldsymbol{T}\boldsymbol{\Lambda}\boldsymbol{T}^{-1} \,. \tag{10.80}$$

In the PPCA model, the data covariance matrix is the covariance matrix of the Gaussian likelihood $p(\boldsymbol{x} \mid \boldsymbol{B}, \boldsymbol{\mu}, \sigma^2)$, which is $\boldsymbol{B}\boldsymbol{B}^\top + \sigma^2\boldsymbol{I}$, see (10.70b). For

$\sigma \to 0$, we obtain $\boldsymbol{B}\boldsymbol{B}^\top$ so that this data covariance must equal the PCA data covariance (and its factorization given in (10.80)) so that

$$\operatorname{Cov}[\mathcal{X}] = \boldsymbol{T}\boldsymbol{\Lambda}\boldsymbol{T}^{-1} = \boldsymbol{B}\boldsymbol{B}^\top \iff \boldsymbol{B} = \boldsymbol{T}\boldsymbol{\Lambda}^{\frac{1}{2}}\boldsymbol{R}, \tag{10.81}$$

i.e., we obtain the maximum likelihood estimate in (10.78) for $\sigma = 0$. From (10.78) and (10.80), it becomes clear that (P)PCA performs a decomposition of the data covariance matrix.

In a streaming setting, where data arrives sequentially, it is recommended to use the iterative expectation maximization (EM) algorithm for maximum likelihood estimation (Roweis, 1998).

To determine the dimensionality of the latent variables (the length of the code, the dimensionality of the lower-dimensional subspace onto which we project the data), Gavish and Donoho (2014) suggest the heuristic that, if we can estimate the noise variance σ^2 of the data, we should discard all singular values smaller than $\frac{4\sigma\sqrt{D}}{\sqrt{3}}$. Alternatively, we can use (nested) cross-validation (Section 8.6.1) or Bayesian model selection criteria (discussed in Section 8.6.2) to determine a good estimate of the intrinsic dimensionality of the data (Minka, 2001b).

Similar to our discussion on linear regression in Chapter 9, we can place a prior distribution on the parameters of the model and integrate them out. By doing so, we (a) avoid point estimates of the parameters and the issues that come with these point estimates (see Section 8.6) and (b) allow for an automatic selection of the appropriate dimensionality M of the latent space. In this *Bayesian PCA*, which was proposed by Bishop (1999), a prior $p(\boldsymbol{\mu}, \boldsymbol{B}, \sigma^2)$ is placed on the model parameters. The generative process allows us to integrate the model parameters out instead of conditioning on them, which addresses overfitting issues. Since this integration is analytically intractable, Bishop (1999) proposes to use approximate inference methods, such as MCMC or variational inference. We refer to the work by Gilks et al. (1996) and Blei et al. (2017) for more details on these approximate inference techniques.

Bayesian PCA

In PPCA, we considered the linear model $p(\boldsymbol{x}_n \,|\, \boldsymbol{z}_n) = \mathcal{N}\big(\boldsymbol{x}_n \,|\, \boldsymbol{B}\boldsymbol{z}_n + \boldsymbol{\mu},\, \sigma^2\boldsymbol{I}\big)$ with prior $p(\boldsymbol{z}_n) = \mathcal{N}\big(\boldsymbol{0},\, \boldsymbol{I}\big)$, where all observation dimensions are affected by the same amount of noise. If we allow each observation dimension d to have a different variance σ_d^2, we obtain *factor analysis* (FA) (Spearman, 1904; Bartholomew et al., 2011). This means that FA gives the likelihood some more flexibility than PPCA, but still forces the data to be explained by the model parameters $\boldsymbol{B}$, $\boldsymbol{\mu}$.However, FA no longer allows for a closed-form maximum likelihood solution so that we need to use an iterative scheme, such as the expectation maximization algorithm, to estimate the model parameters. While in PPCA all stationary points are global optima, this no longer holds for FA. Compared to PPCA, FA does not change if we scale the data, but it does return different solutions if we rotate the data.

factor analysis

An overly flexible likelihood would be able to explain more than just the noise.

An algorithm that is also closely related to PCA is *independent component analysis* (*ICA* (Hyvarinen et al., 2001)). Starting again with the latent-variable perspective $p(\boldsymbol{x}_n \,|\, \boldsymbol{z}_n) = \mathcal{N}\big(\boldsymbol{x}_n \,|\, \boldsymbol{B}\boldsymbol{z}_n + \boldsymbol{\mu},\, \sigma^2\boldsymbol{I}\big)$ we now change the prior on $\boldsymbol{z}_n$ to non-Gaussian distributions. ICA can be used for *blind-source separation*. Imagine you are in a busy train station with many people talking. Your ears play

independent component analysis (ICA)

blind-source separation

the role of microphones, and they linearly mix different speech signals in the train station. The goal of blind-source separation is to identify the constituent parts of the mixed signals. As discussed previously in the context of maximum likelihood estimation for PPCA, the original PCA solution is invariant to any rotation. Therefore, PCA can identify the best lower-dimensional subspace in which the signals live, but not the signals themselves (Murphy, 2012). ICA addresses this issue by modifying the prior distribution $p(\boldsymbol{z})$ on the latent sources to require non-Gaussian priors $p(\boldsymbol{z})$. We refer to the books by Hyvarinen et al. (2001) and Murphy (2012) for more details on ICA.

PCA, factor analysis, and ICA are three examples for dimensionality reduction with linear models. Cunningham and Ghahramani (2015) provide a broader survey of linear dimensionality reduction.

The (P)PCA model we discussed here allows for several important extensions. In Section 10.5, we explained how to do PCA when the input dimensionality D is significantly greater than the number N of data points. By exploiting the insight that PCA can be performed by computing (many) inner products, this idea can be pushed to the extreme by considering infinite-dimensional features. The *kernel trick* is the basis of *kernel PCA* and allows us to implicitly compute inner products between infinite-dimensional features (Schölkopf et al., 1998; Schölkopf and Smola, 2002).

kernel trick
kernel PCA

There are nonlinear dimensionality reduction techniques that are derived from PCA (Burges, 2010, provides a good overview). The auto-encoder perspective of PCA that we discussed previously in this section can be used to render PCA as a special case of a *deep auto-encoder*. In the deep auto-encoder, both the encoder and the decoder are represented by multilayer feedforward neural networks, which themselves are nonlinear mappings. If we set the activation functions in these neural networks to be the identity, the model becomes equivalent to PCA. A different approach to nonlinear dimensionality reduction is the *Gaussian process latent-variable model* (*GP-LVM*) proposed by Lawrence (2005). The GP-LVM starts off with the latent-variable perspective that we used to derive PPCA and replaces the linear relationship between the latent variables $\boldsymbol{z}$ and the observations $\boldsymbol{x}$ with a Gaussian process (GP). Instead of estimating the parameters of the mapping (as we do in PPCA), the GP-LVM marginalizes out the model parameters and makes point estimates of the latent variables $\boldsymbol{z}$. Similar to Bayesian PCA, the *Bayesian GP-LVM* proposed by Titsias and Lawrence (2010) maintains a distribution on the latent variables $\boldsymbol{z}$ and uses approximate inference to integrate them out as well.

deep auto-encoder

Gaussian process latent-variable model
GP-LVM

Bayesian GP-LVM

11

Density Estimation with Gaussian Mixture Models

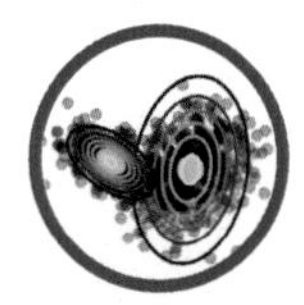

In earlier chapters, we covered already two fundamental problems in machine learning: regression (Chapter 9) and dimensionality reduction (Chapter 10). In this chapter, we will have a look at a third pillar of machine learning: density estimation. On our journey, we introduce important concepts, such as the expectation maximization (EM) algorithm and a latent variable perspective of density estimation with mixture models.

When we apply machine learning to data we often aim to represent data in some way. A straightforward way is to take the data points themselves as the representation of the data; see Figure 11.1 for an example. However, this approach may be unhelpful if the dataset is huge or if we are interested in representing characteristics of the data. In density estimation, we represent the data compactly using a density from a parametric family, e.g., a Gaussian or Beta distribution. For example, we may be looking for the mean and variance of a dataset in order to represent the data compactly using a Gaussian distribution. The mean and variance can be found using tools we discussed in Section 8.3: maximum likelihood or maximum a posteriori estimation. We can then use the mean and variance of this Gaussian to represent the distribution underlying the data, i.e., we think of the dataset to be a typical realization from this distribution if we were to sample from it.

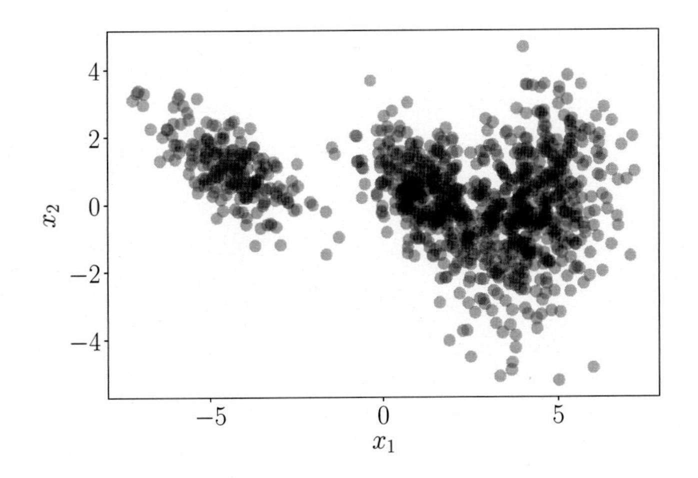

Figure 11.1 Two-dimensional dataset that cannot be meaningfully represented by a Gaussian.

In practice, the Gaussian (or similarly all other distributions we encountered so far) have limited modeling capabilities. For example, a Gaussian approximation of the density that generated the data in Figure 11.1 would be a poor approximation. In the following, we will look at a more expressive family of distributions, which we can use for density estimation: *mixture models*.

mixture model

Mixture models can be used to describe a distribution $p(\boldsymbol{x})$ by a convex combination of K simple (base) distributions

$$p(\boldsymbol{x}) = \sum_{k=1}^{K} \pi_k p_k(\boldsymbol{x}) \tag{11.1}$$

$$0 \leqslant \pi_k \leqslant 1, \quad \sum_{k=1}^{K} \pi_k = 1, \tag{11.2}$$

where the components p_k are members of a family of basic distributions, e.g., Gaussians, Bernoullis, or Gammas, and the π_k are *mixture weights*. Mixture models are more expressive than the corresponding base distributions because they allow for multimodal data representations, i.e., they can describe datasets with multiple "clusters," such as the example in Figure 11.1.

mixture weight

We will focus on Gaussian mixture models (GMMs), where the basic distributions are Gaussians. For a given dataset, we aim to maximize the likelihood of the model parameters to train the GMM. For this purpose, we will use results from Chapter 5, Chapter 6, and Section 7.2. However, unlike other applications we discussed earlier (linear regression or PCA), we will not find a closed-form maximum likelihood solution. Instead, we will arrive at a set of dependent simultaneous equations, which we can only solve iteratively.

11.1 Gaussian Mixture Model

A *Gaussian mixture model* is a density model where we combine a finite number of K Gaussian distributions $\mathcal{N}(\boldsymbol{x} \,|\, \boldsymbol{\mu}_k, \boldsymbol{\Sigma}_k)$ so that

Gaussian mixture model

$$p(\boldsymbol{x} \,|\, \boldsymbol{\theta}) = \sum_{k=1}^{K} \pi_k \mathcal{N}(\boldsymbol{x} \,|\, \boldsymbol{\mu}_k, \boldsymbol{\Sigma}_k) \tag{11.3}$$

$$0 \leqslant \pi_k \leqslant 1, \quad \sum_{k=1}^{K} \pi_k = 1, \tag{11.4}$$

where we defined $\boldsymbol{\theta} := \{\boldsymbol{\mu}_k, \boldsymbol{\Sigma}_k, \pi_k : k = 1, \ldots, K\}$ as the collection of all parameters of the model. This convex combination of Gaussian distribution gives us significantly more flexibility for modeling complex densities than a simple Gaussian distribution (which we recover from (11.3) for $K = 1$). An illustration is given in Figure 11.2, displaying the weighted components and the mixture density, which is given as

$$p(x \,|\, \boldsymbol{\theta}) = 0.5\mathcal{N}\big(x \,|\, -2, \tfrac{1}{2}\big) + 0.2\mathcal{N}\big(x \,|\, 1, 2\big) + 0.3\mathcal{N}\big(x \,|\, 4, 1\big). \tag{11.5}$$

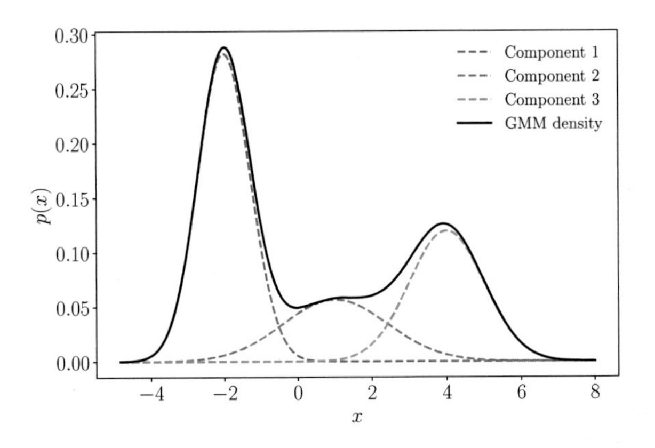

Figure 11.2 Gaussian mixture model. The Gaussian mixture distribution (black) is composed of a convex combination of Gaussian distributions and is more expressive than any individual component. Dashed lines represent the weighted Gaussian components.

11.2 Parameter Learning via Maximum Likelihood

Assume we are given a dataset $\mathcal{X} = \{\boldsymbol{x}_1, \dots, \boldsymbol{x}_N\}$, where $\boldsymbol{x}_n$, $n = 1, \dots, N$, are drawn i.i.d. from an unknown distribution $p(\boldsymbol{x})$. Our objective is to find a good approximation/representation of this unknown distribution $p(\boldsymbol{x})$ by means of a GMM with K mixture components. The parameters of the GMM are the K means $\boldsymbol{\mu}_k$, the covariances $\boldsymbol{\Sigma}_k$, and mixture weights π_k. We summarize all these free parameters in $\boldsymbol{\theta} := \{\pi_k, \boldsymbol{\mu}_k, \boldsymbol{\Sigma}_k : k = 1, \dots, K\}$.

Example 11.1 (Initial Setting)

Throughout this chapter, we will have a simple running example that helps us illustrate and visualize important concepts.

We consider a one-dimensional dataset $\mathcal{X} = \{-3, -2.5, -1, 0, 2, 4, 5\}$ consisting of seven data points and wish to find a GMM with $K = 3$ components that models the density of the data. We initialize the mixture components as

$$p_1(x) = \mathcal{N}(x \mid -4,\ 1) \tag{11.6}$$

$$p_2(x) = \mathcal{N}(x \mid 0,\ 0.2) \tag{11.7}$$

$$p_3(x) = \mathcal{N}(x \mid 8,\ 3) \tag{11.8}$$

and assign them equal weights $\pi_1 = \pi_2 = \pi_3 = \frac{1}{3}$. The corresponding model (and the data points) are shown in Figure 11.3.

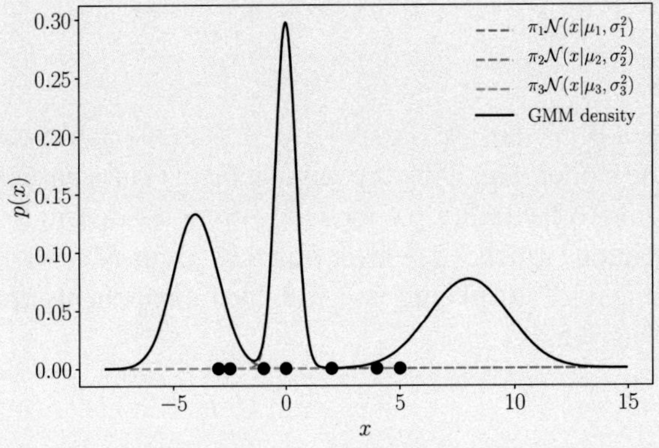

Figure 11.3 Initial setting: GMM (black) with mixture three mixture components (dashed) and seven data points (discs).

In the following, we detail how to obtain a maximum likelihood estimate $\boldsymbol{\theta}_{\text{ML}}$ of the model parameters $\boldsymbol{\theta}$. We start by writing down the likelihood, i.e., the predictive distribution of the training data given the parameters. We exploit our i.i.d. assumption, which leads to the factorized likelihood

$$p(\mathcal{X} \mid \boldsymbol{\theta}) = \prod_{n=1}^{N} p(\boldsymbol{x}_n \mid \boldsymbol{\theta}), \quad p(\boldsymbol{x}_n \mid \boldsymbol{\theta}) = \sum_{k=1}^{K} \pi_k \mathcal{N}\big(\boldsymbol{x}_n \mid \boldsymbol{\mu}_k, \boldsymbol{\Sigma}_k\big), \tag{11.9}$$

where every individual likelihood term $p(\boldsymbol{x}_n \mid \boldsymbol{\theta})$ is a Gaussian mixture density. Then we obtain the log-likelihood as

$$\log p(\mathcal{X} \mid \boldsymbol{\theta}) = \sum_{n=1}^{N} \log p(\boldsymbol{x}_n \mid \boldsymbol{\theta}) = \underbrace{\sum_{n=1}^{N} \log \sum_{k=1}^{K} \pi_k \mathcal{N}\big(\boldsymbol{x}_n \mid \boldsymbol{\mu}_k, \boldsymbol{\Sigma}_k\big)}_{=:\mathcal{L}}. \tag{11.10}$$

We aim to find parameters $\boldsymbol{\theta}_{\text{ML}}^*$ that maximize the log-likelihood $\mathcal{L}$ defined in (11.10). Our "normal" procedure would be to compute the gradient $\mathrm{d}\mathcal{L}/\mathrm{d}\boldsymbol{\theta}$ of the log-likelihood with respect to the model parameters $\boldsymbol{\theta}$, set it to $\boldsymbol{0}$, and solve for $\boldsymbol{\theta}$. However, unlike our previous examples for maximum likelihood estimation (e.g., when we discussed linear regression in Section 9.2), we cannot obtain a closed-form solution. However, we can exploit an iterative scheme to find good model parameters $\boldsymbol{\theta}_{\text{ML}}$, which will turnout to be the EM algorithm for GMMs. The key idea is to update one model parameter at a time while keeping the others fixed.

Remark. If we were to consider a single Gaussian as the desired density, the sum over k in (11.10) vanishes, and the $\log$ can be applied directly to the Gaussian component, such that we get

$$\log \mathcal{N}\big(\boldsymbol{x} \mid \boldsymbol{\mu}, \boldsymbol{\Sigma}\big) = -\tfrac{D}{2}\log(2\pi) - \tfrac{1}{2}\log\det(\boldsymbol{\Sigma}) - \tfrac{1}{2}(\boldsymbol{x}-\boldsymbol{\mu})^\top \boldsymbol{\Sigma}^{-1}(\boldsymbol{x}-\boldsymbol{\mu}). \tag{11.11}$$

This simple form allows us to find closed-form maximum likelihood estimates of $\boldsymbol{\mu}$ and $\boldsymbol{\Sigma}$, as discussed in Chapter 8. In (11.10), we cannot move the $\log$ into the sum over k so that we cannot obtain a simple closed-form maximum likelihood solution. ◇

Any local optimum of a function exhibits the property that its gradient with respect to the parameters must vanish (necessary condition); see Chapter 7. In our case, we obtain the following necessary conditions when we optimize the log-likelihood in (11.10) with respect to the GMM parameters $\boldsymbol{\mu}_k, \boldsymbol{\Sigma}_k, \pi_k$:

$$\frac{\partial \mathcal{L}}{\partial \boldsymbol{\mu}_k} = \boldsymbol{0}^\top \iff \sum_{n=1}^{N} \frac{\partial \log p(\boldsymbol{x}_n \mid \boldsymbol{\theta})}{\partial \boldsymbol{\mu}_k} = \boldsymbol{0}^\top, \tag{11.12}$$

$$\frac{\partial \mathcal{L}}{\partial \boldsymbol{\Sigma}_k} = \boldsymbol{0} \iff \sum_{n=1}^{N} \frac{\partial \log p(\boldsymbol{x}_n \mid \boldsymbol{\theta})}{\partial \boldsymbol{\Sigma}_k} = \boldsymbol{0}, \tag{11.13}$$

$$\frac{\partial \mathcal{L}}{\partial \pi_k} = 0 \iff \sum_{n=1}^{N} \frac{\partial \log p(\boldsymbol{x}_n \mid \boldsymbol{\theta})}{\partial \pi_k} = 0. \tag{11.14}$$

For all three necessary conditions, by applying the chain rule (see Section 5.2.2), we require partial derivatives of the form

$$\frac{\partial \log p(\boldsymbol{x}_n \,|\, \boldsymbol{\theta})}{\partial \boldsymbol{\theta}} = \frac{1}{p(\boldsymbol{x}_n \,|\, \boldsymbol{\theta})} \frac{\partial p(\boldsymbol{x}_n \,|\, \boldsymbol{\theta})}{\partial \boldsymbol{\theta}}, \tag{11.15}$$

where $\boldsymbol{\theta} = \{\boldsymbol{\mu}_k, \boldsymbol{\Sigma}_k, \pi_k, k = 1, \ldots, K\}$ are the model parameters and

$$\frac{1}{p(\boldsymbol{x}_n \,|\, \boldsymbol{\theta})} = \frac{1}{\sum_{j=1}^{K} \pi_j \mathcal{N}\big(\boldsymbol{x}_n \,|\, \boldsymbol{\mu}_j, \boldsymbol{\Sigma}_j\big)}. \tag{11.16}$$

In the following, we will compute the partial derivatives (11.12) through (11.14). But before we do this, we introduce a quantity that will play a central role in the remainder of this chapter: responsibilities.

11.2.1 Responsibilities

We define the quantity

$$r_{nk} := \frac{\pi_k \mathcal{N}\big(\boldsymbol{x}_n \,|\, \boldsymbol{\mu}_k, \boldsymbol{\Sigma}_k\big)}{\sum_{j=1}^{K} \pi_j \mathcal{N}\big(\boldsymbol{x}_n \,|\, \boldsymbol{\mu}_j, \boldsymbol{\Sigma}_j\big)} \tag{11.17}$$

responsibility

as the *responsibility* of the kth mixture component for the nth data point. The responsibility r_{nk} of the kth mixture component for data point $\boldsymbol{x}_n$ is proportional to the likelihood

$$p(\boldsymbol{x}_n \,|\, \pi_k, \boldsymbol{\mu}_k, \boldsymbol{\Sigma}_k) = \pi_k \mathcal{N}\big(\boldsymbol{x}_n \,|\, \boldsymbol{\mu}_k, \boldsymbol{\Sigma}_k\big) \tag{11.18}$$

$\boldsymbol{r}_n$ follows a Boltzmann/Gibbs distribution.

The responsibility r_{nk} is the probability that the kth mixture component generated the nth data point.

of the mixture component given the data point. Therefore, mixture components have a high responsibility for a data point when the data point could be a plausible sample from that mixture component. Note that $\boldsymbol{r}_n := [r_{n1}, \ldots, r_{nK}]^\top \in \mathbb{R}^K$ is a (normalized) probability vector, i.e., $\sum_k r_{nk} = 1$ with $r_{nk} \geqslant 0$. This probability vector distributes probability mass among the K mixture components, and we can think of $\boldsymbol{r}_n$ as a "soft assignment" of $\boldsymbol{x}_n$ to the K mixture components. Therefore, the responsibility r_{nk} from (11.17) represents the probability that $\boldsymbol{x}_n$ has been generated by the kth mixture component.

Example 11.2 (Responsibilities)

For our example from Figure 11.3, we compute the responsibilities r_{nk}

$$\begin{bmatrix} 1.0 & 0.0 & 0.0 \\ 1.0 & 0.0 & 0.0 \\ 0.057 & 0.943 & 0.0 \\ 0.001 & 0.999 & 0.0 \\ 0.0 & 0.066 & 0.934 \\ 0.0 & 0.0 & 1.0 \\ 0.0 & 0.0 & 1.0 \end{bmatrix} \in \mathbb{R}^{N \times K}. \tag{11.19}$$

Here the nth row tells us the responsibilities of all mixture components for x_n. The sum of all K responsibilities for a data point (sum of every row) is 1. The kth column gives us an overview of the responsibility of the kth mixture component. We can see that the third mixture component (third column) is not responsible for any of the first four data points, but takes much responsibility of the remaining data points. The sum of all entries of a column gives us the values N_k, i.e., the total responsibility of the kth mixture component. In our example, we get $N_1 = 2.058,\ N_2 = 2.008,\ N_3 = 2.934$.

In the following, we determine the updates of the model parameters $\boldsymbol{\mu}_k, \boldsymbol{\Sigma}_k, \pi_k$ for given responsibilities. We will see that the update equations all depend on the responsibilities, which makes a closed-form solution to the maximum likelihood estimation problem impossible. However, for given responsibilities we will be updating one model parameter at a time, while keeping the others fixed. After this, we will recompute the responsibilities. Iterating these two steps will eventually converge to a local optimum and is a specific instantiation of the EM algorithm. We will discuss this in some more detail in Section 11.3.

11.2.2 Updating the Means

Theorem 11.1 (Update of the GMM Means)**.** *The update of the mean parameters $\boldsymbol{\mu}_k$, $k = 1, \dots, K$, of the GMM is given by*

$$\boldsymbol{\mu}_k^{new} = \frac{\sum_{n=1}^N r_{nk}\boldsymbol{x}_n}{\sum_{n=1}^N r_{nk}}, \tag{11.20}$$

where the responsibilities r_{nk} are defined in (11.17).

Remark. The update of the means $\boldsymbol{\mu}_k$ of the individual mixture components in (11.20) depends on all means, covariance matrices $\boldsymbol{\Sigma}_k$, and mixture weights π_k via r_{nk} given in (11.17). Therefore, we cannot obtain a closed-form solution for all $\boldsymbol{\mu}_k$ at once. ♢

Proof From (11.15), we see that the gradient of the log-likelihood with respect to the mean parameters $\boldsymbol{\mu}_k$, $k = 1, \dots, K$, requires us to compute the partial derivative

$$\frac{\partial p(\boldsymbol{x}_n \mid \boldsymbol{\theta})}{\partial \boldsymbol{\mu}_k} = \sum_{j=1}^K \pi_j \frac{\partial \mathcal{N}\big(\boldsymbol{x}_n \mid \boldsymbol{\mu}_j, \boldsymbol{\Sigma}_j\big)}{\partial \boldsymbol{\mu}_k} = \pi_k \frac{\partial \mathcal{N}\big(\boldsymbol{x}_n \mid \boldsymbol{\mu}_k, \boldsymbol{\Sigma}_k\big)}{\partial \boldsymbol{\mu}_k} \tag{11.21a}$$

$$= \pi_k (\boldsymbol{x}_n - \boldsymbol{\mu}_k)^\top \boldsymbol{\Sigma}_k^{-1} \mathcal{N}\big(\boldsymbol{x}_n \mid \boldsymbol{\mu}_k, \boldsymbol{\Sigma}_k\big), \tag{11.21b}$$

where we exploited that only the kth mixture component depends on $\boldsymbol{\mu}_k$.

We use our result from (11.21b) in (11.15) and put everything together so that the desired partial derivative of $\mathcal{L}$ with respect to $\boldsymbol{\mu}_k$ is given as

$$\frac{\partial \mathcal{L}}{\partial \boldsymbol{\mu}_k} = \sum_{n=1}^N \frac{\partial \log p(\boldsymbol{x}_n \mid \boldsymbol{\theta})}{\partial \boldsymbol{\mu}_k} = \sum_{n=1}^N \frac{1}{p(\boldsymbol{x}_n \mid \boldsymbol{\theta})} \frac{\partial p(\boldsymbol{x}_n \mid \boldsymbol{\theta})}{\partial \boldsymbol{\mu}_k} \tag{11.22a}$$

$$= \sum_{n=1}^{N} (\boldsymbol{x}_n - \boldsymbol{\mu}_k)^\top \boldsymbol{\Sigma}_k^{-1} \underbrace{\boxed{\frac{\pi_k \mathcal{N}(\boldsymbol{x}_n \mid \boldsymbol{\mu}_k, \boldsymbol{\Sigma}_k)}{\sum_{j=1}^{K} \pi_j \mathcal{N}(\boldsymbol{x}_n \mid \boldsymbol{\mu}_j, \boldsymbol{\Sigma}_j)}}}_{=r_{nk}} \tag{11.22b}$$

$$= \sum_{n=1}^{N} r_{nk} (\boldsymbol{x}_n - \boldsymbol{\mu}_k)^\top \boldsymbol{\Sigma}_k^{-1}. \tag{11.22c}$$

Here we used the identity from (11.16) and the result of the partial derivative in (11.21b) to get to (11.22b). The values r_{nk} are the responsibilities we defined in (11.17).

We now solve (11.22c) for $\boldsymbol{\mu}_k^{\text{new}}$ so that $\frac{\partial \mathcal{L}(\boldsymbol{\mu}_k^{\text{new}})}{\partial \boldsymbol{\mu}_k} = \boldsymbol{0}^\top$ and obtain

$$\sum_{n=1}^{N} r_{nk} \boldsymbol{x}_n = \sum_{n=1}^{N} r_{nk} \boldsymbol{\mu}_k^{\text{new}} \iff \boldsymbol{\mu}_k^{\text{new}} = \frac{\sum_{n=1}^{N} r_{nk} \boldsymbol{x}_n}{\boxed{\sum_{n=1}^{N} r_{nk}}} = \frac{1}{\boxed{N_k}} \sum_{n=1}^{N} r_{nk} \boldsymbol{x}_n, \tag{11.23}$$

where we defined

$$N_k := \sum_{n=1}^{N} r_{nk} \tag{11.24}$$

as the total responsibility of the kth mixture component for the entire dataset. This concludes the proof of Theorem 11.1. □

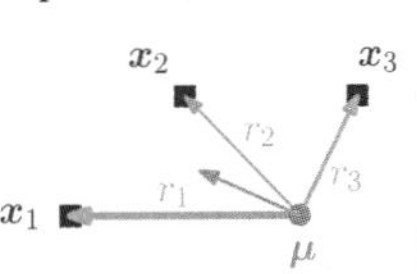

Figure 11.4 Update of the mean parameter of mixture component in a GMM. The mean $\boldsymbol{\mu}$ is being pulled toward individual data points with the weights given by the corresponding responsibilities.

Intuitively, (11.20) can be interpreted as an importance-weighted Monte Carlo estimate of the mean, where the importance weights of data point $\boldsymbol{x}_n$ are the responsibilities r_{nk} of the kth cluster for $\boldsymbol{x}_n$, $k = 1, \ldots, K$. Therefore, the mean $\boldsymbol{\mu}_k$ is pulled toward a data point $\boldsymbol{x}_n$ with strength given by r_{nk}. The means are pulled stronger toward data points for which the corresponding mixture component has a high responsibility, i.e., a high likelihood. Figure 11.4 illustrates this. We can also interpret the mean update in (11.20) as the expected value of all data points under the distribution given by

$$\boldsymbol{r}_k := [r_{1k}, \ldots, r_{Nk}]^\top / N_k, \tag{11.25}$$

which is a normalized probability vector, i.e.,

$$\boldsymbol{\mu}_k \leftarrow \mathbb{E}_{\boldsymbol{r}_k}[\mathcal{X}]. \tag{11.26}$$

Example 11.3 (Mean Updates)

In our example from Figure 11.3, the mean values are updated as follows:

$$\mu_1 : -4 \rightarrow -2.7 \tag{11.27}$$

$$\mu_2 : 0 \rightarrow -0.4 \tag{11.28}$$

$$\mu_3 : 8 \rightarrow 3.7 \tag{11.29}$$

Here we see that the means of the first and third mixture component move toward the regime of the data, whereas the mean of the second component does not change so dramatically. Figure 11.5 illustrates this change, where Figure 11.5(a) shows the GMM density prior to updating the means and Figure 11.5(b) shows the GMM density after updating the mean values μ_k.

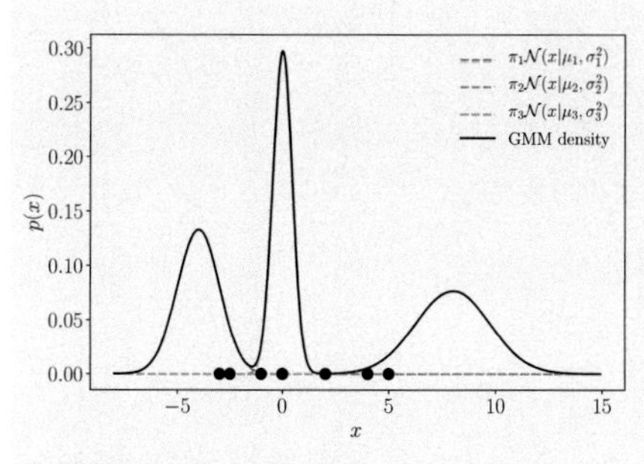

(a) GMM density and individual components prior to updating the mean values

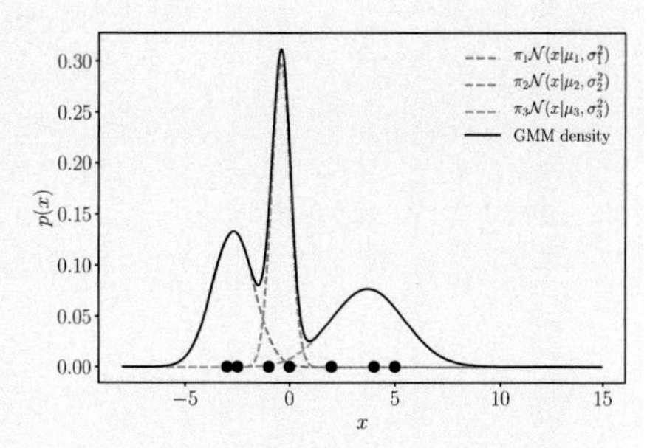

(b) GMM density and individual components after updating the mean values

Figure 11.5 Effect of updating the mean values in a GMM. (a) GMM before updating the mean values; (b) GMM after updating the mean values μ_k while retaining the variances and mixture weights.

The update of the mean parameters in (11.20) look fairly straightforward. However, note that the responsibilities r_{nk} are a function of $\pi_j, \boldsymbol{\mu}_j, \boldsymbol{\Sigma}_j$ for all $j = 1, \dots, K$, such that the updates in (11.20) depend on all parameters of the GMM, and a closed-form solution, which we obtained for linear regression in Section 9.2 or PCA in Chapter 10, cannot be obtained.

11.2.3 Updating the Covariances

Theorem 11.2 (Updates of the GMM Covariances)**.** *The update of the covariance parameters $\boldsymbol{\Sigma}_k$, $k = 1, \dots, K$ of the GMM is given by*

$$\boldsymbol{\Sigma}_k^{new} = \frac{1}{N_k} \sum_{n=1}^{N} r_{nk} (\boldsymbol{x}_n - \boldsymbol{\mu}_k)(\boldsymbol{x}_n - \boldsymbol{\mu}_k)^\top, \tag{11.30}$$

where r_{nk} and N_k are defined in (11.17) *and* (11.24), *respectively.*

Proof To prove Theorem 11.2, our approach is to compute the partial derivatives of the log-likelihood $\mathcal{L}$ with respect to the covariances $\boldsymbol{\Sigma}_k$, set them to $\mathbf{0}$, and solve for $\boldsymbol{\Sigma}_k$. We start with our general approach

$$\frac{\partial \mathcal{L}}{\partial \boldsymbol{\Sigma}_k} = \sum_{n=1}^{N} \frac{\partial \log p(\boldsymbol{x}_n \,|\, \boldsymbol{\theta})}{\partial \boldsymbol{\Sigma}_k} = \sum_{n=1}^{N} \frac{1}{p(\boldsymbol{x}_n \,|\, \boldsymbol{\theta})} \frac{\partial p(\boldsymbol{x}_n \,|\, \boldsymbol{\theta})}{\partial \boldsymbol{\Sigma}_k}. \tag{11.31}$$

We already know $1/p(\boldsymbol{x}_n \,|\, \boldsymbol{\theta})$ from (11.16). To obtain the remaining partial derivative $\partial p(\boldsymbol{x}_n \,|\, \boldsymbol{\theta})/\partial \boldsymbol{\Sigma}_k$, we write down the definition of the Gaussian distribution $p(\boldsymbol{x}_n \,|\, \boldsymbol{\theta})$ (see (11.9)) and drop all terms but the kth. We then obtain

$$\frac{\partial p(\boldsymbol{x}_n \,|\, \boldsymbol{\theta})}{\partial \boldsymbol{\Sigma}_k} \tag{11.32a}$$

$$= \frac{\partial}{\partial \boldsymbol{\Sigma}_k} \left(\pi_k (2\pi)^{-\frac{D}{2}} \det(\boldsymbol{\Sigma}_k)^{-\frac{1}{2}} \exp\left(-\tfrac{1}{2}(\boldsymbol{x}_n - \boldsymbol{\mu}_k)^\top \boldsymbol{\Sigma}_k^{-1} (\boldsymbol{x}_n - \boldsymbol{\mu}_k)\right) \right) \tag{11.32b}$$

$$= \pi_k (2\pi)^{-\frac{D}{2}} \left[\frac{\partial}{\partial \boldsymbol{\Sigma}_k} \det(\boldsymbol{\Sigma}_k)^{-\frac{1}{2}} \exp\left(-\tfrac{1}{2}(\boldsymbol{x}_n - \boldsymbol{\mu}_k)^\top \boldsymbol{\Sigma}_k^{-1}(\boldsymbol{x}_n - \boldsymbol{\mu}_k)\right) + \det(\boldsymbol{\Sigma}_k)^{-\frac{1}{2}} \frac{\partial}{\partial \boldsymbol{\Sigma}_k} \exp\left(-\tfrac{1}{2}(\boldsymbol{x}_n - \boldsymbol{\mu}_k)^\top \boldsymbol{\Sigma}_k^{-1}(\boldsymbol{x}_n - \boldsymbol{\mu}_k)\right)\right]. \tag{11.32c}$$

We now use the identities

$$\frac{\partial}{\partial \boldsymbol{\Sigma}_k} \det(\boldsymbol{\Sigma}_k)^{-\frac{1}{2}} \overset{(5.101)}{=} -\frac{1}{2} \det(\boldsymbol{\Sigma}_k)^{-\frac{1}{2}} \boldsymbol{\Sigma}_k^{-1}, \tag{11.33}$$

$$\frac{\partial}{\partial \boldsymbol{\Sigma}_k} (\boldsymbol{x}_n - \boldsymbol{\mu}_k)^\top \boldsymbol{\Sigma}_k^{-1}(\boldsymbol{x}_n - \boldsymbol{\mu}_k) \overset{(5.106)}{=} -\boldsymbol{\Sigma}_k^{-1}(\boldsymbol{x}_n - \boldsymbol{\mu}_k)(\boldsymbol{x}_n - \boldsymbol{\mu}_k)^\top \boldsymbol{\Sigma}_k^{-1} \tag{11.34}$$

and obtain (after some rearranging) the desired partial derivative required in (11.31) as

$$\frac{\partial p(\boldsymbol{x}_n \,|\, \boldsymbol{\theta})}{\partial \boldsymbol{\Sigma}_k} = \pi_k \mathcal{N}\big(\boldsymbol{x}_n \,|\, \boldsymbol{\mu}_k, \boldsymbol{\Sigma}_k\big) \cdot \left[-\tfrac{1}{2}\big(\boldsymbol{\Sigma}_k^{-1} - \boldsymbol{\Sigma}_k^{-1}(\boldsymbol{x}_n - \boldsymbol{\mu}_k)(\boldsymbol{x}_n - \boldsymbol{\mu}_k)^\top \boldsymbol{\Sigma}_k^{-1}\big)\right]. \tag{11.35}$$

Putting everything together, the partial derivative of the log-likelihood with respect to $\boldsymbol{\Sigma}_k$ is given by

$$\frac{\partial \mathcal{L}}{\partial \boldsymbol{\Sigma}_k} = \sum_{n=1}^{N} \frac{\partial \log p(\boldsymbol{x}_n \,|\, \boldsymbol{\theta})}{\partial \boldsymbol{\Sigma}_k} = \sum_{n=1}^{N} \frac{1}{p(\boldsymbol{x}_n \,|\, \boldsymbol{\theta})} \frac{\partial p(\boldsymbol{x}_n \,|\, \boldsymbol{\theta})}{\partial \boldsymbol{\Sigma}_k} \tag{11.36a}$$

$$= \sum_{n=1}^{N} \underbrace{\frac{\pi_k \mathcal{N}\big(\boldsymbol{x}_n \,|\, \boldsymbol{\mu}_k, \boldsymbol{\Sigma}_k\big)}{\sum_{j=1}^{K} \pi_j \mathcal{N}\big(\boldsymbol{x}_n \,|\, \boldsymbol{\mu}_j, \boldsymbol{\Sigma}_j\big)}}_{=r_{nk}} \cdot \left[-\tfrac{1}{2}\big(\boldsymbol{\Sigma}_k^{-1} - \boldsymbol{\Sigma}_k^{-1}(\boldsymbol{x}_n - \boldsymbol{\mu}_k)(\boldsymbol{x}_n - \boldsymbol{\mu}_k)^\top \boldsymbol{\Sigma}_k^{-1}\big)\right] \tag{11.36b}$$

$$= -\frac{1}{2} \sum_{n=1}^{N} r_{nk} \big(\boldsymbol{\Sigma}_k^{-1} - \boldsymbol{\Sigma}_k^{-1}(\boldsymbol{x}_n - \boldsymbol{\mu}_k)(\boldsymbol{x}_n - \boldsymbol{\mu}_k)^\top \boldsymbol{\Sigma}_k^{-1}\big) \tag{11.36c}$$

$$= -\frac{1}{2} \boldsymbol{\Sigma}_k^{-1} \underbrace{\sum_{n=1}^{N} r_{nk}}_{=N_k} + \frac{1}{2} \boldsymbol{\Sigma}_k^{-1} \left(\sum_{n=1}^{N} r_{nk} (\boldsymbol{x}_n - \boldsymbol{\mu}_k)(\boldsymbol{x}_n - \boldsymbol{\mu}_k)^\top \right) \boldsymbol{\Sigma}_k^{-1}. \tag{11.36d}$$

We see that the responsibilities r_{nk} also appear in this partial derivative. Setting this partial derivative to $\boldsymbol{0}$, we obtain the necessary optimality condition

$$N_k \boldsymbol{\Sigma}_k^{-1} = \boldsymbol{\Sigma}_k^{-1} \left(\sum_{n=1}^{N} r_{nk} (\boldsymbol{x}_n - \boldsymbol{\mu}_k)(\boldsymbol{x}_n - \boldsymbol{\mu}_k)^\top \right) \boldsymbol{\Sigma}_k^{-1} \tag{11.37a}$$

$$\iff N_k \boldsymbol{I} = \left(\sum_{n=1}^{N} r_{nk} (\boldsymbol{x}_n - \boldsymbol{\mu}_k)(\boldsymbol{x}_n - \boldsymbol{\mu}_k)^\top \right) \boldsymbol{\Sigma}_k^{-1}. \tag{11.37b}$$

By solving for $\boldsymbol{\Sigma}_k$, we obtain

$$\boldsymbol{\Sigma}_k^{\text{new}} = \frac{1}{N_k} \sum_{n=1}^{N} r_{nk} (\boldsymbol{x}_n - \boldsymbol{\mu}_k)(\boldsymbol{x}_n - \boldsymbol{\mu}_k)^\top, \tag{11.38}$$

where $\boldsymbol{r}_k$ is the probability vector defined in (11.25). This gives us a simple update rule for $\boldsymbol{\Sigma}_k$ for $k = 1, \ldots, K$ and proves Theorem 11.2. $\square$

Similar to the update of $\boldsymbol{\mu}_k$ in (11.20), we can interpret the update of the covariance in (11.30) as an importance-weighted expected value of the square of the centered data $\tilde{\mathcal{X}}_k := \{\boldsymbol{x}_1 - \boldsymbol{\mu}_k, \dots, \boldsymbol{x}_N - \boldsymbol{\mu}_k\}$.

Example 11.4 (Variance Updates)
In our example from Figure 11.3, the variances are updated as follows:

$$\sigma_1^2 : 1 \to 0.14 \tag{11.39}$$

$$\sigma_2^2 : 0.2 \to 0.44 \tag{11.40}$$

$$\sigma_3^2 : 3 \to 1.53 \tag{11.41}$$

Here we see that the variances of the first and third component shrink significantly, whereas the variance of the second component increases slightly.

Figure 11.6 illustrates this setting. Figure 11.6(a) is identical (but zoomed in) to Figure 11.5(b) and shows the GMM density and its individual components prior to updating the variances. Figure 11.6(b) shows the GMM density after updating the variances.

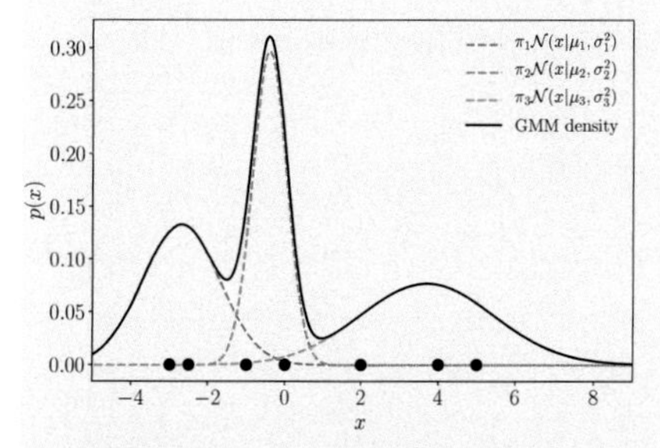

(a) GMM density and individual components prior to updating the variances

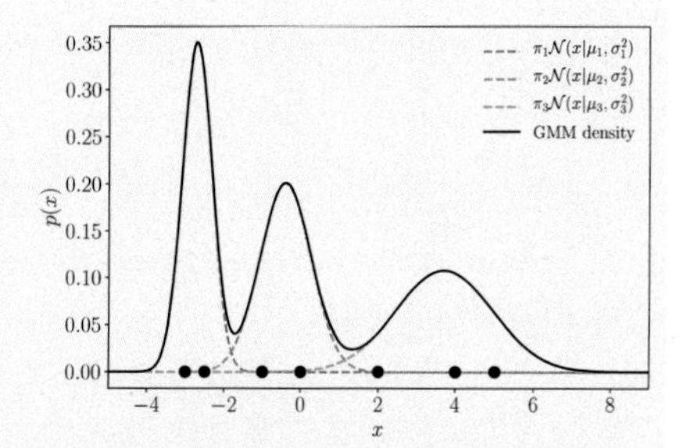

(b) GMM density and individual components after updating the variances

Figure 11.6 Effect of updating the variances in a GMM. (a) GMM before updating the variances; (b) GMM after updating the variances while retaining the means and mixture weights.

Similar to the update of the mean parameters, we can interpret (11.30) as a Monte Carlo estimate of the weighted covariance of data points $\boldsymbol{x}_n$ associated with the kth mixture component, where the weights are the responsibilities r_{nk}. As with the updates of the mean parameters, this update depends on all $\pi_j, \boldsymbol{\mu}_j, \boldsymbol{\Sigma}_j,\ j = 1, \dots, K$, through the responsibilities r_{nk}, which prohibits a closed-form solution.

11.2.4 Updating the Mixture Weights

Theorem 11.3 (Update of the GMM Mixture Weights)**.** *The mixture weights of the GMM are updated as*

$$\pi_k^{new} = \frac{N_k}{N}, \quad k = 1, \dots, K, \tag{11.42}$$

where N is the number of data points and N_k is defined in (11.24).

Proof To find the partial derivative of the log-likelihood with respect to the weight parameters π_k, $k = 1, \ldots, K$, we account for the constraint $\sum_k \pi_k = 1$ by using Lagrange multipliers (see Section 7.2). The Lagrangian is

$$\mathfrak{L} = \mathcal{L} + \lambda \left(\sum_{k=1}^{K} \pi_k - 1 \right) \tag{11.43a}$$

$$= \sum_{n=1}^{N} \log \sum_{k=1}^{K} \pi_k \mathcal{N}\big(\boldsymbol{x}_n \,|\, \boldsymbol{\mu}_k, \boldsymbol{\Sigma}_k\big) + \lambda \left(\sum_{k=1}^{K} \pi_k - 1 \right), \tag{11.43b}$$

where $\mathcal{L}$ is the log-likelihood from (11.10) and the second term encodes for the equality constraint that all the mixture weights need to sum up to 1. We obtain the partial derivative with respect to π_k as

$$\frac{\partial \mathfrak{L}}{\partial \pi_k} = \sum_{n=1}^{N} \frac{\mathcal{N}\big(\boldsymbol{x}_n \,|\, \boldsymbol{\mu}_k, \boldsymbol{\Sigma}_k\big)}{\sum_{j=1}^{K} \pi_j \mathcal{N}\big(\boldsymbol{x}_n \,|\, \boldsymbol{\mu}_j, \boldsymbol{\Sigma}_j\big)} + \lambda \tag{11.44a}$$

$$= \frac{1}{\pi_k} \underbrace{\sum_{n=1}^{N} \frac{\pi_k \mathcal{N}\big(\boldsymbol{x}_n \,|\, \boldsymbol{\mu}_k, \boldsymbol{\Sigma}_k\big)}{\sum_{j=1}^{K} \pi_j \mathcal{N}\big(\boldsymbol{x}_n \,|\, \boldsymbol{\mu}_j, \boldsymbol{\Sigma}_j\big)}}_{=N_k} + \lambda = \frac{N_k}{\pi_k} + \lambda, \tag{11.44b}$$

and the partial derivative with respect to the Lagrange multiplier λ as

$$\frac{\partial \mathfrak{L}}{\partial \lambda} = \sum_{k=1}^{K} \pi_k - 1. \tag{11.45}$$

Setting both partial derivatives to $\mathbf{0}$ (necessary condition for optimum) yields the system of equations

$$\pi_k = -\frac{N_k}{\lambda}, \tag{11.46}$$

$$1 = \sum_{k=1}^{K} \pi_k. \tag{11.47}$$

Using (11.46) in (11.47) and solving for π_k, we obtain

$$\sum_{k=1}^{K} \pi_k = 1 \iff -\sum_{k=1}^{K} \frac{N_k}{\lambda} = 1 \iff -\frac{N}{\lambda} = 1 \iff \lambda = -N. \tag{11.48}$$

This allows us to substitute $-N$ for λ in (11.46) to obtain

$$\pi_k^{\text{new}} = \frac{N_k}{N}, \tag{11.49}$$

which gives us the update for the weight parameters π_k and proves Theorem 11.3. □

We can identify the mixture weight in (11.42) as the ratio of the total responsibility of the kth cluster and the number of data points. Since $N = \sum_k N_k$, the number of data points can also be interpreted as the total responsibility of all mixture components together, such that π_k is the relative importance of the kth mixture component for the dataset.

Remark. Since $N_k = \sum_{i=1}^{N} r_{nk}$, the update equation (11.42) for the mixture weights π_k also depends on all $\pi_j, \boldsymbol{\mu}_j, \boldsymbol{\Sigma}_j, j = 1, \ldots, K$ via the responsibilities r_{nk}. ◇

Example 11.5 (Weight Parameter Updates)
In our running example from Figure 11.3, the mixture weights are updated as follows:

$$\pi_1 : \tfrac{1}{3} \to 0.29 \tag{11.50}$$

$$\pi_2 : \tfrac{1}{3} \to 0.29 \tag{11.51}$$

$$\pi_3 : \tfrac{1}{3} \to 0.42 \tag{11.52}$$

Here we see that the third component gets more weight/importance, while the other components become slightly less important. Figure 11.7 illustrates the effect of updating the mixture weights. Figure 11.7(a) is identical to Figure 11.6(b) and shows the GMM density and its individual components prior to updating the mixture weights. Figure 11.7(b) shows the GMM density after updating the mixture weights.

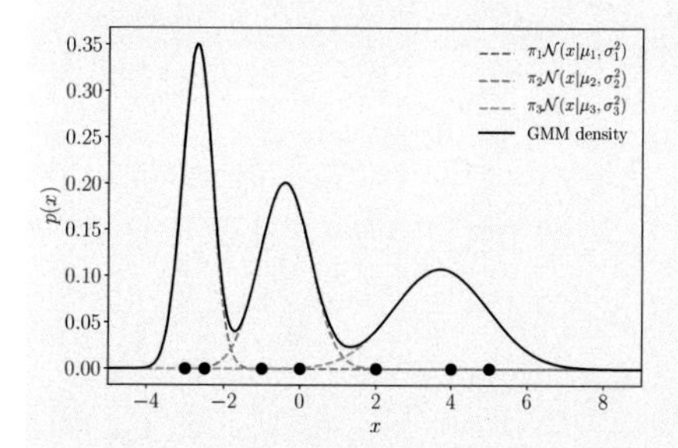

(a) GMM density and individual components prior to updating the mixture weights

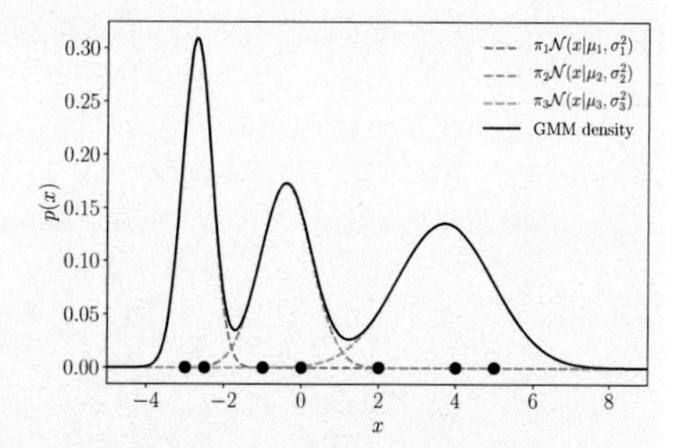

(b) GMM density and individual components after updating the mixture weights

Figure 11.7 Effect of updating the mixture weights in a GMM. (a) GMM before updating the mixture weights; (b) GMM after updating the mixture weights while retaining the means and variances. Note the different scales of the vertical axes.

Overall, having updated the means, the variances, and the weights once, we obtain the GMM shown in Figure 11.7(b). Compared with the initialization shown in Figure 11.3, we can see that the parameter updates caused the GMM density to shift some of its mass toward the data points.

After updating the means, variances, and weights once, the GMM fit in Figure 11.7(b) is already remarkably better than its initialization from Figure 11.3. This is also evidenced by the log-likelihood values, which increased from 28.3 (initialization) to 14.4 after one complete update cycle.

11.3 EM Algorithm

Unfortunately, the updates in (11.20), (11.30), and (11.42) do not constitute a closed-form solution for the updates of the parameters $\boldsymbol{\mu}_k, \boldsymbol{\Sigma}_k, \pi_k$ of the mixture model because the responsibilities r_{nk} depend on those parameters in a complex way. However, the results suggest a simple *iterative scheme* for finding a solution to the parameters estimation problem via maximum likelihood. The expectation maximization algorithm (*EM algorithm*) was proposed by Dempster et al. (1977) and is a general iterative scheme for learning parameters (maximum likelihood or MAP) in mixture models and, more generally, latent-variable models.

EM algorithm

In our example of the Gaussian mixture model, we choose initial values for $\boldsymbol{\mu}_k, \boldsymbol{\Sigma}_k, \pi_k$ and alternate until convergence between

- *E-step:* Evaluate the responsibilities r_{nk} (posterior probability of data point n belonging to mixture component k).
- *M-step:* Use the updated responsibilities to reestimate the parameters $\boldsymbol{\mu}_k, \boldsymbol{\Sigma}_k, \pi_k$.

Every step in the EM algorithm increases the log-likelihood function (Neal and Hinton, 1999). For convergence, we can check the log-likelihood or the parameters directly. A concrete instantiation of the EM algorithm for estimating the parameters of a GMM is as follows:

1. Initialize $\boldsymbol{\mu}_k, \boldsymbol{\Sigma}_k, \pi_k$.
2. *E-step:* Evaluate responsibilities r_{nk} for every data point $\boldsymbol{x}_n$ using current parameters $\pi_k, \boldsymbol{\mu}_k, \boldsymbol{\Sigma}_k$:

$$r_{nk} = \frac{\pi_k \mathcal{N}\big(\boldsymbol{x}_n \,|\, \boldsymbol{\mu}_k, \boldsymbol{\Sigma}_k\big)}{\sum_j \pi_j \mathcal{N}\big(\boldsymbol{x}_n \,|\, \boldsymbol{\mu}_j, \boldsymbol{\Sigma}_j\big)}. \tag{11.53}$$

3. *M-step:* Reestimate parameters $\pi_k, \boldsymbol{\mu}_k, \boldsymbol{\Sigma}_k$ using the current responsibilities r_{nk} (from E-step):

$$\boldsymbol{\mu}_k = \frac{1}{N_k} \sum_{n=1}^{N} r_{nk} \boldsymbol{x}_n, \tag{11.54}$$

$$\boldsymbol{\Sigma}_k = \frac{1}{N_k} \sum_{n=1}^{N} r_{nk} (\boldsymbol{x}_n - \boldsymbol{\mu}_k)(\boldsymbol{x}_n - \boldsymbol{\mu}_k)^\top, \tag{11.55}$$

$$\pi_k = \frac{N_k}{N}. \tag{11.56}$$

Example 11.6 (GMM Fit)

When we run EM on our example from Figure 11.3, we obtain the final result shown in Figure 11.8(a) after five iterations, and Figure 11.8(b) shows how the negative log-likelihood evolves as a function of the EM iterations. The final GMM is given as

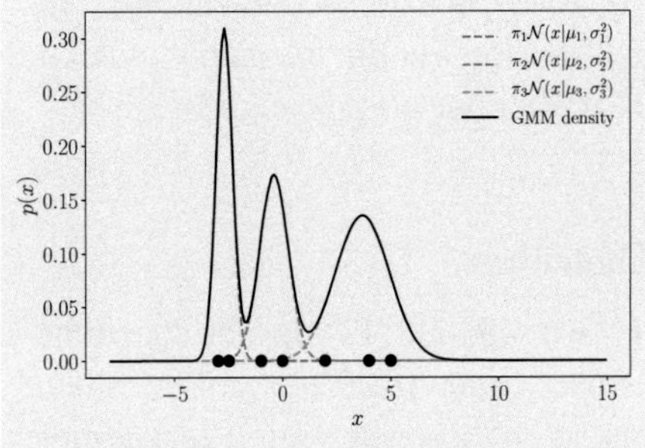

(a) Final GMM fit. After five iterations, the EM algorithm converges and returns this GMM

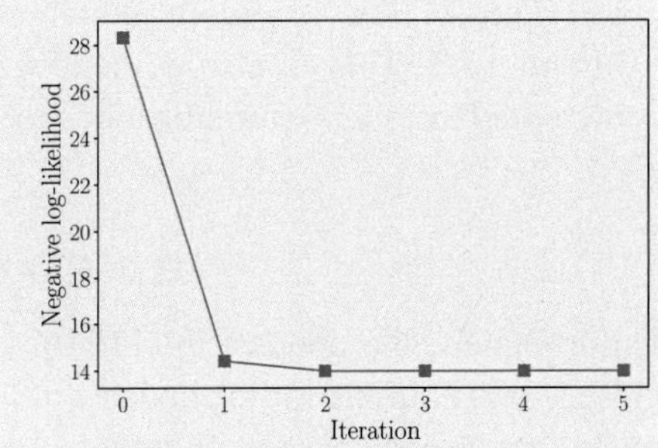

(b) Negative log-likelihood as a function of the EM iterations

Figure 11.8 EM algorithm applied to the GMM from Figure 11.2. (a) Final GMM fit; (b) negative log-likelihood as a function of the EM Iternations.

$$p(x) = 0.29\mathcal{N}(x \mid -2.75,\, 0.06) + 0.28\mathcal{N}(x \mid -0.50,\, 0.25) + 0.43\mathcal{N}(x \mid 3.64,\, 1.63). \tag{11.57}$$

We applied the EM algorithm to the two-dimensional dataset shown in Figure 11.1 with $K = 3$ mixture components. Figure 11.9 illustrates some steps of the EM algorithm and shows the negative log-likelihood as a function of the EM iteration (Figure 11.9(b)). Figure 11.10(a) shows the corresponding final GMM fit. Figure 11.10(b) visualizes the final responsibilities of the mixture components for the data points. The dataset is colored according to the responsibilities of the mixture components when EM converges. While a single mixture component is clearly responsible for the data on the left, the overlap of the two data clusters on the right could have been generated by two mixture components. It becomes clear that there are data points that cannot be uniquely assigned to a single component (either blue or yellow), such that the responsibilities of these two clusters for those points are around 0.5.

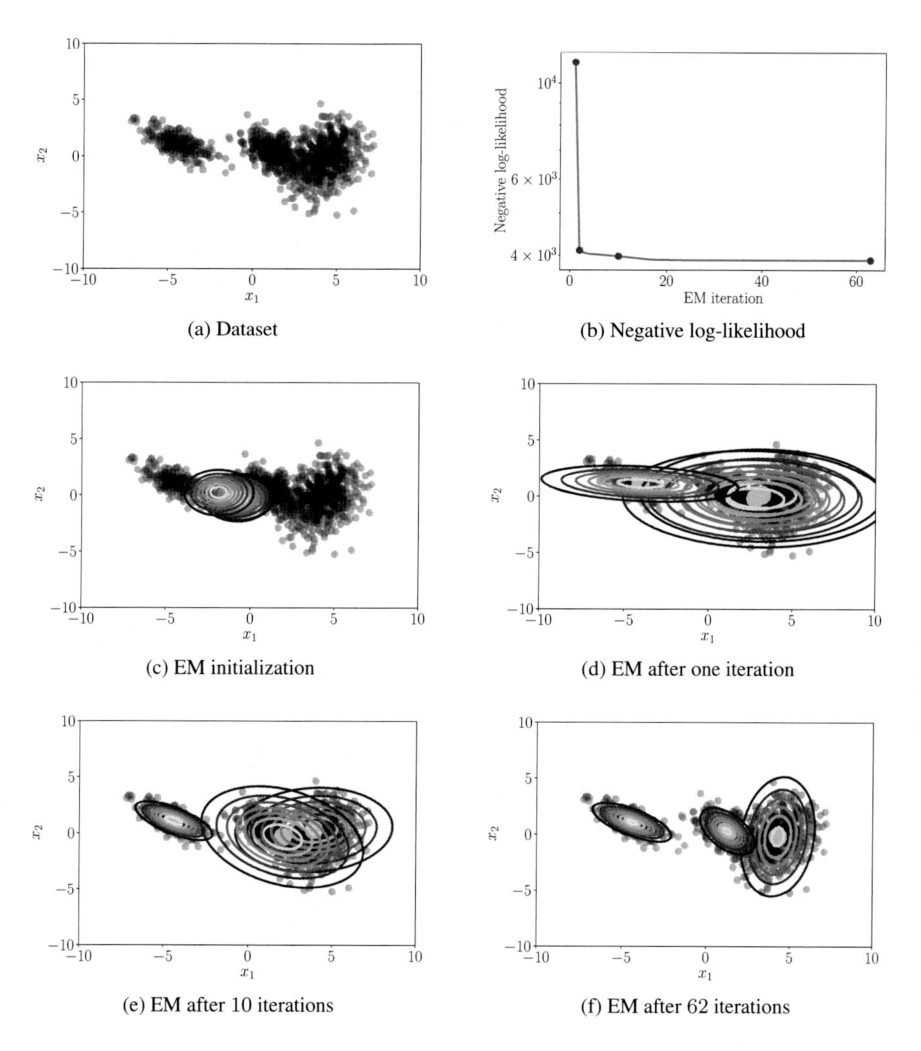

Figure 11.9 Illustration of the EM algorithm for fitting a Gaussian mixture model with three components to a two-dimensional dataset. (a) Dataset; (b) negative log-likelihood (lower is better) as a function of the EM iterations. The red dots indicate the iterations for which the mixture components of the corresponding GMM fits are shown in (c) through (f). The yellow discs indicate the means of the Gaussian mixture components. Figure 11.11(a) shows the final GMM fit.

Figure 11.10 GMM fit and responsibilities when EM converges. (a) GMM fit when EM converges; (b) each data point is colored according to the responsibilities of the mixture components.

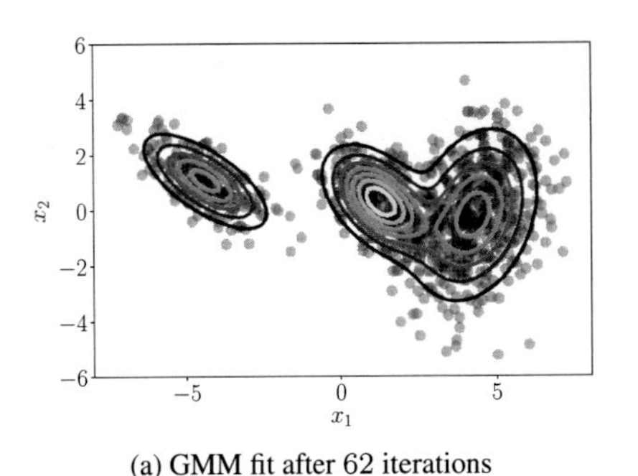

(a) GMM fit after 62 iterations

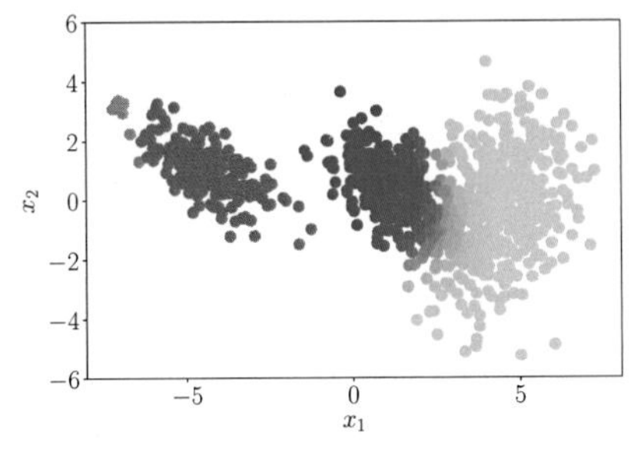

(b) Dataset colored according to the responsibilities of the mixture components

11.4 Latent-Variable Perspective

We can look at the GMM from the perspective of a discrete latent-variable model, i.e., where the latent variable $\boldsymbol{z}$ can attain only a finite set of values. This is in contrast to PCA, where the latent variables were continuous-valued numbers in $\mathbb{R}^M$.

The advantages of the probabilistic perspective are that (i) it will justify some ad hoc decisions we made in the previous sections, (ii) it allows for a concrete interpretation of the responsibilities as posterior probabilities, and (iii) the iterative algorithm for updating the model parameters can be derived in a principled manner as the EM algorithm for maximum likelihood parameter estimation in latent-variable models.

11.4.1 Generative Process and Probabilistic Model

To derive the probabilistic model for GMMs, it is useful to think about the generative process, i.e., the process that allows us to generate data, using a probabilistic model.

We assume a mixture model with K components and that a data point $\boldsymbol{x}$ can be generated by exactly one mixture component. We introduce a binary indicator variable $z_k \in \{0, 1\}$ with two states (see Section 6.2) that indicates whether the kth mixture component generated that data point so that

$$p(\boldsymbol{x} \mid z_k = 1) = \mathcal{N}\big(\boldsymbol{x} \mid \boldsymbol{\mu}_k,\, \boldsymbol{\Sigma}_k\big). \tag{11.58}$$

We define $\boldsymbol{z} := [z_1, \ldots, z_K]^\top \in \mathbb{R}^K$ as a probability vector consisting of $K - 1$ many 0s and exactly one 1. For example, for $K = 3$, a valid $\boldsymbol{z}$ would be $\boldsymbol{z} = [z_1, z_2, z_3]^\top = [0, 1, 0]^\top$, which would select the second mixture component since $z_2 = 1$.

Remark. Sometimes this kind of probability distribution is called "multinoulli," a generalization of the Bernoulli distribution to more than two values (Murphy, 2012). ◇

one-hot encoding
1-of-K representation

The properties of $\boldsymbol{z}$ imply that $\sum_{k=1}^{K} z_k = 1$. Therefore, $\boldsymbol{z}$ is a *one-hot encoding* (also: *1-of-K representation*).

Thus far, we assumed that the indicator variables z_k are known. However, in practice, this is not the case, and we place a prior distribution

$$p(\boldsymbol{z}) = \boldsymbol{\pi} = [\pi_1, \ldots, \pi_K]^\top, \quad \sum_{k=1}^{K} \pi_k = 1, \tag{11.59}$$

on the latent variable $\boldsymbol{z}$. Then the kth entry

$$\pi_k = p(z_k = 1) \tag{11.60}$$

of this probability vector describes the probability that the kth mixture component generated data point $\boldsymbol{x}$.

Remark (Sampling from a GMM). The construction of this latent-variable model (see the corresponding graphical model in Figure 11.11) lends itself to a very simple sampling procedure (generative process) to generate data:

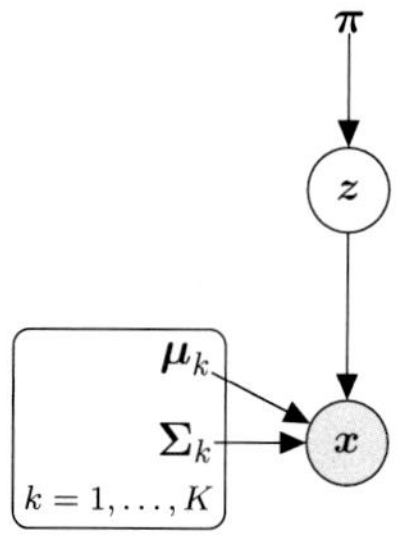

1. Sample $z^{(i)} \sim p(\boldsymbol{z})$.
2. Sample $\boldsymbol{x}^{(i)} \sim p(\boldsymbol{x} \mid z^{(i)} = 1)$.

Figure 11.11 Graphical model for a GMM with a single data point.

$\boldsymbol{\pi}$ $\boldsymbol{z}$ $\boldsymbol{\mu}_k$ $\boldsymbol{\Sigma}_k$ $\boldsymbol{x}$ $k = 1, \ldots, K$

In the first step, we select a mixture component i (via the one-hot encoding $\boldsymbol{z}$) at random according to $p(\boldsymbol{z}) = \boldsymbol{\pi}$; in the second step we draw a sample from the corresponding mixture component. When we discard the samples of the latent variable so that we are left with the $\boldsymbol{x}^{(i)}$, we have valid samples from the GMM. This kind of sampling, where samples of random variables depend on samples from the variable's parents in the graphical model, is called *ancestral sampling*. ◇

ancestral sampling

Generally, a probabilistic model is defined by the joint distribution of the data and the latent variables (see Section 8.4). With the prior $p(\boldsymbol{z})$ defined in (11.59) and (11.60) and the conditional $p(\boldsymbol{x} \mid \boldsymbol{z})$ from (11.58), we obtain all K components of this joint distribution via

$$p(\boldsymbol{x}, z_k = 1) = p(\boldsymbol{x} \mid z_k = 1)p(z_k = 1) = \pi_k \mathcal{N}(\boldsymbol{x} \mid \boldsymbol{\mu}_k, \boldsymbol{\Sigma}_k) \tag{11.61}$$

for $k = 1, \ldots, K$, so that

$$p(\boldsymbol{x}, \boldsymbol{z}) = \begin{bmatrix} p(\boldsymbol{x}, z_1 = 1) \\ \vdots \\ p(\boldsymbol{x}, z_K = 1) \end{bmatrix} = \begin{bmatrix} \pi_1 \mathcal{N}(\boldsymbol{x} \mid \boldsymbol{\mu}_1, \boldsymbol{\Sigma}_1) \\ \vdots \\ \pi_K \mathcal{N}(\boldsymbol{x} \mid \boldsymbol{\mu}_K, \boldsymbol{\Sigma}_K) \end{bmatrix}, \tag{11.62}$$

which fully specifies the probabilistic model.

11.4.2 Likelihood

To obtain the likelihood $p(\boldsymbol{x} \mid \boldsymbol{\theta})$ in a latent-variable model, we need to marginalize out the latent variables (see Section 8.4.3). In our case, this can be done by summing out all latent variables from the joint $p(\boldsymbol{x}, \boldsymbol{z})$ in (11.62) so that

$$p(\boldsymbol{x} \mid \boldsymbol{\theta}) = \sum_{\boldsymbol{z}} p(\boldsymbol{x} \mid \boldsymbol{\theta}, \boldsymbol{z}) p(\boldsymbol{z} \mid \boldsymbol{\theta}), \quad \boldsymbol{\theta} := \{\boldsymbol{\mu}_k, \boldsymbol{\Sigma}_k, \pi_k : k = 1, \ldots, K\}. \tag{11.63}$$

We now explicitly condition on the parameters $\boldsymbol{\theta}$ of the probabilistic model, which we previously omitted. In (11.63), we sum over all K possible one-hot encodings of $\boldsymbol{z}$, which is denoted by $\sum_{\boldsymbol{z}}$. Since there is only a single nonzero single entry in each $\boldsymbol{z}$ there are only K possible configurations/settings of $\boldsymbol{z}$. For example, if $K = 3$, then $\boldsymbol{z}$ can have the configurations

$$\begin{bmatrix}1\\0\\0\end{bmatrix}, \begin{bmatrix}0\\1\\0\end{bmatrix}, \begin{bmatrix}0\\0\\1\end{bmatrix}. \tag{11.64}$$

Summing over all possible configurations of $\boldsymbol{z}$ in (11.63) is equivalent to looking at the nonzero entry of the $\boldsymbol{z}$-vector and writing

$$p(\boldsymbol{x} \mid \boldsymbol{\theta}) = \sum_{\boldsymbol{z}} p(\boldsymbol{x} \mid \boldsymbol{\theta}, \boldsymbol{z}) p(\boldsymbol{z} \mid \boldsymbol{\theta}) \tag{11.65a}$$

$$= \sum_{k=1}^{K} p(\boldsymbol{x} \mid \boldsymbol{\theta}, z_k = 1) p(z_k = 1 \mid \boldsymbol{\theta}) \tag{11.65b}$$

so that the desired marginal distribution is given as

$$p(\boldsymbol{x} \mid \boldsymbol{\theta}) \overset{(11.65\text{b})}{=} \sum_{k=1}^{K} p(\boldsymbol{x} \mid \boldsymbol{\theta}, z_k = 1) p(z_k = 1 | \boldsymbol{\theta}) \tag{11.66a}$$

$$= \sum_{k=1}^{K} \pi_k \mathcal{N}\big(\boldsymbol{x} \mid \boldsymbol{\mu}_k, \boldsymbol{\Sigma}_k\big), \tag{11.66b}$$

which we identify as the GMM model from (11.3). Given a dataset $\mathcal{X}$, we immediately obtain the likelihood

$$p(\mathcal{X} \mid \boldsymbol{\theta}) = \prod_{n=1}^{N} p(\boldsymbol{x}_n \mid \boldsymbol{\theta}) \overset{(11.66\text{b})}{=} \prod_{n=1}^{N} \sum_{k=1}^{K} \pi_k \mathcal{N}\big(\boldsymbol{x}_n \mid \boldsymbol{\mu}_k, \boldsymbol{\Sigma}_k\big), \tag{11.67}$$

which is exactly the GMM likelihood from (11.9). Therefore, the latent-variable model with latent indicators z_k is an equivalent way of thinking about a Gaussian mixture model.

11.4.3 Posterior Distribution

Let us have a brief look at the posterior distribution on the latent variable $\boldsymbol{z}$. According to Bayes' theorem, the posterior of the kth component having generated data point $\boldsymbol{x}$

$$p(z_k = 1 \mid \boldsymbol{x}) = \frac{p(z_k = 1) p(\boldsymbol{x} \mid z_k = 1)}{p(\boldsymbol{x})}, \tag{11.68}$$

where the marginal $p(\boldsymbol{x})$ is given in (11.66b). This yields the posterior distribution for the kth indicator variable z_k

$$p(z_k = 1 \mid \boldsymbol{x}) = \frac{p(z_k = 1) p(\boldsymbol{x} \mid z_k = 1)}{\sum_{j=1}^{K} p(z_j = 1) p(\boldsymbol{x} \mid z_j = 1)} = \frac{\pi_k \mathcal{N}\big(\boldsymbol{x} \mid \boldsymbol{\mu}_k, \boldsymbol{\Sigma}_k\big)}{\sum_{j=1}^{K} \pi_j \mathcal{N}\big(\boldsymbol{x} \mid \boldsymbol{\mu}_j, \boldsymbol{\Sigma}_j\big)}, \tag{11.69}$$

which we identify as the responsibility of the kth mixture component for data point $\boldsymbol{x}$. Note that we omitted the explicit conditioning on the GMM parameters $\pi_k, \boldsymbol{\mu}_k, \boldsymbol{\Sigma}_k$ where $k = 1, \ldots, K$.

11.4.4 Extension to a Full Dataset

Thus far, we have only discussed the case where the dataset consists only of a single data point $\boldsymbol{x}$. However, the concepts of the prior and posterior can be directly extended to the case of N data points $\mathcal{X} := \{\boldsymbol{x}_1, \ldots, \boldsymbol{x}_N\}$.

In the probabilistic interpretation of the GMM, every data point $\boldsymbol{x}_n$ possesses its own latent variable

$$\boldsymbol{z}_n = \left[z_{n1}, \ldots, z_{nK}\right]^\top \in \mathbb{R}^K. \tag{11.70}$$

Previously (when we only considered a single data point $\boldsymbol{x}$), we omitted the index n, but now this becomes important.

We share the same prior distribution $\boldsymbol{\pi}$ across all latent variables $\boldsymbol{z}_n$. The corresponding graphical model is shown in Figure 11.12, where we use the plate notation.

The conditional distribution $p(\boldsymbol{x}_1, \ldots, \boldsymbol{x}_N \mid \boldsymbol{z}_1, \ldots, \boldsymbol{z}_N)$ factorizes over the data points and is given as

$$p(\boldsymbol{x}_1, \ldots, \boldsymbol{x}_N \mid \boldsymbol{z}_1, \ldots, \boldsymbol{z}_N) = \prod_{n=1}^{N} p(\boldsymbol{x}_n \mid \boldsymbol{z}_n). \tag{11.71}$$

To obtain the posterior distribution $p(z_{nk} = 1 \mid \boldsymbol{x}_n)$, we follow the same reasoning as in Section 11.4.3 and apply Bayes' theorem to obtain

$$p(z_{nk} = 1 \mid \boldsymbol{x}_n) = \frac{p(\boldsymbol{x}_n \mid z_{nk} = 1)p(z_{nk} = 1)}{\sum_{j=1}^{K} p(\boldsymbol{x}_n \mid z_{nj} = 1)p(z_{nj} = 1)} \tag{11.72a}$$

$$= \frac{\pi_k \mathcal{N}\big(\boldsymbol{x}_n \mid \boldsymbol{\mu}_k, \boldsymbol{\Sigma}_k\big)}{\sum_{j=1}^{K} \pi_j \mathcal{N}\big(\boldsymbol{x}_n \mid \boldsymbol{\mu}_j, \boldsymbol{\Sigma}_j\big)} = r_{nk}. \tag{11.72b}$$

This means that $p(z_k = 1 \mid \boldsymbol{x}_n)$ is the (posterior) probability that the kth mixture component generated data point $\boldsymbol{x}_n$ and corresponds to the responsibility r_{nk} we introduced in (11.17). Now the responsibilities also have not only an intuitive but also a mathematically justified interpretation as posterior probabilities.

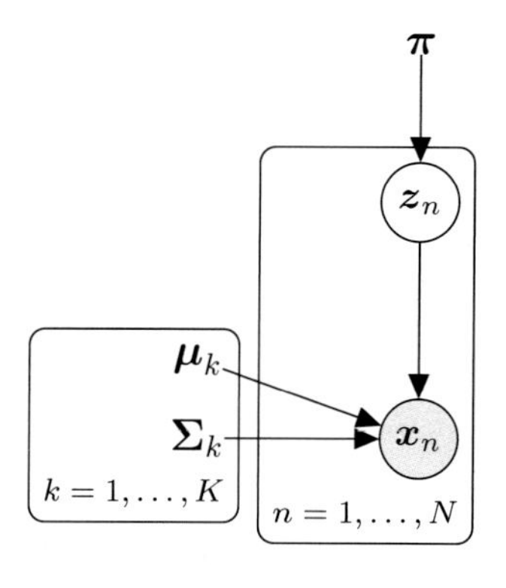

Figure 11.12 Graphical model for a GMM with N data points.

11.4.5 EM Algorithm Revisited

The EM algorithm that we introduced as an iterative scheme for maximum likelihood estimation can be derived in a principled way from the latent-variable perspective. Given a current setting $\boldsymbol{\theta}^{(t)}$ of model parameters, the E-step calculates the expected log-likelihood

$$Q(\boldsymbol{\theta} \mid \boldsymbol{\theta}^{(t)}) = \mathbb{E}_{\boldsymbol{z} \mid \boldsymbol{x}, \boldsymbol{\theta}^{(t)}}[\log p(\boldsymbol{x}, \boldsymbol{z} \mid \boldsymbol{\theta})] \tag{11.73a}$$

$$= \int \log p(\boldsymbol{x}, \boldsymbol{z} \mid \boldsymbol{\theta}) p(\boldsymbol{z} \mid \boldsymbol{x}, \boldsymbol{\theta}^{(t)}) \mathrm{d}\boldsymbol{z}, \tag{11.73b}$$

where the expectation of $\log p(\boldsymbol{x}, \boldsymbol{z} \mid \boldsymbol{\theta})$ is taken with respect to the posterior $p(\boldsymbol{z} \mid \boldsymbol{x}, \boldsymbol{\theta}^{(t)})$ of the latent variables. The M-step selects an updated set of model parameters $\boldsymbol{\theta}^{(t+1)}$ by maximizing (11.73b).

Although an EM iteration does increase the log-likelihood, there are no guarantees that EM converges to the maximum likelihood solution. It is possible that the EM algorithm converges to a local maximum of the log-likelihood. Different initializations of the parameters $\boldsymbol{\theta}$ could be used in multiple EM runs to reduce the risk of ending up in a bad local optimum. We do not go into further details here, but refer to the excellent expositions by Rogers and Girolami (2016) and Bishop (2006).

11.5 Further Reading

The GMM can be considered a generative model in the sense that it is straightforward to generate new data using ancestral sampling (Bishop, 2006). For given GMM parameters $\pi_k, \boldsymbol{\mu}_k, \boldsymbol{\Sigma}_k$, $k = 1, \dots, K$, we sample an index k from the probability vector $[\pi_1, \dots, \pi_K]^\top$ and then sample a data point $\boldsymbol{x} \sim \mathcal{N}(\boldsymbol{\mu}_k, \boldsymbol{\Sigma}_k)$. If we repeat this N times, we obtain a dataset that has been generated by a GMM. Figure 11.1 was generated using this procedure.

Throughout this chapter, we assumed that the number of components K is known. In practice, this is often not the case. However, we could use nested cross-validation, as discussed in Section 8.6.1, to find good models.

Gaussian mixture models are closely related to the K-means clustering algorithm. K-means also uses the EM algorithm to assign data points to clusters. If we treat the means in the GMM as cluster centers and ignore the covariances (or set them to $\boldsymbol{I}$), we arrive at K-means. As also nicely described by MacKay (2003), K-means makes a "hard" assignment of data points to cluster centers $\boldsymbol{\mu}_k$, whereas a GMM makes a "soft" assignment via the responsibilities.

We only touched upon the latent-variable perspective of GMMs and the EM algorithm. Note that EM can be used for parameter learning in general latent-variable models, e.g., nonlinear state-space models (Ghahramani and Roweis, 1999; Roweis and Ghahramani, 1999) and for reinforcement learning as discussed by Barber (2012). Therefore, the latent variable perspective of a GMM is useful to derive the corresponding EM algorithm in a principled way (Bishop, 2006; Barber, 2012; Murphy, 2012).

We only discussed maximum likelihood estimation (via the EM algorithm) for finding GMM parameters. The standard criticisms of maximum likelihood also apply here:

- As in linear regression, maximum likelihood can suffer from severe overfitting. In the GMM case, this happens when the mean of a mixture component is identical to a data point and the covariance tends to $\mathbf{0}$. Then, the likelihood approaches infinity. Bishop (2006) and Barber (2012) discuss this issue in detail.
- We only obtain a point estimate of the parameters $\pi_k, \boldsymbol{\mu}_k, \boldsymbol{\Sigma}_k$ for $k = 1, \ldots, K$, which does not give any indication of uncertainty in the parameter values. A Bayesian approach would place a prior on the parameters, which can be used to obtain a posterior distribution on the parameters. This posterior allows us to compute the model evidence (marginal likelihood), which can be used for model comparison, which gives us a principled way to determine the number of mixture components. Unfortunately, closed-form inference is not possible in this setting because there is no conjugate prior for this model. However, approximations, such as variational inference, can be used to obtain an approximate posterior (Bishop, 2006).

In this chapter, we discussed mixture models for density estimation. There is a plethora of density estimation techniques available. In practice, we often use histograms and kernel density estimation.

histogram

Histograms provide a nonparametric way to represent continuous densities and have been proposed by Pearson (1895). A histogram is constructed by "binning" the data space and counting how many data points fall into each bin. Then a bar is drawn at the center of each bin, and the height of the bar is proportional to the number of data points within that bin. The bin size is a critical hyperparameter, and a bad choice can lead to overfitting and underfitting. Cross-validation, as discussed in Section 8.2.4, can be used to determine a good bin size.

kernel density estimation

Kernel density estimation, independently proposed by Rosenblatt (1956) and Parzen (1962), is a nonparametric way for density estimation. Given N i.i.d. samples, the kernel density estimator represents the underlying distribution as

$$p(\boldsymbol{x}) = \frac{1}{Nh} \sum_{n=1}^{N} k\left(\frac{\boldsymbol{x} - \boldsymbol{x}_n}{h}\right), \tag{11.74}$$

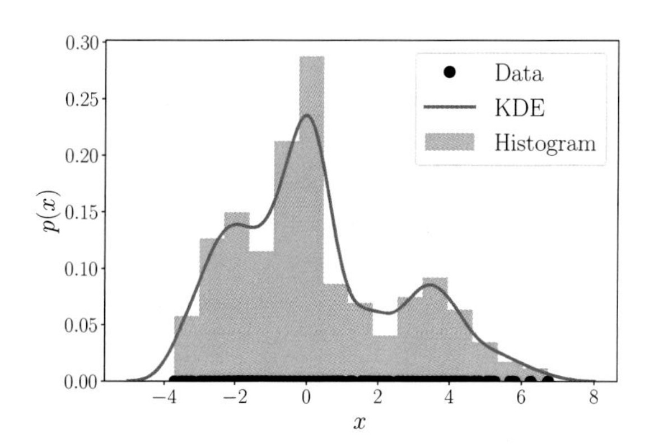

Figure 11.13 Histogram (orange bars) and kernel density estimation (blue line). The kernel density estimator produces a smooth estimate of the underlying density, whereas the histogram is an unsmoothed count measure of how many data points (black) fall into a single bin.

where k is a kernel function, i.e., a nonnegative function that integrates to 1 and $h > 0$ is a smoothing/bandwidth parameter, which plays a similar role as the bin size in histograms. Note that we place a kernel on every single data point $\boldsymbol{x}_n$ in the dataset. Commonly used kernel functions are the uniform distribution and the Gaussian distribution. Kernel density estimates are closely related to histograms, but by choosing a suitable kernel, we can guarantee smoothness of the density estimate. Figure 11.13 illustrates the difference between a histogram and a kernel density estimator (with a Gaussian-shaped kernel) for a given dataset of 250 data points.

12

Classification with Support Vector Machines

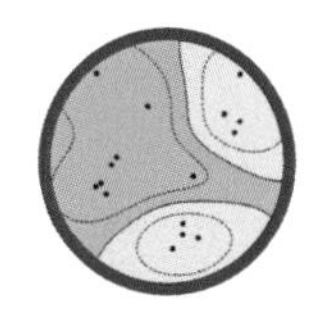

In many situations, we want our machine learning algorithm to predict one of a number of (discrete) outcomes. For example, an email client sorts mail into personal mail and junk mail, which has two outcomes. Another example is a telescope that identifies whether an object in the night sky is a galaxy, star, or planet. There are usually a small number of outcomes, and more importantly there is usually no additional structure on these outcomes. In this chapter, we consider predictors that output binary values, i.e., there are only two possible outcomes. This machine learning task is called *binary classification*. This is in contrast to Chapter 9, where we considered a prediction problem with continuous-valued outputs.

An example of structure is if the outcomes were ordered, like in the case of small, medium, and large t-shirts.

binary classification

For binary classification, the set of possible values that the label/output can attain is binary, and for this chapter we denote them by $\{+1, -1\}$. In other words, we consider predictors of the form

$$f : \mathbb{R}^D \rightarrow \{+1, -1\}. \tag{12.1}$$

Recall from Chapter 8 that we represent each example (data point) $\boldsymbol{x}_n$ as a feature vector of D real numbers. The labels are often referred to as the positive and negative *classes*, respectively. One should be careful not to infer intuitive attributes of positiveness of the $+1$ class. For example, in a cancer detection task, a patient with cancer is often labeled $+1$. In principle, any two distinct values can be used, e.g., $\{\text{True}, \text{False}\}$, $\{0, 1\}$ or $\{\text{red}, \text{blue}\}$. The problem of binary classification is well studied, and we defer a survey of other approaches to Section 12.6.

Input example $\boldsymbol{x}_n$ may also be referred to as inputs, data points, features, or instances.

class

For probabilistic models, it is mathematically convenient to use $\{0, 1\}$ as a binary representation; see the remark after Example 6.12.

We present an approach known as the support vector machine (SVM), which solves the binary classification task. As in regression, we have a supervised learning task, where we have a set of examples $\boldsymbol{x}_n \in \mathbb{R}^D$ along with their corresponding (binary) labels $y_n \in \{+1, -1\}$. Given a training dataset consisting of example–label pairs $\{(\boldsymbol{x}_1, y_1), \ldots, (\boldsymbol{x}_N, y_N)\}$, we would like to estimate parameters of the model that will give the smallest classification error. Similar to Chapter 9, we consider a linear model, and hide away the nonlinearity in a transformation ϕ of the examples (9.13). We will revisit ϕ in Section 12.4.

The SVM provides state-of-the-art results in many applications, with sound theoretical guarantees (Steinwart and Christmann, 2008). There are two main reasons why we chose to illustrate binary classification using SVMs. First, the SVM allows for a geometric way to think about supervised machine learning.

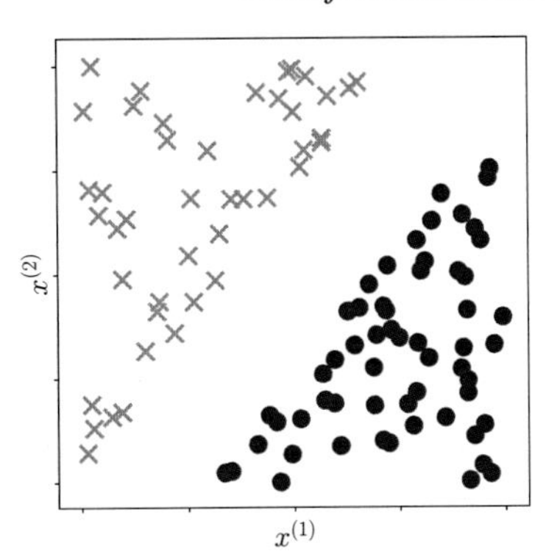

Figure 12.1 Example 2D data, illustrating the intuition of data where we can find a linear classifier that separates red crosses from blue dots.

While in Chapter 9 we considered the machine learning problem in terms of probabilistic models and attacked it using maximum likelihood estimation and Bayesian inference, here we will consider an alternative approach where we reason geometrically about the machine learning task. It relies heavily on concepts, such as inner products and projections, which we discussed in Chapter 3. The second reason why we find SVMs instructive is that in contrast to Chapter 9, the optimization problem for SVM does not admit an analytic solution so that we need to resort to a variety of optimization tools introduced in Chapter 7.

The SVM view of machine learning is subtly different from the maximum likelihood view of Chapter 9. The maximum likelihood view proposes a model based on a probabilistic view of the data distribution, from which an optimization problem is derived. In contrast, the SVM view starts by designing a particular function that is to be optimized during training, based on geometric intuitions. We have seen something similar already in Chapter 10, where we derived PCA from geometric principles. In the SVM case, we start by designing an objective function that is to be minimized on training data, following the principles of empirical risk minimization (Section 8.1). This can also be understood as designing a particular loss function.

Let us derive the optimization problem corresponding to training an SVM on example–label pairs. Intuitively, we imagine binary classification data, which can be separated by a hyperplane as illustrated in Figure 12.1. Here, every example $\boldsymbol{x}_n$ (a vector of dimension 2) is a two-dimensional location ($x_n^{(1)}$ and $x_n^{(2)}$), and the corresponding binary label y_n is one of two different symbols (red cross or blue disc). "Hyperplane" is a word that is commonly used in machine learning, and we encountered hyperplanes already in Section 2.8. A hyperplane is an affine subspace of dimension $D - 1$ (if the corresponding vector space is of dimension D). The examples consist of two classes (there are two possible labels) that have features (the components of the vector representing the example) arranged in such a way as to allow us to separate/classify them by drawing a straight line.

In the following, we formalize the idea of finding a linear separator of the two classes. We introduce the idea of the margin and then extend linear separators to allow for examples to fall on the "wrong" side, incurring a classification error. We present two equivalent ways of formalizing the SVM: the geometric view (Section 12.2.4) and the loss function view (Section 12.2.5). We derive

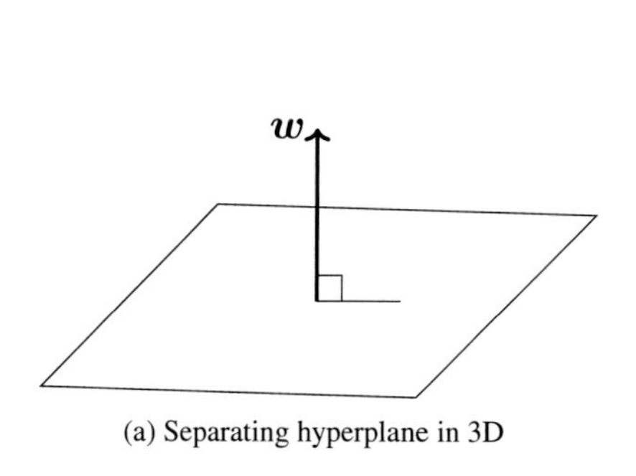

(a) Separating hyperplane in 3D

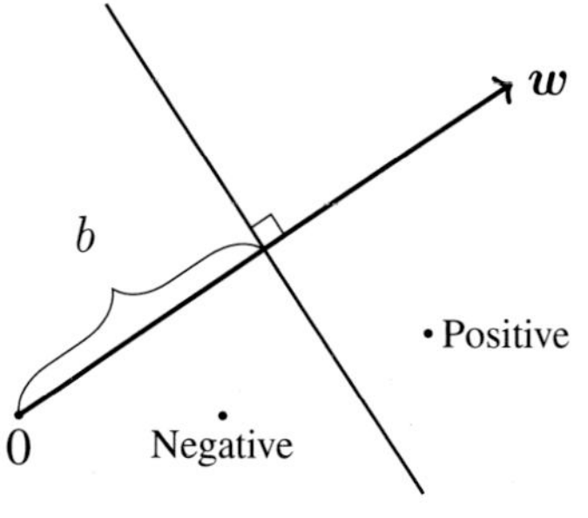

(b) Projection of the setting in (a) onto a plane

Figure 12.2 Equation of a separating hyperplane (12.3). (a) The standard way of representing the equation in 3D. (b) For ease of drawing, we look at the hyperplane edge on.

the dual version of the SVM using Lagrange multipliers (Section 7.2). The dual SVM allows us to observe a third way of formalizing the SVM: in terms of the convex hulls of the examples of each class (Section 12.3.2). We conclude by briefly describing kernels and how to numerically solve the nonlinear kernel-SVM optimization problem.

12.1 Separating Hyperplanes

Given two examples represented as vectors $\boldsymbol{x}_i$ and $\boldsymbol{x}_j$, one way to compute the similarity between them is using an inner product $\langle \boldsymbol{x}_i, \boldsymbol{x}_j \rangle$. Recall from Section 3.2 that inner products are closely related to the angle between two vectors. The value of the inner product between two vectors depends on the length (norm) of each vector. Furthermore, inner products allow us to rigorously define geometric concepts such as orthogonality and projections.

The main idea behind many classification algorithms is to represent data in $\mathbb{R}^D$ and then partition this space, ideally in a way that examples with the same label (and no other examples) are in the same partition. In the case of binary classification, the space would be divided into two parts corresponding to the positive and negative classes, respectively. We consider a particularly convenient partition, which is to (linearly) split the space into two halves using a hyperplane. Let example $\boldsymbol{x} \in \mathbb{R}^D$ be an element of the data space. Consider a function

$$f : \mathbb{R}^D \to \mathbb{R} \tag{12.2a}$$

$$\boldsymbol{x} \mapsto \langle \boldsymbol{w}, \boldsymbol{x} \rangle + b, \tag{12.2b}$$

parametrized by $\boldsymbol{w} \in \mathbb{R}^D$ and $b \in \mathbb{R}$. Recall from Section 2.8 that hyperplanes are affine subspaces. Therefore, we define the hyperplane that separates the two classes in our binary classification problem as

$$\left\{\boldsymbol{x} \in \mathbb{R}^D : f(\boldsymbol{x}) = 0\right\}. \tag{12.3}$$

An illustration of the hyperplane is shown in Figure 12.2, where the vector $\boldsymbol{w}$ is a vector normal to the hyperplane and b the intercept. We can derive that $\boldsymbol{w}$ is a normal vector to the hyperplane in (12.3) by choosing any two examples $\boldsymbol{x}_a$ and $\boldsymbol{x}_b$ on the hyperplane and showing that the vector between them is orthogonal to $\boldsymbol{w}$. In the form of an equation,

$$f(\boldsymbol{x}_a) - f(\boldsymbol{x}_b) = \langle \boldsymbol{w}, \boldsymbol{x}_a \rangle + b - (\langle \boldsymbol{w}, \boldsymbol{x}_b \rangle + b) \tag{12.4a}$$
$$= \langle \boldsymbol{w}, \boldsymbol{x}_a - \boldsymbol{x}_b \rangle , \tag{12.4b}$$

where the second line is obtained by the linearity of the inner product (Section 3.2). Since we have chosen $\boldsymbol{x}_a$ and $\boldsymbol{x}_b$ to be on the hyperplane, this implies that $f(\boldsymbol{x}_a) = 0$ and $f(\boldsymbol{x}_b) = 0$ and hence $\langle \boldsymbol{w}, \boldsymbol{x}_a - \boldsymbol{x}_b \rangle = 0$. Recall that two vectors are orthogonal when their inner product is zero. Therefore, we obtain that $\boldsymbol{w}$ is orthogonal to any vector on the hyperplane.

$\boldsymbol{w}$ is orthogonal to any vector on the hyperplane.

Remark. Recall from Chapter 2 that we can think of vectors in different ways. In this chapter, we think of the parameter vector $\boldsymbol{w}$ as an arrow indicating a direction, i.e., we consider $\boldsymbol{w}$ to be a geometric vector. In contrast, we think of the example vector $\boldsymbol{x}$ as a data point (as indicated by its coordinates), i.e., we consider $\boldsymbol{x}$ to be the coordinates of a vector with respect to the standard basis. ♢

When presented with a test example, we classify the example as positive or negative depending on the side of the hyperplane on which it occurs. Note that (12.3) not only defines a hyperplane; it additionally defines a direction. In other words, it defines the positive and negative side of the hyperplane. Therefore, to classify a test example $\boldsymbol{x}_{\text{test}}$, we calculate the value of the function $f(\boldsymbol{x}_{\text{test}})$ and classify the example as $+1$ if $f(\boldsymbol{x}_{\text{test}}) \geqslant 0$ and -1 otherwise. Thinking geometrically, the positive examples lie "above" the hyperplane and the negative examples "below" the hyperplane.

When training the classifier, we want to ensure that the examples with positive labels are on the positive side of the hyperplane, i.e.,

$$\langle \boldsymbol{w}, \boldsymbol{x}_n \rangle + b \geqslant 0 \quad \text{when} \quad y_n = +1 \tag{12.5}$$

and the examples with negative labels are on the negative side, i.e.,

$$\langle \boldsymbol{w}, \boldsymbol{x}_n \rangle + b < 0 \quad \text{when} \quad y_n = -1. \tag{12.6}$$

Refer to Figure 12.2 for a geometric intuition of positive and negative examples. These two conditions are often presented in a single equation

$$y_n(\langle \boldsymbol{w}, \boldsymbol{x}_n \rangle + b) \geqslant 0. \tag{12.7}$$

Equation (12.7) is equivalent to (12.5) and (12.6) when we multiply both sides of (12.5) and (12.6) with $y_n = 1$ and $y_n = -1$, respectively.

12.2 Primal Support Vector Machine

Based on the concept of distances from points to a hyperplane, we now are in a position to discuss the support vector machine. For a dataset $\{(\boldsymbol{x}_1, y_1), \ldots, (\boldsymbol{x}_N, y_N)\}$ that is linearly separable, we have infinitely many candidate hyperplanes (refer to Figure 12.3), and therefore classifiers, that solve our classification problem without any (training) errors. To find a unique solution, one idea is to choose the separating hyperplane that maximizes the margin between the positive and negative examples. In other words, we want the positive and negative examples to be separated by a large margin (Section 12.2.1). In the following, we compute the distance between an example and a hyperplane to derive the margin. Recall that the closest point on the hyperplane to a given point (example $\boldsymbol{x}_n$) is obtained by the orthogonal projection (Section 3.8).

A classifier with large margin turns out to generalize well (Steinwart and Christmann, 2008).

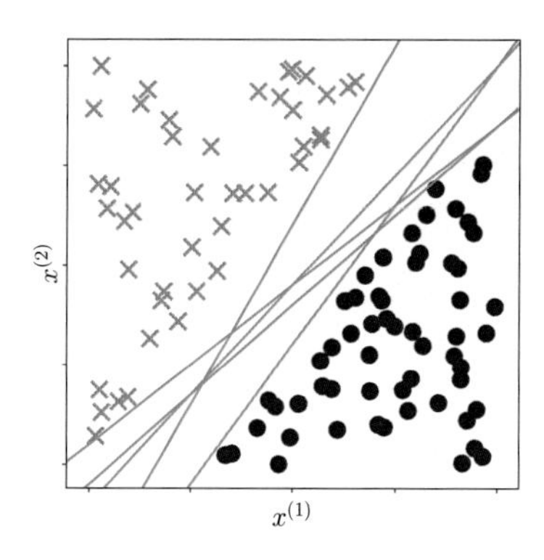

Figure 12.3 Possible separating hyperplanes. There are many linear classifiers (green lines) that separate red crosses from blue dots.

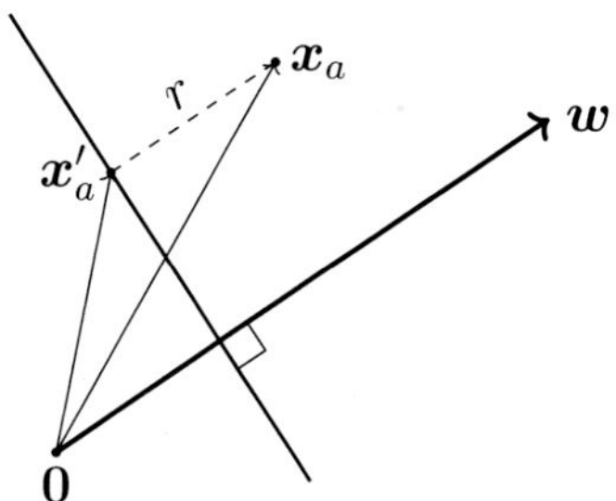

Figure 12.4 Vector addition to express distance to hyperplane: $x_a = x'_a + r\frac{w}{\|w\|}$.

12.2.1 Concept of the Margin

The concept of the *margin* is intuitively simple: It is the distance of the separating hyperplane to the closest examples in the dataset, assuming that the dataset is linearly separable. However, when trying to formalize this distance, there is a technical wrinkle that may be confusing. The technical wrinkle is that we need to define a scale at which to measure the distance. A potential scale is to consider the scale of the data, i.e., the raw values of x_n. There are problems with this, as we could change the units of measurement of x_n and change the values in x_n, and hence change the distance to the hyperplane. As we will see shortly, we define the scale based on the equation of the hyperplane (12.3) itself.

margin

There could be two or more closest examples to a hyperplane.

Consider a hyperplane $\langle w, x\rangle + b$, and an example x_a as illustrated in Figure 12.4. Without loss of generality, we can consider the example x_a to be on the positive side of the hyperplane, i.e., $\langle w, x_a\rangle + b > 0$. We would like to compute the distance $r > 0$ of x_a from the hyperplane. We do so by considering the orthogonal projection (Section 3.8) of x_a onto the hyperplane, which we denote by x'_a. Since w is orthogonal to the hyperplane, we know that the distance r is just a scaling of this vector w. If the length of w is known, then we can use this scaling factor r factor to work out the absolute distance between x_a and x'_a. For convenience, we choose to use a vector of unit length (its norm is 1) and obtain this by dividing w by its norm, $\frac{w}{\|w\|}$. Using vector addition (Section 2.4), we obtain

$$x_a = x'_a + r\frac{w}{\|w\|}. \tag{12.8}$$

Another way of thinking about r is that it is the coordinate of x_a in the subspace spanned by $w/\|w\|$. We have now expressed the distance of x_a from the hyperplane as r, and if we choose $\boldsymbol{x}_a$ to be the point closest to the hyperplane, this distance r is the margin.

Recall that we would like the positive examples to be further than r from the hyperplane, and the negative examples to be further than distance r (in the negative direction) from the hyperplane. Analogously to the combination of (12.5) and (12.6) into (12.7), we formulate this objective as

$$y_n(\langle \boldsymbol{w}, \boldsymbol{x}_n \rangle + b) \geqslant r. \tag{12.9}$$

In other words, we combine the requirements that examples are at least r away from the hyperplane (in the positive and negative direction) into one single inequality.

Since we are interested only in the direction, we add an assumption to our model that the parameter vector $\boldsymbol{w}$ is of unit length, i.e., $\|\boldsymbol{w}\| = 1$, where we use the Euclidean norm $\|\boldsymbol{w}\| = \sqrt{\boldsymbol{w}^\top \boldsymbol{w}}$ (Section 3.1). This assumption also allows a more intuitive interpretation of the distance r (12.8) since it is the scaling factor of a vector of length 1.

We will see other choices of inner products (Section 3.2) in Section 12.4.

Remark. A reader familiar with other presentations of the margin would notice that our definition of $\|\boldsymbol{w}\| = 1$ is different from the standard presentation if the SVM was the one provided by Schölkopf and Smola (2002), for example. In Section 12.2.3, we will show the equivalence of both approaches. ◇

Collecting the three requirements into a single constrained optimization problem, we obtain the objective

$$\begin{aligned} \max_{\boldsymbol{w},b,r} \quad & \underbrace{r}_{\text{margin}} \\ \text{subject to} \quad & \underbrace{y_n(\langle \boldsymbol{w}, \boldsymbol{x}_n \rangle + b) \geqslant r}_{\text{data fitting}}, \quad \underbrace{\|\boldsymbol{w}\| = 1}_{\text{normalization}}, \quad r > 0, \end{aligned} \tag{12.10}$$

which says that we want to maximize the margin r while ensuring that the data lies on the correct side of the hyperplane.

Remark. The concept of the margin turns out to be highly pervasive in machine learning. It was used by Vladimir Vapnik and Alexey Chervonenkis to show that when the margin is large, the "complexity" of the function class is low, and hence learning is possible (Vapnik, 2000). It turns out that the concept is useful for various different approaches for theoretically analyzing generalization error (Steinwart and Christmann, 2008; Shalev-Shwartz and Ben-David, 2014). ◇

12.2.2 Traditional Derivation of the Margin

In the previous section, we derived (12.10) by making the observation that we are only interested in the direction of $\boldsymbol{w}$ and not its length, leading to the assumption that $\|\boldsymbol{w}\| = 1$. In this section, we derive the margin maximization problem by making a different assumption. Instead of choosing that the parameter vector is normalized, we choose a scale for the data. We choose this scale such that the

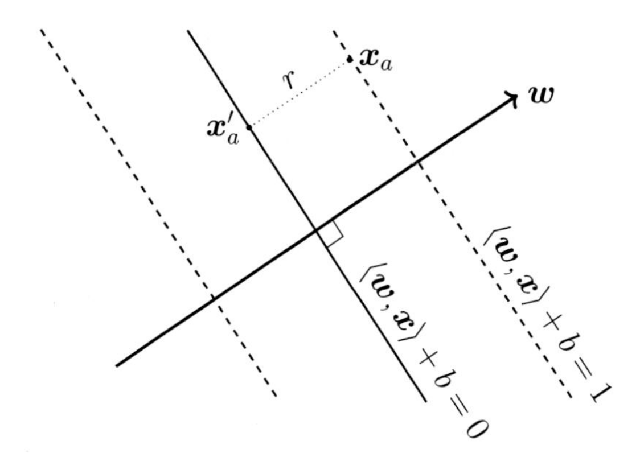

Figure 12.5 Derivation of the margin: $r = \frac{1}{\|\boldsymbol{w}\|}$.

value of the predictor $\langle \boldsymbol{w}, \boldsymbol{x} \rangle + b$ is 1 at the closest example. Let us also denote the example in the dataset that is closest to the hyperplane by $\boldsymbol{x}_a$.

Recall that we currently consider linearly separable data.

Figure 12.5 is identical to Figure 12.4, except that now we rescaled the axes, such that the example $\boldsymbol{x}_a$ lies exactly on the margin, i.e., $\langle \boldsymbol{w}, \boldsymbol{x}_a \rangle + b = 1$. Since $\boldsymbol{x}'_a$ is the orthogonal projection of $\boldsymbol{x}_a$ onto the hyperplane, it must by definition lie on the hyperplane, i.e.,

$$\langle \boldsymbol{w}, \boldsymbol{x}'_a \rangle + b = 0. \tag{12.11}$$

By substituting (12.8) into (12.11), we obtain

$$\left\langle \boldsymbol{w}, \boldsymbol{x}_a - r\frac{\boldsymbol{w}}{\|\boldsymbol{w}\|} \right\rangle + b = 0. \tag{12.12}$$

Exploiting the bilinearity of the inner product (see Section 3.2), we get

$$\langle \boldsymbol{w}, \boldsymbol{x}_a \rangle + b - r\frac{\langle \boldsymbol{w}, \boldsymbol{w} \rangle}{\|\boldsymbol{w}\|} = 0. \tag{12.13}$$

Observe that the first term is 1 by our assumption of scale, i.e., $\langle \boldsymbol{w}, \boldsymbol{x}_a \rangle + b = 1$. From (3.16) in Section 3.1, we know that $\langle \boldsymbol{w}, \boldsymbol{w} \rangle = \|\boldsymbol{w}\|^2$. Hence, the second term reduces to $r\|\boldsymbol{w}\|$. Using these simplifications, we obtain

$$r = \frac{1}{\|\boldsymbol{w}\|}. \tag{12.14}$$

This means we derived the distance r in terms of the normal vector $\boldsymbol{w}$ of the hyperplane. At first glance, this equation is counterintuitive as we seem to have derived the distance from the hyperplane in terms of the length of the vector $\boldsymbol{w}$, but we do not yet know this vector. One way to think about it is to consider the distance r to be a temporary variable that we only use for this derivation. Therefore, for the rest of this section we will denote the distance to the hyperplane by $\frac{1}{\|\boldsymbol{w}\|}$. In Section 12.2.3, we will see that the choice that the margin equals 1 is equivalent to our previous assumption of $\|\boldsymbol{w}\| = 1$ in Section 12.2.1.

We can also think of the distance as the projection error that incurs when projecting $\boldsymbol{x}_a$ onto the hyperplane.

Similar to the argument to obtain (12.9), we want the positive and negative examples to be at least 1 away from the hyperplane, which yields the condition

$$y_n(\langle \boldsymbol{w}, x_n \rangle + b) \geqslant 1. \tag{12.15}$$

Combining the margin maximization with the fact that examples need to be on the correct side of the hyperplane (based on their labels) gives us

$$\max_{\boldsymbol{w},b} \quad \frac{1}{\|\boldsymbol{w}\|} \tag{12.16}$$

$$\text{subject to } y_n(\langle \boldsymbol{w}, \boldsymbol{x}_n \rangle + b) \geqslant 1 \quad \text{for all} \quad n = 1, \dots, N. \tag{12.17}$$

The squared norm results in a convex quadratic programming problem for the SVM (Section 12.5).

Instead of maximizing the reciprocal of the norm as in (12.16), we often minimize the squared norm. We also often include a constant $\frac{1}{2}$ that does not affect the optimal $\boldsymbol{w}, b$ but yields a tidier form when we compute the gradient. Then our objective becomes

$$\min_{\boldsymbol{w},b} \quad \frac{1}{2}\|\boldsymbol{w}\|^2 \tag{12.18}$$

$$\text{subject to } y_n(\langle \boldsymbol{w}, \boldsymbol{x}_n \rangle + b) \geqslant 1 \quad \text{for all} \quad n = 1, \dots, N. \tag{12.19}$$

hard margin SVM

Equation (12.18) is known as the *hard margin SVM*. The reason for the expression "hard" is because the preceding formulation does not allow for any violations of the margin condition. We will see in Section 12.2.4 that this "hard" condition can be relaxed to accommodate violations if the data is not linearly separable.

12.2.3 Why We Can Set the Margin to 1

In Section 12.2.1, we argued that we would like to maximize some value r, which represents the distance of the closest example to the hyperplane. In Section 12.2.2, we scaled the data such that the closest example is of distance 1 to the hyperplane. In this section, we relate the two derivations and show that they are equivalent.

Theorem 12.1. *Maximizing the margin r, where we consider normalized weights as in* (12.10),

$$\begin{aligned} &\max_{\boldsymbol{w},b,r} \quad \underbrace{r}_{\text{margin}} \\ &\textit{subject to} \quad \underbrace{y_n(\langle \boldsymbol{w}, \boldsymbol{x}_n \rangle + b) \geqslant r}_{\text{data fitting}}, \quad \underbrace{\|\boldsymbol{w}\| = 1}_{\text{normalization}}, \quad r > 0, \end{aligned} \tag{12.20}$$

is equivalent to scaling the data, such that the margin is unity:

$$\begin{aligned} &\min_{\boldsymbol{w},b} \quad \underbrace{\frac{1}{2}\|\boldsymbol{w}\|^2}_{\text{margin}} \\ &\textit{subject to} \quad \underbrace{y_n(\langle \boldsymbol{w}, \boldsymbol{x}_n \rangle + b) \geqslant 1}_{\text{data fitting}}. \end{aligned} \tag{12.21}$$

Proof Consider (12.20). Since the square is a strictly monotonic transformation for nonnegative arguments, the maximum stays the same if we consider r^2 in the objective. Since $\|\boldsymbol{w}\| = 1$, we can reparametrize the equation with a new weight vector $\boldsymbol{w}'$ that is not normalized by explicitly using $\frac{\boldsymbol{w}'}{\|\boldsymbol{w}'\|}$. We obtain

$$\max_{\boldsymbol{w}',b,r} \quad r^2$$
$$\text{subject to} \quad y_n \left(\left\langle \frac{\boldsymbol{w}'}{\|\boldsymbol{w}'\|}, \boldsymbol{x}_n \right\rangle + b \right) \geqslant r, \quad r > 0. \tag{12.22}$$

Equation (12.22) explicitly states that the distance r is positive. Therefore, we can divide the first constraint by r, which yields

Note that $r > 0$ because we assumed linear separability, and hence there is no issue to divide by r.

$$\max_{\boldsymbol{w}',b,r} \quad r^2$$
$$\text{subject to} \quad y_n \left(\left\langle \underbrace{\frac{\boldsymbol{w}'}{\|\boldsymbol{w}'\| \, r}}_{\boldsymbol{w}''}, \boldsymbol{x}_n \right\rangle + \underbrace{\frac{b}{r}}_{b''} \right) \geqslant 1, \quad r > 0 \tag{12.23}$$

renaming the parameters to $\boldsymbol{w}''$ and b''. Since $\boldsymbol{w}'' = \frac{\boldsymbol{w}'}{\|\boldsymbol{w}'\| r}$, rearranging for r gives

$$\|\boldsymbol{w}''\| = \left\| \frac{\boldsymbol{w}'}{\|\boldsymbol{w}'\| \, r} \right\| = \frac{1}{r} \cdot \left\| \frac{\boldsymbol{w}'}{\|\boldsymbol{w}'\|} \right\| = \frac{1}{r}. \tag{12.24}$$

By substituting this result into (12.23), we obtain

$$\max_{\boldsymbol{w}'',b''} \quad \frac{1}{\|\boldsymbol{w}''\|^2}$$
$$\text{subject to} \quad y_n (\langle \boldsymbol{w}'', \boldsymbol{x}_n \rangle + b'') \geqslant 1. \tag{12.25}$$

The final step is to observe that maximizing $\frac{1}{\|\boldsymbol{w}''\|^2}$ yields the same solution as minimizing $\frac{1}{2} \|\boldsymbol{w}''\|^2$, which concludes the proof of Theorem 12.1. $\square$

12.2.4 Soft Margin SVM: Geometric View

In the case where data is not linearly separable, we may wish to allow some examples to fall within the margin region, or even to be on the wrong side of the hyperplane as illustrated in Figure 12.6.

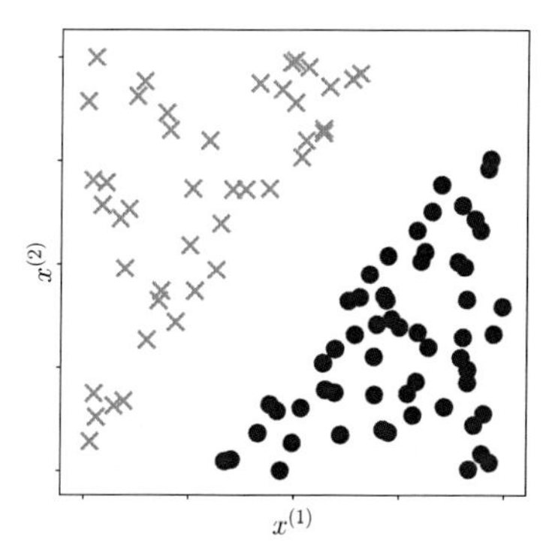

(a) Linearly separable data, with a large margin

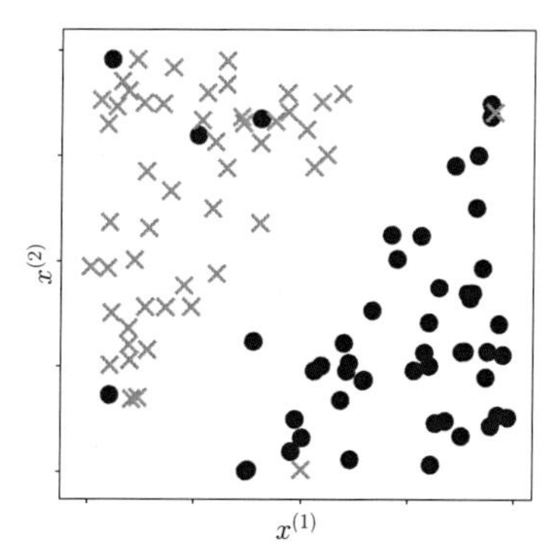

(b) Nonlinearly separable data

Figure 12.6 (a) Linearly separable and (b) nonlinearly separable data.

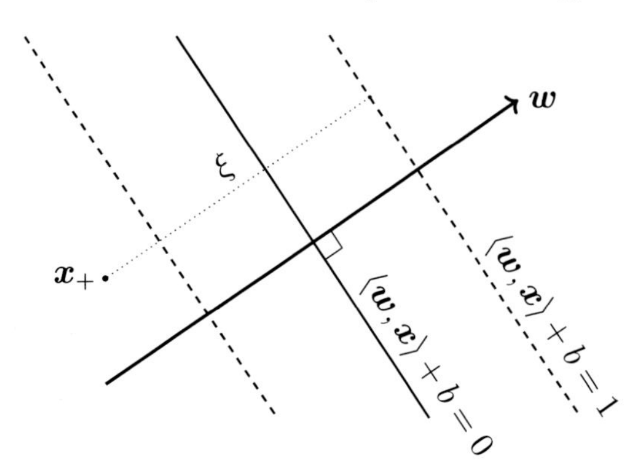

Figure 12.7 Soft margin SVM allows examples to be within the margin or on the wrong side of the hyperplane. The slack variable ξ measures the distance of a positive example $\boldsymbol{x}_+$ to the positive margin hyperplane $\langle \boldsymbol{w}, \boldsymbol{x} \rangle + b = 1$ when $\boldsymbol{x}_+$ is on the wrong side.

soft margin SVM

The model that allows for some classification errors is called the *soft margin SVM*. In this section, we derive the resulting optimization problem using geometric arguments. In Section 12.2.5, we will derive an equivalent optimization problem using the idea of a loss function. Using Lagrange multipliers (Section 7.2), we will derive the dual optimization problem of the SVM in Section 12.3. This dual optimization problem allows us to observe a third interpretation of the SVM: as a hyperplane that bisects the line between convex hulls corresponding to the positive and negative data examples (Section 12.3.2).

slack variable

The key geometric idea is to introduce a *slack variable* ξ_n corresponding to each example–label pair $(\boldsymbol{x}_n, y_n)$ that allows a particular example to be within the margin or even on the wrong side of the hyperplane (refer to Figure 12.7). We subtract the value of ξ_n from the margin, constraining ξ_n to be nonnegative. To encourage correct classification of the samples, we add ξ_n to the objective

$$\min_{\boldsymbol{w}, b, \boldsymbol{\xi}} \quad \frac{1}{2}\|\boldsymbol{w}\|^2 + C\sum_{n=1}^{N} \xi_n \tag{12.26a}$$

$$\text{subject to} \quad y_n(\langle \boldsymbol{w}, \boldsymbol{x}_n \rangle + b) \geqslant 1 - \xi_n \tag{12.26b}$$

$$\xi_n \geqslant 0 \tag{12.26c}$$

soft margin SVM

regularization parameter

regularizer

for $n = 1, \ldots, N$. In contrast to the optimization problem (12.18) for the hard margin SVM, this one is called the *soft margin SVM*. The parameter $C > 0$ trades off the size of the margin and the total amount of slack that we have. This parameter is called the *regularization parameter* since, as we will see in the following section, the margin term in the objective function (12.26a) is a regularization term. The margin term $\|\boldsymbol{w}\|^2$ is called the *regularizer*, and in many books on numerical optimization, the regularization parameter is multiplied with this term (Section 8.2.3). This is in contrast to our formulation in this section. Here a large value of C implies low regularization, as we give the slack variables larger weight, hence giving more priority to examples that do not lie on the correct side of the margin.

There are alternative parametrizations of this regularization, which is why (12.26a) is also often referred to as the C-SVM.

Remark. In the formulation of the soft margin SVM (12.26a) $\boldsymbol{w}$ is regularized, but b is not regularized. We can see this by observing that the regularization term does not contain b. The unregularized term b complicates theoretical analysis

(Steinwart and Christmann, 2008, chapter 1) and decreases computational efficiency (Fan et al., 2008). ◊

12.2.5 Soft Margin SVM: Loss Function View

Let us consider a different approach for deriving the SVM, following the principle of empirical risk minimization (Section 8.2). For the SVM, we choose hyperplanes as the hypothesis class, that is

$$f(\boldsymbol{x}) = \langle \boldsymbol{w}, \boldsymbol{x} \rangle + b. \tag{12.27}$$

We will see in this section that the margin corresponds to the regularization term. The remaining question is, what is the *loss function*? In contrast to Chapter 9, where we consider regression problems (the output of the predictor is a real number), in this chapter, we consider binary classification problems (the output of the predictor is one of two labels $\{+1, -1\}$). Therefore, the error/loss function for each single example–label pair needs to be appropriate for binary classification. For example, the squared loss that is used for regression (9.10b) is not suitable for binary classification.

loss function

Remark. The ideal loss function between binary labels is to count the number of mismatches between the prediction and the label. This means that for a predictor f applied to an example $\boldsymbol{x}_n$, we compare the output $f(\boldsymbol{x}_n)$ with the label y_n. We define the loss to be zero if they match, and one if they do not match. This is denoted by $\mathbf{1}(f(\boldsymbol{x}_n) \neq y_n)$ and is called the *zero-one loss*. Unfortunately, the zero-one loss results in a combinatorial optimization problem for finding the best parameters $\boldsymbol{w}, b$. Combinatorial optimization problems (in contrast to continuous optimization problems discussed in Chapter 7) are in general more challenging to solve. ◊

zero-one loss

What is the loss function corresponding to the SVM? Consider the error between the output of a predictor $f(\boldsymbol{x}_n)$ and the label y_n. The loss describes the error that is made on the training data. An equivalent way to derive (12.26a) is to use the *hinge loss*

hinge loss

$$\ell(t) = \max\{0, 1 - t\} \quad \text{where} \quad t = yf(\boldsymbol{x}) = y(\langle \boldsymbol{w}, \boldsymbol{x} \rangle + b). \tag{12.28}$$

If $f(\boldsymbol{x})$ is on the correct side (based on the corresponding label y) of the hyperplane, and further than distance 1, this means that $t \geqslant 1$ and the hinge loss returns a value of zero. If $f(\boldsymbol{x})$ is on the correct side but too close to the hyperplane $(0 < t < 1)$, the example $\boldsymbol{x}$ is within the margin, and the hinge loss returns a positive value. When the example is on the wrong side of the hyperplane $(t < 0)$, the hinge loss returns an even larger value, which increases linearly. In other words, we pay a penalty once we are closer to than the margin to the hyperplane, even if the prediction is correct, and the penalty increases linearly. An alternative way to express the hinge loss is by considering it as two linear pieces

$$\ell(t) = \begin{cases} 0 & \text{if} \quad t \geqslant 1 \\ 1 - t & \text{if} \quad t < 1 \end{cases}, \tag{12.29}$$

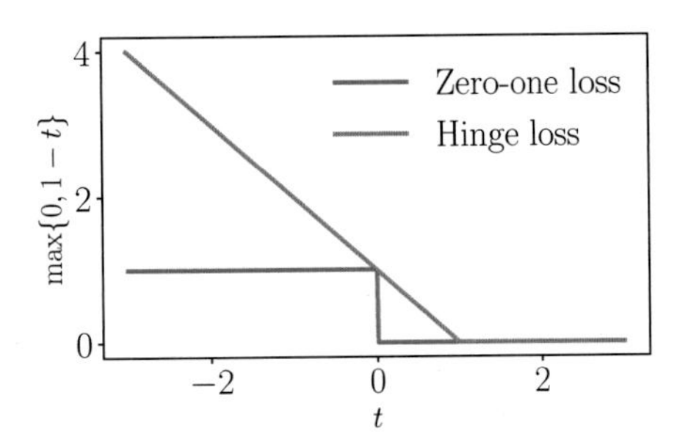

Figure 12.8 The hinge loss is a convex upper bound of zero-one loss.

as illustrated in Figure 12.8. The loss corresponding to the hard margin SVM 12.18 is defined as

$$\ell(t) = \begin{cases} 0 & \text{if} \quad t \geqslant 1 \\ \infty & \text{if} \quad t < 1 \end{cases}. \tag{12.30}$$

This loss can be interpreted as never allowing any examples inside the margin.

For a given training set $\{(\boldsymbol{x}_1, y_1), \dots, (\boldsymbol{x}_N, y_N)\}$, we seek to minimize the total loss, while regularizing the objective with ℓ_2-regularization (see Section 8.2.3). Using the hinge loss (12.28) gives us the unconstrained optimization problem

$$\min_{\boldsymbol{w}, b} \quad \underbrace{\frac{1}{2}\|\boldsymbol{w}\|^2}_{\text{regularizer}} + \underbrace{C \sum_{n=1}^{N} \max\{0, 1 - y_n(\langle \boldsymbol{w}, \boldsymbol{x}_n \rangle + b)\}}_{\text{error term}}. \tag{12.31}$$

regularizer
loss term
error term
regularization

The first term in (12.31) is called the regularization term or the *regularizer* (see Section 9.2.3), and the second term is called the *loss term* or the *error term*. Recall from Section 12.2.4 that the term $\frac{1}{2}\|\boldsymbol{w}\|^2$ arises directly from the margin. In other words, margin maximization can be interpreted as *regularization*.

In principle, the unconstrained optimization problem in (12.31) can be directly solved with (sub-)gradient descent methods as described in Section 7.1. To see that (12.31) and (12.26a) are equivalent, observe that the hinge loss (12.28) essentially consists of two linear parts, as expressed in (12.29). Consider the hinge loss for a single example–label pair (12.28). We can equivalently replace minimization of the hinge loss over t with a minimization of a slack variable ξ with two constraints. In equation form,

$$\min_{t} \max\{0, 1 - t\} \tag{12.32}$$

is equivalent to

$$\begin{aligned} \min_{\xi, t} \quad & \xi \\ \text{subject to} \quad & \xi \geqslant 0, \quad \xi \geqslant 1 - t. \end{aligned} \tag{12.33}$$

By substituting this expression into (12.31) and rearranging one of the constraints, we obtain exactly the soft margin SVM (12.26a).

Remark. Let us contrast our choice of the loss function in this section to the loss function for linear regression in Chapter 9. Recall from Section 9.2.1 that for finding maximum likelihood estimators, we usually minimize the negative log-likelihood. Furthermore, since the likelihood term for linear regression with Gaussian noise is Gaussian, the negative log-likelihood for each example is a squared error function. The squared error function is the loss function that is minimized when looking for the maximum likelihood solution. ◊

12.3 Dual Support Vector Machine

The description of the SVM in the previous sections, in terms of the variables $\boldsymbol{w}$ and b, is known as the primal SVM. Recall that we consider inputs $\boldsymbol{x} \in \mathbb{R}^D$ with D features. Since $\boldsymbol{w}$ is of the same dimension as $\boldsymbol{x}$, this means that the number of parameters (the dimension of $\boldsymbol{w}$) of the optimization problem grows linearly with the number of features.

In the following, we consider an equivalent optimization problem (the so-called dual view), which is independent of the number of features. Instead, the number of parameters increases with the number of examples in the training set. We saw a similar idea appear in Chapter 10, where we expressed the learning problem in a way that does not scale with the number of features. This is useful for problems where we have more features than the number of examples in the training dataset. The dual SVM also has the additional advantage that it easily allows kernels to be applied, as we shall see at the end of this chapter. The word "dual" appears often in mathematical literature, and in this particular case it refers to convex duality. The following subsections are essentially an application of convex duality, which we discussed in Section 7.2.

12.3.1 Convex Duality via Lagrange Multipliers

Recall the primal soft margin SVM (12.26a). We call the variables $\boldsymbol{w}$, b, and ξ corresponding to the primal SVM the primal variables. We use $\alpha_n \geqslant 0$ as the Lagrange multiplier corresponding to the constraint (12.26b) that the examples are classified correctly and $\gamma_n \geqslant 0$ as the Lagrange multiplier corresponding to the nonnegativity constraint of the slack variable; see (12.26c). The Lagrangian is then given by

In Chapter 7, we used λ as Lagrange multipliers. In this section, we follow the notation commonly chosen in SVM literature, and use α and γ.

$$\mathfrak{L}(\boldsymbol{w}, b, \xi, \alpha, \gamma) = \frac{1}{2}\|\boldsymbol{w}\|^2 + C\sum_{n=1}^{N} \xi_n \tag{12.34}$$

$$\underbrace{-\sum_{n=1}^{N} \alpha_n (y_n(\langle \boldsymbol{w}, \boldsymbol{x}_n \rangle + b) - 1 + \xi_n)}_{\text{constraint (12.26b)}} \underbrace{-\sum_{n=1}^{N} \gamma_n \xi_n}_{\text{constraint (12.26c)}} .$$

By differentiating the Lagrangian (12.34) with respect to the three primal variables $\boldsymbol{w}$, b, and ξ respectively, we obtain

$$\frac{\partial \mathfrak{L}}{\partial \boldsymbol{w}} = \boldsymbol{w}^\top - \sum_{n=1}^{N} \alpha_n y_n \boldsymbol{x}_n{}^\top , \tag{12.35}$$

$$\frac{\partial \mathfrak{L}}{\partial b} = \sum_{n=1}^{N} \alpha_n y_n, \tag{12.36}$$

$$\frac{\partial \mathfrak{L}}{\partial \xi_n} = C - \alpha_n - \gamma_n. \tag{12.37}$$

We now find the maximum of the Lagrangian by setting each of these partial derivatives to zero. By setting (12.35) to zero, we find

$$\boldsymbol{w} = \sum_{n=1}^{N} \alpha_n y_n \boldsymbol{x}_n, \tag{12.38}$$

representer theorem

which is a particular instance of the *representer theorem* (Kimeldorf and Wahba, 1970). Equation (12.38) states that the optimal weight vector in the primal is a linear combination of the examples $\boldsymbol{x}_n$. Recall from Section 2.6.1 that this means that the solution of the optimization problem lies in the span of training data. Additionally, the constraint obtained by setting (12.36) to zero implies that the optimal weight vector is an affine combination of the examples. The representer theorem turns out to hold for very general settings of regularized empirical risk minimization (Hofmann et al., 2008; Argyriou and Dinuzzo, 2014). The theorem has more general versions (Schölkopf et al., 2001), and necessary and sufficient conditions on its existence can be found in Yu et al. (2013).

The representer theorem is actually a collection of theorems saying that the solution of minimizing empirical risk lies in the subspace (Section 2.4.3) defined by the examples.

support vector

Remark. The representer theorem (12.38) also provides an explanation of the name "support vector machine." The examples $\boldsymbol{x}_n$, for which the corresponding parameters $\alpha_n = 0$, do not contribute to the solution $\boldsymbol{w}$ at all. The other examples, where $\alpha_n > 0$, are called *support vectors* since they "support" the hyperplane. ◇

By substituting the expression for $\boldsymbol{w}$ into the Lagrangian (12.34), we obtain the dual

$$\begin{aligned}\mathfrak{D}(\xi, \alpha, \gamma) = \frac{1}{2}\sum_{i=1}^{N}\sum_{j=1}^{N} y_i y_j \alpha_i \alpha_j \langle \boldsymbol{x}_i, \boldsymbol{x}_j \rangle - \sum_{i=1}^{N} y_i \alpha_i \left\langle \sum_{j=1}^{N} y_j \alpha_j \boldsymbol{x}_j, \boldsymbol{x}_i \right\rangle \\ + C\sum_{i=1}^{N} \xi_i - b\sum_{i=1}^{N} y_i \alpha_i + \sum_{i=1}^{N} \alpha_i - \sum_{i=1}^{N} \alpha_i \xi_i - \sum_{i=1}^{N} \gamma_i \xi_i.\end{aligned} \tag{12.39}$$

Note that there are no longer any terms involving the primal variable $\boldsymbol{w}$. By setting (12.36) to zero, we obtain $\sum_{n=1}^{N} y_n \alpha_n = 0$. Therefore, the term involving b also vanishes. Recall that inner products are symmetric and bilinear (see Section 3.2). Therefore, the first two terms in (12.39) are over the same objects. These terms (colored blue) can be simplified, and we obtain the Lagrangian

$$\mathfrak{D}(\xi, \alpha, \gamma) = -\frac{1}{2}\sum_{i=1}^{N}\sum_{j=1}^{N} y_i y_j \alpha_i \alpha_j \langle \boldsymbol{x}_i, \boldsymbol{x}_j \rangle + \sum_{i=1}^{N} \alpha_i + \sum_{i=1}^{N} (C - \alpha_i - \gamma_i)\xi_i. \tag{12.40}$$

The last term in this equation is a collection of all terms that contain slack variables ξ_i. By setting (12.37) to zero, we see that the last term in (12.40) is also zero. Furthermore, by using the same equation and recalling that the Lagrange multipliers γ_i are nonnegative, we conclude that $\alpha_i \leqslant C$. We now obtain the dual optimization problem of the SVM, which is expressed exclusively in terms of the Lagrange multipliers α_i. Recall from Lagrangian duality (Definition 7.1) that we maximize the dual problem. This is equivalent to minimizing the negative dual problem, such that we end up with the *dual SVM*

dual SVM

$$\begin{aligned} \min_{\boldsymbol{\alpha}} \quad & \frac{1}{2}\sum_{i=1}^{N}\sum_{j=1}^{N} y_i y_j \alpha_i \alpha_j \langle \boldsymbol{x}_i, \boldsymbol{x}_j \rangle - \sum_{i=1}^{N} \alpha_i \\ \text{subject to} \quad & \sum_{i=1}^{N} y_i \alpha_i = 0 \\ & 0 \leqslant \alpha_i \leqslant C \quad \text{for all} \quad i = 1, \dots, N. \end{aligned} \tag{12.41}$$

The equality constraint in (12.41) is obtained from setting (12.36) to zero. The inequality constraint $\alpha_i \geqslant 0$ is the condition imposed on Lagrange multipliers of inequality constraints (Section 7.2). The inequality constraint $\alpha_i \leqslant C$ is discussed in the previous paragraph.

The set of inequality constraints in the SVM is called "box constraints" because they limit the vector $\boldsymbol{\alpha} = [\alpha_1, \dots, \alpha_N]^\top \in \mathbb{R}^N$ of Lagrange multipliers to be inside the box defined by 0 and C on each axis. These axis-aligned boxes are particularly efficient to implement in numerical solvers (Dostál, 2009, chapter 5).

It turns out that examples that lie exactly on the margin are examples whose dual parameters lie strictly inside the box constraints, $0 < \alpha_i < C$. This is derived using the Karush Kuhn Tucker conditions, for example in Schölkopf and Smola (2002).

Once we obtain the dual parameters $\boldsymbol{\alpha}$, we can recover the primal parameters $\boldsymbol{w}$ by using the representer theorem (12.38). Let us call the optimal primal parameter $\boldsymbol{w}^*$. However, there remains the question on how to obtain the parameter b^*. Consider an example $\boldsymbol{x}_n$ that lies exactly on the margin's boundary, i.e., $\langle \boldsymbol{w}^*, \boldsymbol{x}_n \rangle + b = y_n$. Recall that y_n is either $+1$ or -1. Therefore, the only unknown is b, which can be computed by

$$b^* = y_n - \langle \boldsymbol{w}^*, \boldsymbol{x}_n \rangle . \tag{12.42}$$

Remark. In principle, there may be no examples that lie exactly on the margin. In this case, we should compute $|y_n - \langle \boldsymbol{w}^*, \boldsymbol{x}_n \rangle|$ for all support vectors and take the median value of this absolute value difference to be the value of b^*. A derivation of this can be found at `http://fouryears.eu/2012/06/07/the-svm-bias-term-conspiracy/`. ◇

12.3.2 Dual SVM: Convex Hull View

Another approach to obtain the dual SVM is to consider an alternative geometric argument. Consider the set of examples $\boldsymbol{x}_n$ with the same label. We would like to build a convex set that contains all the examples such that it is the smallest possible set. This is called the convex hull and is illustrated in Figure 12.9.

Let us first build some intuition about a convex combination of points. Consider two points $\boldsymbol{x}_1$ and $\boldsymbol{x}_2$ and corresponding nonnegative weights $\alpha_1, \alpha_2 \geqslant 0$ such that $\alpha_1 + \alpha_2 = 1$. The equation $\alpha_1 \boldsymbol{x}_1 + \alpha_2 \boldsymbol{x}_2$ describes each point on a line between $\boldsymbol{x}_1$ and $\boldsymbol{x}_2$. Consider what happens when we add a third point $\boldsymbol{x}_3$ along with a weight $\alpha_3 \geqslant 0$ such that $\sum_{n=1}^{3} \alpha_n = 1$. The convex combination

Figure 12.9 Convex hulls. (a) Convex hull of points, some of which lie within the boundary; (b) convex hulls around positive and negative examples.

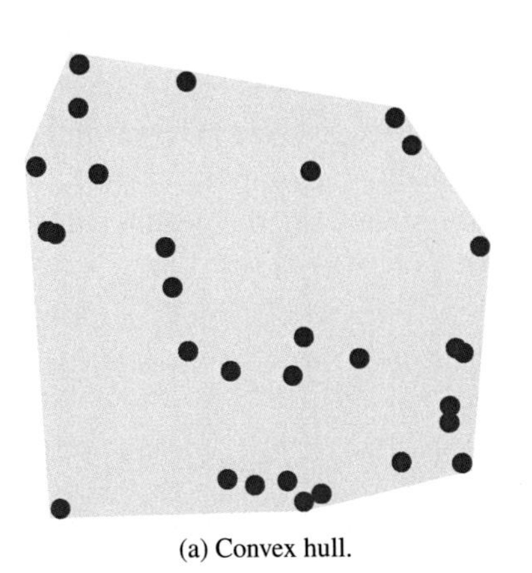

(a) Convex hull.

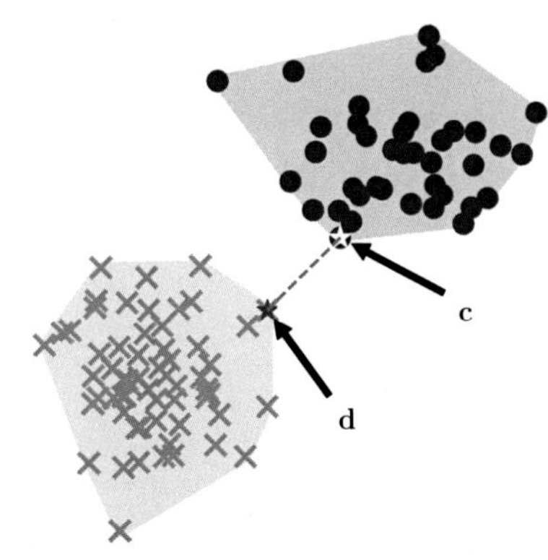

(b) Convex hulls around positive (blue) and negative (red) examples. The distance between the two convex sets is the length of the difference vector $\boldsymbol{c} - \boldsymbol{d}$.

convex hull

of these three points $\boldsymbol{x}_1, \boldsymbol{x}_2, \boldsymbol{x}_3$ spans a two-dimensional area. The *convex hull* of this area is the triangle formed by the edges corresponding to each pair of of points. As we add more points, and the number of points becomes greater than the number of dimensions, some of the points will be inside the convex hull, as we can see in Figure 12.9(a).

In general, building a convex convex hull can be done by introducing nonnegative weights $\alpha_n \geqslant 0$ corresponding to each example $\boldsymbol{x}_n$. Then the convex hull can be described as the set

$$\operatorname{conv}(\boldsymbol{X}) = \left\{ \sum_{n=1}^{N} \alpha_n \boldsymbol{x}_n \right\} \quad \text{with} \quad \sum_{n=1}^{N} \alpha_n = 1 \quad \text{and} \quad \alpha_n \geqslant 0, \tag{12.43}$$

for all $n = 1, \ldots, N$. If the two clouds of points corresponding to the positive and negative classes are separated, then the convex hulls do not overlap. Given the training data $(\boldsymbol{x}_1, y_1), \ldots, (\boldsymbol{x}_N, y_N)$, we form two convex hulls, corresponding to the positive and negative classes respectively. We pick a point $\boldsymbol{c}$, which is in the convex hull of the set of positive examples and is closest to the negative class distribution. Similarly, we pick a point $\boldsymbol{d}$ in the convex hull of the set of negative examples and is closest to the positive class distribution; see Figure 12.9(b). We define a difference vector between $\boldsymbol{d}$ and $\boldsymbol{c}$ as

$$\boldsymbol{w} := \boldsymbol{c} - \boldsymbol{d}. \tag{12.44}$$

Picking the points $\boldsymbol{c}$ and $\boldsymbol{d}$ as in the preceding cases, and requiring them to be closest to each other is equivalent to minimizing the length/norm of $\boldsymbol{w}$, so that we end up with the corresponding optimization problem

$$\arg\min_{\boldsymbol{w}} \|\boldsymbol{w}\| = \arg\min_{\boldsymbol{w}} \frac{1}{2} \|\boldsymbol{w}\|^2 . \tag{12.45}$$

Since $\boldsymbol{c}$ must be in the positive convex hull, it can be expressed as a convex combination of the positive examples, i.e., for nonnegative coefficients α_n^+

$$\boldsymbol{c} = \sum_{n:y_n=+1} \alpha_n^+ \boldsymbol{x}_n . \tag{12.46}$$

In (12.46), we use the notation $n : y_n = +1$ to indicate the set of indices n for which $y_n = +1$. Similarly, for the examples with negative labels, we obtain

$$\boldsymbol{d} = \sum_{n:y_n=-1} \alpha_n^- \boldsymbol{x}_n. \tag{12.47}$$

By substituting (12.44), (12.46), and (12.47) into (12.45), we obtain the objective

$$\min_{\boldsymbol{\alpha}} \frac{1}{2} \left\| \sum_{n:y_n=+1} \alpha_n^+ \boldsymbol{x}_n - \sum_{n:y_n=-1} \alpha_n^- \boldsymbol{x}_n \right\|^2. \tag{12.48}$$

Let $\boldsymbol{\alpha}$ be the set of all coefficients, i.e., the concatenation of $\boldsymbol{\alpha}^+$ and $\boldsymbol{\alpha}^-$. Recall that we require that for each convex hull that their coefficients sum to one,

$$\sum_{n:y_n=+1} \alpha_n^+ = 1 \quad \text{and} \quad \sum_{n:y_n=-1} \alpha_n^- = 1. \tag{12.49}$$

This implies the constraint

$$\sum_{n=1}^{N} y_n \alpha_n = 0. \tag{12.50}$$

This result can be seen by multiplying out the individual classes

$$\sum_{n=1}^{N} y_n \alpha_n = \sum_{n:y_n=+1} (+1)\alpha_n^+ + \sum_{n:y_n=-1} (-1)\alpha_n^- \tag{12.51a}$$

$$= \sum_{n:y_n=+1} \alpha_n^+ - \sum_{n:y_n=-1} \alpha_n^- = 1 - 1 = 0. \tag{12.51b}$$

The objective function (12.48) and the constraint (12.50), along with the assumption that $\boldsymbol{\alpha} \geqslant \boldsymbol{0}$, give us a constrained (convex) optimization problem. This optimization problem can be shown to be the same as that of the dual hard margin SVM (Bennett and Bredensteiner, 2000a).

Remark. To obtain the soft margin dual, we consider the reduced hull. The *reduced hull* is similar to the convex hull but has an upper bound to the size of the coefficients $\boldsymbol{\alpha}$. The maximum possible value of the elements of $\boldsymbol{\alpha}$ restricts the size that the convex hull can take. In other words, the bound on $\boldsymbol{\alpha}$ shrinks the convex hull to a smaller volume (Bennett and Bredensteiner, 2000b). ◊

reduced hull

12.4 Kernels

Consider the formulation of the dual SVM (12.41). Notice that the inner product in the objective occurs only between examples $\boldsymbol{x}_i$ and $\boldsymbol{x}_j$. There are no inner products between the examples and the parameters. Therefore, if we consider a set of features $\boldsymbol{\phi}(\boldsymbol{x}_i)$ to represent $\boldsymbol{x}_i$, the only change in the dual SVM will be to replace the inner product. This modularity, where the choice of the classification method (the SVM) and the choice of the feature representation $\boldsymbol{\phi}(\boldsymbol{x})$ can be considered separately, provides flexibility for us to explore the two problems independently. In this section, we discuss the representation $\boldsymbol{\phi}(\boldsymbol{x})$ and briefly introduce the idea of kernels, but do not go into the technical details.

Since $\phi(\boldsymbol{x})$ could be a nonlinear function, we can use the SVM (which assumes a linear classifier) to construct classifiers that are nonlinear in the examples $\boldsymbol{x}_n$. This provides a second avenue, in addition to the soft margin, for users to deal with a dataset that is not linearly separable. It turns out that there are many algorithms and statistical methods that have this property that we observed in the dual SVM: the only inner products are those that occur between examples. Instead of *explicitly* defining a nonlinear feature map $\phi(\cdot)$ and computing the resulting inner product between examples $\boldsymbol{x}_i$ and $\boldsymbol{x}_j$, we define a similarity function $k(\boldsymbol{x}_i, \boldsymbol{x}_j)$ between $\boldsymbol{x}_i$ and $\boldsymbol{x}_j$. For a certain class of similarity functions, called *kernels*, the similarity function *implicitly* defines a nonlinear feature map $\phi(\cdot)$. Kernels are by definition functions $k : \mathcal{X} \times \mathcal{X} \to \mathbb{R}$ for which there exists a Hilbert space $\mathcal{H}$ and $\phi : \mathcal{X} \to \mathcal{H}$ a feature map such that

kernel

The inputs $\mathcal{X}$ of the kernel function can be very general, and are not necessarily restricted to $\mathbb{R}^D$.

$$k(\boldsymbol{x}_i, \boldsymbol{x}_j) = \langle \phi(\boldsymbol{x}_i), \phi(\boldsymbol{x}_j) \rangle_{\mathcal{H}} . \tag{12.52}$$

There is a unique reproducing kernel Hilbert space associated with every kernel k (Aronszajn, 1950; Berlinet and Thomas-Agnan, 2004). In this unique association, $\phi(\boldsymbol{x}) = k(\cdot, \boldsymbol{x})$ is called the *canonical feature map*. The generalization from an inner product to a kernel function (12.52) is known as the *kernel trick* (Schölkopf and Smola, 2002; Shawe-Taylor and Cristianini, 2004), as it hides away the explicit nonlinear feature map.

canonical feature map

kernel trick

The matrix $\boldsymbol{K} \in \mathbb{R}^{N \times N}$, resulting from the inner products or the application of $k(\cdot, \cdot)$ to a dataset, is called the *Gram matrix*, and is often just referred to as the *kernel matrix*. Kernels must be symmetric and positive semidefinite functions so that every kernel matrix $\boldsymbol{K}$ is symmetric and positive semidefinite (Section 3.2.3):

Gram matrix

kernel matrix

$$\forall \boldsymbol{z} \in \mathbb{R}^N : \boldsymbol{z}^\top \boldsymbol{K} \boldsymbol{z} \geqslant 0. \tag{12.53}$$

Some popular examples of kernels for multivariate real-valued data $\boldsymbol{x}_i \in \mathbb{R}^D$ are the polynomial kernel, the Gaussian radial basis function kernel, and the rational quadratic kernel (Schölkopf and Smola, 2002; Rasmussen and Williams, 2006). Figure 12.10 illustrates the effect of different kernels on separating hyperplanes on an example dataset. Note that we are still solving for hyperplanes, that is, the hypothesis class of functions is still linear. The nonlinear surfaces are due to the kernel function.

Remark. Unfortunately for the fledgling machine learner, there are multiple meanings of the word "kernel." In this chapter, the word "kernel" comes from the idea of the reproducing kernel Hilbert space (RKHS) (Aronszajn, 1950; Saitoh, 1988). We have discussed the idea of the kernel in linear algebra (Section 2.7.3), where the kernel is another word for the null space. The third common use of the word "kernel" in machine learning is the smoothing kernel in kernel density estimation (Section 11.5). ◇

Since the explicit representation $\phi(\boldsymbol{x})$ is mathematically equivalent to the kernel representation $k(\boldsymbol{x}_i, \boldsymbol{x}_j)$, a practitioner will often design the kernel function such that it can be computed more efficiently than the inner product between explicit feature maps. For example, consider the polynomial kernel (Schölkopf and Smola, 2002), where the number of terms in the explicit expansion grows very quickly (even for polynomials of low degree) when the input dimension is

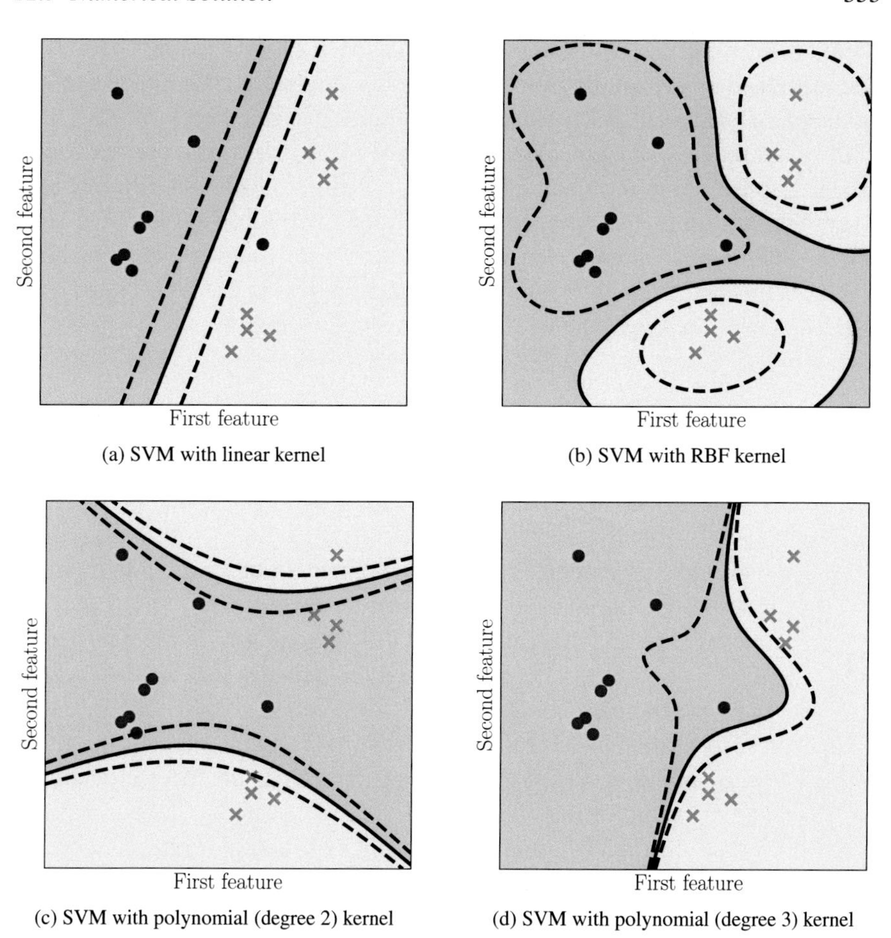

Figure 12.10 SVM with different kernels. Note that while the decision boundary is nonlinear, the underlying problem being solved is for a linear separating hyperplane (albeit with a nonlinear kernel).

large. The kernel function only requires one multiplication per input dimension, which can provide significant computational savings. Another example is the Gaussian radial basis function kernel (Schölkopf and Smola, 2002; Rasmussen and Williams, 2006), where the corresponding feature space is infinite dimensional. In this case, we cannot explicitly represent the feature space but can still compute similarities between a pair of examples using the kernel.

The choice of kernel, as well as the parameters of the kernel, is often chosen using nested cross-validation (Section 8.6.1).

Another useful aspect of the kernel trick is that there is no need for the original data to be already represented as multivariate real-valued data. Note that the inner product is defined on the output of the function $\phi(\cdot)$, but does not restrict the input to real numbers. Hence, the function $\phi(\cdot)$ and the kernel function $k(\cdot, \cdot)$ can be defined on any object, e.g., sets, sequences, strings, graphs, and distributions (Ben-Hur et al., 2008; Gärtner, 2008; Shi et al., 2009; Sriperumbudur et al., 2010; Vishwanathan et al., 2010).

12.5 Numerical Solution

We conclude our discussion of SVMs by looking at how to express the problems derived in this chapter in terms of the concepts presented in Chapter 7. We consider two different approaches for finding the optimal solution for the

SVM. First we consider the loss view of SVM 8.2.2 and express this as an unconstrained optimization problem. Then we express the constrained versions of the primal and dual SVMs as quadratic programs in standard form 7.3.2.

Consider the loss function view of the SVM (12.31). This is a convex unconstrained optimization problem, but the hinge loss (12.28) is not differentiable. Therefore, we apply a subgradient approach for solving it. However, the hinge loss is differentiable almost everywhere, except for one single point at the hinge $t = 1$. At this point, the gradient is a set of possible values that lie between 0 and -1. Therefore, the subgradient g of the hinge loss is given by

$$g(t) = \begin{cases} -1 & t < 1 \\ [-1, 0] & t = 1 \\ 0 & t > 1 \end{cases}. \tag{12.54}$$

Using this subgradient, we can apply the optimization methods presented in Section 7.1.

Both the primal and the dual SVM result in a convex quadratic programming problem (constrained optimization). Note that the primal SVM in (12.26a) has optimization variables that have the size of the dimension D of the input examples. The dual SVM in (12.41) has optimization variables that have the size of the number N of examples.

To express the primal SVM in the standard form (7.45) for quadratic programming, let us assume that we use the dot product (3.5) as the inner product. We rearrange the equation for the primal SVM (12.26a), such that the optimization variables are all on the right and the inequality of the constraint matches the standard form. This yields the optimization

Recall from Section 3.2 that we use the phrase dot product to mean the inner product on Euclidean vector space.

$$\begin{aligned} \min_{\boldsymbol{w}, b, \boldsymbol{\xi}} \quad & \frac{1}{2}\|\boldsymbol{w}\|^2 + C \sum_{n=1}^{N} \xi_n \\ \text{subject to} \quad & -y_n \boldsymbol{x}_n^\top \boldsymbol{w} - y_n b - \xi_n \leqslant -1 \\ & -\xi_n \leqslant 0 \end{aligned} \tag{12.55}$$

$n = 1, \ldots, N$. By concatenating the variables $\boldsymbol{w}, b, \boldsymbol{x}_n$ into a single vector, and carefully collecting the terms, we obtain the following matrix form of the soft margin SVM:

$$\begin{aligned} \min_{\boldsymbol{w}, b, \boldsymbol{\xi}} \quad & \frac{1}{2} \begin{bmatrix} \boldsymbol{w} \\ b \\ \boldsymbol{\xi} \end{bmatrix}^\top \begin{bmatrix} \boldsymbol{I}_D & \boldsymbol{0}_{D,N+1} \\ \boldsymbol{0}_{N+1,D} & \boldsymbol{0}_{N+1,N+1} \end{bmatrix} \begin{bmatrix} \boldsymbol{w} \\ b \\ \boldsymbol{\xi} \end{bmatrix} + \begin{bmatrix} \boldsymbol{0}_{D+1,1} & C\boldsymbol{1}_{N,1} \end{bmatrix}^\top \begin{bmatrix} \boldsymbol{w} \\ b \\ \boldsymbol{\xi} \end{bmatrix} \\ \text{subject to} \quad & \begin{bmatrix} -\boldsymbol{Y}\boldsymbol{X} & -\boldsymbol{y} & -\boldsymbol{I}_N \\ \boldsymbol{0}_{N,D+1} & & -\boldsymbol{I}_N \end{bmatrix} \begin{bmatrix} \boldsymbol{w} \\ b \\ \boldsymbol{\xi} \end{bmatrix} \leqslant \begin{bmatrix} -\boldsymbol{1}_{N,1} \\ \boldsymbol{0}_{N,1} \end{bmatrix}. \end{aligned} \tag{12.56}$$

In the preceding optimization problem, the minimization is over $[\boldsymbol{w}^\top, b, \boldsymbol{\xi}^\top]^\top \in \mathbb{R}^{D+1+N}$, and we use the notation: $\boldsymbol{I}_m$ to represent the identity matrix of size $m \times m$, $\boldsymbol{0}_{m,n}$ to represent the matrix of zeros of size $m \times n$, and $\boldsymbol{1}_{m,n}$ to represent the matrix of ones of size $m \times n$. In addition, $\boldsymbol{y}$ is the vector of labels $[y_1, \ldots, y_N]^\top$, $\boldsymbol{Y} = \text{diag}(\boldsymbol{y})$ is an N by N matrix where the elements of the diagonal are from $\boldsymbol{y}$, and $\boldsymbol{X} \in \mathbb{R}^{N \times D}$ is the matrix obtained by concatenating all the examples.

We can similarly perform a collection of terms for the dual version of the SVM (12.41). To express the dual SVM in standard form, we first have to express the kernel matrix $\boldsymbol{K}$ such that each entry is $K_{ij} = k(\boldsymbol{x}_i, \boldsymbol{x}_j)$. If we have an explicit feature representation $\boldsymbol{x}_i$, then we define $K_{ij} = \langle \boldsymbol{x}_i, \boldsymbol{x}_j \rangle$. For convenience of notation, we introduce a matrix with zeros everywhere except on the diagonal, where we store the labels, that is, $\boldsymbol{Y} = \operatorname{diag}(\boldsymbol{y})$. The dual SVM can be written as

$$\begin{aligned} \min_{\boldsymbol{\alpha}} \quad & \frac{1}{2}\boldsymbol{\alpha}^\top \boldsymbol{Y}\boldsymbol{K}\boldsymbol{Y}\boldsymbol{\alpha} - \boldsymbol{1}_{N,1}^\top \boldsymbol{\alpha} \\ \text{subject to} \quad & \begin{bmatrix} \boldsymbol{y}^\top \\ -\boldsymbol{y}^\top \\ -\boldsymbol{I}_N \\ \boldsymbol{I}_N \end{bmatrix} \boldsymbol{\alpha} \leqslant \begin{bmatrix} \boldsymbol{0}_{N+2,1} \\ C\boldsymbol{1}_{N,1} \end{bmatrix}. \end{aligned} \tag{12.57}$$

Remark. In Sections 7.3.1 and 7.3.2, we introduced the standard forms of the constraints to be inequality constraints. We will express the dual SVM's equality constraint as two inequality constraints, i.e.,

$$\boldsymbol{A}\boldsymbol{x} = \boldsymbol{b} \quad \text{is replaced by} \quad \boldsymbol{A}\boldsymbol{x} \leqslant \boldsymbol{b} \quad \text{and} \quad \boldsymbol{A}\boldsymbol{x} \geqslant \boldsymbol{b}. \tag{12.58}$$

Particular software implementations of convex optimization methods may provide the ability to express equality constraints. ◇

Since there are many different possible views of the SVM, there are many approaches for solving the resulting optimization problem. The approach presented here, expressing the SVM problem in standard convex optimization form, is not often used in practice. The two main implementations of SVM solvers are Chang and Lin (2011) (which is open source) and Joachims (1999). Since SVMs have a clear and well-defined optimization problem, many approaches based on numerical optimization techniques (Nocedal and Wright, 2006) can be applied (Shawe-Taylor and Sun, 2011).

12.6 Further Reading

The SVM is one of many approaches for studying binary classification. Other approaches include the perceptron, logistic regression, Fisher discriminant, nearest neighbor, naive Bayes, and random forest (Bishop, 2006; Murphy, 2012). A short tutorial on SVMs and kernels on discrete sequences can be found in Ben-Hur et al. (2008). The development of SVMs is closely linked to empirical risk minimization, discussed in Section 8.2. Hence, the SVM has strong theoretical properties (Vapnik, 2000; Steinwart and Christmann, 2008). The book about kernel methods (Schölkopf and Smola, 2002) includes many details of support vector machines and how to optimize them. A broader book about kernel methods (Shawe-Taylor and Cristianini, 2004) also includes many linear algebra approaches for different machine learning problems.

An alternative derivation of the dual SVM can be obtained using the idea of the Legendre–Fenchel transform (Section 7.3.3). The derivation considers each term of the unconstrained formulation of the SVM (12.31) separately and calculates their convex conjugates (Rifkin and Lippert, 2007). Readers interested in the functional analysis view (also the regularization methods view) of SVMs are referred to the work by Wahba (1990). Theoretical exposition of kernels (Aronszajn, 1950; Schwartz, 1964; Saitoh, 1988; Manton and Amblard, 2015) requires a basic grounding in linear operators (Akhiezer and Glazman, 1993). The idea of kernels have been generalized to Banach spaces (Zhang et al., 2009) and Kreĭn spaces (Ong et al., 2004; Loosli et al., 2016).

Observe that the hinge loss has three equivalent representations, as shown in (12.28) and (12.29), as well as the constrained optimization problem in (12.33). The formulation (12.28) is often used when comparing the SVM loss function with other loss functions (Steinwart, 2007). The two-piece formulation (12.29) is convenient for computing subgradients, as each piece is linear. The third formulation (12.33), as seen in Section 12.5, enables the use of convex quadratic programming (Section 7.3.2) tools.

Since binary classification is a well-studied task in machine learning, other words are also sometimes used, such as discrimination, separation, and decision. Furthermore, there are three quantities that can be the output of a binary classifier. First is the output of the linear function itself (often called the score), which can take any real value. This output can be used for ranking the examples, and binary classification can be thought of as picking a threshold on the ranked examples (Shawe-Taylor and Cristianini, 2004). The second quantity that is often considered the output of a binary classifier is the output determined after it is passed through a nonlinear function to constrain its value to a bounded range, for example in the interval $[0, 1]$. A common nonlinear function is the sigmoid function (Bishop, 2006). When the nonlinearity results in well-calibrated probabilities (Gneiting and Raftery, 2007; Reid and Williamson, 2011), this is called class probability estimation. The third output of a binary classifier is the final binary decision $\{+1, -1\}$, which is the one most commonly assumed to be the output of the classifier.

The SVM is a binary classifier that does not naturally lend itself to a probabilistic interpretation. There are several approaches for converting the raw output of the linear function (the score) into a calibrated class probability estimate ($P(Y = 1|X = \boldsymbol{x})$) that involve an additional calibration step (Platt, 2000; Zadrozny and Elkan, 2001; Lin et al., 2007). From the training perspective, there are many related probabilistic approches. We mentioned at the end of Section 12.2.5 that there is a relationship between loss function and the likelihood (also compare Sections 8.2 and 8.3). The maximum likelihood approach corresponding to a well-calibrated transformation during training is called logistic regression, which comes from a class of methods called generalized linear models. Details of logistic regression from this point of view can be found in Agresti (2002, chapter 5) and McCullagh and Nelder (1989, chapter 4). Naturally, one could take a more Bayesian view of the classifier output by estimating a posterior distribution using Bayesian logistic regression. The Bayesian view also includes the specification of the prior, which includes design choices such as conjugacy (Section 6.6.1) with the likelihood. Additionally, one could consider latent functions as priors, which results in Gaussian process classification (Rasmussen and Williams, 2006, chapter 3).

References

Abel, Niels H. 1826. *Démonstration de l'Impossibilité de la Résolution Algébrique des Équations Générales qui Passent le Quatrième Degré*. Grøndahl & Søn.

Adhikari, Ani, and DeNero, John. 2018. *Computational and Inferential Thinking: The Foundations of Data Science*. Gitbooks.

Agarwal, Arvind, and Daumé III, Hal. 2010. A Geometric View of Conjugate Priors. *Machine Learning*, **81**(1), 99–113.

Agresti, A. 2002. *Categorical Data Analysis*. Wiley.

Akaike, Hirotugu. 1974. A New Look at the Statistical Model Identification. *IEEE Transactions on Automatic Control*, **19**(6), 716–723.

Akhiezer, Naum I., and Glazman, Izrail M. 1993. *Theory of Linear Operators in Hilbert Space*. Dover Publications.

Alpaydin, Ethem. 2010. *Introduction to Machine Learning*. MIT Press.

Amari, Shun-ichi. 2016. *Information Geometry and Its Applications*. Springer.

Argyriou, Andreas, and Dinuzzo, Francesco. 2014. A Unifying View of Representer Theorems. In: *Proceedings of the International Conference on Machine Learning*.

Aronszajn, Nachman. 1950. Theory of Reproducing Kernels. *Transactions of the American Mathematical Society*, **68**, 337–404.

Axler, Sheldon. 2015. *Linear Algebra Done Right*. Springer.

Bakir, Gökhan, Hofmann, Thomas, Schölkopf, Bernhard, Smola, Alexander J., Taskar, Ben, and Vishwanathan, S. V. N. (eds). 2007. *Predicting Structured Data*. MIT Press.

Barber, David. 2012. *Bayesian Reasoning and Machine Learning*. Cambridge University Press.

Barndorff-Nielsen, Ole. 2014. *Information and Exponential Families: In Statistical Theory*. Wiley.

Bartholomew, David, Knott, Martin, and Moustaki, Irini. 2011. *Latent Variable Models and Factor Analysis: A Unified Approach*. Wiley.

Baydin, Atılım G., Pearlmutter, Barak A., Radul, Alexey A., and Siskind, Jeffrey M. 2018. Automatic Differentiation in Machine Learning: A Survey. *Journal of Machine Learning Research*, **18**, 1–43.

Beck, Amir, and Teboulle, Marc. 2003. Mirror Descent and Nonlinear Projected Subgradient Methods for Convex Optimization. *Operations Research Letters*, **31**(3), 167–175.

Belabbas, Mohamed-Ali, and Wolfe, Patrick J. 2009. Spectral Methods in Machine Learning and New Strategies for Very Large Datasets. *Proceedings of the National Academy of Sciences*, 0810600105.

Belkin, Mikhail, and Niyogi, Partha. 2003. Laplacian Eigenmaps for Dimensionality Reduction and Data Representation. *Neural Computation*, **15**(6), 1373–1396.

Ben-Hur, Asa, Ong, Cheng Soon, Sonnenburg, Sören, Schölkopf, Bernhard, and Rätsch, Gunnar. 2008. Support Vector Machines and Kernels for Computational Biology. *PLoS Computational Biology*, **4**(10), e1000173.

Bennett, Kristin P., and Bredensteiner, Erin J. 2000a. Duality and Geometry in SVM Classifiers. In: *Proceedings of the International Conference on Machine Learning*.

Bennett, Kristin P., and Bredensteiner, Erin J. 2000b. Geometry in Learning. Pages 132–145 in *Geometry at Work*. Mathematical Association of America.

Berlinet, Alain, and Thomas-Agnan, Christine. 2004. *Reproducing Kernel Hilbert Spaces in Probability and Statistics*. Springer.

Bertsekas, Dimitri P. 1999. *Nonlinear Programming*. Athena Scientific.

Bertsekas, Dimitri P. 2009. *Convex Optimization Theory*. Athena Scientific.

Bickel, Peter J., and Doksum, Kjell. 2006. *Mathematical Statistics, Basic Ideas and Selected Topics*. Vol. 1. Prentice Hall.

Bickson, Danny, Dolev, Danny, Shental, Ori, Siegel, Paul H., and Wolf, Jack K. 2007. Linear Detection via Belief Propagation. In: *Proceedings of the Annual Allerton Conference on Communication, Control, and Computing*.

Billingsley, Patrick. 1995. *Probability and Measure*. Wiley.

Bishop, Christopher M. 1995. *Neural Networks for Pattern Recognition*. Clarendon Press.

Bishop, Christopher M. 1999. Bayesian PCA. In: *Advances in Neural Information Processing Systems*.

Bishop, Christopher M. 2006. *Pattern Recognition and Machine Learning*. Springer.

Blei, David M., Kucukelbir, Alp, and McAuliffe, Jon D. 2017. Variational Inference: A Review for Statisticians. *Journal of the American Statistical Association*, **112**(518), 859–877.

Blum, Arvim, and Hardt, Moritz. 2015. The Ladder: A Reliable Leaderboard for Machine Learning Competitions. In: *International Conference on Machine Learning*.

Bonnans, J. Frédéric, Gilbert, J. Charles, Lemaréchal, Claude, and Sagastizábal, Claudia A. 2006. *Numerical Optimization: Theoretical and Practical Aspects*. Springer.

Borwein, Jonathan M., and Lewis, Adrian S. 2006. *Convex Analysis and Nonlinear Optimization*. 2nd edn. Canadian Mathematical Society.

Bottou, Léon. 1998. Online Algorithms and Stochastic Approximations. Pages 9–42 in *Online Learning and Neural Networks*. Cambridge University Press.

Bottou, Léon, Curtis, Frank E., and Nocedal, Jorge. 2018. Optimization Methods for Large-Scale Machine Learning. *SIAM Review*, **60**(2), 223–311.

Boucheron, Stephane, Lugosi, Gabor, and Massart, Pascal. 2013. *Concentration Inequalities: A Nonasymptotic Theory of Independence*. Oxford University Press.

Boyd, Stephen, and Vandenberghe, Lieven. 2004. *Convex Optimization*. Cambridge University Press.

Boyd, Stephen, and Vandenberghe, Lieven. 2018. *Introduction to Applied Linear Algebra*. Cambridge University Press.

Brochu, Eric, Cora, Vlad M., and de Freitas, Nando. 2009. *A Tutorial on Bayesian Optimization of Expensive Cost Functions, with Application to Active User Modeling and Hierarchical Reinforcement Learning*. Tech. rept. TR-2009-023. Department of Computer Science, University of British Columbia.

Brooks, Steve, Gelman, Andrew, Jones, Galin L., and Meng, Xiao-Li (eds). 2011. *Handbook of Markov Chain Monte Carlo*. Chapman and Hall/CRC.

Brown, Lawrence D. 1986. *Fundamentals of Statistical Exponential Families: With Applications in Statistical Decision Theory*. Institute of Mathematical Statistics.

Bryson, Arthur E. 1961. A Gradient Method for Optimizing Multi-Stage Allocation Processes. In: *Proceedings of the Harvard University Symposium on Digital Computers and Their Applications*.

Bubeck, Sébastien. 2015. Convex Optimization: Algorithms and Complexity. *Foundations and Trends in Machine Learning*, **8**(3-4), 231–357.

Bühlmann, Peter, and Van De Geer, Sara. 2011. *Statistics for High-Dimensional Data*. Springer.

Burges, Christopher. 2010. Dimension Reduction: A Guided Tour. *Foundations and Trends in Machine Learning*, **2**(4), 275–365.

Carroll, J Douglas, and Chang, Jih-Jie. 1970. Analysis of Individual Differences in Multidimensional Scaling via an N-Way Generalization of "Eckart–Young" Decomposition. *Psychometrika*, **35**(3), 283–319.

Casella, George, and Berger, Roger L. 2002. *Statistical Inference*. Duxbury.

Çinlar, Erhan. 2011. *Probability and Stochastics*. Springer.

Chang, Chih-Chung, and Lin, Chih-Jen. 2011. LIBSVM: A Library for Support Vector Machines. *ACM Transactions on Intelligent Systems and Technology*, **2**, 27:1–27:27. Software available at `www.csie.ntu.edu.tw/~cjlin/libsvm`.

Cheeseman, Peter. 1985. In Defense of Probability. In: *Proceedings of the International Joint Conference on Artificial Intelligence*.

Chollet, Francois, and Allaire, J. J. 2018. *Deep Learning with R*. Manning Publications.

Codd, Edgar F. 1990. *The Relational Model for Database Management*. Addison-Wesley Longman Publishing.

Cunningham, John P., and Ghahramani, Zoubin. 2015. Linear Dimensionality Reduction: Survey, Insights, and Generalizations. *Journal of Machine Learning Research*, **16**, 2859–2900.

Datta, Biswa N. 2010. *Numerical Linear Algebra and Applications*. SIAM.

Davidson, Anthony C., and Hinkley, David V. 1997. *Bootstrap Methods and Their Application*. Cambridge University Press.

Dean, Jeffrey, Corrado, Greg S., Monga, Rajat, et al. 2012. Large Scale Distributed Deep Networks. In: *Advances in Neural Information Processing Systems*.

Deisenroth, Marc P., and Mohamed, Shakir. 2012. Expectation Propagation in Gaussian Process Dynamical Systems. In: *Advances in Neural Information Processing Systems*.

Deisenroth, Marc P., and Ohlsson, Henrik. 2011. A General Perspective on Gaussian Filtering and Smoothing: Explaining Current and Deriving New Algorithms. In: *Proceedings of the American Control Conference*.

Deisenroth, Marc P., Fox, Dieter, and Rasmussen, Carl E. 2015. Gaussian Processes for Data-Efficient Learning in Robotics and Control. *IEEE Transactions on Pattern Analysis and Machine Intelligence*, **37**(2), 408–423.

Dempster, Arthur P., Laird, Nan M., and Rubin, Donald B. 1977. Maximum Likelihood from Incomplete Data via the EM Algorithm. *Journal of the Royal Statistical Society*, **39**(1), 1–38.

Deng, Li, Seltzer, Michael L., Yu, Dong, Acero, Alex, Mohamed, Abdel-rahman, and Hinton, Geoffrey E. 2010. Binary Coding of Speech Spectrograms Using a Deep Auto-Encoder. *Proceedings of Interspeech*.

Devroye, Luc. 1986. *Non-Uniform Random Variate Generation*. Springer.

Donoho, David L., and Grimes, Carrie. 2003. Hessian Eigenmaps: Locally Linear Embedding Techniques for High-Dimensional Data. *Proceedings of the National Academy of Sciences*, **100**(10), 5591–5596.

Dostál, Zdeněk. 2009. *Optimal Quadratic Programming Algorithms: With Applications to Variational Inequalities*. Springer.

Douven, Igor. 2017. Abduction. In: *The Stanford Encyclopedia of Philosophy*. Metaphysics Research Lab, Stanford University. `https://plato.stanford.edu/cgi-bin/encyclopedia/archinfo.cgi?entry=abduction&archive=sum2017`

Downey, Allen B. 2014. *Think Stats: Exploratory Data Analysis*. 2nd edn. O'Reilly Media.

Dreyfus, Stuart. 1962. The Numerical Solution of Variational Problems. *Journal of Mathematical Analysis and Applications*, **5**(1), 30–45.

Drumm, Volker, and Weil, Wolfgang. 2001. *Lineare Algebra und Analytische Geometrie*. Lecture Notes, Universität Karlsruhe (TH).

Dudley, Richard M. 2002. *Real Analysis and Probability*. Cambridge University Press.

Eaton, Morris L. 2007. *Multivariate Statistics: A Vector Space Approach*. Institute of Mathematical Statistics Lecture Notes.

Eckart, Carl, and Young, Gale. 1936. The Approximation of One Matrix by Another of Lower Rank. *Psychometrika*, **1**(3), 211–218.

Efron, Bradley, and Hastie, Trevor. 2016. *Computer Age Statistical Inference: Algorithms, Evidence and Data Science*. Cambridge University Press.

Efron, Bradley, and Tibshirani, Robert J. 1993. *An Introduction to the Bootstrap.* Chapman and Hall/CRC.

Elliott, Conal. 2009. Beautiful Differentiation. In: *International Conference on Functional Programming.*

Evgeniou, Theodoros, Pontil, Massimiliano, and Poggio, Tomaso. 2000. Statistical Learning Theory: A Primer. *International Journal of Computer Vision*, **38**(1), 9–13.

Fan, Rong-En, Chang, Kai-Wei, Hsieh, Cho-Jui, Wang, Xiang-Rui, and Lin, Chih-Jen. 2008. LIBLINEAR: A Library for Large Linear Classification. *Journal of Machine Learning Research*, **9**, 1871–1874.

Gal, Yarin, van der Wilk, Mark, and Rasmussen, Carl E. 2014. Distributed Variational Inference in Sparse Gaussian Process Regression and Latent Variable Models. In: *Advances in Neural Information Processing Systems.*

Gärtner, Thomas. 2008. *Kernels for Structured Data.* World Scientific.

Gavish, Matan, and Donoho, David L. 2014. The Optimal Hard Threshold for Singular Values is $4\sqrt{3}$. *IEEE Transactions on Information Theory*, **60**(8), 5040–5053.

Gelman, Andrew, Carlin, John B., Stern, Hal S., and Rubin, Donald B. 2004. *Bayesian Data Analysis.* Chapman & Hall/CRC.

Gentle, James E. 2004. *Random Number Generation and Monte Carlo Methods.* Springer.

Ghahramani, Zoubin. 2015. Probabilistic Machine Learning and Artificial Intelligence. *Nature*, **521**, 452–459.

Ghahramani, Zoubin, and Roweis, Sam T. 1999. Learning Nonlinear Dynamical Systems Using an EM Algorithm. In: *Advances in Neural Information Processing Systems.*

Gilks, Walter R., Richardson, Sylvia, and Spiegelhalter, David J. 1996. *Markov Chain Monte Carlo in Practice.* Chapman and Hall/CRC.

Gneiting, Tilmann, and Raftery, Adrian E. 2007. Strictly Proper Scoring Rules, Prediction, and Estimation. *Journal of the American Statistical Association*, **102**(477), 359–378.

Goh, Gabriel. 2017. Why Momentum Really Works. *Distill.*

Gohberg, Israel, Goldberg, Seymour, and Krupnik, Nahum. 2012. *Traces and Determinants of Linear Operators.* Birkhäuser.

Golan, Jonathan S. 2007. *The Linear Algebra a Beginning Graduate Student Ought to Know.* Springer.

Golub, Gene H., and Van Loan, Charles F. 2012. *Matrix Computations.* JHU Press.

Goodfellow, Ian, Bengio, Yoshua, and Courville, Aaron. 2016. *Deep Learning.* MIT Press.

Graepel, Thore, Candela, Joaquin Quiñonero-Candela, Borchert, Thomas, and Herbrich, Ralf. 2010. Web-Scale Bayesian Click-through Rate Prediction for Sponsored Search Advertising in Microsoft's Bing Search Engine. In: *Proceedings of the International Conference on Machine Learning.*

Griewank, Andreas, and Walther, Andrea. 2003. Introduction to Automatic Differentiation. In: *Proceedings in Applied Mathematics and Mechanics.*

Griewank, Andreas, and Walther, Andrea. 2008. *Evaluating Derivatives, Principles and Techniques of Algorithmic Differentiation.* SIAM.

Grimmett, Geoffrey R., and Welsh, Dominic. 2014. *Probability: An Introduction.* Oxford University Press.

Grinstead, Charles M., and Snell, J. Laurie. 1997. *Introduction to Probability.* American Mathematical Society.

Hacking, Ian. 2001. *Probability and Inductive Logic.* Cambridge University Press.

Hall, Peter. 1992. *The Bootstrap and Edgeworth Expansion.* Springer.

Hallin, Marc, Paindaveine, Davy, and Šiman, Miroslav. 2010. Multivariate Quantiles and Multiple-Output Regression Quantiles: From ℓ_1 Optimization to Halfspace Depth. *Annals of Statistics*, **38**, 635–669.

Hasselblatt, Boris, and Katok, Anatole. 2003. *A First Course in Dynamics with a Panorama of Recent Developments.* Cambridge University Press.

Hastie, Trevor, Tibshirani, Robert, and Friedman, Jerome. 2001. *The Elements of Statistical Learning – Data Mining, Inference, and Prediction.* Springer.

Hausman, Karol, Springenberg, Jost T., Wang, Ziyu, Heess, Nicolas, and Riedmiller, Martin. 2018. Learning an Embedding Space for Transferable Robot Skills. In: *Proceedings of the International Conference on Learning Representations*.

Hazan, Elad. 2015. Introduction to Online Convex Optimization. *Foundations and Trends in Optimization*, **2**(3–4), 157–325.

Hensman, James, Fusi, Nicolò, and Lawrence, Neil D. 2013. Gaussian Processes for Big Data. In: *Proceedings of the Conference on Uncertainty in Artificial Intelligence*.

Herbrich, Ralf, Minka, Tom, and Graepel, Thore. 2007. TrueSkill(TM): A Bayesian Skill Rating System. In: *Advances in Neural Information Processing Systems*.

Hiriart-Urruty, Jean-Baptiste, and Lemaréchal, Claude. 2001. *Fundamentals of Convex Analysis*. Springer.

Hoffman, Matthew D., Blei, David M., and Bach, Francis. 2010. Online Learning for Latent Dirichlet Allocation. In: *Advances in Neural Information Processing Systems*.

Hoffman, Matthew D., Blei, David M., Wang, Chong, and Paisley, John. 2013. Stochastic Variational Inference. *Journal of Machine Learning Research*, **14**(1), 1303–1347.

Hofmann, Thomas, Schölkopf, Bernhard, and Smola, Alexander J. 2008. Kernel Methods in Machine Learning. *Annals of Statistics*, **36**(3), 1171–1220.

Hogben, Leslie. 2013. *Handbook of Linear Algebra*. Chapman and Hall/CRC.

Horn, Roger A., and Johnson, Charles R. 2013. *Matrix Analysis*. Cambridge University Press.

Hotelling, Harold. 1933. Analysis of a Complex of Statistical Variables into Principal Components. *Journal of Educational Psychology*, **24**, 417–441.

Hyvarinen, Aapo, Oja, Erkki, and Karhunen, Juha. 2001. *Independent Component Analysis*. Wiley.

Imbens, Guido W., and Rubin, Donald B. 2015. *Causal Inference for Statistics, Social and Biomedical Sciences*. Cambridge University Press.

Jacod, Jean, and Protter, Philip. 2004. *Probability Essentials*. Springer.

Jaynes, Edwin T. 2003. *Probability Theory: The Logic of Science*. Cambridge University Press.

Jefferys, William H., and Berger, James O. 1992. Ockham's Razor and Bayesian Analysis. *American Scientist*, **80**, 64–72.

Jeffreys, Harold. 1961. *Theory of Probability*. Oxford University Press.

Jimenez Rezende, Danilo, and Mohamed, Shakir. 2015. Variational Inference with Normalizing Flows. In: *Proceedings of the International Conference on Machine Learning*.

Jimenez Rezende, Danilo, Mohamed, Shakir, and Wierstra, Daan. 2014. Stochastic Backpropagation and Approximate Inference in Deep Generative Models. In: *Proceedings of the International Conference on Machine Learning*.

Joachims, Thorsten. 1999. Making Large-Scale SVM Learning Practical. Pages 169–184 in: *Advances in Kernel Methods – Support Vector Learning*. MIT Press.

Jordan, Michael I., Ghahramani, Zoubin, Jaakkola, Tommi S., and Saul, Lawrence K. 1999. An Introduction to Variational Methods for Graphical Models. *Machine Learning*, **37**, 183–233.

Julier, Simon J., and Uhlmann, Jeffrey K. 1997. A New Extension of the Kalman Filter to Nonlinear Systems. In: *Proceedings of AeroSense Symposium on Aerospace/Defense Sensing, Simulation and Controls*.

Kaiser, Marcus, and Hilgetag, Claus C. 2006. Nonoptimal Component Placement, but Short Processing Paths, Due to Long-Distance Projections in Neural Systems. *PLoS Computational Biology*, **2**(7), e95.

Kalman, Dan. 1996. A Singularly Valuable Decomposition: The SVD of a Matrix. *College Mathematics Journal*, **27**(1), 2–23.

Kalman, Rudolf E. 1960. A New Approach to Linear Filtering and Prediction Problems. *Transactions of the ASME – Journal of Basic Engineering*, **82**(Series D), 35–45.

Kamthe, Sanket, and Deisenroth, Marc P. 2018. Data-Efficient Reinforcement Learning with Probabilistic Model Predictive Control. In: *Proceedings of the International Conference on Artificial Intelligence and Statistics*.

Katz, Victor J. 2004. *A History of Mathematics*. Pearson/Addison-Wesley.

Kelley, Henry J. 1960. Gradient Theory of Optimal Flight Paths. *Ars Journal*, **30**(10), 947–954.

Kimeldorf, George S., and Wahba, Grace. 1970. A Correspondence between Bayesian Estimation on Stochastic Processes and Smoothing by Splines. *Annals of Mathematical Statistics*, **41**(2), 495–502.

Kingma, Diederik P., and Ba, Jimmy. 2014. Adam: A Method for Stochastic Optimization. In: *Proceedings of the International Conference on Learning Representations*.

Kingma, Diederik P., and Welling, Max. 2014. Auto-Encoding Variational Bayes. In: *Proceedings of the International Conference on Learning Representations*.

Kittler, Josef, and Föglein, Janos. 1984. Contextual Classification of Multispectral Pixel Data. *Image and Vision Computing*, **2**(1), 13–29.

Kolda, Tamara G., and Bader, Brett W. 2009. Tensor Decompositions and Applications. *SIAM Review*, **51**(3), 455–500.

Koller, Daphne, and Friedman, Nir. 2009. *Probabilistic Graphical Models*. MIT Press.

Kong, Linglong, and Mizera, Ivan. 2012. Quantile Tomography: Using Quantiles with Multivariate Data. *Statistica Sinica*, **22**, 1598–1610.

Lang, Serge. 1987. *Linear Algebra*. Springer.

Lawrence, Neil D. 2005. Probabilistic Non-Linear Principal Component Analysis with Gaussian Process Latent Variable Models. *Journal of Machine Learning Research*, **6**(Nov.), 1783–1816.

Leemis, Lawrence M., and McQueston, Jacquelyn T. 2008. Univariate Distribution Relationships. *American Statistician*, **62**(1), 45–53.

Lehmann, Erich L., and Romano, Joseph P. 2005. *Testing Statistical Hypotheses*. Springer.

Lehmann, Erich Leo, and Casella, George. 1998. *Theory of Point Estimation*. Springer.

Liesen, Jörg, and Mehrmann, Volker. 2015. *Linear Algebra*. Springer.

Lin, Hsuan-Tien, Lin, Chih-Jen, and Weng, Ruby C. 2007. A Note on Platt's Probabilistic Outputs for Support Vector Machines. *Machine Learning*, **68**, 267–276.

Ljung, Lennart. 1999. *System Identification: Theory for the User*. Prentice Hall.

Loosli, Gaëlle, Canu, Stéphane, and Ong, Cheng Soon. 2016. Learning SVM in Kreĭn Spaces. *IEEE Transactions of Pattern Analysis and Machine Intelligence*, **38**(6), 1204–1216.

Luenberger, David G. 1969. *Optimization by Vector Space Methods*. Wiley.

MacKay, David J. C. 1992. Bayesian Interpolation. *Neural Computation*, **4**, 415–447.

MacKay, David J. C. 1998. Introduction to Gaussian Processes. Pages 133–165 of: *Neural Networks and Machine Learning*. Springer.

MacKay, David J. C. 2003. *Information Theory, Inference, and Learning Algorithms*. Cambridge University Press.

Magnus, Jan R., and Neudecker, Heinz. 2007. *Matrix Differential Calculus with Applications in Statistics and Econometrics*. Wiley.

Manton, Jonathan H., and Amblard, Pierre-Olivier. 2015. A Primer on Reproducing Kernel Hilbert Spaces. *Foundations and Trends in Signal Processing*, **8**(1–2), 1–126.

Markovsky, Ivan. 2011. *Low Rank Approximation: Algorithms, Implementation, Applications*. Springer.

Maybeck, Peter S. 1979. *Stochastic Models, Estimation, and Control*. Academic Press.

McCullagh, Peter, and Nelder, John A. 1989. *Generalized Linear Models*. CRC Press.

McEliece, Robert J., MacKay, David J. C., and Cheng, Jung-Fu. 1998. Turbo Decoding as an Instance of Pearl's "Belief Propagation" Algorithm. *IEEE Journal on Selected Areas in Communications*, **16**(2), 140–152.

Mika, Sebastian, Rätsch, Gunnar, Weston, Jason, Schölkopf, Bernhard, and Müller, Klaus-Robert. 1999. Fisher Discriminant Analysis with Kernels. Pages 41–48 of: *Proceedings of the Workshop on Neural Networks for Signal Processing*.

Minka, Thomas P. 2001a. *A Family of Algorithms for Approximate Bayesian Inference*. Ph.D. thesis, Massachusetts Institute of Technology.

Minka, Tom. 2001b. Automatic Choice of Dimensionality of PCA. In: *Advances in Neural Information Processing Systems*.

Mitchell, Tom. 1997. *Machine Learning*. McGraw-Hill.

Mnih, Volodymyr, Kavukcuoglu, Koray, & Silver, David, et al. 2015. Human-Level Control through Deep Reinforcement Learning. *Nature*, **518**, 529–533.

Moonen, Marc, and De Moor, Bart. 1995. *SVD and Signal Processing, III: Algorithms, Architectures and Applications*. Elsevier.

Moustaki, Irini, Knott, Martin, and Bartholomew, David J. 2015. *Latent-Variable Modeling*. American Cancer Society. Pages 1–10.

Müller, Andreas C., and Guido, Sarah. 2016. *Introduction to Machine Learning with Python: A Guide for Data Scientists*. O'Reilly Publishing.

Murphy, Kevin P. 2012. *Machine Learning: A Probabilistic Perspective*. MIT Press.

Neal, Radford M. 1996. *Bayesian Learning for Neural Networks*. Ph.D. thesis, Department of Computer Science, University of Toronto.

Neal, Radford M., and Hinton, Geoffrey E. 1999. A View of the EM Algorithm that Justifies Incremental, Sparse, and Other Variants. Pages 355–368 of: *Learning in Graphical Models*. MIT Press.

Nelsen, Roger. 2006. *An Introduction to Copulas*. Springer.

Nesterov, Yuri. 2018. *Lectures on Convex Optimization*. Springer.

Neumaier, Arnold. 1998. Solving Ill-Conditioned and Singular Linear Systems: A Tutorial on Regularization. *SIAM Review*, **40**, 636–666.

Nocedal, Jorge, and Wright, Stephen J. 2006. *Numerical Optimization*. Springer.

Nowozin, Sebastian, Gehler, Peter V., Jancsary, Jeremy, and Lampert, Christoph H. (eds). 2014. *Advanced Structured Prediction*. MIT Press.

O'Hagan, Anthony. 1991. Bayes–Hermite Quadrature. *Journal of Statistical Planning and Inference*, **29**, 245–260.

Ong, Cheng Soon, Mary, Xavier, Canu, Stéphane, and Smola, Alexander J. 2004. Learning with Non-Positive Kernels. In: *Proceedings of the International Conference on Machine Learning*.

Ormoneit, Dirk, Sidenbladh, Hedvig, Black, Michael J., and Hastie, Trevor. 2001. Learning and Tracking Cyclic Human Motion. In: *Advances in Neural Information Processing Systems*.

Page, Lawrence, Brin, Sergey, Motwani, Rajeev, and Winograd, Terry. 1999. *The PageRank Citation Ranking: Bringing Order to the Web*. Tech. rept. Stanford InfoLab.

Paquet, Ulrich. 2008. *Bayesian Inference for Latent Variable Models*. Ph.D. thesis, University of Cambridge.

Parzen, Emanuel. 1962. On Estimation of a Probability Density Function and Mode. *Annals of Mathematical Statistics*, **33**(3), 1065–1076.

Pearl, Judea. 1988. *Probabilistic Reasoning in Intelligent Systems: Networks of Plausible Inference*. Morgan Kaufmann.

Pearl, Judea. 2009. *Causality: Models, Reasoning and Inference*. 2nd edn. Cambridge University Press.

Pearson, Karl. 1895. Contributions to the Mathematical Theory of Evolution. II. Skew Variation in Homogeneous Material. *Philosophical Transactions of the Royal Society A: Mathematical, Physical and Engineering Sciences*, **186**, 343–414.

Pearson, Karl. 1901. On Lines and Planes of Closest Fit to Systems of Points in Space. *Philosophical Magazine*, **2**(11), 559–572.

Peters, Jonas, Janzing, Dominik, and Schölkopf, Bernhard. 2017. *Elements of Causal Inference: Foundations and Learning Algorithms*. MIT Press.

Petersen, Kaare B., and Pedersen, Michael S. 2012. *The Matrix Cookbook*. Tech. rept. Technical University of Denmark.

Platt, John C. 2000. Probabilistic Outputs for Support Vector Machines and Comparisons to Regularized Likelihood Methods. In: *Advances in Large Margin Classifiers*.

Pollard, David. 2002. *A User's Guide to Measure Theoretic Probability*. Cambridge University Press.

Polyak, Roman A. 2016. The Legendre Transformation in Modern Optimization. Pages 437–507 of: *Optimization and Its Applications in Control and Data Sciences*. Springer.

Press, William H., Teukolsky, Saul A., Vetterling, William T., and Flannery, Brian P. 2007. *Numerical Recipes: The Art of Scientific Computing*. Cambridge University Press.

Proschan, Michael A., and Presnell, Brett. 1998. Expect the Unexpected from Conditional Expectation. *American Statistician*, **52**(3), 248–252.

Raschka, Sebastian, and Mirjalili, Vahid. 2017. *Python Machine Learning: Machine Learning and Deep Learning with Python, scikit-learn, and TensorFlow*. Packt Publishing.

Rasmussen, Carl E., and Ghahramani, Zoubin. 2001. Occam's Razor. In: *Advances in Neural Information Processing Systems*.

Rasmussen, Carl E., and Ghahramani, Zoubin. 2003. Bayesian Monte Carlo. In: *Advances in Neural Information Processing Systems*.

Rasmussen, Carl E., and Williams, Christopher K. I. 2006. *Gaussian Processes for Machine Learning*. MIT Press.

Reid, Mark, and Williamson, Robert C. 2011. Information, Divergence and Risk for Binary Experiments. *Journal of Machine Learning Research*, **12**, 731–817.

Rifkin, Ryan M., and Lippert, Ross A. 2007. Value Regularization and Fenchel Duality. *Journal of Machine Learning Research*, **8**, 441–479.

Rockafellar, Ralph T. 1970. *Convex Analysis*. Princeton University Press.

Rogers, Simon, and Girolami, Mark. 2016. *A First Course in Machine Learning*. Chapman and Hall/CRC.

Rosenbaum, Paul R. 2017. *Observation and Experiment: An Introduction to Causal Inference*. Harvard University Press.

Rosenblatt, Murray. 1956. Remarks on Some Nonparametric Estimates of a Density Function. *Annals of Mathematical Statistics*, **27**(3), 832–837.

Roweis, Sam T. 1998. EM Algorithms for PCA and SPCA. In: *Advances in Neural Information Processing Systems*.

Roweis, Sam T., and Ghahramani, Zoubin. 1999. A Unifying Review of Linear Gaussian Models. *Neural Computation*, **11**(2), 305–345.

Roy, Anindya, and Banerjee, Sudipto. 2014. *Linear Algebra and Matrix Analysis for Statistics*. Chapman and Hall/CRC.

Rubinstein, Reuven Y., and Kroese, Dirk P. 2016. *Simulation and the Monte Carlo Method*. Wiley.

Ruffini, Paolo. 1799. *Teoria Generale delle Equazioni, in cui si Dimostra Impossibile la Soluzione Algebraica delle Equazioni Generali di Grado Superiore al Quarto*. Stamperia di S. Tommaso d'Aquino.

Rumelhart, David E., Hinton, Geoffrey E., and Williams, Ronald J. 1986. Learning Representations by Back-Propagating Errors. *Nature*, **323**(6088), 533–536.

Sæmundsson, Steindór, Hofmann, Katja, and Deisenroth, Marc P. 2018. Meta Reinforcement Learning with Latent Variable Gaussian Processes. In: *Proceedings of the Conference on Uncertainty in Artificial Intelligence*.

Saitoh, Saburou. 1988. *Theory of Reproducing Kernels and Its Applications*. Longman Scientific & Technical.

Särkkä, Simo. 2013. *Bayesian Filtering and Smoothing*. Cambridge University Press.

Schölkopf, Bernhard, Herbrich, Ralf, and Smola, Alexander J. 2001. A Generalized Representer Theorem. In: *Proceedings of the International Conference on Computational Learning Theory*.

Schölkopf, Bernhard, and Smola, Alexander J. 2002. *Learning with Kernels – Support Vector Machines, Regularization, Optimization, and Beyond*. MIT Press.

Schölkopf, Bernhard, Smola, Alexander J., and Müller, Klaus-Robert. 1997. Kernel Principal Component Analysis. In: *Proceedings of the International Conference on Artificial Neural Networks*.

Schölkopf, Bernhard, Smola, Alexander J., and Müller, Klaus-Robert. 1998. Nonlinear Component Analysis as a Kernel Eigenvalue Problem. *Neural Computation*, **10**(5), 1299–1319.

Schwartz, Laurent. 1964. Sous Espaces Hilbertiens d'Espaces Vectoriels Topologiques et Noyaux Associés. *Journal d'Analyse Mathématique*, **13**, 115–256.

Schwarz, Gideon E. 1978. Estimating the Dimension of a Model. *Annals of Statistics*, **6**(2), 461–464.

Shahriari, Bobak, Swersky, Kevin, Wang, Ziyu, Adams, Ryan P., and De Freitas, Nando. 2016. Taking the Human out of the Loop: A Review of Bayesian Optimization. *Proceedings of the IEEE*, **104**(1), 148–175.

Shalev-Shwartz, Shai, and Ben-David, Shai. 2014. *Understanding Machine Learning: From Theory to Algorithms*. Cambridge University Press.

Shawe-Taylor, John, and Cristianini, Nello. 2004. *Kernel Methods for Pattern Analysis*. Cambridge University Press.

Shawe-Taylor, John, and Sun, Shiliang. 2011. A Review of Optimization Methodologies in Support Vector Machines. *Neurocomputing*, **74**(17), 3609–3618.

Shental, Ori, Siegel, Paul H., Wolf, Jack K., Bickson, Danny, and Dolev, Danny. 2008. Gaussian Belief Propagation Solver for Systems of Linear Equations. In: *Proceedings of the International Symposium on Information Theory*.

Shewchuk, Jonathan R. 1994. *An Introduction to the Conjugate Gradient Method without the Agonizing Pain*.

Shi, Jianbo, and Malik, Jitendra. 2000. Normalized Cuts and Image Segmentation. *IEEE Transactions on Pattern Analysis and Machine Intelligence*, **22**(8), 888–905.

Shi, Qinfeng, Petterson, James, Dror, Gideon, Langford, John, Smola, Alexander J., and Vishwanathan, S. V. N. 2009. Hash Kernels for Structured Data. *Journal of Machine Learning Research*, 2615–2637.

Shiryayev, Albert N. 1984. *Probability*. Springer.

Shor, Naum Z. 1985. *Minimization Methods for Non-Differentiable Functions*. Springer.

Shotton, Jamie, Winn, John, Rother, Carsten, and Criminisi, Antonio. 2006. TextonBoost: Joint Appearance, Shape and Context Modeling for Multi-Class Object Recognition and Segmentation. In: *Proceedings of the European Conference on Computer Vision*.

Smith, Adrian F. M., and Spiegelhalter, David. 1980. Bayes Factors and Choice Criteria for Linear Models. *Journal of the Royal Statistical Society B*, **42**(2), 213–220.

Snoek, Jasper, Larochelle, Hugo, and Adams, Ryan P. 2012. Practical Bayesian Optimization of Machine Learning Algorithms. In: *Advances in Neural Information Processing Systems*.

Spearman, Charles. 1904. "General Intelligence," Objectively Determined and Measured. *American Journal of Psychology*, **15**(2), 201–292.

Sriperumbudur, Bharath K., Gretton, Arthur, Fukumizu, Kenji, Schölkopf, Bernhard, and Lanckriet, Gert R. G. 2010. Hilbert Space Embeddings and Metrics on Probability Measures. *Journal of Machine Learning Research*, **11**, 1517–1561.

Steinwart, Ingo. 2007. How to Compare Different Loss Functions and Their Risks. *Constructive Approximation*, **26**, 225–287.

Steinwart, Ingo, and Christmann, Andreas. 2008. *Support Vector Machines*. Springer.

Stoer, Josef, and Burlirsch, Roland. 2002. *Introduction to Numerical Analysis*. Springer.

Strang, Gilbert. 1993. The Fundamental Theorem of Linear Algebra. *American Mathematical Monthly*, **100**(9), 848–855.

Strang, Gilbert. 2003. *Introduction to Linear Algebra*. Wellesley–Cambridge Press.

Stray, Jonathan. 2016. *The Curious Journalist's Guide to Data*. Tow Center for Digital Journalism at Columbia's Graduate School of Journalism.

Strogatz, Steven. 2014. Writing about Math for the Perplexed and the Traumatized. *Notices of the American Mathematical Society*, **61**(3), 286–291.

Sucar, Luis E., and Gillies, Duncan F. 1994. Probabilistic Reasoning in High-Level Vision. *Image and Vision Computing*, **12**(1), 42–60.

Szeliski, Richard, Zabih, Ramin, Scharstein, Daniel, et al. 2008. A Comparative Study of Energy Minimization Methods for Markov Random Fields with Smoothness-Based Priors. *IEEE Transactions on Pattern Analysis and Machine Intelligence*, **30**(6), 1068–1080.

Tandra, Haryono. 2014. The Relationship between the Change of Variable Theorem and the Fundamental Theorem of Calculus for the Lebesgue Integral. *Teaching of Mathematics*, **17**(2), 76–83.

Tenenbaum, Joshua B., De Silva, Vin, and Langford, John C. 2000. A Global Geometric Framework for Nonlinear Dimensionality Reduction. *Science*, **290**(5500), 2319–2323.

Tibshirani, Robert. 1996. Regression Selection and Shrinkage via the Lasso. *Journal of the Royal Statistical Society B*, **58**(1), 267–288.

Tipping, Michael E., and Bishop, Christopher M. 1999. Probabilistic Principal Component Analysis. *Journal of the Royal Statistical Society: Series B*, **61**(3), 611–622.

Titsias, Michalis K., and Lawrence, Neil D. 2010. Bayesian Gaussian Process Latent Variable Model. In: *Proceedings of the International Conference on Artificial Intelligence and Statistics*.

Toussaint, Marc. 2012. *Some Notes on Gradient Descent*. `https://ipvs.informatik.uni-stuttgart.de/mlr/marc/notes/gradientDescent.pdf`.

Trefethen, Lloyd N., and Bau III, David. 1997. *Numerical Linear Algebra*. SIAM.

Tucker, Ledyard R. 1966. Some Mathematical Notes on Three-Mode Factor Analysis. *Psychometrika*, **31**(3), 279–311.

Vapnik, Vladimir N. 1998. *Statistical Learning Theory*. Wiley.

Vapnik, Vladimir N. 1999. An Overview of Statistical Learning Theory. *IEEE Transactions on Neural Networks*, **10**(5), 988–999.

Vapnik, Vladimir N. 2000. *The Nature of Statistical Learning Theory*. Springer.

Vishwanathan, S. V. N., Schraudolph, Nicol N., Kondor, Risi, and Borgwardt, Karsten M. 2010. Graph Kernels. *Journal of Machine Learning Research*, **11**, 1201–1242.

von Luxburg, Ulrike, and Schölkopf, Bernhard. 2011. Statistical Learning Theory: Models, Concepts, and Results. Pages 651–706 of: *Handbook of the History of Logic*, vol. 10. Elsevier.

Wahba, Grace. 1990. *Spline Models for Observational Data*. Society for Industrial and Applied Mathematics.

Walpole, Ronald E., Myers, Raymond H., Myers, Sharon L., and Ye, Keying. 2011. *Probability and Statistics for Engineers and Scientists*. Prentice Hall.

Wasserman, Larry. 2004. *All of Statistics*. Springer.

Wasserman, Larry. 2007. *All of Nonparametric Statistics*. Springer.

Whittle, Peter. 2000. *Probability via Expectation*. Springer.

Wickham, Hadley. 2014. Tidy Data. *Journal of Statistical Software*, **59**, 1–23.

Williams, Christopher K. I. 1997. Computing with Infinite Networks. In: *Advances in Neural Information Processing Systems*.

Yu, Yaoliang, Cheng, Hao, Schuurmans, Dale, and Szepesvári, Csaba. 2013. Characterizing the Representer Theorem. In: *Proceedings of the International Conference on Machine Learning*.

Zadrozny, Bianca, and Elkan, Charles. 2001. Obtaining Calibrated Probability Estimates from Decision Trees and Naive Bayesian Classifiers. In: *Proceedings of the International Conference on Machine Learning*.

Zhang, Haizhang, Xu, Yuesheng, and Zhang, Jun. 2009. Reproducing Kernel Banach Spaces for Machine Learning. *Journal of Machine Learning Research*, **10**, 2741–2775.

Zia, Royce K. P., Redish, Edward F., and McKay, Susan R. 2009. Making Sense of the Legendre Transform. *American Journal of Physics*, **77**(7), 614–622.

Index

Printed in the United States
By Bookmasters